Graduate Texts in Mathematics 32

Nathan Jacobson

Lectures in Abstract Algebra

III. Theory of Fields and Galois Theory

Springer-Verlag New York Heidelberg Berlin

Nathan Jacobson
Yale University
Department of Mathematics
New Haven, Connecticut 06520

AMS Subject Classification
12-01

Library of Congress Cataloging in Publication Data

Jacobson, Nathan, 1910–
Lectures in abstract algebra.
(Graduate texts in mathematics; v. 32)
Reprint of the 1951–1964 ed. published by Van Nostrand, New York in The University series in higher mathematics.
Bibliography: v. 3, p.
Includes indexes.
CONTENTS: 2. Linear algebra. 3. Theory of fields and Galois theory. 1. Algebra, Abstract. I. Title. II. Series.
QA162.J3 1975 512′.02 75–15564

Second corrected printing

Softcover reprint of the hardcover 1st edition 1964

Originally published in the University Series in Higher Mathematics (D. Van Nostrand Company): edited by M. H. Stone, L. Nirenberg and S. S. Chern.

ISBN-13: 978-0-387-90124-4 e-ISBN-13: 978-1-4612-9872-4
DOI: 10.1007/978-1-4612-9872-4

TO

POLLY

PREFACE

The present volume completes the series of texts on algebra which the author began more than ten years ago. The account of field theory and Galois theory which we give here is based on the notions and results of general algebra which appear in our first volume and on the more elementary parts of the second volume, dealing with linear algebra. The level of the present work is roughly the same as that of Volume II.

In preparing this book we have had a number of objectives in mind. First and foremost has been that of presenting the basic field theory which is essential for an understanding of modern algebraic number theory, ring theory, and algebraic geometry. The parts of the book concerned with this aspect of the subject are Chapters I, IV, and V dealing respectively with finite dimensional field extensions and Galois theory, general structure theory of fields, and valuation theory. Also the results of Chapter III on abelian extensions, although of a somewhat specialized nature, are of interest in number theory. A second objective of our account has been to indicate the links between the present theory of fields and the classical problems which led to its development. This purpose has been carried out in Chapter II, which gives Galois' theory of solvability of equations by radicals, and in Chapter VI, which gives Artin's application of the theory of real closed fields to the solution of Hilbert's problem on positive definite rational functions. Finally, we have wanted to present the parts of field theory which are of importance to analysis. Particularly noteworthy here is the Tarski-Seidenberg decision method for polynomial equations and inequalities in real closed fields which we treat in Chapter VI.

As in the case of our other two volumes, the exercises form an important part of the text. Also we are willing to admit that quite a few of these are intentionally quite difficult.

Again, it is a pleasure for me to acknowledge my great indebtedness to my friends, Professors Paul Cohn and George Seligman, for their care in reading a preliminary version of this material. Many of their suggestions have been incorporated in the present volume. I am indebted also to Professors Cohn and James Reid and to my wife for help with the proof reading. Finally, I wish to acknowledge my appreciation to the U. S. Air Force Office of Scientific Development whose support during a summer and half of an academic year permitted the completion of this work at an earlier date than would have been possible otherwise.

N. J.

New Haven, Conn.
January 20, 1964

CONTENTS

Introduction

In this book we shall assume that the reader is familiar with the general notions of algebra and the results on fields which appear in Vol. I, and with the more elementary parts of Vol. II. In particular, we presuppose a knowledge of the characteristic of a field, prime field, construction of the field of fractions of a commutative integral domain, construction of simple algebraic and transcendental extensions of a field. These ideas appear in Chaps. II and III of Vol. I. We shall need also the elementary factorization theory of Chap. IV. From Vol. II we require the basic notions of vector space over a field, dimensionality, linear transformation, linear function, compositions of linear transformations, bilinear form. On the other hand, the deeper results on canonical forms of linear transformations and bilinear forms will not be needed.

In this Introduction we shall re-do some things we have done before. Our motivation for this is twofold. In the first place, it will be useful for the applications that we shall make to sharpen some of the earlier results. In the second place, it will be convenient to list for easy reference some of the results that will be used frequently in the sequel. The topics that we shall treat here are: extension of homomorphisms (cf. Vol. I, Chap. III), algebras (Vol. II, Chap. VII), and tensor products * of vector spaces and algebras (Vol. II, Chap. VII). The notion of extension of homomorphism is one of the main tools in the theory of fields. The concept of an algebra arises naturally when one studies a field relative to a selected subfield as base field. The concept of tensor product is of lesser importance in field theory and it per-

* In Vol. II this notion was called the Kronecker product. Current usage favors the term tensor product, so we shall adopt this in the present volume. Also we shall use the currently standard notation $\otimes$ for the $\times$ of Vol. II.

haps could be avoided altogether. However, this notion has attained enormous importance throughout algebra and algebraic topology in recent years. For this broader reason it is a good idea for the student to become adept in handling tensor products, and we shall use these freely when it seems appropriate.

1. Extension of homomorphisms. Throughout this book we shall adopt the convention that the rings we consider all have identity elements $1 \neq 0$. The term subring will therefore mean subring in the old sense (as in Vol. I) containing 1, and by a homomorphism of a ring $\mathfrak{A}$ into a ring $\mathfrak{B}$ we shall understand a homomorphism in the old sense sending the 1 of $\mathfrak{A}$ into the 1 of $\mathfrak{B}$.

Now let $\mathfrak{o}$ be a subring of a field P and let Φ be the subfield of P generated by $\mathfrak{o}$. We recall that the elements of Φ can be expressed as simple fractions $\alpha\beta^{-1}$ of elements $\alpha, \beta \in \mathfrak{o}$ $(\beta \neq 0)$. Hence Φ is the subring of P generated by $\mathfrak{o}$ and the inverses of the elements of the set $\mathfrak{o}^*$ of non-zero elements of $\mathfrak{o}$. The set $\mathfrak{o}^*$ contains 1 and is closed under the multiplication of $\mathfrak{o}$. It is sometimes useful to generalize this situation in the following way: We are given a subring $\mathfrak{o}$ of P and a subset M of $\mathfrak{o}^*$ containing 1 and closed under multiplication. We shall refer to such a subset as a sub-semigroup of the multiplicative group of the field. We are interested in the subring $\mathfrak{o}_M$ generated by $\mathfrak{o}$ and the inverses of the elements of M. For example, we could take P to be the field R_0 of rational numbers and $M = \{2^k | k = 0, 1, 2, \cdots\}$. Then $\mathfrak{o}_M$ is the subring of rational numbers whose denominators are powers of 2. In the general case,

$$\mathfrak{o}_M = \{\alpha\beta^{-1} | \alpha \in \mathfrak{o}, \quad \beta \in M\};$$

for, if we denote the set on the right-hand side of this equation by $\mathfrak{o}'$, then clearly $\mathfrak{o}' \subseteq \mathfrak{o}_M$ and $\mathfrak{o}'$ contains $\mathfrak{o} = \{\alpha = \alpha 1^{-1}\}$. Also $\mathfrak{o}'$ contains every $\beta^{-1} = 1\beta^{-1}$ for $\beta \in M$. One checks directly that $\mathfrak{o}'$ is a subring of P. Then it follows that $\mathfrak{o}' = \mathfrak{o}_M$.

Now suppose P$'$ is a second field and we have a homomorphism s of $\mathfrak{o}$ into P$'$ such that $\beta^s \neq 0$ for every $\beta \in M$. Our first homomorphism extension theorem concerns this situation. This is the following result.

I. *Let* $\mathfrak{o}$ *be a subring* (*with* 1) *of a field* P, M *a subset of non-zero elements of* $\mathfrak{o}$ *containing* 1 *and closed under multiplication,* $\mathfrak{o}_M$ *the subring of* P *generated by* $\mathfrak{o}$ *and the inverses of the elements of* M. *Let* s *be a homomorphism of* $\mathfrak{o}$ *into a field* P′ *such that* $\beta^s \neq 0$ *for every* $\beta \in M$. *Then* s *has a unique extension to a homomorphism* S *of* $\mathfrak{o}_M$ *into* P′. *Moreover,* S *is an isomorphism if and only if* s *is an isomorphism.*

Proof. Let $\alpha_1\beta_1{}^{-1} = \alpha_2\beta_2{}^{-1}$, $\alpha_i \in \mathfrak{o}$, $\beta_i \in M$. Then $\alpha_1\beta_2 = \alpha_2\beta_1$ and consequently $\alpha_1{}^s\beta_2{}^s = \alpha_2{}^s\beta_1{}^s$. This relation in P′ gives $\alpha_1{}^s(\beta_1{}^s)^{-1} = \alpha_2{}^s(\beta_2{}^s)^{-1}$. Hence the mapping

$$S\colon \alpha\beta^{-1} \to \alpha^s(\beta^s)^{-1}, \quad \alpha \in \mathfrak{o}, \quad \beta \in M$$

which is defined on the whole of $\mathfrak{o}_M = \{\alpha\beta^{-1}\}$ is single-valued. One checks that S is a homomorphism (Vol. I, p. 92). If $\alpha \in \mathfrak{o}$, then $\alpha^S = (\alpha 1^{-1})^S = \alpha^s 1^s = \alpha^s$, so S is the same as s on $\mathfrak{o}$. Hence S is a homomorphism of $\mathfrak{o}_M$ which extends the given homomorphism of $\mathfrak{o}$. Now let S' be any such extension. Then the relation $\beta\beta^{-1} = 1$ for $\beta \in M$ gives $\beta^{S'}(\beta^{-1})^{S'} = 1$, so $(\beta^{-1})^{S'} = (\beta^{S'})^{-1}$. If $\alpha \in \mathfrak{o}$, then we have $(\alpha\beta^{-1})^{S'} = \alpha^{S'}(\beta^{S'})^{-1} = \alpha^s(\beta^s)^{-1} = (\alpha\beta^{-1})^S$. Hence $S' = S$ and S is unique. Clearly, if S is an isomorphism, then its restriction s to $\mathfrak{o}$ is an isomorphism. Now assume s is an isomorphism and let $\alpha\beta^{-1}$ be in the kernel of the homomorphism S: $0 = (\alpha\beta^{-1})^S = \alpha^s(\beta^s)^{-1}$. Then $\alpha^s = 0$, $\alpha = 0$, and $\alpha\beta^{-1} = 0$. This shows that the kernel of S is 0; hence S is an isomorphism.

We consider next an arbitrary commutative ring $\mathfrak{A}$ and the polynomial ring $\mathfrak{A}[x]$, x an element which is transcendental relative to $\mathfrak{A}$ (Vol. I, p. 93). The elements of $\mathfrak{A}[x]$ have the form $a_0 + a_1x + a_2x^2 + \cdots + a_nx^n$ where the $a_i \in \mathfrak{A}$ and $a_0 + a_1x + \cdots + a_nx^n = 0$ only if all the $a_i = 0$. We now have the following homomorphism theorem.

II. *Let* $\mathfrak{A}$ *be a commutative ring,* $\mathfrak{A}[x]$ *the polynomial ring over* $\mathfrak{A}$ *in a transcendental element* x *and let* s *be a homomorphism of* $\mathfrak{A}$ *into a commutative ring* $\mathfrak{B}$. *If* u *is any element of* $\mathfrak{B}$ *there exists a unique homomorphism* S *of* $\mathfrak{A}[x]$ *into* $\mathfrak{B}$ *such that:* $a^S = a^s$, $a \in \mathfrak{A}$, $x^S = u$.

The reader is referred to Vol. I, p. 97, for the proof. This result has an immediate extension to a polynomial ring $\mathfrak{A}[x_1, x_2, \cdots, x_r]$ where the x_i are algebraically independent elements. We recall that the algebraic independence of the x_i means the following:

If $(m_1, m_2, \cdots, m_r)$ is an r-tuple of non-negative integers m_i, then a relation $\sum_{m_i} a_{m_1\cdots m_r} x_1^{m_1} \cdots x_r^{m_r} = 0$, $a_{m_1\cdots m_r} \varepsilon\, \mathfrak{A}$, can hold only if every $a_{m_1\cdots m_r} = 0$. From now on we shall refer to elements x_i which belong to a commutative ring and are algebraically independent relative to a subring $\mathfrak{A}$ as *indeterminates* (relative to $\mathfrak{A}$). Then we have

III. *Let* $\mathfrak{A}[x_1, \cdots, x_r]$ *be a commutative polynomial ring in* x_i *which are indeterminates (relative to* $\mathfrak{A}$*) and let* s *be a homomorphism of* $\mathfrak{A}$ *into a commutative ring* $\mathfrak{B}$. *If* $u_1, u_2, \cdots, u_r$ *are arbitrary elements of* $\mathfrak{B}$, *then there exists a unique homomorphism* S *of* $\mathfrak{A}[x_i]$ *into* $\mathfrak{B}$ *such that* 1) $a^S = a^s$, $a \varepsilon\, \mathfrak{A}$; 2) $x_i^S = u_i$, $i = 1, 2, \cdots, r$.

We now suppose we have a commutative ring $\mathfrak{C}$, $\mathfrak{A}$ a subring, s a homomorphism of $\mathfrak{A}$ into another commutative ring $\mathfrak{B}$. Let $t_1, t_2, \cdots, t_r$ be elements of $\mathfrak{C}$ and let $\mathfrak{A}[t_1, t_2, \cdots, t_r]$ be the subring of $\mathfrak{C}$ generated by $\mathfrak{A}$ and the t_i. Under what conditions can s be extended to a homomorphism S of $\mathfrak{A}[t_i] \equiv \mathfrak{A}[t_1, t_2, \cdots, t_r]$ into $\mathfrak{B}$ so that $t_i^S = u_i$, $1 \leq i \leq r$, where the u_i are prescribed elements of $\mathfrak{B}$? The answer to this basic question is

IV. *Let* $\mathfrak{B}$ *and* $\mathfrak{C}$ *be commutative rings,* $\mathfrak{A}$ *a subring of* $\mathfrak{C}$, s *a homomorphism of* $\mathfrak{A}$ *into* $\mathfrak{B}$. *Let* $t_1, \cdots, t_r$ *be elements of* $\mathfrak{C}$, $u_1, \cdots, u_r$ *elements of* $\mathfrak{B}$. *Then there exists a homomorphism* S *of* $\mathfrak{A}[t_1, \cdots, t_r]$ *into* $\mathfrak{B}$ *such that* $a^S = a^s$, $a \varepsilon\, \mathfrak{A}$ *and* $t_i^S = u_i$, $i = 1, 2, \cdots, r$, *if and only if for every polynomial* $f(x_1, \cdots, x_r) \varepsilon\, \mathfrak{A}[x_i]$, x_i *indeterminates, such that* $f(t_1, \cdots, t_r) = 0$ *we have* $f^s(u_1, \cdots, u_r) = 0$. *Here* $f^s(x_1, \cdots, x_r)$ *is obtained by applying* s *to the coefficients of* $f(x_1, \cdots, x_r)$. *If* S *exists, it is unique.*

Proof. The set $\mathfrak{R}$ of polynomials $f(x_1, \cdots, x_r)$ such that $f(t_1, \cdots, t_r) = 0$ is the kernel of the homomorphism $h(x_1, \cdots, x_r) \rightarrow h(t_1, \cdots, t_r)$ of $\mathfrak{A}[x_i]$ into $\mathfrak{A}[t_i]$. Hence we have the isomorphism $\tau: h(t_1, \cdots, t_r) \rightarrow h(x_1, \cdots, x_r) + \mathfrak{R}$ of $\mathfrak{A}[t_i]$ onto the difference ring $\mathfrak{A}[x_i]/\mathfrak{R}$. Next we consider the homomorphism $h(x_1, \cdots, x_r) \rightarrow h^s(u_1, \cdots, u_r)$ of $\mathfrak{A}[x_i]$ into $\mathfrak{B}$ (cf. III). Assume that $f^s(u_1, \cdots, u_r) = 0$ for every $f \varepsilon\, \mathfrak{R}$. Then every $f \varepsilon\, \mathfrak{R}$ is mapped into 0 by the homomorphism $h(x_1, \cdots, x_r) \rightarrow h^s(u_1, \cdots, u_r)$ so $\mathfrak{R}$ is contained in the kernel of this homomorphism. It follows (Vol. I, p. 70) that we have the homomorphism $h(x_1, \cdots, x_r) +$

$\mathfrak{K} \to h^s(u_1, \cdots, u_r)$ of $\mathfrak{A}[x_i]/\mathfrak{K}$ into $\mathfrak{B}$. Combining this with the isomorphism τ we obtain the homomorphism

$$(1) \qquad S: h(t_1, \cdots, t_r) \to h^s(u_1, \cdots, u_r)$$

of $\mathfrak{A}[t_i]$ into $\mathfrak{B}$. This is the required extension of s. If S' is any extension of s to a homomorphism of $\mathfrak{A}[t_i]$ into $\mathfrak{B}$ such that $a^{S'} = a^s$ and $t_i^{S'} = u_i$, then $h(t_1, \cdots, t_r)^{S'} = h^s(u_1, \cdots, u_r)$; hence $S' = S$ and S is unique. Also, it is trivial that, if $f(t_1, \cdots, t_r) = 0$, then $0 = f(t_1, \cdots, t_r)^S = f^s(u_1, \cdots, u_r)$ if S is a homomorphism of $\mathfrak{A}[t_1, \cdots, t_r]$ satisfying our conditions. Hence it is clear that the condition stated in the theorem is necessary for the existence of the extension S.

We have noted in the proof that the set $\mathfrak{K}$ of polynomials $f(x_1, \cdots, x_r)$ such that $f(t_1, \cdots, t_r) = 0$ is the kernel of a homomorphism. Hence this is an ideal in the polynomial ring $\mathfrak{A}[x_1, x_2, \cdots, x_r]$. Now let $X = \{g\}$ be a set of generators of $\mathfrak{K}$: $X \subseteq \mathfrak{K}$ and every element $f \in \mathfrak{K}$ has the form $\Sigma a_i(x_1, \cdots, x_r)g_i(x_1, \cdots, x_r)$ where the $a_i\,(x_1, \cdots, x_r) \in \mathfrak{A}[x_1, x_2, \cdots, x_r]$ and the $g_i(x_1, \cdots, x_r) \in X$. It is clear that, if $g^s(u_1, \cdots, u_r) = 0$ holds for every $g \in X$, then also $f^s(u_1, \cdots, u_r) = 0$ for every $f \in \mathfrak{K}$. Hence we can obtain from IV the following result which is often easier to apply than IV itself:

IV′. *Let $\mathfrak{B}$ and $\mathfrak{C}$ be commutative rings, $\mathfrak{A}$ a subring of $\mathfrak{C}$, and s a homomorphism of $\mathfrak{A}$ into $\mathfrak{B}$. Let X be a set of generators of the ideal $\mathfrak{K}$ of polynomials f in $\mathfrak{A}[x_1, x_2, \cdots, x_r]$, x_i indeterminates, such that $f(t_1, t_2, \cdots, t_r) = 0$. Then there exists a homomorphism S of $\mathfrak{A}[t_1, t_2, \cdots, t_r]$ into $\mathfrak{B}$ such that $a^S = a^s$, $a \in \mathfrak{A}$, and $t_i^S = u_i$, $1 \leq i \leq r$, if and only if $g^s(u_1, \cdots, u_r) = 0$ for every $g \in X$. If S exists, then it is unique.*

We now consider the important special case of IV′ in which $\mathfrak{A} = \Phi$ a field and $r = 1$. Then we know that $\Phi[x]$ is a principal ideal domain (Vol. I, p. 100). Hence the ideal $\mathfrak{K} = (f(x))$, where $(f(x))$ denotes the ideal of polynomial multiples of the polynomial $f(x) \in \mathfrak{K}$. It is clear that $\mathfrak{K} \neq (1) = \Phi[x]$ since, otherwise, $0 = \Phi[x]/\mathfrak{K} \cong \Phi[t] \supseteq \Phi$ which contradicts $1 \neq 0$. Since $(\alpha) = (1)$ if α is a non-zero element of Φ, it is clear that the possibilities for $\mathfrak{K}$ are $\mathfrak{K} = (0)$ or $\mathfrak{K} = (f(x))$ where $f(x)$ is a non-zero polynomial in $\Phi[x]$ of positive degree. In the first case we have $\Phi[x] \cong \Phi[t]$ and t is transcendental. Then II (or IV) is applicable and shows that s can be extended to a homomorphism S sending t into any $u \in \mathfrak{B}$. Now suppose that $f(x) \neq 0$. In this case we call the element $t \in \mathfrak{C}$ *algebraic over* Φ since we have a non-zero polynomial $f(x)$ such that $f(t) = 0$. The ideal $\mathfrak{K}$ is, by definition, the set of polynomials $g(x)$ such that $g(t) = 0$. The polynomial $f(x)$ is a polynomial of least degree in $\mathfrak{K}$ and every other polynomial contained in $\mathfrak{K} = (f(x))$ has the form $g(x)f(x)$. We can normalize $f(x)$ by multiplying it by the inverse of its leading coefficient to obtain a polynomial with leading coefficient 1. If we let $f(x)$ be this polynomial, then clearly f can be characterized by the properties that it is the polynomial of least degree belonging to $\Phi[x]$ with leading coefficient 1 satisfying $f(t) = 0$. We shall call $f(x)$ the *minimum polynomial* (*over* Φ) of the algebraic element $t \in \mathfrak{C}$. We can now state the following result which is a special case of IV′.

V. *Let $\mathfrak{B}$ and $\mathfrak{C}$ be commutative rings, Φ a subfield of $\mathfrak{C}$, t an element of $\mathfrak{C}$ which is algebraic over Φ, and s an isomorphism of Φ into $\mathfrak{B}$:*

$$\begin{array}{ccccc} \mathfrak{C} \supseteq & \Phi[t] & \supseteq & \Phi & \\ & S \searrow & & \downarrow s & \\ & & & \mathfrak{B} & \end{array}$$

Then s can be extended to a homomorphism S of $\Phi[t]$ into $\mathfrak{B}$ so that $t^S = u$, if and only if $f^s(u) = 0$ for the minimum polynomial $f(x)$ of t over Φ. When the extension exists it is unique.

Remarks. The condition one has to put on u to insure the existence of S can be stated also in the following way: u is algebraic over the image Φ^s of Φ and its minimum polynomial over Φ^s is a factor of $f^s(x)$. The equation (1) giving the form of S now becomes

$$S: g(t) \rightarrow g^s(u). \tag{2}$$

It is immediate from this that S is an isomorphism if and only if $f^s(x)$ is the minimum polynomial of u.

2. Algebras. We recall the definition of an algebra $\mathfrak{A}$ over a field Φ (Vol. II, p. 36 and p. 225): $\mathfrak{A}$ is a vector space over Φ in which a product $xy \varepsilon \mathfrak{A}$ is defined for x, y in $\mathfrak{A}$ such that

$$(3) \qquad \begin{gathered} (x_1 + x_2)y = x_1y + x_2y, \quad x(y_1 + y_2) = xy_1 + xy_2 \\ \alpha(xy) = (\alpha x)y = x(\alpha y), \quad \alpha \varepsilon \Phi. \end{gathered}$$

We shall be interested only in algebras which have identities 1 and which are associative; hence in this volume "algebra" will always mean just this.

We shall usually encounter algebras in the following way: We are given a ring $\mathfrak{A}$ and a subfield Φ of the center of $\mathfrak{A}$. Then we can consider $\mathfrak{A}$ as a vector space over Φ by taking αx, $\alpha \varepsilon \Phi$, $x \varepsilon \mathfrak{A}$, to be the ring product of α and x in $\mathfrak{A}$. Clearly this makes $\mathfrak{A}$ a vector space over Φ. Also (3) is clear since α is in the center. Hence we have an algebra $\mathfrak{A}/\Phi$ ($\mathfrak{A}$ over Φ).* This procedure for defining an algebra will be used in studying a field P relative to a subfield Φ. Then we obtain the algebra P$/\Phi$.

Another algebra which is basic is the algebra $\mathfrak{L}_\Phi(\mathfrak{M})$ of linear transformations of a vector space $\mathfrak{M}$ over a field Φ. Here $A + B$, AB and αA for $A, B \varepsilon \mathfrak{L}_\Phi(\mathfrak{M})$ and $\alpha \varepsilon \Phi$ are defined by $x(A + B) = xA + xB$, $x(AB) = (xA)B$, $x(\alpha A) = \alpha(xA) = (\alpha x)A$. The dimensionality $[\mathfrak{L}_\Phi(\mathfrak{M}):\Phi]$ of $\mathfrak{L}_\Phi(\mathfrak{M})$ over Φ is finite if and only if $[\mathfrak{M}:\Phi]$ is finite. If $[\mathfrak{M}:\Phi] = m$, then $[\mathfrak{L}_\Phi(\mathfrak{M}):\Phi] = m^2$ (Vol. II, p. 41).

Evidently an algebra is a ring relative to the + of the vector space and the multiplication ab. A *subalgebra* $\mathfrak{B}$ of an algebra $\mathfrak{A}$ over Φ is a subspace of $\mathfrak{A}$ which is also a subring. An *ideal* of $\mathfrak{A}/\Phi$ is a subspace which is an ideal of $\mathfrak{A}$ as a ring. A *homomorphism* s of the algebra $\mathfrak{A}/\Phi$ into the algebra $\mathfrak{B}/\Phi$ is a mapping of $\mathfrak{A}$ into $\mathfrak{B}$ which is Φ-linear and a ring homomorphism. Isomorphisms and automorphisms are defined in a similar fashion. If $\mathfrak{K}$ is an ideal in $\mathfrak{A}/\Phi$, then the factor space $\mathfrak{A}/\mathfrak{K}$ is an algebra over Φ relative to its vector space compositions and the multiplication $(a + \mathfrak{K})(b + \mathfrak{K}) = ab + \mathfrak{K}$. We have the algebra homomorphism $a \to a + \mathfrak{K}$ of $\mathfrak{A}/\Phi$ onto $\mathfrak{A}/\mathfrak{K}$ over Φ. If s is a homomorphism of $\mathfrak{A}/\Phi$ into $\mathfrak{B}/\Phi$, then the image $\mathfrak{A}^s$ is a subalgebra of $\mathfrak{B}$ and the

* We shall use the notation $\mathfrak{A}/\mathfrak{B}$ also for the difference ring of $\mathfrak{A}$ relative to the ideal $\mathfrak{B}$. Which of these meanings is intended will always be clear from the context.

kernel $\mathfrak{K}$ of s is an ideal in $\mathfrak{A}$. We have the isomorphism $a + \mathfrak{K} \rightarrow a^s$ of $\mathfrak{A}/\mathfrak{K}$ onto $\mathfrak{A}^s$. The basic results on ring homomorphisms extend to algebras and we shall use these without comment.

We shall now record some elementary results on finite dimensional algebras which will be used frequently in the sequel. The first concerns a dimensionality relation for $\mathfrak{A}/\Phi$ and $\mathfrak{A}/\mathrm{E}$, where E is a subfield of Φ. Evidently if E is a subfield of Φ, then we can restrict the multiplication αx, $\alpha \in \Phi$, $x \in \mathfrak{A}$ to α in E. This turns $\mathfrak{A}$ into an algebra $\mathfrak{A}$ over E. Also since E is a subfield of Φ we can define the algebra Φ/E. We now have

VI. *Let $\mathfrak{A}$ be an algebra over Φ, E a subfield of Φ. Suppose $[\mathfrak{A}:\Phi] < \infty$ and $[\Phi:\mathrm{E}] < \infty$. Then*

$$[\mathfrak{A}:\mathrm{E}] = [\mathfrak{A}:\Phi][\Phi:\mathrm{E}]. \tag{4}$$

Proof. Let (u_i), $1 \leq i \leq n$, be a basis for $\mathfrak{A}/\Phi$, (γ_j), $1 \leq j \leq m$, a basis for Φ/E. Then (4) will follow if we can show that $(\gamma_j u_i)$ is a basis for $\mathfrak{A}/\mathrm{E}$. First let $a \in \mathfrak{A}$. Then $a = \sum_1^n \alpha_i u_i$, $\alpha_i \in \Phi$, and $\alpha_i = \sum_{j=1}^m \epsilon_{ij}\gamma_j$ where $\epsilon_{ij} \in \mathrm{E}$. Then $a = \Sigma\epsilon_{ij}\gamma_j u_i$ is a linear combination of the elements $\gamma_j u_i$ with coefficients ϵ_{ij} in E. Now suppose $\Sigma\epsilon_{ij}\gamma_j u_i = 0$ where the $\epsilon_{ij} \in \mathrm{E}$. Then we have $\Sigma\alpha_i u_i = 0$ for $\alpha_i = \sum_j \epsilon_{ij}\gamma_j$ in Φ. Since the u_i are Φ-independent, this gives $\alpha_i = 0$, $1 \leq i \leq n$. Then the formulas $\alpha_i = \Sigma\epsilon_{ij}\gamma_j$ and the E-independence of the γ_j give $\epsilon_{ij} = 0$ for all i, j. This proves that the elements $\gamma_j u_i$ are E-independent and so these form a basis for $\mathfrak{A}/\mathrm{E}$.

VII. *Let $\mathfrak{A}$ be a finite dimensional algebra over a field Φ. Then $\mathfrak{A}$ is a division ring if and only if $\mathfrak{A}$ is an integral domain.*

Proof. We know that division rings are integral domains (Vol. I, p. 54). Now suppose $\mathfrak{A}$ is an integral domain and let a be any non-zero element of $\mathfrak{A}$. Consider the right multiplication a_R: $x \rightarrow xa$ determined by a. This is a linear transformation in $\mathfrak{A}/\Phi$ and, since $ba = 0$ in $\mathfrak{A}$ implies $b = 0$, the null space of a_R is 0. It follows that a_R is surjective (that is, maps $\mathfrak{A}$ onto $\mathfrak{A}$). Hence there exists an element a' such that $a'a = a'a_R = 1$. Thus a

has a left inverse. A similar argument using the left multiplication a_L shows that a has a right inverse. Hence every non-zero element of $\mathfrak{A}$ is a unit and $\mathfrak{A}$ is a division ring.

We consider next algebras $\mathfrak{A} = \Phi[t]$ which have a single generator t (cf. § 1). We have the homomorphism $g(x) \rightarrow g(t)$ of $\Phi[x]$, x an indeterminate, onto $\mathfrak{A}$. If $\mathfrak{K}$ is the kernel, then $\mathfrak{A} \cong \Phi[x]/\mathfrak{K}$. Also we have seen in § 1 that $\mathfrak{K} = (f(x))$ where $f(x) = 0$ or is a non-zero polynomial with leading coefficient 1. In the first case, t is transcendental and the homomorphism we indicated is an isomorphism. In the second case, t is algebraic and $f(x)$ is its minimum polynomial. Then we have

VIII. *Let $\mathfrak{A} = \Phi[t]$ be an algebra over Φ generated by a single algebraic element t whose minimum polynomial is $f(x)$. Then*

$$[\mathfrak{A}:\Phi] = \deg f(x), \tag{5}$$

the degree of $f(x)$.

Proof. Let $n = \deg f(x)$. Then we assert that $(1, t, \cdots, t^{n-1})$ is a basis for $\mathfrak{A}/\Phi$. Thus let a be any element of $\mathfrak{A} = \Phi[t]$. This has the form $g(t)$, $g(x)$ in $\Phi[x]$. By the division process in $\Phi[x]$ we can write $g(x) = f(x)q(x) + r(x)$ where $\deg r(x) < \deg f(x)$. Then if we apply the homomorphism of $\Phi[x]/\Phi$ onto $\Phi[t]/\Phi$ sending x into t, we obtain $a = g(t) = 0q(t) + r(t)$. Since $\deg r(x) < n$, this shows that $a = r(t)$ is a Φ-linear combination of $1, t, \cdots, t^{n-1}$. Next we note that $1, t, \cdots, t^{n-1}$ are linearly independent over Φ since otherwise we would have a polynomial $g(x) \neq 0$ of degree $< n$ such that $g(t) = 0$. This contradicts the hypothesis that $f(x)$ is the minimum polynomial. Hence $(1, t, \cdots, t^{n-1})$ is a basis and (5) holds.

We recall that $\Phi[t] \cong \Phi[x]/(f(x))$, $f(x)$ a polynomial of positive degree, is a field if and only if $f(x)$ is irreducible (Vol. I, p. 101). Otherwise, $\Phi[t]$ is not an integral domain. It is useful to have a more complete analysis of the structure of $\Phi[t]$ in terms of the minimum polynomial $f(x)$. We shall indicate the results in the following exercises.

EXERCISES

1. An algebra $\mathfrak{A}$ is a *direct sum* of ideals $\mathfrak{A}_i$ if $\mathfrak{A}$ is a vector space direct sum of the subspaces $\mathfrak{A}_i$. Let $\mathfrak{A} = \Phi[t]$, t algebraic with minimum polynomial $f(x)$.

Suppose $f(x) = f_1(x)f_2(x) \cdots f_r(x)$ where $(f_i(x), f_j(x)) = 1$ if $i \neq j$. Set $q_i(x) = f(x)/f_i(x)$. Show that there exist polynomials $a_i(x)$ such that

$$\sum_1^r a_i(x)q_i(x) = 1.$$

Set $e_i = a_i(t)q_i(t)$ and show that

$$e_1 + e_2 + \cdots + e_r = 1, \quad e_i^2 = e_i, \quad e_ie_j = 0, \quad i \neq j.$$

Show that $\mathfrak{A} = \mathfrak{A}e_1 \oplus \mathfrak{A}e_2 \oplus \cdots \oplus \mathfrak{A}e_r$ and that the ideal $\mathfrak{A}e_i = \{ae_i | a \in \mathfrak{A}\}$ considered as an algebra with identity e_i has the form $\Phi[te_i]$ and is isomorphic to $\Phi[x]/(f_i(x))$.

2. Let $\mathfrak{A} = \Phi[t]$, t algebraic with minimum polynomial $f(x)$. Let $f(x) = p_1(x)^{k_1}p_2(x)^{k_2} \cdots p_r(x)^{k_r}$, $p_i(x)$ irreducible, $p_i(x) \neq p_j(x)$, $i \neq j$. Show that if $z = p_1(t)p_2(t) \cdots p_r(t)$, then the ideal $\mathfrak{N} = \mathfrak{A}z$ in $\mathfrak{A}$ is nilpotent in the sense that there exists an integer k such that every product of k elements of $\mathfrak{N}$ is 0. Show that $\overline{\mathfrak{A}} = \mathfrak{A}/\mathfrak{N} = \Phi[\bar{t}]$, $\bar{t} = t + \mathfrak{N}$, and $\bar{t}$ is algebraic with minimum polynomial $g(x) = p_1(x)p_2(x) \cdots p_r(x)$. Show that $\overline{\mathfrak{A}} = \overline{\mathfrak{A}}_1 \oplus \overline{\mathfrak{A}}_2 \oplus \cdots \oplus \overline{\mathfrak{A}}_r$ where $\overline{\mathfrak{A}}_i$ is an ideal which as an algebra is isomorphic to the field $\Phi[x]/(p_i(x))$.

3. Let $\mathfrak{A}/\Phi$ be an *algebraic* algebra in the sense that every element of $\mathfrak{A}$ is algebraic. Prove that, if $\mathfrak{A}$ is an integral domain, then $\mathfrak{A}$ is a division ring.

3. Tensor products of vector spaces. Let $\mathfrak{M}$, $\mathfrak{N}$ and $\mathfrak{P}$ be vector spaces over the same field Φ. Then a *bilinear mapping* of $\mathfrak{M}$, $\mathfrak{N}$ into $\mathfrak{P}$ is a mapping of the product set $\mathfrak{M} \times \mathfrak{N}$ into $\mathfrak{P}$ such that, if $x \times y$ denotes the image of the pair (x, y), $x \in \mathfrak{M}$, $y \in \mathfrak{N}$, then

$$\begin{aligned} (x_1 + x_2) \times y &= x_1 \times y + x_2 \times y, \\ x \times (y_1 + y_2) &= x \times y_1 + x \times y_2 \\ \alpha(x \times y) &= \alpha x \times y = x \times \alpha y, \quad \alpha \in \Phi. \end{aligned} \tag{6}$$

It is clear that the product xy in any algebra $\mathfrak{A}$ is bilinear from $\mathfrak{A}$, $\mathfrak{A}$ to $\mathfrak{A}$. We shall say that a vector space $\mathfrak{P}$ and a bilinear mapping $\otimes$ of $\mathfrak{M}$, $\mathfrak{N}$ into $\mathfrak{P}$ is a *tensor product* of $\mathfrak{M}$ and $\mathfrak{N}$ and we write $\mathfrak{P} = \mathfrak{M} \otimes \mathfrak{N}$ if the pair $(\otimes, \mathfrak{P})$ is "universal" for bilinear mappings in the sense that the following condition is fulfilled:

If $\mathfrak{P}'$ is any vector space and $\times'$ is a bilinear mapping of $\mathfrak{M}$, $\mathfrak{N}$ into $\mathfrak{P}'$, then there exists a unique linear mapping π of $\mathfrak{P}$ into $\mathfrak{P}'$ such that $(x \otimes y)\pi = x \times' y$.

This notion is a special case of the general concept of the tensor product of a right module $\mathfrak{M}$ over a ring $\mathfrak{A}$ and a left module $\mathfrak{N}$ over $\mathfrak{A}$. The special case we have defined for vector spaces is treated under slightly different but equivalent hypotheses in Vol.

II, Chap. VII. In particular, a proof of the existence of a tensor product of vector spaces and nearly all the basic properties we shall require were given in Vol. II. At this point we shall give another derivation of some of these basic results which is more in keeping with the spirit of the now standard treatment of the module case.

We first give a construction of a tensor product. To do this one begins with a vector space $\mathfrak{F}$ having as basis the product set $\mathfrak{M} \times \mathfrak{N}$ of pairs (x, y), $x \in \mathfrak{M}$, $y \in \mathfrak{N}$. Thus the elements of $\mathfrak{F}$ are the expressions $\xi_1(x_1, y_1) + \xi_2(x_2, y_2) + \cdots + \xi_m(x_m, y_m)$ where $\xi_i \in \Phi$, $x_i \in \mathfrak{M}$, $y_i \in \mathfrak{N}$, and the pairs (x_i, y_i) are distinct. If two elements are given we can introduce terms with 0 coefficients and thus suppose that the elements are $\sum_1^m \xi_i(x_i, y_i)$ and $\sum_1^m \eta_i(x_i, y_i)$. Then equality holds if and only if $\xi_i = \eta_i$, $i = 1, 2, \cdots, m$. Addition is defined by $\sum_1^m \xi_i(x_i, y_i) + \sum_1^m \eta_i(x_i, y_i) = \sum_1^m (\xi_i + \eta_i)(x_i, y_i)$ and multiplication by α in Φ by $\alpha\Sigma\xi_i(x_i, y_i) = \Sigma(\alpha\xi_i)(x_i, y_i)$. It is immediate that $\mathfrak{F}$ is a vector space over Φ. Since $\mathfrak{M} \times \mathfrak{N}$ is usually infinite, $\mathfrak{F}$ is usually an infinite dimensional space. Now let $\mathfrak{R}$ be the subspace of $\mathfrak{F}$ spanned by all the vectors of the following forms:

$$
(7)\qquad
\begin{aligned}
&(x_1 + x_2, y) - (x_1, y) - (x_2, y)\\
&(x, y_1 + y_2) - (x, y_1) - (x, y_2)\\
&(\alpha x, y) - (x, \alpha y)\\
&\alpha(x, y) - (\alpha x, y),
\end{aligned}
$$

$x \in \mathfrak{M}$, $y \in \mathfrak{N}$, $\alpha \in \Phi$. Let $\mathfrak{P}$ be the factor space $\mathfrak{F}/\mathfrak{R}$ and set $x \otimes y \equiv (x, y) + \mathfrak{R}$, the coset of (x, y) in $\mathfrak{F}/\mathfrak{R}$. Then we have:

$$
\begin{aligned}
&(x_1 + x_2) \otimes y - x_1 \otimes y - x_2 \otimes y\\
&\qquad = (x_1 + x_2, y) - (x_1, y) - (x_2, y) + \mathfrak{R} = \mathfrak{R}\\
&x \otimes (y_1 + y_2) - x \otimes y_1 - x \otimes y_2\\
&\qquad = (x, y_1 + y_2) - (x, y_1) - (x, y_2) + \mathfrak{R} = \mathfrak{R}
\end{aligned}
$$

$$\alpha(x \otimes y) - \alpha x \otimes y = \alpha(x, y) - (\alpha x, y) + \mathfrak{R} = \mathfrak{R}$$

$$\alpha x \otimes y - x \otimes \alpha y = (\alpha x, y) - (x, \alpha y) + \mathfrak{R} = \mathfrak{R}.$$

Hence $x \otimes y$ is bilinear. Since the vectors (x, y) generate $\mathfrak{F}$, the cosets $x \otimes y$ generate $\mathfrak{P} = \mathfrak{F}/\mathfrak{N}$.

Now let $\times'$ be a bilinear mapping of $\mathfrak{M}, \mathfrak{N}$ into the vector space $\mathfrak{P}'$. Since the vectors (x, y) form a basis for $\mathfrak{F}$, there exists a linear mapping π' of $\mathfrak{F}$ into $\mathfrak{P}'$ such that $(x, y)\pi' = x \times' y$. Let $\mathfrak{K}$ be the kernel of π'. Then $((x_1 + x_2, y) - (x_1, y) - (x_2, y))\pi' = (x_1 + x_2) \times' y - x_1 \times' y - x_2 \times' y = 0$; so $(x_1 + x_2, y) - (x, y) - (x_2, y) \in \mathfrak{K}$. Similarly, $(x, y_1 + y_2) - (x, y_1) - (x, y_2) \in \mathfrak{K}$, $(\alpha x, y) - \alpha(x, y) \in \mathfrak{K}$, and $(\alpha x, y) - (x, \alpha y) \in \mathfrak{K}$. This implies that $\mathfrak{N} \subseteq \mathfrak{K}$ and, consequently, we have the linear mapping π of $\mathfrak{P} = \mathfrak{F}/\mathfrak{N}$ into $\mathfrak{P}'$ such that $(x \otimes y)\pi \equiv ((x, y) + \mathfrak{N})\pi = x \times' y$. Since the space $\mathfrak{P} = \mathfrak{F}/\mathfrak{N}$ is generated by the elements $x \otimes y$, it is clear that π is uniquely determined by the linearity property and $(x \otimes y)\pi = x \times' y$. We have therefore shown that $(\mathfrak{P}, \otimes)$ is a tensor product of $\mathfrak{M}$ and $\mathfrak{N}$ and accordingly we shall write $\mathfrak{P} = \mathfrak{M} \otimes \mathfrak{N}$ (or $\mathfrak{M} \otimes_\Phi \mathfrak{N}$, if it is necessary to indicate the base field Φ). It is immediate from the definition that if $(\mathfrak{P}_1, \otimes_1)$ and $(\mathfrak{P}_2, \otimes_2)$ are two tensor products, then we have a linear mapping of $\mathfrak{P}_1$ into $\mathfrak{P}_2$ such that $x \otimes_1 y \rightarrow x \otimes_2 y$ and we have a linear mapping of $\mathfrak{P}_2$ into $\mathfrak{P}_1$ such that $x \otimes_2 y \rightarrow x \otimes_1 y$. Since the $x \otimes_i y$ generate $\mathfrak{P}_i$, the products in both orders of the two linear mappings are identity mappings. It follows that both mappings are surjective (onto) linear isomorphisms. In this sense the tensor product is uniquely determined and so we may speak of *the* tensor product of $\mathfrak{M}$ and $\mathfrak{N}$.

Let $\{e_\alpha\}$ and $\{f_\beta\}$ be sets of generators for $\mathfrak{M}$ and $\mathfrak{N}$ respectively. Then any $x \in \mathfrak{M}$ has the form $x = \sum_1^m \xi_i e_i$ where $\{e_i\}$ is a finite subset of $\{e_\alpha\}$ and any $y \in \mathfrak{N}$ has the form $y = \sum_1^n \eta_j f_j$, $\{f_j\} \subseteq \{f_\beta\}$. Hence, by the bilinearity of $\otimes$ we have $x \otimes y = \Sigma \xi_i \eta_j e_i \otimes f_j$. Since the elements $x \otimes y$ generate $\mathfrak{M} \otimes \mathfrak{N}$, we see that the products $e_\alpha \otimes f_\beta$ generate $\mathfrak{M} \otimes \mathfrak{N}$. Now suppose that the $\{e_\alpha\}$ and $\{f_\beta\}$ are independent as well as generators, that is, these form bases for their respective spaces. We assert that the set of products $\{e_\alpha \otimes f_\beta\}$ is a basis for $\mathfrak{M} \otimes \mathfrak{N}$. Since these are generators we just need to show that they are linearly independent. For this purpose we form a vector space $\mathfrak{P}'$ with basis $g_{\alpha\beta}$ in 1–1 correspondence with the product set (α, β) of the index sets of α and of β. If $x = \Sigma\xi_i e_i$ and $y = \Sigma\eta_j f_j$, then we define $x \times' y = \Sigma\xi_i\eta_j g_{ij}$. It is easy to check that the product $\times'$ is bilinear, so we have the linear mapping π of $\mathfrak{M} \otimes \mathfrak{N}$ into $\mathfrak{P}'$ sending $x \otimes y \rightarrow x \times' y$. In particular, $e_\alpha \otimes f_\beta \rightarrow e_\alpha \times' f_\beta = g_{\alpha\beta}$. Since the $g_{\alpha\beta}$ are linearly independent, the same holds for the $e_\alpha \otimes f_\beta$ and we have proved

IX. *Let $\{e_\alpha\}$ and $\{f_\beta\}$ be generators for $\mathfrak{M}$ over Φ and $\mathfrak{N}$ over Φ respectively. Then the set $\{e_\alpha \otimes f_\beta\}$ generates $\mathfrak{M} \otimes \mathfrak{N}$. Moreover, if the $\{e_\alpha\}$ and $\{f_\beta\}$ are bases, then the same holds for $\{e_\alpha \otimes f_\beta\}$.*

The second property actually characterizes the tensor product among the bilinear mappings of $\mathfrak{M}$ and $\mathfrak{N}$. More precisely, let $\times'$ be a bilinear mapping from $\mathfrak{M}$ and $\mathfrak{N}$ to a space $\mathfrak{P}'$ and suppose there exists a basis (e_α) for $\mathfrak{M}$ over Φ and a basis (f_β) for $\mathfrak{N}$ over Φ such that $(e_\alpha \times' f_\beta)$ is a basis for $\mathfrak{P}'$. Then $(\mathfrak{P}', \times')$ is a tensor product. Thus we have the linear mapping of $\mathfrak{M} \otimes \mathfrak{N}$ into $\mathfrak{P}'$ sending $e_\alpha \otimes f_\beta$ into $e_\alpha \times' f_\beta$. Since the $e_\alpha \times' f_\beta$ generate $\mathfrak{P}'$, the mapping is surjective and, since the $e_\alpha \times' f_\beta$ are linearly independent, the mapping is 1–1. Thus we have a linear isomorphism of $\mathfrak{M} \otimes \mathfrak{N}$ onto $\mathfrak{P}'$, mapping $x \otimes y$ into $x \times' y$. This implies that $(\mathfrak{P}', \times')$ is a tensor product.

In the case of finite dimensional spaces we have the following simple criterion.

X. *Let $\times'$ be a bilinear mapping of the finite dimensional spaces $\mathfrak{M}$ and $\mathfrak{N}$ into $\mathfrak{P}'$ and suppose that $\mathfrak{P}'$ is generated by the products $x \times' y$. Then the dimensionality $[\mathfrak{P}':\Phi] \leq [\mathfrak{M}:\Phi][\mathfrak{N}:\Phi]$ and equality holds if and only if $(\mathfrak{P}', \times')$ is a tensor product of $\mathfrak{M}$ and $\mathfrak{N}$.*

Proof. Let (e_i), (f_j) be bases for $\mathfrak{M}$ and $\mathfrak{N}$ respectively. Then every $x \times' y$ is a linear combination of the elements $e_i \times' f_j$ and so every element of $\mathfrak{P}'$ is a linear combination of these elements. This implies $[\mathfrak{P}':\Phi] \leq [\mathfrak{M}:\Phi][\mathfrak{N}:\Phi]$. $(\mathfrak{P}', \times')$ is the tensor product if and only if the set $(e_i \times' f_j)$ is a basis. This is the case if and only if the equality holds in the dimensionality relation.

We recall that, if A is a linear mapping of $\mathfrak{M}$ into $\mathfrak{M}_1$ and B is a linear mapping of $\mathfrak{N}$ into $\mathfrak{N}_1$, then there exists a uniquely determined linear mapping $A \otimes B$ of $\mathfrak{M} \otimes \mathfrak{N}$ into $\mathfrak{M}_1 \otimes \mathfrak{N}_1$ such that $(x \otimes y)(A \otimes B) = xA \otimes yB$ (Vol. II, p. 211). We recall also that, if P is an extension field of the field Φ so that P is a vector space over Φ and $\mathfrak{M}$ is any vector space over Φ, then $P \otimes_\Phi \mathfrak{M}$ can be considered as a vector space over P by means of the product $\rho(\Sigma\rho_i \otimes x_i) = \Sigma\rho\rho_i \otimes x_i$, $\rho, \rho_i \in P$, $x_i \in \mathfrak{M}$ (Vol. II, p. 221). We denote this vector space as $\mathfrak{M}_P$ and we refer to it as the space obtained from $\mathfrak{M}$ by extending the base field to P. If A is a linear transformation in $\mathfrak{M}$ over Φ, then $1 \otimes A$ (defined by $(\Sigma\rho_i \otimes x_i)\ (1 \otimes A) = \Sigma\rho_i \otimes x_iA)$ is a linear transformation in $\mathfrak{M}_P$ over P which may be considered as the extension of A to $\mathfrak{M}_P$. We shall use the same letter A to denote this extension. If (e_α) is a basis for $\mathfrak{M}$ over Φ, then $(1 \otimes e_\alpha)$ is a basis for $\mathfrak{M}_P$ over P, so $\mathfrak{M}$ over Φ and $\mathfrak{M}_P$ over P have the same dimensionality. If $\mathfrak{M}$ is finite dimensional with basis (e_i), $1 \leq i \leq n$, and A is the linear transformation with matrix (α_{ij}) relative to this basis, then $e_iA = \Sigma\alpha_{ij}e_j$ and $(1 \otimes e_i)A = \Sigma\alpha_{ij}(1 \otimes e_j)$. Hence the extension A has the same matrix relative to the basis $(1 \otimes e_i)$.

We recall also that the tensor product is commutative in the sense that there exists a 1–1 linear transformation such that $x \otimes y \to y \otimes x$ of $\mathfrak{M} \otimes \mathfrak{N}$ onto $\mathfrak{N} \otimes \mathfrak{M}$. Moreover, associativity holds in the sense that there is a linear isomorphism of $(\mathfrak{M} \otimes \mathfrak{N}) \otimes \mathfrak{S}$ onto $\mathfrak{M} \otimes (\mathfrak{N} \otimes \mathfrak{S})$ mapping $(x \otimes y) \otimes z$ into $x \otimes (y \otimes z)$. These results have been established in Vol. II, pp. 209–210. We shall indicate alternative proofs in some of the following exercises.

EXERCISES

1. Show that, if $\{f_\beta\}$ is a set of generators for $\mathfrak{N}$, then every element of $\mathfrak{M} \otimes \mathfrak{N}$ has the form $\Sigma x_i \otimes f_i$, $\{f_i\}$ a finite subset of $\{f_\beta\}$ and $x_i \in \mathfrak{M}$. Show that, if the $\{f_\beta\}$ are linearly independent, then $\Sigma x_i \otimes f_i = 0$ if and only if every $x_i = 0$.

2. Show that, if $\mathfrak{M}_1$ is a subspace of $\mathfrak{M}$, then the subspace $\mathfrak{M}_1 \otimes \mathfrak{N}$ generated by all vectors $x_1 \otimes y$, $x_1 \in \mathfrak{M}_1$, $y \in \mathfrak{N}$ is the tensor product of $\mathfrak{M}_1$ and $\mathfrak{N}$ relative to the $\otimes$ defined in $\mathfrak{M} \otimes \mathfrak{N}$.

3. Let $\mathfrak{K}$ be a subspace of $\mathfrak{M}$, $\mathfrak{L}$ a subspace of $\mathfrak{N}$. Show that $(\mathfrak{M}/\mathfrak{K}) \otimes (\mathfrak{N}/\mathfrak{L})$ and $(\mathfrak{M} \otimes \mathfrak{N})/(\mathfrak{K} \otimes \mathfrak{N} + \mathfrak{M} \otimes \mathfrak{L})$ are isomorphic under a linear mapping such that $(x + \mathfrak{K}) \otimes (y + \mathfrak{L}) \to x \otimes y + (\mathfrak{K} \otimes \mathfrak{N} + \mathfrak{M} \otimes \mathfrak{L})$.

4. Let $\mathfrak{M}_1, \mathfrak{M}_2, \cdots, \mathfrak{M}_r$ and $\mathfrak{P}$ be vector spaces over Φ. Define an r-linear mapping $(x_1, \cdots, x_r) \to x_1 \times x_2 \times \cdots \times x_r \in \mathfrak{P}$, $x_i \in \mathfrak{M}_i$, by the properties:

$$x_1 \times \cdots \times (x_i' + x_i'') \times \cdots \times x_r = x_1 \times \cdots \times x_i' \times \cdots \times x_r + x_1 \times \cdots \times x_i'' \times \cdots \times x_r$$

$$\alpha(x_1 \times \cdots \times x_r) = x_1 \times \cdots \times \alpha x_i \times \cdots \times x_r.$$

Show that there exists a $\mathfrak{P}$ and an r-linear mapping of $\mathfrak{M}_1, \cdots, \mathfrak{M}_r$ into $\mathfrak{P}$ such that: if $(x_1, \cdots, x_r) \rightarrow x_1 \times' x_2 \times' \cdots \times' x_r$ is an r-linear mapping of $\mathfrak{M}_1, \cdots, \mathfrak{M}_r$ into $\mathfrak{P}'$, then there exists a unique linear mapping π of $\mathfrak{P}$ into $\mathfrak{P}'$ such that $(x_1 \otimes \cdots \otimes x_r)\pi = x_1 \times' \cdots \times' x_r$. Denote this $\mathfrak{P}$ together with its product as the tensor product $\mathfrak{M}_1 \otimes \mathfrak{M}_2 \otimes \cdots \otimes \mathfrak{M}_r$.

5. Show that $\mathfrak{M} \otimes \mathfrak{N} \otimes \mathfrak{P}$ is isomorphic to $\mathfrak{M} \otimes (\mathfrak{N} \otimes \mathfrak{P})$ and $(\mathfrak{M} \otimes \mathfrak{N}) \otimes \mathfrak{P}$ by means of linear mappings such that $x \otimes y \otimes z \rightarrow x \otimes (y \otimes z)$ and $(x \otimes y) \otimes z$ respectively. Generalize to r factors.

6. Show that $\mathfrak{M} \otimes \mathfrak{N}$ is isomorphic to $\mathfrak{N} \otimes \mathfrak{M}$ under a linear mapping sending $x \otimes y \rightarrow y \otimes x$. (Hint: Given $\mathfrak{N} \otimes \mathfrak{M}$, define $x \times' y = y \otimes x$, $x \in \mathfrak{M}$, $y \in \mathfrak{N}$. Show that this gives a bilinear mapping of $\mathfrak{M}, \mathfrak{N}$ into $\mathfrak{N} \otimes \mathfrak{M}$ and apply the defining property of $\mathfrak{M} \otimes \mathfrak{N}$. Then reverse the roles of $\mathfrak{M}$ and $\mathfrak{N}$.)

4. Tensor product of algebras. We recall that, if $\mathfrak{A}_1$ and $\mathfrak{A}_2$ are algebras over Φ, then the vector space $\mathfrak{A} = \mathfrak{A}_1 \otimes \mathfrak{A}_2$ is an algebra relative to its vector space compositions and the multiplication

$$(8) \qquad \left(\sum_i a_{1i} \otimes a_{2i}\right)\left(\sum_j b_{1j} \otimes b_{2j}\right) = \sum_{i,j} a_{1i}b_{1j} \otimes a_{2i}b_{2j},$$

$a_{1i}, b_{1j} \in \mathfrak{A}_1$, $a_{2i}, b_{2j} \in \mathfrak{A}_2$ (Vol. II, p. 225). The associativity of $\mathfrak{A}_1$ and $\mathfrak{A}_2$ implies associativity of $\mathfrak{A}_1 \otimes \mathfrak{A}_2$ and $1_1 \otimes 1_2$ is the identity 1 of $\mathfrak{A} = \mathfrak{A}_1 \otimes \mathfrak{A}_2$ if 1_i is the identity of $\mathfrak{A}_i$. Also $\mathfrak{A}$ is commutative if the $\mathfrak{A}_i$ are commutative. The basic property of the tensor product of algebras is the following homomorphism theorem.

XI. *Let $\mathfrak{A}_i$, $i = 1, 2$, be algebras over Φ, s_i a homomorphism of $\mathfrak{A}_i$ into an algebra $\mathfrak{B}$ such that $a_1^{s_1}a_2^{s_2} = a_2^{s_2}a_1^{s_1}$, $a_1 \in \mathfrak{A}_1$, $a_2 \in \mathfrak{A}_2$. Then there exists a homomorphism s of $\mathfrak{A} = \mathfrak{A}_1 \otimes \mathfrak{A}_2$ into $\mathfrak{B}$ such that*

$$(9) \qquad (\Sigma a_{1i} \otimes a_{2i})^s = \Sigma a_{1i}^{s_1} a_{2i}^{s_2}.$$

Proof. The algebra product $a_1 \times' a_2 \equiv a_1^{s_1}a_2^{s_2} \in \mathfrak{B}$ defines a bilinear mapping of $\mathfrak{A}_1, \mathfrak{A}_2$ into $\mathfrak{B}$. This is clear from the linearity of the s_i and the properties of the multiplication composition in $\mathfrak{B}$. Hence the definition implies that we have a linear mapping s of $\mathfrak{A}_1 \otimes \mathfrak{A}_2$ into $\mathfrak{B}$ such that $(a_1 \otimes a_2)^s = a_1^s a_2^s$. Then s has the form (9). We have $((a_1 \otimes a_2)(b_1 \otimes b_2))^s = (a_1b_1 \otimes a_2b_2)^s = (a_1b_1)^{s_1}(a_2b_2)^{s_2} = a_1^{s_1}b_1^{s_1}a_2^{s_2}b_2^{s_2} = a_1^{s_1}a_2^{s_2}b_1^{s_1}b_2^{s_2} = ((a_1 \otimes a_2)^s (b_1 \otimes b_2)^s)$. This implies that s is an algebra homomorphism.

Suppose now that the following condition holds in $\mathfrak{B}$:

(i) If (e_α) is a basis for $\mathfrak{A}_1$ over Φ and (f_β) is a basis for $\mathfrak{A}_2$ over Φ, then the set $\{e_\alpha^{s_1} f_\beta^{s_2}\}$ is linearly independent.

An equivalent condition for this which we shall sometimes find more convenient is

(i′) If (f_β) is a basis for $\mathfrak{A}_2$ over Φ, then a relation $a_1^{s_1} f_1^{s_2} + a_2^{s_1} f_2^{s_2} + \cdots + a_m^{s_1} f_m^{s_2} = 0$ for $a_i \in \mathfrak{A}_1$ and $f_i \in (f_\beta)$ implies that every $a_i = 0$ (cf. ex. 1 of § 3).

Now we have seen that, if (i) or (i′) holds, then the mapping s given by (9) is an isomorphism of $\mathfrak{A} = \mathfrak{A}_1 \otimes \mathfrak{A}_2$ as vector space into $\mathfrak{B}$. Since this is an algebra homomorphism, clearly it is an algebra isomorphism. We remark that (i) cannot hold unless s_1 and s_2 are isomorphisms.

The result we have obtained actually gives an internal characterization of $\mathfrak{A}_1 \otimes \mathfrak{A}_2$. For this we note that $a_1 \rightarrow a_1^{s_1} \equiv a_1 \otimes 1_2$ and $a_2 \rightarrow a_2^{s_2} \equiv 1_1 \otimes a_2$ are homomorphisms of $\mathfrak{A}_1$ and $\mathfrak{A}_2$ respectively into $\mathfrak{A}_1 \otimes \mathfrak{A}_2$, since the linearity of the mappings we have indicated follows from the bilinearity of $a_1 \otimes a_2$, and the homomorphism for multiplication is clear from (9). The commutativity condition: $a_1^{s_1} a_2^{s_2} = a_2^{s_2} a_1^{s_1}$ is clear, since $a_1^{s_1} a_2^{s_2} = (a_1 \otimes 1_2)(1_1 \otimes a_2) = a_1 \otimes a_2 = (1_1 \otimes a_2)(a_1 \otimes 1_2) = a_2^{s_2} a_1^{s_1}$. Finally, if (e_α) and (f_β) are bases for $\mathfrak{A}_1$ and $\mathfrak{A}_2$ respectively, then the set $\{e_\alpha^{s_1} f_\beta^{s_2}\} = \{e_\alpha \otimes f_\beta\}$ is linearly independent. It follows that $(e_\alpha^{s_1})$ is a basis for $\mathfrak{A}_1^{s_1} = \{a_1 \otimes 1\}$ and $(f_\beta^{s_2})$ is a basis for $\mathfrak{A}_2^{s_2}$. Also s_1 and s_2 are isomorphisms and we can identify $\mathfrak{A}_1^{s_1}$ with $\mathfrak{A}_1$, $\mathfrak{A}_2^{s_2}$ with $\mathfrak{A}_2$. Our results evidently lead to the following internal characterization of the tensor product of algebras:

XII. *Let $\mathfrak{A}$ be an algebra, $\mathfrak{A}_1$ and $\mathfrak{A}_2$ subalgebras such that*

(i) $a_1 a_2 = a_2 a_1$, $a_i \in \mathfrak{A}_i$.

(ii) *If (e_α) is a basis for $\mathfrak{A}_1$ and (f_β) is a basis for $\mathfrak{A}_2$, then $\{e_\alpha f_\beta\}$ is a linearly independent set.*

(iii) *$\mathfrak{A}$ is generated by $\mathfrak{A}_1$ and $\mathfrak{A}_2$.*

Then $\Sigma a_{1i} \otimes a_{2i} \rightarrow \Sigma a_{1i} a_{2i}$ is an isomorphism of $\mathfrak{A}_1 \otimes \mathfrak{A}_2$ onto $\mathfrak{A}$.

Because of this result and the situation we noted in $\mathfrak{A}_1 \otimes \mathfrak{A}_2$ itself, we shall say that $\mathfrak{A}$ *is the tensor product of its subalgebras* $\mathfrak{A}_1$ *and* $\mathfrak{A}_2$ if the above conditions (i)–(iii) are fulfilled. As we have seen, the condition (ii) can be replaced by the equivalent condition:

(ii′) *If* (f_β) *is a basis for* $\mathfrak{A}_2$, *then* $a_1 f_1 + a_2 f_2 + \cdots + a_m f_m = 0$ *for* $a_i \in \mathfrak{A}_1$, $f_i \in (f_\beta)$ *implies every* $a_i = 0$.

Of course, the roles of $\mathfrak{A}_1$ and $\mathfrak{A}_2$ can be interchanged in this. We remark also that (ii) and (iii) can be combined in a single condition: *If* (e_α) *is a basis for* $\mathfrak{A}_1$ *and* (f_β) *is a basis for* $\mathfrak{A}_2$, *then* $(e_\alpha f_\beta)$ *is a basis for* $\mathfrak{A}$. For finite dimensional algebras this is equivalent to the dimensionality condition: $[\mathfrak{A}:\Phi] = [\mathfrak{A}_1:\Phi][\mathfrak{A}_2:\Phi]$ (cf. X).

EXERCISES

1. Let $\mathfrak{A}$ be an algebra over the field Φ and let $\mathfrak{A}[x]$ be the algebra of polynomials in an indeterminate x over $\mathfrak{A}$. Show that $\mathfrak{A}[x]$ is the tensor product of its subalgebra $\mathfrak{A}$ (constants of $\mathfrak{A}[x]$) and its subalgebra $\Phi[x]$ of polynomials in x with coefficients in Φ. Use this to prove that $\Phi[x, y]$, x, y indeterminates, is the tensor product of its subalgebras $\Phi[x]$ and $\Phi[y]$.

2. Let $\Phi(x, y)$ be the field of rational expressions in the indeterminates x, y, that is, the field of fractions of $\Phi[x, y]$. Let $\mathfrak{A}$ be the subset of fractions with denominators of the form $f(x)g(y)$, $f(x) \in \Phi[x]$, $g(y) \in \Phi[y]$. Show that $\mathfrak{A}$ is a subalgebra of $\Phi(x, y)$ which contains the subalgebras $\Phi(x)$, $\Phi(y)$ where these are the fields of fractions of $\Phi[x]$ and $\Phi[y]$ respectively. Show that $\mathfrak{A}$ is the tensor product of these subalgebras and that $\mathfrak{A}$ is not a field.

Chapter I

FINITE DIMENSIONAL EXTENSION FIELDS

If Φ is a subfield of a field P, then we have seen that we can consider P as an algebra over Φ. In this chapter we shall be concerned primarily with the situation in which P is finite dimensional over the subfield Φ. We shall be concerned particularly with the general results of Galois theory that are of importance throughout algebra and especially in the theory of algebraic numbers. We shall consider the notions of normality, separability, and pure inseparability for extension fields, Galois cohomology, regular representations, traces, and norms. Also the basic results on finite fields will be derived and the notion of composites of two extension fields will be considered.

In most of our considerations, and indeed throughout this book, we shall usually be given a field Φ and we shall be concerned with extension fields P/Φ. The ways of obtaining such extensions have already been indicated in Vol. I, pp. 100–104. At the beginning of this chapter we adopt a different point of view. Here we are given the top field P and we look down at its various subfields; moreover, we do not insist that these contain any particular subfield (except, of course, the prime field). The treatment here will be abstract in the sense that no knowledge of the structure of an extension is required. In spite of this we can give a survey of the subfields which are of finite co-dimension in the given field P and those which are Galois in P. These surveys are given in two general "Galois correspondences." After these rather abstract considerations we shall go down to Φ and we shall apply the general results to the extension P/Φ in terms of polynomial equations with coefficients in P.

1. Some vector spaces associated with mappings of fields. Let E and P be two fields, and let $\mathfrak{L}(\mathrm{E}, \mathrm{P})$ denote the set of homomorphisms of the additive group $(\mathrm{E}, +)$ of E into $(\mathrm{P}, +)$. The set $\mathfrak{L}(\mathrm{E}, \mathrm{P})$ is a group relative to the composition $A + B$ defined by $\epsilon(A + B) = \epsilon A + \epsilon B$ for ϵ in E. One checks that $A + B \varepsilon \mathfrak{L}(\mathrm{E}, \mathrm{P})$ and that the group conditions hold. The 0 of $\mathfrak{L}(\mathrm{E}, \mathrm{P})$ is the mapping 0 such that $\epsilon 0 = 0$, the 0 of P, for all ϵ in E, and $-A$ is given by $\epsilon(-A) = -\epsilon A$ (cf. Vol. I, § 2.13 and Vol. II, § 2.2). If Δ is a third field and $A \varepsilon \mathfrak{L}(\mathrm{E}, \mathrm{P})$ and $B \varepsilon \mathfrak{L}(\mathrm{P}, \Delta)$, then the resultant AB defined by $\epsilon(AB) = (\epsilon A)B$ is an element of $\mathfrak{L}(\mathrm{E}, \Delta)$. Both distributive laws hold for this composition. In combined form they say that, if $A_1, A_2 \varepsilon \mathfrak{L}(\mathrm{E}, \mathrm{P})$ and $B_1, B_2 \varepsilon \mathfrak{L}(\mathrm{P}, \Delta)$, then $(A_1 + A_2)(B_1 + B_2) = A_1B_1 + A_1B_2 + A_2B_1 + A_2B_2$. Finally, we note that the associative law of multiplication holds: If Γ is another field and $A \varepsilon \mathfrak{L}(\mathrm{E}, \mathrm{P})$, $B \varepsilon \mathfrak{L}(\mathrm{P}, \Delta)$, $C \varepsilon \mathfrak{L}(\Delta, \Gamma)$, then $(AB)C = A(BC) \varepsilon \mathfrak{L}(\mathrm{E}, \Gamma)$. All of these assertions are readily verified and they are very similar to facts about composition of linear mappings which we have considered in Vol. II, § 2.2. We leave it to the reader to carry out the verifications.

The results we have indicated imply that $\mathfrak{L}(\mathrm{E}, \mathrm{E})$ is a ring under the compositions of addition and multiplication. This is just the ring of endomorphisms of the additive group $(\mathrm{E}, +)$ which has been considered in the general case in Vol. I, § 2.13. If $\rho \varepsilon \mathrm{P}$, then the mapping $\rho_R : \xi \to \xi\rho (= \rho\xi)$ in P belongs to $\mathfrak{L}(\mathrm{P}, \mathrm{P})$. Since $AB \varepsilon \mathfrak{L}(\mathrm{E}, \mathrm{P})$ for A in $\mathfrak{L}(\mathrm{E}, \mathrm{P})$ and B in $\mathfrak{L}(\mathrm{P}, \mathrm{P})$, we see that $A\rho_R \varepsilon \mathfrak{L}(\mathrm{E}, \mathrm{P})$. This observation permits us to convert $\mathfrak{L}(\mathrm{E}, \mathrm{P})$ into a right vector space over the field P. For this purpose we *define* $A\rho = A\rho_R$ for $A \varepsilon \mathfrak{L}(\mathrm{E}, \mathrm{P})$ and $\rho \varepsilon \mathrm{P}$. Then we have

$$\begin{aligned}
(A + B)\rho &= (A + B)\rho_R = A\rho_R + B\rho_R = A\rho + B\rho \\
A(\rho + \sigma) &= A(\rho + \sigma)_R = A(\rho_R + \sigma_R) \\
&= A\rho_R + A\sigma_R = A\rho + A\sigma \\
A(\rho\sigma) &= A(\rho\sigma)_R = A(\rho_R\sigma_R) = (A\rho_R)\sigma_R = (A\rho)\sigma \\
A1 &= A1_R = A,
\end{aligned}$$

which shows that $\mathfrak{L}(\mathrm{E}, \mathrm{P})$ is a right vector space over P.

We note next that if ϵ_R denotes the mapping $\eta \to \eta\epsilon$ in E, then $\epsilon_R \varepsilon \mathfrak{L}(\mathrm{E}, \mathrm{E})$. Hence, if $A \varepsilon \mathfrak{L}(\mathrm{E}, \mathrm{P})$, then $\epsilon_R A \varepsilon \mathfrak{L}(\mathrm{E}, \mathrm{P})$. We can now consider $\mathfrak{L}(E, P)$ also as a left vector space over E by defining $\epsilon A = \epsilon_R A$. It should be remarked that, if we do this, then there is an ambiguity in writing ϵA which can mean either the image of ϵ under A or the endomorphism $\epsilon_R A$. For this reason we shall avoid considering $\mathfrak{L}(E, P)$ as a left vector space over E and use instead the product $\epsilon_R A$ when this will be needed.

All that we have just said applies also to fields over a given field Φ. Consider the fields E/Φ and P/Φ. In this connection it is natural to consider the subset $\mathfrak{L}_\Phi(E, P)$ of $\mathfrak{L}(E, P)$ of linear transformations of E as vector space over Φ into P over Φ. If $\alpha \in \Phi$ and $\xi, \rho \in P$, then $(\alpha\xi)\rho_R = (\alpha\xi)\rho = \alpha(\xi\rho) = \alpha(\xi\rho_R)$, which implies that $\rho_R \in \mathfrak{L}_\Phi(P, P)$. If $A \in \mathfrak{L}_\Phi(E, P)$, then $A\rho \equiv A\rho_R \in \mathfrak{L}_\Phi(E, P)$; so it is clear that $\mathfrak{L}_\Phi(E, P)$ is a subspace of the right vector space $\mathfrak{L}(E, P)$ over P. If $\mathfrak{A}$ is any right vector space over P, we denote its dimensionality over P as $[\mathfrak{A}:P]_R$. Then we have the following important result on $[\mathfrak{L}_\Phi(E, P):P]_R$.

Theorem 1. *Let* $E/\Phi, P/\Phi$ *be fields over* Φ *and let* $\mathfrak{L}_\Phi(E, P)$ *be the right vector space over* P *of linear mappings of* E/Φ *into* P/Φ. *Then* $[E:\Phi]$ *is finite if and only if* $[\mathfrak{L}_\Phi(E, P):P]_R$ *is finite and when both are finite then*

$$[E:\Phi] = [\mathfrak{L}_\Phi(E, P):P]_R. \tag{1}$$

Proof. Let $\eta_1, \eta_2, \cdots, \eta_n$ be elements of E which are linearly independent over Φ. Then we may imbed this set in a basis $\{\eta_\alpha\}$ for E over Φ (Vol. II, p. 239). If we choose a correspondent $\tau_\alpha \in P$ for each η_α, then there exists a unique element $A \in \mathfrak{L}_\Phi(E, P)$ such that $\eta_\alpha A = \tau_\alpha$ for every η_α. This implies that for each $i = 1, 2, \cdots, n$, there exists a linear mapping E_i (not necessarily unique) such that $\eta_i E_i = 1$, $\eta_j E_i = 0$ if $j \neq i$. Then if $\rho_i \in P$,

$$\eta_j\left(\sum_1^n E_i\rho_i\right) = \sum_{i=1}^n (\eta_j E_i)\rho_i = \rho_j.$$

Hence $\sum_1^n E_i\rho_i = 0$ implies every $\rho_i = 0$, which shows that, if $[E:\Phi]$ is infinite, then for every n there exist n right P-independent elements of $\mathfrak{L}_\Phi(E, P)$. Then $[\mathfrak{L}_\Phi(E, P):P]_R \geq n$ for every n, so this dimensionality is infinite. Next suppose $[E:\Phi] = n <$

∞ and that the η's constitute a basis. Let $A \in \mathfrak{L}_\Phi(E, P)$ and set $\eta_i A = \rho_i$. Then $\eta_j\left(\sum_1^n E_i\rho_i\right) = \rho_j = \eta_j A$. Thus A and $\Sigma E_i\rho_i$ have the same effect on the basis $(\eta_1, \eta_2, \cdots, \eta_n)$ for E/Φ. It follows that $A = \sum_1^n E_i\rho_i$ and, since the E_i are right independent over P, these form a basis for $\mathfrak{L}_\Phi(E, P)$ over P. Hence $[\mathfrak{L}_\Phi(E, P):P]_R = n = [E:\Phi]$. This completes the proof.

We now drop Φ and consider again E and P arbitrary fields and $\mathfrak{L}(E, P)$ the group of homomorphisms of $(E, +)$ into $(P, +)$. We consider this as a right vector space over P as before. Let $\mathfrak{A}$ be a subspace of this space. Let ϵ be a fixed element of E. Then ϵ determines a mapping f_ϵ of $\mathfrak{A}$ into P by the rule that $f_\epsilon(A) = \epsilon A \in P$. We have $f_\epsilon(A + B) = \epsilon(A + B) = \epsilon A + \epsilon B = f_\epsilon(A) + f_\epsilon(B)$ and, if $\rho \in P$, then $f_\epsilon(A\rho) = \epsilon(A\rho) = (\epsilon A)\rho = f_\epsilon(A)\rho$. Thus we see that f_ϵ is a P-linear mapping of the right vector space $\mathfrak{A}$ over P into the one dimensional space P over P, that is, $f_\epsilon \in \mathfrak{A}^*$, the conjugate space of $\mathfrak{A}$. Of course, $\mathfrak{A}^*$ is a left vector space over P. The process we have just indicated produces a collection $\{f_\epsilon | \epsilon \in E\}$ of linear functions. This collection is "total" in the sense that, if $f_\epsilon(A) = 0$ for all ϵ, then $A = 0$. This is clear since the requirement is that $\epsilon A = 0$ for all ϵ and this is just the definition of $A = 0$. We can now prove the following useful

Lemma. *Let $\mathfrak{A}$ be a subspace of $\mathfrak{L}(E, P)$ over P such that $[\mathfrak{A}:P]_R = n < \infty$. Then there exist elements $\epsilon_1, \epsilon_2, \cdots, \epsilon_n \in E$ and a right basis $E_1, E_2, \cdots, E_n$ for $\mathfrak{A}$ over P such that $\epsilon_i E_j = \delta_{ij}$ ($\delta_{ij} = 0$ if $i \neq j$, $\delta_{ii} = 1$).*

Proof. We are given that $[\mathfrak{A}:P]_R = n < \infty$. This implies that the conjugate space $\mathfrak{A}^*$ is n-dimensional. Let $\mathfrak{B}^*$ be the subspace of $\mathfrak{A}^*$ spanned by the linear functions f_ϵ, $\epsilon \in E$. Since $f(A) = 0$ for all $f \in \mathfrak{B}^*$ implies that $A = 0$, it follows that $\mathfrak{B}^* = \mathfrak{A}^*$ (Vol. 2, § 2.10). Hence we can find n linear functions $f_{\epsilon_1}, f_{\epsilon_2}, \cdots, f_{\epsilon_n}$ which form a basis for $\mathfrak{A}^*$. Since $\mathfrak{A}$ can be considered as the conjugate space of $\mathfrak{A}^*$, we can find a basis $E_1, E_2, \cdots, E_n$ for $\mathfrak{A}$ over P such that $f_{\epsilon_i}(E_j) = \delta_{ij}$. Recalling the meaning of f_ϵ we see that we have $\epsilon_i E_j = \delta_{ij}$ as required.

2. The Jacobson-Bourbaki correspondence. Let P be a field and let $\mathfrak{L}(P, P)$ be the ring of endomorphisms of the additive group $(P, +)$. As before, we consider $\mathfrak{L}(P, P)$ as a right vector space over P. If Φ is a subfield, then $\mathfrak{L}_\Phi(P, P)$ the ring of linear transformations of P/Φ is a subring of $\mathfrak{L}(P, P)$ and a subspace of $\mathfrak{L}(P, P)$ over P. Moreover, we have seen (Th. 1) that, if Φ is of *finite co-dimension* in P in the sense that $[P:\Phi] = n < \infty$, then $[\mathfrak{L}_\Phi(P, P):P]_R = n$. These properties of $\mathfrak{L}_\Phi(P, P)$ in no way refer to the subfield Φ. We shall now show that they are characteristic of the sets $\mathfrak{L}_\Phi(P, P)$. This is a consequence of the following

Theorem 2 (Jacobson-Bourbaki). *Let* P *be a field and* $\mathfrak{A}$ *a set of endomorphisms of* $(P, +)$ *such that:*

(i) $\mathfrak{A}$ *is a subring of* $\mathfrak{L}(P, P)$ *the ring of endomorphisms of* $(P, +)$ *(containing the identity mapping, by our convention, Introd. p.* 2).
(ii) $\mathfrak{A}$ *is a subspace of* $\mathfrak{L}(P, P)$ *as right vector space over* P.
(iii) $[\mathfrak{A}:P]_R = n < \infty$.

Let Φ *be the subset of* P *of elements* α *such that* $\alpha_R A = A\alpha_R$ *for all* $A \in \mathfrak{A}$. *Then* Φ *is a subfield of* P, $[P:\Phi] = n$ *and* $\mathfrak{A} = \mathfrak{L}_\Phi(P, P)$ *the complete set of linear transformations of* P/Φ.

Proof (Hochschild). The verification that Φ is a subfield is immediate and will be omitted. Next we apply the lemma of § 1 to obtain elements $\rho_1, \rho_2, \cdots, \rho_n$ in P and a right basis $(E_1, E_2, \cdots, E_n)$ for $\mathfrak{A}$ over P such that $\rho_i E_j = \delta_{ij}$. Since $\rho_R\sigma_R = \sigma_R\rho_R$ for any ρ, σ in P, it is clear that Φ is the set of $\alpha \in P$ satisfying $\alpha_R E_i = E_i\alpha_R$, $i = 1, 2, \cdots, n$. Also it follows from $\rho_i E_j = \delta_{ij}$ that, if we express the element A of $\mathfrak{A}$ as $A = \sum_1^n E_i\sigma_i$, then $\rho_j A = \sum_i (\rho_j E_i)\sigma_i = \sigma_j$. Hence the representation of any A in terms of the basis reads: $A = \sum_1^n E_i(\rho_i A)$ or $A = \sum_1^n E_i(\rho_i A)_R$. We shall now use this formula to show that every E_i maps P into Φ. For this purpose let σ be any element of P and consider the mapping $E_j\sigma_R E_k$, $j, k = 1, 2, \cdots, n$, which belongs to $\mathfrak{A}$, since $\mathfrak{A}$ is a subring of the ring of endomorphisms. The formula we obtained can be applied for $A = E_j\sigma_R E_k$ to give

$$\begin{aligned} E_j\sigma_R E_k &= \sum_i E_i(\rho_i E_j \sigma_R E_k)_R \\ &= E_j(1\sigma_R E_k)_R \\ &= E_j(\sigma E_k)_R. \end{aligned}$$

This means that for arbitrary ρ in P we have

$$\rho E_j \sigma_R E_k = \rho E_j(\sigma E_k)_R.$$

In other words, $((\rho E_j)\sigma)E_k = (\rho E_j)(\sigma E_k)$. Then

$$(\sigma(\rho E_j))E_k = (\sigma E_k)(\rho E_j).$$

If we think of σ as the argument, this gives the operator identity $(\rho E_j)_R E_k = E_k(\rho E_j)_R$, which implies that $\rho E_j \in \Phi$, and this holds for all $\rho \in$ P. We can now show that the ρ_i we started with form a basis for P/Φ. Let $\sigma \in$ P and consider the element $\sigma' = \sigma - \sum_j (\sigma E_j)\rho_j$ in P. Since $\sigma E_j \in \Phi$ and $\alpha_R E_k = E_k \alpha_R$ for α in Φ, we have $\sigma' E_k = \sigma E_k - \left(\sum_j (\sigma E_j)\rho_j\right) E_k = \sigma E_k - \left(\sum_j \rho_j(\sigma E_j)_R\right) E_k = \sigma E_k - \sum_j (\rho_j E_k)(\sigma E_j)_R = \sigma E_k - \sigma E_k = 0$. Since $1 \in \mathfrak{A}$, $1 = \Sigma E_k \lambda_k$ for suitable $\lambda_k \in$ P. Then $\sigma' E_k = 0$ implies $\sigma' 1 = 0$ so $\sigma' = 0$. We therefore see that $\sigma = \Sigma(\sigma E_j)\rho_j$ is a Φ-linear combination of the ρ_j. If $\Sigma \alpha_i \rho_i = 0$, $\alpha_i \in \Phi$, then $\alpha_j = (\Sigma \alpha_i \rho_i)E_j = 0$. Hence $(\rho_1, \rho_2, \cdots, \rho_n)$ is a basis for P over Φ and $[\mathrm{P}:\Phi] = n$. Since $\alpha_R A = A\alpha_R$ for every $\alpha \in \Phi$ and $A \in \mathfrak{A}$, every $A \in \mathfrak{A}$ is a linear transformation of P over Φ. Hence $\mathfrak{A} \subseteq \mathfrak{L}_\Phi(\mathrm{P}, \mathrm{P})$. Since $[\mathfrak{L}_\Phi(\mathrm{P}, \mathrm{P}):\mathrm{P}]_R = n$ by Theorem 1, and $[\mathfrak{A}:\mathrm{P}]_R = n$, we see that $\mathfrak{A} = \mathfrak{L}_\Phi(\mathrm{P}, \mathrm{P})$.

Theorem 2 permits us to establish our first and most general "Galois correspondence" for a field P. This concerns two collections of objects: the collection $\mathscr{F}$ of subfields Φ which are of finite co-dimension in P and the collection $\mathscr{R}$ of sets of endomorphisms of (P, +) having the properties (i), (ii), (iii) of the theorem. To each $\Phi \in \mathscr{F}$ we associate $R(\Phi) \equiv \mathfrak{L}_\Phi(\mathrm{P}, \mathrm{P})$. This is a subring of $\mathfrak{L}(\mathrm{P}, \mathrm{P})$, a subspace of $\mathfrak{L}(\mathrm{P}, \mathrm{P})$ over P and satisfies $[\mathfrak{L}_\Phi(\mathrm{P}, \mathrm{P}):\mathrm{P}]_R < \infty$. Hence $R(\Phi) = \mathfrak{L}_\Phi(\mathrm{P}, \mathrm{P}) \in \mathscr{R}$. On the other hand, if $\mathfrak{A} \varepsilon \mathscr{R}$, then we can associate the subfield $F(\mathfrak{A}) = \Phi = \{\alpha | \alpha \varepsilon \mathrm{P}, \alpha_R A = A\alpha_R, A \varepsilon \mathfrak{A}\}$. This is of finite co-dimension in P and so it belongs to $\mathscr{F}$. By Theorem 2, we have $R(F(\mathfrak{A})) = \mathfrak{A}$. If $\Phi \varepsilon \mathscr{F}$ and $\mathfrak{A} = R(\Phi) = \mathfrak{L}_\Phi(\mathrm{P}, \mathrm{P})$, then $[\mathfrak{A}:\mathrm{P}]_R = [\mathrm{P}:\Phi]$ by Theorem 1 and $[\mathfrak{A}:\mathrm{P}]_R = [\mathrm{P}:F(\mathfrak{A})]$ by Theorem 2. If $\alpha \varepsilon \Phi$, we certainly have $\alpha_R A = A\alpha_R$ for $A \varepsilon \mathfrak{A}$. Hence $\Phi \subseteq F(\mathfrak{A})$ by the definition of F. Since $[\mathrm{P}:\Phi] = [\mathrm{P}:F(\mathfrak{A})][F(\mathfrak{A}):\Phi]$ (VI, Introd.) and $[\mathrm{P}:\Phi] = [\mathrm{P}:F(\mathfrak{A})]$, we have $[F(\mathfrak{A}):\Phi] = 1$ and so $\Phi = F(\mathfrak{A}) = F(R(\Phi))$. The two relations

$$R(F(\mathfrak{A})) = \mathfrak{A}, \quad \mathfrak{A} \varepsilon \mathscr{R}$$
$$F(R(\Phi)) = \Phi, \quad \Phi \varepsilon \mathscr{F}$$

imply that the mappings R and F are inverses and are 1–1 of $\mathscr{F}$ onto $\mathscr{R}$ and $\mathscr{R}$ onto $\mathscr{F}$ respectively. It should be noted that the definitions of R and F show that these mappings are order reversing for the inclusion relation: $\Phi_1 \subseteq \Phi_2$ for subfields implies $R(\Phi_1) \supseteq R(\Phi_2)$ and $\mathfrak{A}_1 \subseteq \mathfrak{A}_2$ for $\mathfrak{A}_i \varepsilon \mathscr{R}$ implies $F(\mathfrak{A}_1) \supseteq F(\mathfrak{A}_2)$.

In § 4 we shall establish a Galois correspondence between finite groups of automorphisms of a field P and certain subfields of finite co-dimension in P. Later (§ 8, Chap. IV) we shall establish a similar correspondence between certain Lie algebras of derivations in P and certain subfields of P. Both of these correspondences will be derived from the general "Jacobson-Bourbaki correspondence" which we have just given. In addition to this we shall need some information on special generators for some of the rings $\mathfrak{A} \varepsilon \mathscr{R}$. For the automorphism theory the generators are automorphisms of P. The results we require for these will be derived in the next section.

EXERCISES

1. Let $\mathfrak{A}$ be a set of endomorphisms of (P, +) satisfying conditions (i) and (ii) of Theorem 2. Show that $\mathfrak{A}$ is an irreducible ring of endomorphisms (Vol. II, p. 259). Apply the density theorem for such rings (Vol. II, p. 274) to show that, if $\rho_1, \rho_2, \cdots, \rho_m$ are Φ-independent elements (Φ as in Th. 2) and $\sigma_1, \sigma_2, \cdots, \sigma_m$ are arbitrary in P, then there exists an $A \varepsilon \mathfrak{A}$ such that $\rho_i A = \sigma_i$, $i = 1, 2, \cdots, m$. Use this result to give another proof of Theorem 2.

2. Let P be an arbitrary extension field of the field Φ. Show that, if $\alpha \varepsilon \mathrm{P}$ satisfies $\alpha_R A = A\alpha_R$ for all $A \varepsilon \mathfrak{L}_\Phi(\mathrm{P}, \mathrm{P})$, then $\alpha \varepsilon \Phi$.

3. Let $(\rho_1, \rho_2, \cdots, \rho_n)$ be a basis of P/Φ, $(A_1, A_2, \cdots, A_n)$ a right basis for $\mathfrak{L}_\Phi(\mathrm{P}, \mathrm{P})$ over P. Show that the $n \times n$ matrix $(\rho_i A_j)$ has an inverse in P_n.

3. Dedekind independence theorem for isomorphisms of a field. Let s be an isomorphism of a field E into a field P. Then s is an isomorphism of the additive group (E, +) of E into (P, +) satisfying the multiplicative condition $(\epsilon\eta)^s = \epsilon^s\eta^s$. We can write this in operator form as:

$$\eta_R s = s(\eta^s)_R \tag{2}$$

where η_R is the multiplication by η in E and $(\eta^s)_R$ is the multiplication by η^s in P. If both E and P are fields over Φ, then an isomorphism of E/Φ into P/Φ is an algebra isomorphism of the first algebra into the second. Hence, in addition to the conditions: $(\epsilon + \eta)^s = \epsilon^s + \eta^s$, $(\epsilon\eta)^s = \epsilon^s\eta^s$, $1^s = 1$, s is 1–1, we have $(\alpha\epsilon)^s = \alpha\epsilon^s$ for $\alpha \in \Phi$. The first and last of these are just the conditions that $s \in \mathfrak{L}_\Phi(\mathrm{E}, \mathrm{P})$. Hence if s is an isomorphism of E/Φ into P/Φ, then $\alpha^s = (\alpha 1)^s = \alpha 1^s = \alpha$ holds for every $\alpha \in \Phi$. Conversely, this condition implies that $(\alpha\epsilon)^s = \alpha\epsilon^s$, $\epsilon \in$ E. Thus an isomorphism of E/Φ into P/Φ is just an isomorphism of E into P which is the identity mapping on Φ.

We shall now derive two basic results on linear relations connecting isomorphisms of E into P (no Φ).

Theorem 3 (Dedekind). *Let* E *and* P *be fields and let* $s_1, s_2, \cdots, s_n$ *be distinct isomorphisms of* E *into* P. *Then the* s_i *are right linearly independent over* P: $\Sigma s_i\rho_i = 0$, $\rho_i \in$ P, *implies every* $\rho_i = 0$. *Here* $s\rho \equiv s\rho_R$.

Proof. If the assertion is false, then we have a shortest relation, which by suitable ordering reads:

$$s_1\rho_1 + s_2\rho_2 + \cdots + s_r\rho_r = 0, \tag{3}$$

where every $\rho_i \neq 0$. Suppose $r > 1$. Since $s_1 \neq s_2$ there exists $\eta \in$ E such that $\eta^{s_1} \neq \eta^{s_2}$. Now multiply (3) on the left by η_R. If we take into account (2), this gives: $s_1\eta^{s_1}\rho_1 + s_2\eta^{s_2}\rho_2 + \cdots + s_r\eta^{s_r}\rho_r = 0$. Next we multiply (3) on the right by η^{s_1} and obtain $s_1\rho_1\eta^{s_1} + s_2\rho_2\eta^{s_1} + \cdots + s_r\rho_r\eta^{s_1} = 0$. Subtraction of the two new relations gives

$$s_2\rho_2(\eta^{s_2} - \eta^{s_1}) + s_3\rho_3(\eta^{s_3} - \eta^{s_1}) + \cdots = 0.$$

Since $\rho_2(\eta^{s_2} - \eta^{s_1}) \neq 0$, this is a non-trivial relation which is shorter than (3). Hence we are forced to conclude that $r = 1$, that is, $s_1\rho_1 = 0$. Since $\rho_{1R}{}^{-1}$ exists, this gives $s_1 = 0$ contrary to the assumption that s_1 is an isomorphism.

We can combine Theorem 1 and Dedekind's theorem to obtain the following

Corollary. *Let* E *and* P *be fields over* Φ *such that* $[\mathrm{E}:\Phi] = n < \infty$. *Then there exist at most* n *distinct isomorphisms of* E/Φ *into* P/Φ.

Proof. Let $s_1, s_2, \cdots, s_r$ be distinct isomorphisms of E/Φ into P/Φ. Then these are elements of $\mathfrak{L}_\Phi(\mathrm{E}, \mathrm{P})$ which are right P-independent. Since $[\mathfrak{L}_\Phi(\mathrm{E}, \mathrm{P}):\mathrm{P}]_R = n$, we must have $r \leq n$.

In the next section we shall be concerned with right P-vector spaces spanned by a finite number of automorphisms of a field. More generally, let $s_1, s_2, \cdots, s_n$ be distinct isomorphisms of E into P and let $\mathfrak{A}$ be the set of endomorphisms of the form

$$s_1\rho_1 + s_2\rho_2 + \cdots + s_n\rho_n, \quad \rho_i \in \mathrm{P}. \tag{4}$$

Evidently, $\mathfrak{A}$ is a subspace of the right P-vector space $\mathfrak{L}(\mathrm{E}, \mathrm{P})$. Moreover, if $\epsilon \in \mathrm{E}$, then $\epsilon_R s_i = s_i(\epsilon^{s_i})_R$, by (2), so

$$\epsilon_R\left(\sum_1^n s_i\rho_i\right) = \sum_1^n s_i(\epsilon^{s_i})_R\rho_i = \sum_1^n s_i(\epsilon^{s_i}\rho_i)\cdot$$

This shows that $\mathfrak{A}$ is closed under left multiplication by arbitrary ϵ_R, $\epsilon \in \mathrm{E}$. We shall require the following

Theorem 4. *Let* E *and* P *be fields,* $s_1, s_2, \cdots, s_n$ *isomorphisms of* E *into* P, *and let* $\mathfrak{A}$ *be the right* P-*subspace of* $\mathfrak{L}(\mathrm{E}, \mathrm{P})$ *of endomorphisms* $\Sigma s_i\rho_i$, $\rho_i \in \mathrm{P}$. *Let* $\mathfrak{B}$ *be a* P-*subspace of* $\mathfrak{A}$ *which is invariant under left multiplication by elements* ϵ_R, $\epsilon \in \mathrm{E}$. *Then* $\mathfrak{B} = s_{i_1}\mathrm{P} + s_{i_2}\mathrm{P} + \cdots + s_{i_r}\mathrm{P}\left(= \left\{\sum_j s_{i_j}\rho_{i_j}\right\}\right)$ *where* $\{s_{i_1}, s_{i_2}, \cdots, s_{i_r}\} = \mathfrak{B} \cap \{s_1, s_2, \cdots, s_n\}$.

Proof. It is clear that $\left\{\sum_{j=1}^r s_{i_j}\rho_{i_j} \middle| \rho_{i_j} \in \mathrm{P}\right\} \subseteq \mathfrak{B}$. To prove the opposite inclusion it suffices to show that, if $\sum_1^n s_i\rho_i \in \mathfrak{B}$, then the s_i for which $\rho_i \neq 0$ are contained in $\mathfrak{B}$. Suppose this is not the case. Then we have an element $s_{k_1}\rho_{k_1} + s_{k_2}\rho_{k_2} + \cdots + s_{k_s}\rho_{k_s}$ in

$\mathfrak{B}$ in which every $\rho_{k_t} \neq 0$ and $s_{k_t} \notin \mathfrak{B}$. We can then argue as in the proof of Dedekind's theorem. We assume s minimal. If $s > 1$, we apply the process we used before to obtain a shorter element of the same type contained in $\mathfrak{B}$. Then $s_{k_1}\rho_{k_1} \in \mathfrak{B}$ which implies that $s_{k_1} \in \mathfrak{B}$ contrary to assumption.

EXERCISE

1. Let $E = \Phi(\theta)$ where θ is algebraic over Φ and $(f(x))$ is the kernel of the homomorphism $g(x) \to g(\theta)$ (Vol. I, p. 103). Then $[E:\Phi] = \deg f$. Use the extension theorem V of Introduction to show that the number of isomorphisms of E/Φ into P/Φ does not exceed $\deg f$. Extend this result to obtain an alternative proof of the Corollary to Theorem 3.

4. Finite groups of automorphisms. Let G be a group of automorphisms of a field P and let Φ be the subset of P of elements α such that $\alpha^s = \alpha$ for every $s \in G$. We shall call Φ the set of G-*invariants* of P. Since the invariants (or fixed elements) of an automorphism form a subfield, Φ is a subfield of P. We denote $\Phi = I(G)$ (or $I_P(G)$ if it is necessary to indicate P) and we call a subfield which has this form, that is, which is the subfield of invariants of a group of automorphisms, *Galois in* P. We shall also say that P is *Galois over* Φ or P/Φ *is Galois*.

The process we have just indicated associates with groups of automorphisms G, subfields $I(G)$, and we have the mapping $G \to I(G)$ of these groups into subfields of P. We now define a mapping in the opposite direction. If Φ is any subfield of P, then we associate with Φ the set $A(\Phi)$ (or $A_P(\Phi)$) consisting of the automorphisms of P/Φ, that is, the automorphisms s of P such that $\alpha^s = \alpha$ for all $\alpha \in \Phi$. Evidently, $A(\Phi)$ is a subgroup of the group A of all the automorphisms of P. We call $A(\Phi)$ *the Galois group* of P/Φ. We have the subfield-group mapping $\Phi \to A(\Phi)$. The following properties of the mappings $G \to I(G)$, $\Phi \to A(\Phi)$ are clear from the definitions:

(α) $G_1 \supseteq G_2 \Rightarrow I(G_1) \subseteq I(G_2)$ ($\Rightarrow$ denotes "implies").

(β) $\Phi_1 \supseteq \Phi_2 \Rightarrow A(\Phi_1) \subseteq A(\Phi_2)$.

(γ) $I(A(\Phi)) \supseteq \Phi$.

(δ) $A(I(G)) \supseteq G$.

These relations have the following consequences:

$$(\epsilon) \qquad I(A(I(G))) = I(G).$$

$$(\eta) \qquad A(I(A(\Phi))) = A(\Phi).$$

The proofs of these two are identical so we consider (ϵ) only. Here we use (γ) for $\Phi = I(G)$ and obtain $I(A(I(G))) \supseteq I(G)$. On the other hand, if we apply I to $A(I(G)) \supseteq G$, we obtain $I(G) \supseteq I(A(I(G)))$. Hence ($\epsilon$) holds. A consequence of (ϵ) is that Φ is Galois in P if and only if Φ is the set of invariants of the Galois group of P/Φ, that is, $\Phi = I(A(\Phi))$. Clearly this condition is sufficient. On the other hand, if $\Phi = I(G)$ for *some* group of automorphisms G, then $\Phi = I(G) = I(A(I(G))) = I(A(\Phi))$.

We shall now study the Galois correspondences $\Phi \rightarrow A(\Phi)$, $G \rightarrow I(G)$ starting with finite groups of automorphisms. We denote the order of a group G by $(G{:}1)$ and, more generally, the index of a subgroup H in G by $(G{:}H)$. We shall deduce all the results on the subfield–group correspondence from the Jacobson-Bourbaki theorem (Th. 2) via the following

Lemma. *Let G be a finite group of automorphisms in the field* P *and let* $\mathfrak{A} = \left\{\sum_1^n s_i\rho_i \,\middle|\, s_i \in G, \rho_i \in \mathrm{P}\right\}$. *Then* $\mathfrak{A}$ *satisfies the hypotheses* (i), (ii), (iii) *of Theorem 2,* $[\mathfrak{A}:\mathrm{P}]_R = (G{:}1)$, *and the subfield* Φ *given in Theorem 2 is the subfield of G-invariants. If* $\mathfrak{B}$ *is a subring of* $\mathfrak{A}$ *and a subspace of* $\mathfrak{A}$ *over* P, *then*

$$\mathfrak{B} = \left\{\sum_1^r t_j\rho_j \,\middle|\, t_j \in H, \rho_i \in \mathrm{P}\right\}$$

where $H = \{t_j\}$ *is a subgroup of G.*

Proof. If $\rho \in \mathrm{P}$ and s is an automorphism, then (2) shows that $\rho_R s = s(\rho^s)_R$. Hence $(s_i\rho_i)(s_j\rho_j) = s_i(\rho_{iR}s_j)\rho_{jR} = s_is_j(\rho_i^{s_j})_R\rho_{jR} = s_is_j\rho_i^{s_j}\rho_j \in \mathfrak{A}$ since $s_is_j \in G$. This implies that $\mathfrak{A}$ is a subring of the ring of endomorphisms $\mathfrak{L}(\mathrm{P}, \mathrm{P})$. Since $1 \in G$ and $G \subseteq \mathfrak{A}$, $1 \in \mathfrak{A}$. It is clear that $\mathfrak{A}$ is a subspace of $\mathfrak{L}(\mathrm{P}, \mathrm{P})$ as right vector space over P. Since the s_i are independent over P by Dedekind's theorem, $[\mathfrak{A}:\mathrm{P}]_R = (G{:}1) < \infty$. The subfield Φ of Theorem 2 is the set of $\alpha \in \mathrm{P}$ such that $\alpha_R A = A\alpha_R$ for all $A \in \mathfrak{A}$. Since $\alpha_R\rho_R =$

$\rho_R\alpha_R$, $\rho \varepsilon$ P, anyhow, the condition is equivalent to $\alpha_R s_i = s_i\alpha_R$, $s_i \varepsilon G$. Since $\alpha_R s_i = s_i(\alpha^{s_i})_R$, this is equivalent to $s_i(\alpha^{s_i})_R = s_i\alpha_R$, $s_i \varepsilon G$. Since s_i^{-1} exists, this becomes $(\alpha^{s_i})_R = \alpha_R$ or $\alpha^{s_i} = \alpha$ which shows that $\alpha_R A = A\alpha_R$, $A \varepsilon \mathfrak{A}$ is equivalent to: α is G-invariant. Now let $\mathfrak{B}$ be a subring of $\mathfrak{A}$ which is a P-subspace. Then $\mathfrak{B} \supseteq 1\mathrm{P} = \{\rho_R | \rho \varepsilon \mathrm{P}\}$ and consequently $\mathfrak{B}$ is invariant under left multiplication by the ρ_R. Hence, by Theorem 4, $\mathfrak{B} = t_1\mathrm{P} + t_2\mathrm{P} + \cdots + t_r\mathrm{P}$ where $H = \{t_j\} = G \cap \mathfrak{B}$. Evidently $H = G \cap \mathfrak{B}$ is closed under multiplication so this is a finite subsemigroup of G. Hence H is a subgroup of G.

The main result on finite groups of automorphisms of a field is

Theorem 5. *Let* P *be a field and let* $\mathscr{A}$ *be the collection of finite groups of automorphisms in* P, $\mathscr{I}$ *the collection of subfields of* P *which are Galois and of finite co-dimension in* P. *If* $\Phi \varepsilon \mathscr{I}$, *let* $A(\Phi)$ *be the Galois group of* P/Φ *and, if* $G \varepsilon \mathscr{A}$, *let* $I(G)$ *be the subfield of* P *of G-invariants. Then:* (i) *If* $\Phi \varepsilon \mathscr{I}$, $A(\Phi) \varepsilon \mathscr{A}$, *and if* $G \varepsilon \mathscr{A}$, $I(G) \varepsilon \mathscr{I}$. *Moreover,* $I(A(\Phi)) = \Phi$ *and* $A(I(G)) = G$. (ii) *If* $G \varepsilon \mathscr{A}$, *then* $(G:1) = [\mathrm{P}:I(G)]$. (iii) *If* $\Phi \varepsilon \mathscr{I}$ *and* E *is a subfield of* P *containing* Φ, *then* $\mathrm{E} \varepsilon \mathscr{I}$. (iv) *In this situation* $H = A(\mathrm{E})$, *which is a subgroup of* $G = A(\Phi)$, *is invariant in G if and only if* E *is Galois over* Φ. *Then the Galois group* $A_{\mathrm{E}}(\Phi)$ *of* E/Φ *is isomorphic to* G/H.

Proof. (i)–(ii). If $G \varepsilon \mathscr{A}$ and $\mathfrak{A} = \{\Sigma s_i\rho_i | s_i \varepsilon G, \rho_i \varepsilon \mathrm{P}\}$, then $[\mathrm{P}:I(G)] = [\mathfrak{A}:\mathrm{P}]_R = (G:1)$, by the lemma and Theorem 2. If we set $\Phi = I(G)$ and $G' = A(\Phi)$ the Galois group of P/Φ, then the corollary to Dedekind's theorem shows that $(G':1) \leq [\mathrm{P}:\Phi] = (G:1)$. Since $G \subseteq G'$ is evident, $G' = G$. Thus $A(I(G)) = G$. Next let Φ be Galois and of finite co-dimension in P. Then $\Phi = I(G)$ where G is the Galois group of P/Φ. This is finite by the corollary to Dedekind's theorem. Hence $A(\Phi) \varepsilon \mathscr{A}$ and $I(A(\Phi)) = \Phi$. This completes the proof of (i) and (ii). (iii) Let $\Phi \varepsilon \mathscr{I}$ and let $\mathfrak{A}$ be the ring of endomorphisms defined by the Galois group G of P/Φ. By Theorem 2, $\mathfrak{A} = \mathfrak{L}_\Phi(\mathrm{P}, \mathrm{P})$. Now let E be a subfield of P containing Φ. Then $\mathfrak{B} = \mathfrak{L}_{\mathrm{E}}(\mathrm{P}, \mathrm{P})$ is a subring of $\mathfrak{A}$ of the sort considered in the lemma. Hence $\mathfrak{B} = t_1\mathrm{P} + \cdots + t_r\mathrm{P}$ where $H = \{t_j\}$ is a subgroup of G. Since $\mathrm{E} = \{\epsilon | \epsilon_R B = B\epsilon_R, B \varepsilon \mathfrak{B}\}$, it follows that E is the subfield of H-invariants. This proves (iii). (iv) If $s \in G$, E^s the image of E under s is another subfield of P containing Φ and it follows directly from the definition that $A(E^s) = s^{-1}Hs$. Hence H is invariant in G if and only if $E^s = E$ for every $s \in G$. We proceed to show that this holds if and only if E is Galois over Φ and then $A_E(\Phi) \cong G/H$. Assume first that $E^s = E$ and let G' be the group of restrictions s' to E of the $s \in G$. Then G' is a finite group of automorphisms in E and $I(G') = \Phi$. Hence Φ is Galois in E and $G' = A_E(\Phi)$ by (i) applied to E. The mapping $s \to s'$ is a homomorphism of G onto G'. The kernel is the set of $s \in G$ such that $s' = 1$ on E. This is H. Hence $G' \cong G/H$. Next let E be Galois over Φ. Then we have $[E:\Phi]$ distinct automorphisms of E over Φ and these can be considered as isomorphisms of E/Φ into P/Φ. On the other hand, by the corollary to Dedekind's theorem there are at most $[E:\Phi]$ isomorphisms of E/Φ into P/Φ so these must coincide with the automorphisms of E/Φ. If $s \in G$, the restriction of s to E is an isomorphism of E/Φ into P/Φ; hence this is an automorphism. This implies that $E^s = E$ for all $s \in G$.

Theorem 5 establishes, in particular, a bijection (1–1, onto mapping) between the collection of subfields E of P which contain a fixed subfield Φ, which is Galois and of finite co-dimension in P, and the collection of subgroups H of the Galois group G of P/Φ. This correspondence satisfies the properties in (iii) and (iv). We remark also that $\{H\}$ is finite, which implies that the collection of fields beween P and Φ is finite. At this point there is one serious gap in our theory: We have given no conditions that P be finite dimensional Galois over Φ. The next three sections will be devoted to filling this gap and to forging the link between the present "abstract" Galois theory and the theory of equations.

EXERCISES

1. Let C be the field of complex numbers and let $P = C(\xi)$, a simple transcendental extension of C (Vol. I, p. 101). Let s be the automorphism of P/C such that $\xi^s = \epsilon\xi$ where ϵ is a primitive n-th root of 1 and let t be the automorphism of P/C such that $\xi^t = \xi^{-1}$. Show that $s^n = 1, t^2 = 1, st = ts^{-1}$ and that the group G of automorphisms generated by s, t is of order $2n$. Show that the subfield of G-invariants is $C(\eta)$, $\eta = \xi^n + \xi^{-n}$.

2. Determine the Galois group of $\Phi(\rho)$ over Φ where Φ is the field of rational numbers and $\rho^4 = 2$.

3. Let P be finite dimensional Galois over Φ and suppose the Galois group $G = G_1 \times G_2$, G_i subgroups of G. Show that, if P_i is the subfield corresponding to G_i, then P_i/Φ is Galois and $P = P_1 \otimes P_2$ (over Φ).

4. Let Φ be a field of characteristic $\neq 2$ and let P be an extension such that $[P:\Phi] = 2$. Show that $P = \Phi(\theta)$ where $\theta^2 = \alpha \in \Phi$. Use this to prove that P is Galois over Φ.

5. Show that, if Φ_1 and Φ_2 are Galois in P, then $\Phi_1 \cap \Phi_2$ is Galois in P. Let Φ be a field of characteristic 0, $P = \Phi(\xi)$, ξ transcendental. Let $\Phi_1 = \Phi(\xi^2)$, $\Phi_2 = \Phi(\xi(\xi + 1))$. Show that $[P:\Phi_1] = 2 = [P:\Phi_2]$ but $[P:\Phi_1 \cap \Phi_2]$ is infinite.

6. Show that, if R_0 is the field of rational numbers, then $R_0(\sqrt[3]{2})$ is not Galois over R_0.

7. Let G be an arbitrary group of automorphisms in a field P and let $\mathfrak{A} = \{\Sigma s_i\rho_i | s_i \in G, \rho_i \in P\}$. Show that $\mathfrak{A}$ satisfies the hypotheses (i) and (ii) of Theorem 2. Show that $\Phi = \{\alpha | \alpha_R A = A\alpha_R, A \in \mathfrak{A}\}$ is the subfield of G-invariants. Use these results and ex. 1, § 2, to prove that, if E is a subfield of P containing Φ such that $[E:\Phi] < \infty$, then any isomorphism of E/Φ into P/Φ can be extended to an automorphism of P/Φ.

8. (Kaplansky). P, Φ, E, and G as in ex. 7. Prove that E is Galois in P. (In other words, if Φ is Galois in P and $E \supseteq \Phi$ satisfies $[E:\Phi] < \infty$, then E is Galois in P.) (Hint: Set $H = G \cap A(E)$. Let Δ be a finite dimensional subspace of P/Φ containing E. Use ex. 1, § 2, to show that $\mathfrak{L}_\Phi(\Delta, P)$ has a right P-basis of the form $(\bar{s}_1, \bar{s}_2, \cdots, \bar{s}_n)$, $\bar{s}_i$ the restriction to Δ of $s_i \in G$. Use Theorem 4 to show that $\mathfrak{L}_E(\Delta, P)$ has a P-basis $(\bar{t}_1, \bar{t}_2, \cdots, \bar{t}_r)$, $t_j \in H$. Use this and ex. 2, § 2, to prove that $E = I(H)$.)

5. Splitting field of a polynomial. Let Φ be a given field and $f(x)$ a non-zero polynomial contained in the polynomial ring $\Phi[x]$, x an indeterminate. We recall that an element ρ of Φ is called a root of $f(x)$ or of the equation $f(x) = 0$ if $f(\rho) = 0$. We know that this is the case if and only if $f(x) = (x - \rho)g(x)$ in $\Phi[x]$ (Vol. I, p. 99); and if deg $f(x) = n$, then $f(x)$ has at most n roots in Φ (Vol. I, p. 104). If $\rho_1, \rho_2, \cdots, \rho_r$ are distinct roots, then

$$f(x) = (x - \rho_1)(x - \rho_2) \cdots (x - \rho_r)g(x).$$

In Vol. I, pp. 101–102, we have given a construction for an extension P/Φ in which a given irreducible polynomial $f(x) \in \Phi[x]$ has a root. If we apply this to an irreducible factor of any non-zero $f(x) \in \Phi[x]$, we obtain an extension P/Φ containing a root of $f(x)$. We shall now establish the existence of a minimal field extension P in which a given polynomial $f(x)$, deg $f(x) > 0$, decomposes as a product of linear factors. Unless otherwise indicated we shall assume our polynomials have leading coefficients 1. Then we require an extension P/Φ such that

$$(5) \qquad f(x) = (x - \rho_1)(x - \rho_2) \cdots (x - \rho_n)$$

in P[x]. If $\Phi(\rho_1, \rho_2, \cdots, \rho_n)$ denotes the subfield of P/Φ generated by the ρ_i, then evidently the factorization (5) is valid also in $\Phi(\rho_1, \rho_2, \cdots, \rho_n)[x]$. Hence, if P/$\Phi$ is to be minimal, then we must have $P = \Phi(\rho_1, \rho_2, \cdots, \rho_n)$. We recall also that the factorization (5) is unique in P[x] apart from factors in Φ (Vol. I, p. 100, p. 123). From this it follows that the set $\{\rho_i\}$ is the complete set of roots of $f(x)$ in P and that, if Σ/Φ is a subfield of $\Phi(\rho_1, \cdots, \rho_n)/\Phi$ such that $f(x)$ is a product of linear factors in $\Sigma[x]$, then $\Sigma = \Phi(\rho_1, \rho_2, \cdots, \rho_n)$. This leads us to give the following

Definition 1. *Let Φ be a field and $f(x)$ a polynomial of positive degree with coefficients in Φ (leading coefficient* 1). *Then an extension field* P/Φ *is called a* splitting field *of $f(x)$ if a factorization* (5) *holds in* P[x] *and* $P = \Phi(\rho_1, \rho_2, \cdots, \rho_n)$.

We shall now state two immediate results which will be used frequently.

Lemma 1. (1) *If* P/Φ *is a splitting field of $f(x)$* ε $\Phi[x]$ *and* Σ/Φ *is a subfield of* P/Φ, *then* P/Σ *is a splitting field of $f(x)$.* (2) *If* P/Σ *is a splitting field for $f(x)$* ε $\Phi[x]$ *and* $\Sigma = \Phi(\sigma_1, \cdots, \sigma_r)$ *where* $f(\sigma_j) = 0$, *then* P/Φ *is a splitting field of $f(x)$.*

Proof. (1) This is an immediate consequence of the definition. (2) By assumption we have $P = \Sigma(\rho_1, \cdots, \rho_n)$ where (5) holds in P[x]. Also $\Sigma = \Phi(\sigma_1, \cdots, \sigma_r)$ and $f(\sigma_j) = 0$. It follows that every σ_j is one of the ρ_i; hence $P = \Phi(\rho_1, \rho_2, \cdots, \rho_n)$.

We can now prove the following existence theorem.

Theorem 6. *Any polynomial $f(x)$* ε $\Phi[x]$ *of positive degree has a splitting field* P/Φ.

Proof. Let $f(x) = f_1(x)f_2(x) \cdots f_k(x)$ be the factorization of $f(x)$ into irreducible factors (with leading coefficients 1). Evidently $k \leq n = \deg f(x)$. We use induction on $n - k$. If $n - k = 0$, the $f_i(x)$ are all of degree 1 and this means that Φ itself is a splitting field. Now assume $n - k > 0$ so that some f_i, say $f_1(x)$, is of degree > 1. Then there exists an extension field E/Φ such that $E = \Phi(\rho)$ and $f_1(\rho) = 0$. Then $f_1(x) = (x - \rho)f_1^*(x)$ in E[x] and so $f(x)$ is a product of $l > k$ irreducible factors in E[x].

Then $n - l < n - k$ and the induction hypothesis permits us to conclude that there exists a splitting field P/E for $f(x)$. Since E $= \Phi(\rho)$, the lemma shows that P/Φ is a splitting field for $f(x)$.

We consider some examples of splitting fields.

(1) $f(x) = x^2 + \alpha x + \beta$. If f is reducible in $\Phi[x]$, then Φ is a splitting field. Otherwise, we let $P = \Phi[x]/(f(x))$, which is a field since f is irreducible. If we set $\rho_1 = x + (f(x))$ where, as usual, $(f(x))$ denotes the principal ideal generated by this polynomial, then $f(\rho_1) = 0$ in P so $f(x) = (x - \rho_1)(x - \rho_2)$ in P$[x]$. Thus $P = \Phi[\rho_1] = \Phi(\rho_1)$ is a splitting field. Since $f(x)$ is the minimum polynomial of ρ_1, $[P:\Phi] = 2$, by VIII of the Introduction.

(2) Let p be a prime and let $f(x) = x^p - 1 = (x - 1)(x^{p-1} + x^{p-2} + \cdots + 1)$, $\Phi = R_0$ the field of rational numbers. Then P is a splitting field of $f(x)$ if and only if it is a splitting field of $g(x) = x^{p-1} + x^{p-2} + \cdots + 1$. It is known that $g(x)$ is irreducible (Vol. I, ex. 2, p. 127), so $P = R_0[x]/(g(x))$ is a field over R_0 and $P = R_0[\rho] = R_0(\rho)$, $\rho = x + (g(x))$. We have $\rho^p = 1$ and $\rho \neq 1$, which implies that ρ is of order p in the multiplicative group P* of P. Hence, $1, \rho, \rho^2, \cdots, \rho^{p-1}$ are distinct and all of these are roots of $x^p - 1 = 0$. It follows that $x^p - 1 = \prod_{i=0}^{p-1} (x - \rho^i)$ and so $P = R_0(\rho)$ is a splitting field.

(3) $f(x) = (x^2 - 2)(x^2 - 3)$, $\Phi = R_0$. We first form $E = R_0(\rho)$ where $\rho^2 = 2$. We know that $x^2 - 2$ is irreducible in $R_0[x]$ (Euclid). In E we have $x^2 - 2 = (x - \rho)(x + \rho)$. However, $x^2 - 3$ is irreducible in E$[x]$. Otherwise, there exists $\eta \in E$ such that $\eta^2 = 3$. But $\eta = \alpha + \beta\rho$, α, β rational and $\eta^2 = (\alpha^2 + 2\beta^2) + 2\alpha\beta\rho$; so, if this $= 3$, then $\alpha\beta = 0$ and $\alpha^2 + 2\beta^2 = 3$. If $\beta = 0$, we must have $\alpha^2 = 3$ and, if $\alpha = 0$, we have $\beta^2 = 3/2$. Both of these are impossible for rational numbers. We now form $P = E(\eta)$ where $\eta^2 = 3$. We have $P = R_0(\rho, \eta)$ and in P$[x]$, $(x^2 - 2)(x^2 - 3) = (x - \rho)(x + \rho)(x - \eta)(x + \eta)$, so P/R_0 is a splitting field. Using VI and VIII of the Introduction, one sees that $[P:\Phi] = 4$.

Before continuing with our discussion of splitting fields it will be well to fix some notations on field extensions and algebra extensions of a field Φ which to some extent have already been used. If S is a subset of a field P/Φ, then we let $\Phi[S]$ and $\Phi(S)$ respectively denote the subalgebra and the subfield over Φ generated by S. By definition, the first of these is the intersection of all subalgebras of P/Φ containing S and the second is the intersection of all subfields of P/Φ containing S. It is clear that $\Phi[S]$ is the subspace of P over Φ spanned by 1 and all monomials $\sigma_1\sigma_2\cdots\sigma_m$, $\sigma_i \in S$, and that $\Phi(S)$ is the set of elements $\alpha\beta^{-1}$, α, β in $\Phi[S]$, $\beta \neq 0$. It follows directly from the definition that, if S_1 and S_2 are subsets of P, then $(\Phi(S_1))(S_2) = \Phi(S_1 \cup S_2)$ where the first of these is, of course, the subfield of P/$\Phi(S_1)$ generated by S_2.

If ρ is an algebraic element of P/Φ, then we know that $[\Phi[\rho]:\Phi]$

$= \deg f(x)$ where $f(x)$ is the minimum polynomial of ρ over Φ (Introd. VIII). In fact, we have seen that, if $\deg f(x) = n$, then $(1, \rho, \rho^2, \cdots, \rho^{n-1})$ is a basis for P/Φ. Since the dimensionality $[\Phi[\rho]:\Phi]$ is finite, we know that $\Phi[\rho]$ is a field (Introd. VII). Hence it is clear that $\Phi(\rho) = \Phi[\rho]$. We shall now generalize these results in proving the following key lemma on successive algebraic extensions.

Lemma 2. *Let* $P = \Phi(\rho_1, \rho_2, \cdots, \rho_m)$ *and assume that* ρ_i *is algebraic over* $\Phi(\rho_1, \rho_2, \cdots, \rho_{i-1})$, $i = 1, 2, \cdots, m$. *Then* $[P:\Phi] < \infty$ *and* $P = \Phi[\rho_1, \rho_2, \cdots, \rho_m]$.

(We shall see later (p. 254) that, conversely, if $P = \Phi[\rho_1, \rho_2, \cdots, \rho_n]$ is a field then the ρ_i are algebraic over Φ.)

Proof. We have seen that this holds for $m = 1$. Suppose $m > 1$ and assume the result holds for $r < m$. Then $\Phi(\rho_1, \cdots, \rho_r) = \Phi[\rho_1, \cdots, \rho_r]$ and this is finite dimensional over Φ. Since ρ_{r+1} is algebraic over $\Phi(\rho_1, \cdots, \rho_r)$, we have $\Phi(\rho_1, \cdots, \rho_r)(\rho_{r+1}) = \Phi(\rho_1, \cdots, \rho_r)[\rho_{r+1}]$ and the dimensionality of this extension over $\Phi(\rho_1, \cdots, \rho_r)$ is finite. It follows that

$$(6)\quad [\Phi(\rho_1, \cdots, \rho_{r+1}):\Phi] = [\Phi(\rho_1, \cdots, \rho_r)(\rho_{r+1}):\Phi]$$

$$= [\Phi(\rho_1, \cdots, \rho_r)(\rho_{r+1}):\Phi(\rho_1, \cdots, \rho_r)][\Phi(\rho_1, \cdots, \rho_r):\Phi]$$

is finite. Also $\Phi(\rho_1, \cdots, \rho_{r+1}) = \Phi(\rho_1, \cdots, \rho_r)[\rho_{r+1}]$ and $\Phi(\rho_1, \cdots, \rho_r) = \Phi[\rho_1, \cdots, \rho_r]$ imply that every element of $\Phi(\rho_1, \cdots, \rho_{r+1})$ is a polynomial in the ρ_i, for $1 \leq i \leq r+1$. Hence $\Phi(\rho_1, \cdots, \rho_{r+1}) = \Phi[\rho_1, \cdots, \rho_{r+1}]$. The lemma now follows by induction.

This result is applicable in particular if the ρ_i are all algebraic over Φ. Since the roots ρ_i of $f(x)$ are algebraic over Φ, it is apparent that if P/Φ is a splitting field of $f(x) \in \Phi[x]$, then $[P:\Phi] < \infty$.

We shall now show that any two splitting fields of a polynomial are isomorphic over Φ. In order to carry out an inductive argument (and for other reasons, too) it is useful to generalize the result as follows. Let Φ and $\bar{\Phi}$ be fields which are isomorphic and let $\alpha \to \bar{\alpha}$ be an isomorphism of Φ onto $\bar{\Phi}$. We know that this can be extended to a unique isomorphism $f(x) \to \bar{f}(x)$ of $\Phi[x]$

onto $\bar{\Phi}[x]$ so that $x \to x$ (Introd. II). We wish to consider a splitting field over Φ of a polynomial $f(x)$ and a splitting field over $\bar{\Phi}$ of the corresponding polynomial $\bar{f}(x)$ in $\bar{\Phi}[x]$. The following theorem will imply uniqueness of the splitting field and gives an important result on the number of isomorphisms of a splitting field.

Theorem 7. *Let $\alpha \to \bar{\alpha}$ be an isomorphism of a field Φ onto the field $\bar{\Phi}$ and let $f(x)$ be a polynomial of positive degree with leading coefficient* 1, *$f(x)$ in $\Phi[x]$, and let $\bar{f}(x)$ be the corresponding polynomial in $\bar{\Phi}[x]$. Let P and $\bar{\text{P}}$ be splitting fields over Φ and $\bar{\Phi}$ of $f(x)$ and $\bar{f}(x)$ respectively. Then there exists an isomorphism of P onto $\bar{\text{P}}$ which coincides with the given isomorphism on Φ. Moreover, if $\bar{f}(x)$ is a product of distinct linear factors in $\bar{\text{P}}[x]$, then the number of extensions of the given isomorphism on Φ to an isomorphism of P into $\bar{\text{P}}$ is* $[\text{P}:\Phi]$.

Proof. Both assertions will be proved by induction on $[\text{P}:\Phi]$. If $[\text{P}:\Phi] = 1$, $\text{P} = \Phi$ and $f(x) = \Pi(x - \rho_i)$ in $\Phi[x]$. Applying the isomorphism $h(x) \to \bar{h}(x)$ of $\Phi[x]$ we obtain $\bar{f}(x) = \Pi(x - \bar{\rho}_i)$ and the $\bar{\rho}_i \in \bar{\Phi}$. It follows that these are the roots of $\bar{f}(x) = 0$ in $\bar{\text{P}}$ so $\bar{\text{P}} = \bar{\Phi}$ and both results hold in this case. Now assume $[\text{P}:\Phi] > 1$. Then $f(x)$ is not a product of linear factors in $\Phi[x]$ and so it has an irreducible factor $g(x)$ of degree $r > 1$. Then $\bar{g}(x)$ is a factor of $\bar{f}(x)$. Also we may assume that $g(x) = \prod_1^r (x - \rho_i)$, $\bar{g}(x) = \prod_1^r (x - \bar{\sigma}_i)$ where $f(x) = \prod_1^n (x - \rho_i)$ and $\bar{f}(x) = \prod_1^n (x - \bar{\sigma}_i)$. Since $g(x)$ is irreducible, this is the minimum polynomial over Φ of ρ_1, and $\text{E} = \Phi(\rho_1)$ is r-dimensional over Φ ($r = \deg g > 1$). The extension theorem V of the Introduction implies that the isomorphism $\alpha \to \bar{\alpha}$ can be extended to a unique isomorphism of $\text{E} = \Phi[\rho_1] = \Phi(\rho_1)$ into $\bar{\text{P}}$ so that $\rho_1 \to \bar{\sigma}_i$, $i = 1, \cdots, r$. We observe next that the indicated isomorphisms of E into P are the only extensions of $\alpha \to \bar{\alpha}$. Again by the extension theorem V, in any isomorphism of E extending the given isomorphism, ρ_1 is mapped into an element $\bar{\sigma}$ such that $\bar{g}(\bar{\sigma}) = 0$. Since $\bar{g}(x) = \prod_1^r (x - \bar{\sigma}_i)$, it follows that $\bar{\sigma} = \bar{\sigma}_i$ for some i,

$1 \leq i \leq r$. Then the isomorphism coincides with one of those we indicated. Thus we see that $\alpha \to \bar{\alpha}$ can be extended to an isomorphism of $E = \Phi(\rho_1)$ into $\bar{P}$ and the number of such extensions is the number of distinct elements in $\{\bar{\sigma}_1, \cdots, \bar{\sigma}_r\}$. In particular, if $\bar{f}(x)$ is a product of distinct linear factors, then this is true also for $\bar{g}(x)$ and the number is then $r = \deg \bar{g}(x) = [E:\Phi]$. We now replace the base field Φ by E and let $\bar{E}$ be its image under one of the chosen extensions of the isomorphism on Φ to an isomorphism of E into $\bar{P}$. We denote this extension by $\epsilon \to \bar{\epsilon}$. Then P/E is a splitting field of $f(x)$ and $\bar{P}/\bar{E}$ is a splitting field of $\bar{f}(x)$. Moreover, $[P:E] < [P:\Phi]$ since $[E:\Phi] = r > 1$. Hence the induction hypothesis shows that $\epsilon \to \bar{\epsilon}$ can be extended to an isomorphism of P onto $\bar{P}$ and the number of such extensions is $[P:E]$ if $\bar{f}(x)$ is a product of distinct linear factors in $\bar{P}[x]$. If we take into account the first result on the extension of $\alpha \to \bar{\alpha}$ to $\epsilon \to \bar{\epsilon}$, we see that there exists an extension of the isomorphism $\alpha \to \bar{\alpha}$ to an isomorphism of P onto $\bar{P}$ and, if $\bar{f}(x)$ splits into distinct linear factors, then we obtain $[P:E][E:\Phi] = [P:\Phi]$ distinct isomorphisms since we have $[E:\Phi]$ extensions to E and each of these has $[P:E]$ extensions to P. Thus we obtain $[P:\Phi] = [P:E][E:\Phi]$ distinct extensions. It is clear that we have accounted for every extension in our enumeration (cf. also Cor. to Th. 3) and so the proof is complete.

We now specialize the result we have just proved by taking $\bar{\Phi} = \Phi$ and the identity mapping $\alpha \to \alpha$ in Φ. Then the conclusion is that, if P/Φ and $\bar{P}/\Phi$ are two splitting fields of the same polynomial $f(x)$, then P/Φ and $\bar{P}/\Phi$ are isomorphic. Moreover, the second part of the result is that, if $f(x)$ has distinct roots, then the number of automorphisms of P/Φ is $[P:\Phi]$. In other words, $(G:1) = [P:\Phi]$ for G the Galois group of P/Φ.

EXERCISES

1. Construct a splitting field over the rationals for $x^5 - 2$. Find the dimensionality.
2. Let P/Φ be a splitting field of $f(x) \neq 0$ in $\Phi[x]$ and let E be a subfield of P/Φ. Show that any isomorphism of E/Φ into P/Φ can be extended to an automorphism of P.
3. Show that the dimensionality of a splitting field P/Φ of a polynomial $f(x)$ of degree n cannot exceed $n!$.

6. Multiple roots. Separable polynomials. Let $f(x)$ be a polynomial of positive degree in $\Phi[x]$ and let P/Φ be a splitting field. We now write

$$(7)\qquad f(x) = (x - \rho_1)^{k_1}(x - \rho_2)^{k_2} \cdots (x - \rho_r)^{k_r},$$

$\rho_i \in P$, $\rho_i \neq \rho_j$ if $i \neq j$, and we say that ρ_i is a root of *multiplicity* k_i of $f(x) = 0$. If $k_i = 1$, then ρ_i is called a *simple root;* otherwise ρ_i is a *multiple root.* If we have a second splitting field $\bar{P}$ over Φ, then $f(x) = \prod_1^r (x - \bar{\rho}_j)^{k_j}$ in $\bar{P}$ where $\rho_j \to \bar{\rho}_j$ in an isomorphism of P/Φ onto $\bar{P}/\Phi$. It is clear that the existence of multiple roots for f is independent of the particular choice of a splitting field. We shall now carry over a classical criterion for multiple roots which can be tested in $\Phi[x]$ itself. For this we need the standard formal derivative (or derivation) in $\Phi[x]$. Thus we define a linear mapping $f \to f'$ in $\Phi[x]$ by specifying that $(x^i)' = ix^{i-1}$, $i = 0, 1, 2, \cdots$, $x^0 \equiv 1$. Since $(1, x, x^2, \cdots)$ is a basis for $\Phi[x]$ over Φ, this defines a unique linear mapping $f \to f'$ in $\Phi[x]$ over Φ. We call f' the (formal) derivative of f and we note the basic rule:

$$(8)\qquad (fg)' = f'g + fg'.$$

Because of the linearity of the derivative, it suffices to check this for $f = x^i$, $g = x^j$ in the basis (x^i) for $\Phi[x]$. Then $fg = x^{i+j}$ so that $(fg)' = (i + j)x^{i+j-1}$, $f'g = ix^{i+j-1}$, $fg' = jx^{i+j-1}$, so (8) is valid. We can now prove

Theorem 8. *If* $f(x) \in \Phi[x]$ *and* $\deg f > 0$, *then all the roots of* f *(in its splitting field) are simple if and only if* $(f, f') = 1$ *(that is,* 1 *is the highest common factor of* f *and* f'*).*

Proof. Let $d(x)$ be the highest common factor (f, f') of f and f' in $\Phi[x]$ (cf. Vol. I, p. 100, p. 122). Suppose $f(x)$ has a multiple root in $P[x]$, so $f(x) = (x - \rho)^k g(x)$, $k > 1$. If we take derivatives in $P[x]$, this gives $f' = (x - \rho)^k g' + k(x - \rho)^{k-1} g$ which is divisible by $x - \rho$ since $k - 1 \geq 1$. Thus $(x - \rho) \mid f$ (i.e. $x - \rho$ is a factor of f) and $(x - \rho) \mid f'$, so $(x - \rho) \mid d$. Hence $d(x) \neq 1$. Next, suppose all the roots of f are simple. Then we have $f(x) = \prod_1^n (x - \rho_i)$, $\rho_i \neq \rho_j$, $i \neq j$. The usual extension of (8) to several factors gives

$$f'(x) = \sum_{j=1}^{n} (x - \rho_1) \cdots (x - \rho_{j-1})(x - \rho_{j+1}) \cdots (x - \rho_n).$$

It is clear from this that $(x - \rho_i) \nmid f'(x)$ and this implies that $(f, f') = 1$.

If f is irreducible in $\Phi[x]$, then $(f, f') \neq 1$ implies that $f | f'$. By degree considerations this can happen only if $f' = 0$. If the characteristic is 0, this evidently implies that f is an element of Φ. If the characteristic is $p \neq 0$ and $f(x) = \alpha_0 x^n + \alpha_1 x^{n-1} + \alpha_2 x^{n-2} + \cdots + \alpha_n$, then $f'(x) = n\alpha_0 x^{n-1} + (n-1)\alpha_1 x^{n-2} + (n-2)\alpha_2 x^{n-3} + \cdots$, so $f'(x) = 0$ implies that $(n - i)\alpha_i = 0$. This implies that $\alpha_i = 0$ if the integer $n - i$ is not divisible by p. Hence we see that $f(x) = \beta_0 x^{mp} + \beta_1 x^{(m-1)p} + \cdots + \beta_m = g(x^p)$ where $g(x) = \beta_0 x^m + \beta_1 x^{m-1} + \cdots + \beta_m$. This condition is also clearly sufficient that $f' = 0$ since $(x^{kp})' = kpx^{kp-1} = 0$. In the characteristic $p \neq 0$ case we shall see that the conditions: f irreducible of positive degree, $f' = 0$, can be fulfilled. This is a basic difference between fields of characteristic 0 and those of characteristic $p \neq 0$ and this is the root of a host of complications in the latter case.

Let us now look more closely at fields of characteristic $p \neq 0$. We recall that, if Φ is of this type, then we have

$$(\alpha + \beta)^p = \alpha^p + \beta^p, \quad (\alpha\beta)^p = \alpha^p \beta^p \tag{9}$$

in Φ (Vol. I, ex. 3, p. 120). The second of these is clear and the first is a consequence of the binomial theorem and the fact that the binomial coefficient $\binom{p}{i} = p!/i!(p-i)!$ is divisible by p for $1 \leq i \leq p - 1$ since this is an integer and p occurs in the numerator of the fraction but not in the denominator. We note also that, if $\alpha^p = \beta^p$, then $(\alpha - \beta)^p = \alpha^p - \beta^p = 0$ so $\alpha = \beta$. Thus we see from this and (9) that the mapping $\alpha \to \alpha^p$ is an isomorphism of Φ into itself. The image $\Phi^p = \{\alpha^p | \alpha \in \Phi\}$ is a subfield, the subfield of p-th powers. We can iterate the mapping $\alpha \to \alpha^p$ and obtain the isomorphism $\alpha \to \alpha^{p^e}$, $e = 1, 2, \cdots$ of Φ onto the subfield Φ^{p^e} of p^e-th powers.

We prove next the following general result which will be useful later on.

Lemma. *If Φ is a field of characteristic $p \neq 0$, then $x^p - \alpha$ is irreducible in $\Phi[x]$ unless $\alpha = \beta^p$, $\beta \in \Phi$, in which case $x^p - \alpha = (x - \beta)^p$.*

Proof. Let P be a splitting field for $x^p - \alpha$. If β is a root of $x^p - \alpha = 0$, then $\alpha = \beta^p$. Hence $x^p - \alpha = x^p - \beta^p = (x - \beta)^p$ in P$[x]$. Now suppose $x^p - \alpha = g(x)h(x)$ in $\Phi[x]$ where $\deg g = k$ and $1 \leq k \leq p - 1$. Then in P$[x]$ we must have $g = (x - \beta)^k = x^k - k\beta x^{k-1} + \cdots$. This implies that $k\beta \in \Phi$; hence $\beta \in \Phi$. Then $x^p - \alpha = (x - \beta)^p$ holds in $\Phi[x]$.

We now consider the following example. We let $I_p \equiv I/(p)$ the field of residues modulo p, and we let $\Phi = I_p(\xi)$, ξ transcendental. Then we claim that $\xi \notin \Phi^p$. Now, if $\gamma \in \Phi$, we can write $\gamma = \alpha(\xi)\beta(\xi)^{-1}$ where $\alpha(\xi)$ and $\beta(\xi)$ are polynomials. Then $\gamma^p = \alpha(\xi^p)\beta(\xi^p)^{-1}$, since $\alpha(\xi) = \alpha_0 + \alpha_1\xi + \cdots$ implies $\alpha(\xi)^p = \alpha_0{}^p + \alpha_1{}^p\xi^p + \cdots = \alpha_0 + \alpha_1\xi^p + \cdots$ (by Fermat's theorem). Hence $\gamma^p = \xi$ implies that $\alpha(\xi^p) = \beta(\xi^p)\xi$ and this is impossible since 1, $\xi, \cdots$ are I_p-independent. Thus we see that $\xi \notin \Phi^p$ and hence, by the lemma, $x^p - \xi$ is irreducible in $\Phi[x]$. On the other hand, we have seen that $x^p - \xi$ has p equal roots in its splitting field. We note also that $(x^p - \xi)' = 0$.

We shall now call a polynomial f (of positive degree) *separable* if it is a product of irreducible polynomials in $\Phi[x]$ all of which have only simple roots in a splitting field. Our discussion shows that, if Φ is of characteristic 0, then every $f(x) \in \Phi[x]$ is separable, whereas for characteristic $p \neq 0$ there exist inseparable polynomials.

EXERCISES

1. Prove the following extension of the lemma: $x^{p^e} - \alpha$ is irreducible unless $\alpha \in \Phi^p$.

2. Let Φ_0 be a finite field of q elements and let P $= \Phi_0(\xi)$, ξ transcendental over Φ_0. Let G be the finite group of the automorphisms of P over Φ_0 such that $\xi \rightarrow \xi + \alpha$, $\alpha \in \Phi_0$. Show that $\Phi = I(G) = \Phi_0(\xi^q - \xi)$.

3. Let Φ be a field of characteristic $p \neq 0$ and let $\xi_1, \xi_2, \cdots, \xi_n$ be indeterminates over Φ, P $= \Phi(\xi_1, \xi_2, \cdots, \xi_n)$ the field of fractions of $\Phi[\xi_1, \xi_2, \cdots, \xi_n]$. Show that $[\mathrm{P}:\Phi(\xi_1{}^p, \xi_2{}^p, \cdots, \xi_n{}^p)] = p^n$. Show also that the Galois group of P over $\Phi(\xi_1{}^p, \xi_2{}^p, \cdots, \xi_n{}^p)$ is the identity.

4. Let $P = \Phi(\xi_1, \xi_2, \cdots, \xi_n)$ of characteristic $p \neq 0$ and suppose that $\xi_i^{p^{e_i}} \varepsilon \Phi$ for $i = 1, 2, \cdots, n$, e_i a positive integer. Show that the Galois group of P over Φ is the identity.

5. Let Φ be a field of characteristic $p \neq 0$. A polynomial with coefficients in Φ is called a *p-polynomial* if it has the form $x^{p^m} + \alpha_1 x^{p^{m-1}} + \alpha_2 x^{p^{m-2}} + \cdots + \alpha_m x$. Show that a polynomial (with leading coefficient 1) is a p-polynomial if and only if its roots form a subgroup of the additive group of the splitting field and all the roots have the same multiplicity p^e. Show that the roots of the displayed p-polynomial are all simple if and only if $\alpha_m \neq 0$.

6. Let $f(x)$ be irreducible in $\Phi[x]$, Φ of characteristic $p \neq 0$. Show that $f(x)$ can be written in the form $g(x^{p^e})$ where $g(x)$ is irreducible and has distinct roots. Use this to show that every root of $f(x)$ has the same multiplicity p^e (in a splitting field).

7. Let Φ be a field of characteristic 0, $f(x)$ a polynomial of positive degree contained in $\Phi[x]$. Show that if $d(x)$ is the highest common factor of $f(x)$ and $f'(x)$, then $g(x) = f(x)d(x)^{-1}$ has simple roots which are the distinct roots of $f(x)$.

7. The "fundamental theorem" of Galois theory. We now take up again the abstract Galois theory of § 4 and we shall answer first the question which we raised at the end of § 4: that of characterizing finite dimensional Galois extensions. The result is the following

Theorem 9. *A field* P/Φ *is finite dimensional Galois over* Φ *if and only* P *is a splitting field over* Φ *of a separable polynomial* $f(x) \varepsilon \Phi[x]$.

Proof. Let $f(x)$ be separable and let $f(x) = f_1(x)^{e_1} \cdots f_l(x)^{e_l}$ where the $f_i(x)$ are irreducible in $\Phi[x]$ and $f_i \neq f_j$ if $i \neq j$. Then $f_i(x)$ has only simple roots. Moreover, since f_i and f_j for $i \neq j$ are distinct irreducible polynomials, their highest common factor is 1. Hence $1 = a(x)f_i(x) + b(x)f_j(x)$ for $a(x)$, $b(x)$ in $\Phi[x]$, and this implies that f_i and f_j have no common roots in any extension field. It follows that $g(x) = f_1(x)f_2(x) \cdots f_l(x)$ has no multiple roots, and it is clear that, if P/Φ is a splitting field of f, then it is a splitting field also for g. Now we know that any splitting field is finite dimensional and Theorem 6 implies that, if G is the Galois group of P/Φ, then $(G:1) = [P:\Phi]$. Let $\Phi' = I(G)$ the set of G-invariants. Then, by Theorem 5 (ii), $(G:1) = [P:\Phi']$. Since $\Phi' \supseteq \Phi$ we have $\Phi = \Phi'$, which shows that Φ is Galois in P or P is Galois over Φ. This proves the sufficiency of the condition. Next assume P is finite dimensional Galois over Φ and let G be the Galois group. We know that G is finite and we indicate it as

$G = \{s_1, s_2, \cdots, s_n\}$. If $\rho \in P$, we shall call the images ρ^{s_i} under $s_i \in G$ the *conjugates* of ρ in P/Φ. We may assume that $\rho^{s_1}, \cdots, \rho^{s_r}$ are distinct and that this set includes all the conjugates. Then we assert that $h(x) = \prod_{j=1}^{r} (x - \rho^{s_j}) \in \Phi[x]$. To see this let $s \in G$ and let s be its extension to $P[x]$ such that $x^s = x$. Then we have $h^s(x) = \prod_{1}^{r} (x - \rho^{s_j s})$. Since the elements $\rho^{s_1 s}, \cdots, \rho^{s_r s}$ are distinct conjugates, this set is the complete set of conjugates and so $h^s(x) = h(x)$, $s \in G$. Hence $h(x) \in \Phi[x]$. This shows that the minimum polynomial over Φ of any $\rho \in P$ (this is actually $h(x)$) is separable and splits as a product of linear factors in $P[x]$. Now let $(\rho_1, \rho_2, \cdots, \rho_n)$ be a basis for P/Φ and let $f_i(x)$ be the minimum polynomial of ρ_i over Φ. Then $f(x) = \Pi f_i(x)$ is separable and clearly P is a splitting field over Φ of f.

The main Galois correspondence (Th. 5) can now be applied to state the following result that is known classically as the

Fundamental Theorem of Galois Theory. *Let* P *be a splitting field over* Φ *of a separable polynomial and let* G *be the Galois group of* P/Φ. *With each subgroup* H *of* G *we associate the subfield* E *of* P *over* Φ *of* H*-invariants and with each subfield* E *over* Φ *we associate the subgroup* H *of* G *of elements* t *such that* $\epsilon^t = \epsilon$ *for all* ϵ *in* E. *Then these two correspondences are inverses and are bijections of the set of subgroups of* G *and the set of subfields of* P *over* Φ. *The correspondences are order inverting relative to inclusion and*

$$(H:1) = [P:E], \quad (G:H) = [E:\Phi]. \tag{10}$$

Moreover, H *is invariant in* G *if and only if the corresponding field* E *is Galois over* Φ *and in this case the Galois group of* E/Φ *is isomorphic to the factor group* G/H.

All of this can be read off directly from Theorem 5 and the remarks which follow it. The only part which has not been made explicit before is (10). Now it is clear from the definition that H is the Galois group of P/E. Hence $(H:1) = [P:E]$. Also $(G:1) = [P:\Phi]$, so $(G:H) = (G:1)/(H:1) = [P:\Phi]/[P:E] = [E:\Phi]$. We note also that $(G:H) = [E:\Phi]$ *is the number of distinct isomor-*

phisms of E/Φ *into* P/Φ. To see this we consider the restrictions $\bar{s}$ to E of the elements $s \varepsilon G$. If $\bar{s} = \bar{t}$ for $s, t \varepsilon G$, then $\overline{st^{-1}} = \bar{1}$ which means that $st^{-1} \varepsilon H$. Then the cosets Hs and Ht are identical. The converse follows by retracing the steps. Hence we see the collection $\{\bar{s} | s \varepsilon G\}$ contains $(G:H)$ distinct isomorphisms of E/Φ into P/Φ. We know also that there are no more than $[\mathrm{E}:\Phi] = (G:H)$ isomorphisms of E/Φ into P/Φ (Cor. to Th. 3). Hence we have caught them all. Incidentally, we have shown also that every isomorphism of E/Φ into P/Φ is a restriction of an automorphism of P/Φ. In other words, any such isomorphism can be extended to an automorphism (cf. ex. 7, § 4).

8. Normal extensions. Normal closures. At the beginning of the last section we gave an abstract characterization of splitting fields of separable polynomials: these are just the finite dimensional Galois extensions. We shall now give two abstract characterizations of arbitrary splitting fields.

Theorem 10. *The following three conditions on a finite dimensional extension* P/Φ *are equivalent:*

(1) P/Φ *is a splitting field of a polynomial* $f(x) \varepsilon \Phi[x]$.
(2) *Any isomorphism s of* P/Φ *into an extension field* Δ/Φ *is an automorphism.*
(3) *Every irreducible polynomial* $g(x) \varepsilon \Phi[x]$ *which has a root in* P *is a product of linear factors in* P[x].

Proof. (1) ⇒ (2) ("⇒" means "implies"): Let $\mathrm{P} = \Phi(\rho_1, \rho_2, \cdots, \rho_n)$ where $f(x) = \Pi(x - \rho_i)$ in P[x] and $f(x) \varepsilon \Phi[x]$. Suppose $\Delta \supseteq \mathrm{P} \supseteq \Phi$ and let s be an isomorphism of P/Φ into Δ/Φ. Since $f(\rho_i) = 0$, we have $f(\rho_i^s) = 0$ and, since $\{\rho_i\}$ is the complete set of roots of $f(x)$ in Δ, ρ_i^s is one of the ρ_j. Hence s maps every generator ρ_i of $\mathrm{P} = \Phi(\rho_1, \rho_2, \cdots, \rho_n)$ into P. Hence $\mathrm{P}^s \subseteq \mathrm{P}$. Since s is 1–1 Φ-linear and $[\mathrm{P}:\Phi] < \infty$, we have $\mathrm{P}^s = \mathrm{P}$ and s is an automorphism. (2) ⇒ (3): Assume every isomorphism of P/Φ into any extension field Δ/Φ is an automorphism. Let $g(x)$ be irreducible in Φ[x] and have a root σ in P. Write $\mathrm{P} = \Phi(\rho_1, \rho_2, \cdots, \rho_m)$ and let $f_i(x)$ be the minimum polynomial of ρ_i over Φ. Set $f(x) = g(x) \prod_1^m f_i(x)$ and let Δ/P be a splitting field over P of $f(x)$. Since $\rho_i \varepsilon$ P and $f(x) \varepsilon \Phi[x]$, Δ is also a splitting field over Φ of $f(x)$ and it contains a splitting field over Φ of $g(x)$. Hence it will follow that $g(x)$ is a product of linear factors in P$[x]$ if we can show that every root σ' of $g(x)$ contained in Δ is contained in P. To prove this we note that, since $g(x)$ is irreducible in $\Phi[x]$ and $g(\sigma) = 0 = g(\sigma')$, there exists an isomorphism s of $\Phi(\sigma)/\Phi$ into $\Phi(\sigma')/\Phi$ such that $\sigma^s = \sigma'$. We now observe that we can consider Δ as a splitting field over $\Phi(\sigma)$ and over $\Phi(\sigma')$ of $f(x) = f^s(x)$. Hence the main isomorphism theorem for splitting fields (Th. 6) shows that s can be extended to an automorphism s of Δ. Since $\alpha^s = \alpha$, $\alpha \varepsilon \Phi$, s is a Φ-automorphism of Δ. Its restriction to P is an isomorphism of P/Φ into Δ/Φ. Hence, by hypothesis, this restriction is an automorphism of P/Φ. Since $\sigma \varepsilon$ P, it follows that $\sigma^s = \sigma' \varepsilon$ P. This proves (3). (3) $\Rightarrow$ (1). Write P $= \Phi(\rho_1, \rho_2, \cdots, \rho_m)$ and let $f_i(x)$ be the minimum polynomial of ρ_i over Φ. If we assume (3), then $f_i(x)$ is a product of linear factors in P$[x]$. Hence P/Φ is a splitting field of $f(x) = \Pi f_i(x)$. This completes the proof.

A finite dimensional extension P/Φ satisfying any one (hence all) of the conditions of Theorem 10 is called a *normal extension*. It is clear from the condition (1) that, if P is normal over Φ and E is a subfield of P/Φ, then P is normal over E. On the other hand, if P $\supseteq$ E $\supseteq \Phi$, then it may well happen that P/E and E/Φ are normal and P/Φ is not (ex. 1 below). Let P $= \Phi(\sigma_1, \cdots, \sigma_m)$ be an arbitrary finite dimensional extension of Φ and let $f(x) = \Pi f_i(x)$ where $f_i(x)$ is the minimum polynomial of σ_i over Φ. Let Δ/P be a splitting field of $f(x) \varepsilon \Phi[x]$. Then Δ/Φ is a splitting field of $f(x)$; hence Δ/Φ is normal. Now let Δ'/Φ be any normal extension of P/Φ. Since Δ' contains σ_i and $f_i(x)$ is irreducible in Φ, condition (2) of Theorem 8 shows that Δ'/Φ contains a splitting field of $f(x)$. Hence we have an isomorphism of Δ/Φ into Δ'/Φ. This implies that no proper subfield of Δ containing P is normal over Φ. We now define a *normal closure* of P/Φ as a normal extension of Φ containing P and having the property that no proper subfield containing P is normal over Φ. Then we can say that Δ/Φ is a normal closure of P/Φ and the remark about Δ and Δ' shows that such an extension is determined up to Φ-isomorphism by P/Φ.

EXERCISE

1. Let $P = R_0(\sqrt[4]{2})$, R_0 the rationals, and let $E = R_0(\sqrt{2}) \subseteq P$. Show that P/E and E/R_0 are normal but P/R_0 is not normal.

9. Structure of algebraic extensions. Separability. The structure theory of fields will be taken up in detail in Chapter IV. However, at this point it is convenient to derive the basic theorems on algebraic extensions and more generally on the set of algebraic elements of any field P/Φ. We have shown in Vol. I (p. 183) that, if P is a field over Φ, then the subset A of elements of P which are algebraic over Φ form a subfield over Φ and every element of P which is algebraic over A is contained in A. The subfield A/Φ is called the *algebraic closure* of Φ in P and Φ is called *algebraically closed in* P if $\Phi = A$. The field P/Φ is *algebraic* if $P = A$, that is, every element of ρ is algebraic over Φ. Thus the second part of the result we have quoted above is that, if A is the algebraic closure of Φ in P, then A is algebraically closed in P. We shall now indicate another proof of these results which is based on the Lemma 2 of § 5: If ρ_i is algebraic over $\Phi(\rho_1, \cdots, \rho_{i-1})$ then $\Phi(\rho_1, \rho_2, \cdots, \rho_n)/\Phi$ is finite dimensional. Now let A be the set of elements of P which are algebraic over Φ and let $\rho, \sigma \in A$. Then $\Phi(\rho, \sigma)$ is finite dimensional. Since $\Phi(\tau)/\Phi$ is infinite dimensional for transcendental τ, it follows that every element of $\Phi(\rho, \sigma)$ is algebraic over Φ. In particular, $\rho \pm \sigma$, $\rho\sigma$, and ρ^{-1} are algebraic if $\rho \neq 0$. Since ρ and σ are arbitrary in P, this implies that A is a subfield of P. Also it is clear that $A \supseteq \Phi$. Now let ρ be an element of P which is algebraic over A and let $f(x) = x^n + \alpha_1 x^{n-1} + \cdots + \alpha_n$ be its minimum polynomial over A. Then the $\alpha_i \in A$ and so are algebraic over Φ. Moreover, it is clear that ρ is algebraic over $\Phi(\alpha_1, \alpha_2, \cdots, \alpha_n)$. It now follows that $\Phi(\alpha_1, \cdots, \alpha_n, \rho)$ is finite dimensional over Φ. Hence ρ is algebraic over Φ and consequently A is algebraically closed in the field P. The result we proved on the algebraic closure of A in P implies the following transitivity property: *if* B/A *is algebraic and* A/Φ *is algebraic, then* B/Φ *is algebraic.* To see this let Γ/Φ be the subfield of B/Φ of elements which are algebraic over Φ. Clearly $\Gamma \supseteq A$ and we have seen that, if $\beta \in B$ is algebraic over Γ, then it belongs to Γ. On the other hand, if β is any element of B, then β is algebraic over A, hence over Γ, so $\beta \in \Gamma$. This shows that $B = \Gamma$ is algebraic over Φ.

There are several other useful remarks on algebraic elements which are worth recording here for future reference. The first of these, which is implicit in what we have proved before, is that, if $\mathfrak{A}/\Phi$ is a subalgebra of a field P/Φ, then every element of $\mathfrak{A}$ is algebraic over Φ if and only if every finite subset X of $\mathfrak{A}$ is contained in a finite dimensional subalgebra $\mathfrak{X}/\Phi$. This implies that every $\xi \in \mathfrak{A}$ is algebraic over Φ since it implies that $\Phi[\xi]$ is finite dimensional. On the other hand, if every element of $\mathfrak{A}$ is algebraic and $X = \{\xi_1, \xi_2, \cdots, \xi_r\}$, then the lemma we quoted shows that $\Phi(\xi_1, \cdots, \xi_r)/\Phi$ is finite dimensional. We recall also that $\Phi[\xi_1, \cdots, \xi_r] = \Phi(\xi_1, \cdots, \xi_r)$, which implies that, if every element of $\mathfrak{A}$ is algebraic, then $\mathfrak{A}$ is a subfield of P/Φ. We note also that, if E/Φ is an algebraic subfield of P/Φ and Δ/Φ is an arbitrary subfield, then the subalgebra $E\Delta/\Phi$ generated by E and Δ is a subfield which is algebraic over Δ. To see this we observe that $E\Delta$ is the set of elements of the form $\Sigma\epsilon_i\delta_i$, $\epsilon_i \in E$, $\delta_i \in \Delta$. Hence, if X is a finite subset of $E\Delta$, then there exists a finite subset $\{\epsilon_i\}$ such that every element of X is a Δ-linear combination of the ϵ_i. Since E/Φ is algebraic, we may imbed the set $\{\epsilon_i\}$ in a finite dimensional subalgebra. If we express the ϵ_i in terms of a basis $\{\eta_j\}$ for this subalgebra, then we see that every element of X has the form $\Sigma\delta_j\eta_j$, $\delta_j \in \Delta$. Since $\eta_j\eta_k = \Sigma\gamma_{jkl}\eta_l$, $\gamma_{jkl} \in \Phi$, it is clear that the set $\Sigma\Delta\eta_j$ of Δ-linear combinations of the η_j is a subalgebra of P/Δ. We have therefore proved that every finite subset of $E\Delta$ is contained in a finite dimensional subalgebra over Δ. Hence every element of $E\Delta$ is algebraic over Δ and $E\Delta$ is a subfield.

An algebraic element $\rho \in P/\Phi$ is called *separable* (*algebraic*) over Φ if its minimum polynomial over Φ is separable. It is clear that ρ is separable over Φ if and only if there exists a polynomial $f(x) \in \Phi[x]$ with distinct roots such that $f(\rho) = 0$. Also ρ is separable if and only if there exists a polynomial $f(x) \in \Phi[x]$ with $(f, f') = 1$ such that $f(\rho) = 0$. If Φ' is an extension field of Φ, we shall again have $(f, f') = 1$ in $\Phi'[x]$ (since $(f', f) = af + bf'$, ex. 3, p. 122 of Vol. I). It follows that, if Φ'/Φ is a subfield of P/Φ and $\rho \in P$ is separable over Φ, then ρ is separable over Φ'. We have seen (§ 6) that every polynomial with coefficients in a field of characteristic 0 is separable. Consequently, the results we shall now consider become trivial in the characteristic 0 case.

An extension P/Φ will be called *separable* (*algebraic*) if every element $\rho \in P$ is separable over Φ. Let A/Φ be algebraic and let Σ be the subset of A of elements which are separable over Φ. We wish to show that Σ is a subfield containing Φ and that every element of A which is separable over Σ is contained in Σ. For this we shall need the following

Lemma 1. *Let* $P \supseteq E \supseteq \Phi$ *where* E *and* Φ *are subfields of* P *and* E/Φ *is finite dimensional Galois. Then any element* $\theta \in P$ *which is separable algebraic over* E *is separable algebraic over* Φ.

Proof. Let $g(x)$ be the minimum polynomial of θ over E. If $s \in G$ the Galois group of E/Φ, then s has a unique extension to $E[x]$ satisfying $x^s = x$. Let $g^{s_1}(x), g^{s_2}(x), \cdots, g^{s_r}(x)$ be the distinct images of $g(x)$ under $s \in G$ and let $f(x) = \prod_1^r g^{s_i}(x)$. Then $f^s(x) = f(x)$ for all $s \in G$, which implies that $f(x) \in \Phi[x]$. Since $g(x)$ is irreducible in $E[x]$ and $(g, g') = 1$, the same is true for every g^{s_i}. Hence every $g^{s_i}(x)$ has distinct roots. We note also that, if $i \neq j$, then g^{s_i} and g^{s_j} are relatively prime, since otherwise $(g^{s_i}, g^{s_j}) = g^{s_i}(x) = g^{s_j}(x)$ because these are irreducible in $E[x]$. This contradicts $g^{s_i} \neq g^{s_j}$ for $i \neq j$. Thus $1 = (g^{s_i}, g^{s_j})$ and consequently these have no common roots in a splitting field for $f(x)$. It is now clear that $f(x)$ has distinct roots. Since $f(\theta) = 0$ and $f \in \Phi[x]$, we see that θ is separable over Φ.

Clearly if $\rho \in E$, then ρ is separable algebraic over E (with minimum polynomial $x - \rho$). Hence Lemma 1 shows that ρ is separable over Φ. In other words, we have the

Corollary. *Any finite dimensional Galois extension is separable.*

We can now prove the main result on separability.

Theorem 11. *If* A/Φ *is algebraic, then the set* Σ *of elements of* A *which are separable over* Φ *is a subfield containing* Φ. *Moreover,* Σ *contains every element of* A *which is separable algebraic over* Σ.

Proof. Let $\rho, \sigma \in \Sigma$ and let $g(x)$ and $h(x)$ be the minimum polynomials over Φ of ρ and σ respectively. Then $f(x) = g(x)h(x)$ is separable. If Δ is a splitting field over $\Phi(\rho, \sigma)$ of $f(x)$, then Δ is also a splitting field over Φ of $f(x)$ (the normal closure of $\Phi(\rho, \sigma)/\Phi$). Hence Δ/Φ is Galois, so by the corollary above, every element of Δ is separable over Φ. In particular, $\rho \pm \sigma$, $\rho\sigma$, ρ^{-1} (if $\rho \neq 0$) and every element of Φ are separable over Φ. This proves that Σ is a subfield containing Φ. Now let θ be an element of A which is separable algebraic over Σ and let $x^n + \rho_1 x^{n-1} + \cdots + \rho_n$, $\rho_i \in \Sigma$, be its minimum polynomial over Σ. The subfield $\Phi(\rho_1, \rho_2, \cdots, \rho_n; \theta)$ is finite dimensional over Φ. Let Δ/Φ be its normal closure. Let $f_i(x)$ be the minimum polynomial of ρ_i over Φ. Then Δ contains a splitting field E/Φ of $f(x) = \prod_1^n f_i(x)$ and this is Galois over Φ since $f(x)$ is separable. Evidently $E \supseteq \Phi(\rho_1, \rho_2, \cdots, \rho_n)$. Also θ is separable over $\Phi(\rho_1, \rho_2, \cdots, \rho_n)$ since $x^n + \rho_1 x^{n-1} + \cdots + \rho_n$ is its minimum polynomial. Hence θ is separable algebraic over E. Then θ is separable algebraic over Φ by Lemma 1. This proves the second statement.

If the only elements of an algebraic extension A/Φ which are separable are the elements of Φ, then we say that A/Φ is *purely inseparable*. Similarly, an algebraic element ρ is *purely inseparable* over Φ if $\Phi(\rho)/\Phi$ is purely inseparable. It is clear from the definitions that, if ρ is at the same time separable and purely inseparable over Φ, then $\rho \in \Phi$. Also, it should be remarked that an element can be inseparable (= not separable) without being purely inseparable (cf. ex. 3 below). If A/Φ is algebraic and Σ/Φ is the maximal separable subfield of A/Φ (that is, the subfield of all the separable elements), then the second half of Theorem 11 states that A/Σ is purely inseparable. This shows that every algebraic extension A/Φ can be built up in two "pure" stages: first, a separable extension Σ/Φ and next a purely inseparable extension A/Σ. The second part of Theorem 11 and the argument we used before for algebraic extensions (p. 44) implies the transitivity: If A/Φ is separable algebraic and B/A is separable algebraic, then B/Φ is separable algebraic. We are going to prove a similar transitivity for purely inseparable extensions. Since everything is trivial for characteristic 0, we shall assume in the rest of this section that the characteristic is $p \neq 0$. We shall need the following important criterion for separable and purely inseparable elements.

Lemma 2. *Let Φ be of characteristic $p \neq 0$. (i) Then an algebraic element ρ of an extension field is separable over Φ if and only if $\Phi(\rho) = \Phi(\rho^p) = \Phi(\rho^{p^2}) = \cdots$. (ii) If ρ is purely inseparable, then its minimum polynomial has the form $x^{p^e} - \alpha$, $\alpha \in \Phi$. On the other hand, if ρ satisfies an equation of the form $x^{p^e} = \alpha \in \Phi$, $e \geq 0$, then ρ is purely inseparable over Φ.*

Proof. Let $g(x)$ be the minimum polynomial of ρ over Φ. (i) Suppose first that ρ is not separable. Then $g(x) = h(x^p)$ and ρ^p is a root of $h(x)$. Hence $[\Phi(\rho^p):\Phi] \leqq \deg h(x) < \deg g(x) = [\Phi(\rho):\Phi]$. Consequently, $\Phi(\rho^p) \subset \Phi(\rho)$. Next suppose ρ is separable so that $g(x)$ has distinct roots. Let $h(x)$ be the minimum polynomial of ρ over $\Phi(\rho^p)$. Then $h(x) \mid g(x)$, so $h(x)$ has distinct roots. Also ρ is a root of the polynomial $x^p - \rho^p \in \Phi(\rho^p)[x]$, so $h(x) \mid x^p - \rho^p = (x - \rho)^p$. Since $h(x)$ has distinct roots, this implies that $h(x) = x - \rho$. Hence $\rho \in \Phi(\rho^p) = \Phi[\rho^p]$ and ρ is a polynomial in ρ^p with coefficients in Φ. Taking p-th powers shows that ρ^p is a polynomial in ρ^{p^2} with coefficients in Φ. Hence $\rho \in \Phi(\rho^{p^2})$. A repetition of the argument shows that $\Phi(\rho) = \Phi(\rho^p) = \Phi(\rho^{p^2}) = \cdots$. This proves (i). (ii) Let ρ be purely inseparable over Φ and write $g(x) = h(x^{p^e})$ where e is maximal for this. Then $h'(x) \neq 0$ since, otherwise, $h(x) = k(x^p)$ and $g(x) = k(x^{p^{e+1}})$ contrary to the choice of e. We have $h(\rho^{p^e}) = 0$, so ρ^{p^e} is a root of a separable polynomial. Since ρ was assumed purely inseparable, this implies that $\rho^{p^e} = \alpha \in \Phi$ and ρ is a root of $x^{p^e} - \alpha$. Since $g(x) = h(x^{p^e})$ is the minimum polynomial of ρ over Φ, it is clear that $g(x) = x^{p^e} - \alpha$. Next assume that $\rho^{p^e} = \alpha \in \Phi$ for some non-negative integer e. Let $\sigma \in \Phi(\rho) = \Phi[\rho]$ so that $\sigma = \alpha_0 + \alpha_1\rho + \cdots + \alpha_m\rho^m$, $\alpha_i \in \Phi$. Then $\sigma^{p^e} = \alpha_0^{p^e} + \alpha_1^{p^e}\rho^{p^e} + \cdots + \alpha_m^{p^e}(\rho^{p^e})^m \in \Phi$. If σ is separable, then $\Phi(\sigma) = \Phi(\sigma^{p^e})$, by (i). Hence $\Phi(\sigma) = \Phi$ and $\sigma \in \Phi$. Thus ρ is purely inseparable.

The second part of this lemma shows that A/Φ is purely inseparable if and only if every element of A satisfies an equation of the form $x^{p^e} = \alpha \in \Phi$. Since $(x^{p^e})^{p^f} = x^{p^{e+f}}$, this implies that if B/A is purely inseparable and A/Φ is purely inseparable, then B/Φ is purely inseparable. Also it is clear from the second part of the lemma that if A is purely inseparable over Φ, then it is purely inseparable over any subfield E of P/Φ.

EXERCISES

1. Let A/Φ be algebraic. Verify that the set of elements $\gamma \varepsilon A$ which are purely inseparable over Φ form a subfield containing Φ.

2. Let I_p be the field $I/(p)$ and let $P = I_p(\xi, \eta)$ where ξ, η are indeterminates (cf. ex. 3, § 6). Let $\Phi = I_p(\xi^p, \eta^p - \eta - \xi)$. Show that $[P:\Phi] = p^2$ and determine the maximal separable subfield of P/Φ.

3. (J. D. Reid). Let P be as in ex. 2 and let $E = P(\zeta)$ where $\zeta^{p^2} = \xi\zeta^p + \xi\eta$. Show that $[E:P] = p^2$ and E/P is inseparable. Show that E/P contains no purely inseparable element over P not contained in P.

4. Let P/Φ be algebraic and let $P(\xi_1, \xi_2, \cdots, \xi_r)$ be the field of fractions of $P[\xi_1, \xi_2, \cdots, \xi_r]$, ξ_i indeterminates. Let A be the set of elements in $P(\xi_i)$ having the form Fg^{-1}, where $F \varepsilon P[\xi_1, \cdots, \xi_r]$, $g \varepsilon \Phi[\xi_1, \cdots, \xi_r]$. Prove that A is a subfield of $P(\xi_1, \cdots, \xi_r)/\Phi(\xi_1, \cdots, \xi_r)$ which is algebraic. Prove that $A = P(\xi_1, \cdots, \xi_r)$. Hence prove that every non-zero polynomial with coefficients in P has a non-zero multiple with coefficients in Φ.

10. Degrees of separability and inseparability. Structure of normal extensions. We assume throughout this section that the characteristic is $p \neq 0$ and we consider finite dimensional extensions. For such an extension P/Φ with maximal separable subfield Σ/Φ we consider the dimensionalities $[\Sigma:\Phi]$ and $[P:\Sigma]$, which we call the *separability degree* and *inseparability degree* respectively of P/Φ. We write $[\Sigma:\Phi] = [P:\Phi]_s$, $[P:\Sigma] = [P:\Phi]_i$. Then we have

$$[P:\Phi] = [P:\Phi]_s[P:\Phi]_i. \tag{11}$$

We shall now show that $[P:\Phi]_i = p^f$, which amounts to saying that the dimensionality of a purely inseparable extension is a power of the characteristic. If $P = \Phi$, this is clear since $[P:\Phi] = 1 = p^0$. Otherwise, let $\rho \varepsilon P$, $\notin \Phi$. Then Lemma 2 of § 9 shows that the minimum polynomial of ρ over Φ has degree p^e, $e > 0$. Then $[\Phi(\rho):\Phi] = p^e$ and $[P:\Phi(\rho)] < [P:\Phi]$. Since P is purely inseparable over $\Phi(\rho)$, we may assume (using induction on the dimensionality) that $[P:\Phi(\rho)] = p^g$. Then $[P:\Phi] = p^e p^g = p^{e+g}$.

We now consider successive finite dimensional extensions: Δ/P is finite dimensional and P/Φ is finite dimensional; hence Δ/Φ is finite dimensional. We have seen that, if P/Φ and Δ/P are separable (purely inseparable), then Δ/Φ is separable (purely inseparable). If P/Φ is separable and Δ/P is purely inseparable, then one sees easily that P/Φ is the maximal separable subfield of Δ/Φ. Then $[\Delta:\Phi]_s = [P:\Phi]$ and $[\Delta:\Phi]_i = [\Delta:P]$. We now consider the interesting combination: P/Φ purely inseparable and Δ/P separable. We shall show that the maximal separable subfield of the result Δ/Φ will have the same dimensionality as Δ/P. A considerably sharper statement is the following

Lemma. *Let* Δ/P *be separable,* P/Φ *purely inseparable. Then* $\Delta/\Phi = \mathrm{P} \otimes_\Phi \Sigma$ *where* Σ/Φ *is the maximal separable subfield of* Δ/Φ*. Moreover,* $[\Delta:\mathrm{P}] = [\Sigma:\Phi]$.

Proof. It follows easily from XII, Introduction that the statement $\Delta = \mathrm{P} \otimes_\Phi \Sigma$ is equivalent to: there exists a Φ-basis for Σ which is at the same time a P-basis for Δ. This implies that $[\Sigma:\Phi] = [\Delta:\mathrm{P}]$. We proceed to determine the required type of basis. First, let $(\delta_1, \delta_2, \cdots, \delta_n)$ be a basis for Δ/P and write $\delta_i\delta_j = \sum_k \rho_{ijk}\delta_k$, $\rho_{ijk} \in \mathrm{P}$. If δ is any element of Δ and $g(x) = h(x^{p^e})$ is its minimum polynomial over Φ such that $h(x)$ is separable, then δ^{p^e} is separable over Φ. Hence also $\delta^{p^{e+f}} = (\delta^{p^e})^{p^f}$ is separable. It follows that we can choose e so that every $\delta_i^{p^e}$ and every $\rho_{ijk}{}^{p^e}$ is separable over Φ. Since P/Φ is purely inseparable, this implies that $\alpha_{ijk} = \rho_{ijk}{}^{p^e} \in \Phi$. We have $\delta_i^{p^e}\delta_j^{p^e} = \Sigma\rho_{ijk}{}^{p^e}\delta_k{}^{p^e}$; hence, if we put $\delta_i^{p^e} = \sigma_i$, then we have $\sigma_i \in \Sigma$ and $\sigma_i\sigma_j = \Sigma\alpha_{ijk}\sigma_k$, $\alpha_{ijk} \in \Phi$. We claim that $(\sigma_1, \sigma_2, \cdots, \sigma_n)$ is a basis for Δ/P and for Σ/Φ. We note first that the multiplication table for the σ_i shows that $\sum_i \Phi\sigma_i$ is a Φ-subalgebra of Σ/Φ and $\sum_i \mathrm{P}\sigma_i$ is a P-subalgebra of Δ/P. Also the number of σ_i is n; so to show that $(\sigma_1, \sigma_2, \cdots, \sigma_n)$ is a P-basis for Δ it is enough to prove that every $\delta \in \Delta$ is a P-linear combination of the σ_i. To prove that $(\sigma_1, \sigma_2, \cdots, \sigma_n)$ is a Φ-basis for Σ it will be enough to show that every $\sigma \in \Sigma$ is a Φ-linear combination of the σ_i since, if these are P-linearly independent, then they are certainly Φ-linearly independent. Now let $\delta \in \Delta$. Then δ is separable over P so $\delta \in \mathrm{P}[\delta^{p^e}]$. We have $\delta = \Sigma\rho_i\delta_i$, $\rho_i \in \mathrm{P}$, since $(\delta_1, \cdots, \delta_n)$ is a P-basis. Then $\delta^{p^e} = \Sigma\rho_i^{p^e}\delta_i^{p^e} = \Sigma\rho_i^{p^e}\sigma_i \in \sum_i \mathrm{P}\sigma_i$. Since $\delta \in \mathrm{P}[\delta^{p^e}]$, this implies that $\delta \in \sum_i \mathrm{P}\sigma_i$. Next let $\sigma \in \Sigma$. Then, as we have just shown, $\sigma = \Sigma\rho_i\sigma_i$, $\rho_i \in \mathrm{P}$. If f is large enough, then $\rho_i^{p^f} \in \Phi$ and $\sigma^{p^f} = \Sigma\rho_i^{p^f}\sigma_i^{p^f} \in \sum_i \Phi\sigma_i$. Since σ is separable over Φ, $\sigma \in \Phi[\sigma^{p^f}] \subseteq \sum_i \Phi\sigma_i$. This completes the proof.

We can now prove

Theorem 12. *If* Δ/P *and* P/Φ *are finite dimensional, then*

(12) $$[\Delta:\Phi]_s = [\Delta:\mathrm{P}]_s[\mathrm{P}:\Phi]_s, \quad [\Delta:\Phi]_i = [\Delta:\mathrm{P}]_i[\mathrm{P}:\Phi]_i.$$

Proof. It is enough to prove the first of these equations since the second will follow from it and (11). Assume first that Δ/P is purely inseparable. Then any element of Δ/Φ which is separable over Φ is separable over P and so belongs to P. Hence the maximal separable subfield Σ/Φ of Δ/Φ is contained in P and so this is the maximal separable subfield of P/Φ. Hence $[\Delta:\Phi]_s = [\mathrm{P}:\Phi]_s$. On the other hand, since Δ/P is purely inseparable, $[\Delta:\mathrm{P}]_s = 1$. Hence (12) holds in this case. Next assume Δ/P is separable. Considering the maximal separable subfield Σ/Φ of P/Φ as base field, we apply the lemma to the separable extension Δ/P and the purely inseparable extension P/Σ. This gives $[\Delta:\mathrm{P}] = [\Sigma':\Sigma]$ where Σ'/Σ is the maximal separable subfield of Δ/Σ. Since separability is transitive, it is clear that Σ'/Φ is the maximal separable subfield of Δ/Φ. Hence, by definition, $[\Delta:\Phi]_s = [\Sigma':\Phi]$ and

$$\begin{aligned}[\Delta:\Phi]_s &= [\Sigma':\Phi] = [\Sigma':\Sigma][\Sigma:\Phi]\\ &= [\Delta:\mathrm{P}][\Sigma:\Phi].\end{aligned}$$

Clearly, $[\Sigma:\Phi] = [\mathrm{P}:\Phi]_s$ and, since Δ/P is separable, $[\Delta:\mathrm{P}] = [\Delta:\mathrm{P}]_s$. Substituting in the above equation gives (12) in this case. Finally, let Δ/P be arbitrary. Let E/P be the maximal separable subfield of Δ/P, so Δ/E is purely inseparable. Then, on considering $\mathrm{E} \supseteq \mathrm{P} \supseteq \Phi$ where E/P is separable, we see that $[\mathrm{E}:\Phi]_s = [\mathrm{E}:\mathrm{P}]_s[\mathrm{P}:\Phi]_s$. Since Δ/E is purely inseparable, the first case applied to $\Delta \supseteq \mathrm{E} \supseteq \Phi$ gives $[\Delta:\Phi]_s = [\Delta:\mathrm{E}]_s[\mathrm{E}:\Phi]_s$. Similarly, considering $\Delta \supseteq \mathrm{E} \supseteq \mathrm{P}$, we obtain $[\Delta:\mathrm{P}]_s = [\Delta:\mathrm{E}]_s[\mathrm{E}:\mathrm{P}]_s$. Combining, we obtain

$$\begin{aligned}[\Delta:\Phi]_s &= [\Delta:\mathrm{E}]_s[\mathrm{E}:\Phi]_s\\ &= [\Delta:\mathrm{E}]_s[\mathrm{E}:\mathrm{P}]_s[\mathrm{P}:\Phi]_s\\ &= [\Delta:\mathrm{P}]_s[\mathrm{P}:\Phi]_s,\end{aligned}$$

which is (12) in the general case.

We have seen that, if P/Φ is Galois, then P/Φ is separable (Cor. to Lemma 1, § 9). Also, since P/Φ is a splitting field, this extension is normal. Conversely, if P/Φ is normal and separable, then P/Φ is a splitting field of a polynomial which is separable. Hence P/Φ is Galois. Thus the condition: P/Φ Galois is equivalent to P/Φ is separable and normal. We claim also that any purely inseparable extension P/Φ is normal. To see this let $g(x) \in \Phi[x]$ be irreducible and suppose $g(\sigma) = 0$ for $\sigma \in P$. Then $g(x)$ is the minimum polynomial of σ over Φ, so $g(x) = x^{p^e} - \alpha$. Since $\sigma^{p^e} = \alpha$, we have the factorization $g(x) = x^{p^e} - \alpha = x^{p^e} - \sigma^{p^e} = (x - \sigma)^{p^e}$, so $g(x)$ is a product of linear factors in $P[x]$. Hence P/Φ is normal. The following theorem gives a rather precise description of the structure of normal extensions.

Theorem 13. *If P/Φ is finite dimensional normal, then $P = \Sigma \otimes_\Phi \Gamma$ where Σ/Φ is Galois and Γ/Φ is purely inseparable. Conversely, if Γ/Φ is a finite dimensional purely inseparable extension and Σ/Φ is finite dimensional separable extension, then the algebra $P/\Phi = \Gamma \otimes_\Phi \Sigma$ is a field and this is normal if Σ/Φ is Galois.*

Proof. Assume P/Φ finite dimensional normal. Let Γ be the set of purely inseparable elements over Φ, so Γ is a subfield over Φ (ex. 1, § 9). Let $\rho \in P$ and let $g(x)$ be the minimum polynomial of ρ over Φ and write $g(x) = h(x^{p^e})$ where $h(x)$ is separable. Since $g(x)$ is irreducible in $\Phi[x]$, it is clear that $h(x)$ is irreducible in $\Phi[x]$. Since $h(\rho^{p^e}) = 0$, the normality of P/Φ implies that $h(x) = \prod_1^r (x - \beta_i)$ in $P[x]$. Also $g(x) = h(x^{p^e}) = \prod_1^r (x^{p^e} - \beta_i)$ is a product of linear factors in $P[x]$. Hence $x^{p^e} - \beta_i$ has a root ρ_i in P and consequently $x^{p^e} - \beta_i = x^{p^e} - \rho_i^{p^e} = (x - \rho_i)^{p^e}$. Then $g(x) = h(x^{p^e}) = \prod_1^r (x^{p^e} - \beta_i) = \prod_1^r (x - \rho_i)^{p^e}$. Now set $k(x) = \prod_1^r (x - \rho_i)$. Since the $\beta_i = \rho_i^{p^e}$ are distinct, the ρ_i are distinct. We have $k(x)^{p^e} = \Pi(x^{p^e} - \rho_i^{p^e}) = \Pi(x^{p^e} - \beta_i) = g(x)$. If $k(x) = x^m + \sigma_1 x^{m-1} + \cdots + \sigma_m$, then the $\sigma_j \in P$ and $k(x)^{p^e} = x^{mp^e} + \sigma_1^{p^e} x^{(m-1)p^e} + \cdots + \sigma_m^{p^e} = g(x)$ which shows that $\sigma_j^{p^e} \in \Phi$; hence the $\sigma_j \in \Gamma$. Hence $k(x) \in \Gamma[x]$. Since ρ is a root of $k(x)$ and $k(x)$ has distinct roots, ρ is separable over Γ. Since ρ was any element of P, this proves that P/Γ is separable. The factorization $P = \Gamma \otimes_\Phi \Sigma$ where Σ is the maximal separable subfield of P/Φ therefore follows from the lemma at the beginning of this section. If $\sigma \in \Sigma$ and $f(x)$ is its minimum polynomial over Φ, then $f(x) = \Pi(x - \sigma_k)$ in P[x]. Evidently the σ_k are separable over Φ so they are contained in Σ. Hence the factorization $f(x) = \Pi(x - \sigma_k)$ holds in Σ[x]. This proves that Σ/Φ is normal as well as separable. Hence it is Galois over Φ. This proves the first statement. Now let Γ/Φ be finite dimensional purely inseparable, Σ/Φ finite dimensional separable. We shall show first that, if $(\sigma_1, \sigma_2, \cdots, \sigma_m)$ is a basis for Σ/Φ, then the same is true for $(\sigma_1^{p^e}, \sigma_2^{p^e}, \cdots, \sigma_m^{p^e})$, $e \geq 1$. Clearly it suffices to show that every element $\sigma \in \Sigma$ is a Φ-linear combination of the $\sigma_i^{p^e}$. Now for any $j \geq 0$, σ^j is a linear combination of the σ_i. Taking p^e-th powers shows that $(\sigma^{p^e})^j$ is a Φ-linear combination of the $\sigma_i^{p^e}$ for $j = 0, 1, 2, \cdots$. Since σ is separable, we know that $\sigma \in \Phi[\sigma^{p^e}]$ (Lemma 2 of § 9). Hence $\sigma \in \sum_i \Phi\sigma_i^{p^e}$ and $(\sigma_1^{p^e}, \sigma_2^{p^e}, \cdots, \sigma_m^{p^e})$ is a basis for Σ/Φ. Now consider $P = \Gamma \otimes_\Phi \Sigma$. This is a commutative algebra and any element of this algebra can be written in the form $\Sigma\gamma_i \otimes \sigma_i$, $\gamma_i \in \Gamma$, and $(\sigma_1, \cdots, \sigma_m)$, the basis for Σ/Φ. Now if $\rho = \Sigma\gamma_i \otimes \sigma_i \neq 0$, then one of the γ_i, say, $\gamma_1 \neq 0$. Since Γ/Φ is purely inseparable, we can choose $e \geq 0$ so that $\gamma_i^{p^e} = \alpha_i \in \Phi$, $1 \leq i \leq m$. Then $\rho^{p^e} = \Sigma\alpha_i \otimes \sigma_i^{p^e} = 1 \otimes \Sigma\alpha_i\sigma_i^{p^e}$. Since $(\sigma_1^{p^e}, \cdots, \sigma_m^{p^e})$ is a basis for Σ/Φ and $\alpha_1 \neq 0$, $\Sigma\alpha_i\sigma_i^{p^e}$ is a nonzero element of Σ. Since this has an inverse in Σ, ρ has an inverse in P. Thus P is a field. Now assume Σ/Φ is Galois. Then Σ/Φ is a splitting field of a polynomial $g(x) \in \Phi[x]$ and Γ/Φ is a splitting field of $h(x) \in \Phi[x]$. Since P is generated by subfields isomorphic to Γ and Σ, it follows that P/Φ is a splitting field of $g(x)h(x)$. Hence P/Φ is normal.

EXERCISES

1. Let E/Φ be finite dimensional, P/Φ an arbitrary field extension. Show that the number of distinct isomorphisms of E/Φ into P/Φ does not exceed $[E:\Phi]_s$. Show that this number is attained if P/Φ is the normal closure of E/Φ.

2. Let P/Φ be finite dimensional normal, Σ/Φ the maximal separable subfield, G the Galois group of P/Φ. Show that G maps Σ into itself and that the mapping $s \to \bar{s}$ the restriction of $s \in G$ to Σ is an isomorphism of G onto the Galois group of Σ/Φ. Show that $I(G) = \Gamma/\Phi$, the maximal purely inseparable subfield of P/Φ.

11. Primitive elements. In this section and the next we shall obtain some special generations of finite dimensional extensions P/Φ. The results are valid for arbitrary Φ. However, the proofs in these two sections will require Φ to be infinite; the validity of the results for finite Φ will be established in § 13.

If $P = \Phi(\theta)$, that is, P is generated over Φ by θ, then we have called P a simple extension of Φ (Vol. I, p. 101). We shall now say also that θ is a *primitive* element of P/Φ. We shall prove two results on existence of primitive elements.

Theorem 14. *Let Φ be an infinite field and let $P = \Phi(\xi, \eta)$ be a field generated over Φ by a separable algebraic element ξ and an algebraic element η. Then P/Φ has a primitive element.*

Proof. Let $f(x)$ and $g(x)$ be the minimum polynomial over Φ of ξ and η respectively and let Δ/P be a splitting field of $f(x)g(x)$. Then Δ/Φ is a splitting field of $f(x)g(x)$ containing P. Let $\xi_1 = \xi, \xi_2, \cdots, \xi_m$ be the distinct roots of $f(x)$, $\eta_1, \eta_2, \cdots, \eta_r$ those of $g(x)$. Then the ξ_i are all the roots of $f(x)$ and we may assume $m > 1$ since, otherwise, $\xi \varepsilon \Phi$ and $P = \Phi(\eta)$. Consider one of the linear equations $x\xi_1 + \eta_1 = x\xi_i + \eta_j$, $i = 2, \cdots, m, j = 1, \cdots, r$. This has at most one solution in Φ. Hence, since Φ is infinite, we can avoid the finite set of solutions of these equations and choose $x = \gamma \varepsilon \Phi$ so that $\gamma\xi_1 + \eta_1 \neq \gamma\xi_i + \eta_j$, $i = 2, \cdots, m; j = 1, \cdots, r$. We assert that $\theta = \gamma\xi_1 + \eta_1$ is a primitive element of P. Thus consider the polynomial $g(\theta - \gamma x)$ which evidently belongs to $\Phi(\theta)[x]$. We have $g(\theta - \gamma\xi_1) = g(\eta_1) = 0$ and, since $\theta - \gamma\xi_i \neq \eta_j$ for $i = 2, \cdots, m$ and $j = 1, 2, \cdots, r$, $g(\theta - \gamma\xi_i) \neq 0$. Hence the highest common factor of $g(\theta - \gamma x)$ and $f(x) = \prod_1^m (x - \xi_i)$ is $x - \xi = x - \xi_1$. Hence there exist $A(x), B(x) \varepsilon \Phi(\theta)[x]$ so that $(x - \xi) = A(x)f(x) + B(x)g(\theta - \gamma x)$, which implies that $\xi \varepsilon \Phi(\theta)$. Then $\eta = \theta - \gamma\xi \varepsilon \Phi(\theta)$ and θ is a primitive element.

The result just proved has an immediate extension, by induction on k to show that, if $P = \Phi(\xi_1, \cdots, \xi_k, \eta)$ where the ξ_i are separable algebraic and η is algebraic, then P has a primitive element. In particular, we see that any finite dimensional separable extension has a primitive element. We note that the number of intermediate fields of such an extension is finite. This is clear since P can be imbedded in an extension Δ/Φ which is finite dimensional Galois, and the set of intermediate fields between Δ and Φ are in 1–1 correspondence with the set of subgroups of a finite group—the Galois group of Δ/Φ. The theorem on primitive elements for finite dimensional separable extensions is therefore also a consequence of the following

Theorem 15 (Artin). *Let Φ be an infinite field and P a finite dimensional extension field of Φ. Then P/Φ is a simple extension if and only if there are only a finite number of intermediate fields between P and Φ.*

Proof. Suppose first that $P = \Phi(\theta)$ and let E be an intermediate field. Let $g(x)$ be the minimum polynomial of θ over E and let E'/Φ be the field generated by the coefficients of $g(x)$. Then $E' \subseteq E$, but $g(x)$ is also the minimum polynomial over E' of θ. Hence $[P:E'] = \deg g(x) = [P:E]$. Hence $E = E'$ is generated by the coefficients of $g(x)$. Now $g(x)$ is a factor of the minimum polynomial $f(x)$ of θ over Φ and both $g(x)$, $f(x) \varepsilon P[x]$. Since $f(x)$ has only a finite number of distinct factors in $P[x]$ with leading coefficients 1, the number of E is finite. Next assume that there are only a finite number of intermediate fields between P and Φ. It suffices to show that, if $\xi, \eta \varepsilon P$, then $\Phi(\xi, \eta)$ is simple. Now let $\alpha \varepsilon \Phi$ and consider the subfield $P_\alpha = \Phi(\xi + \alpha\eta)$. We have an infinite number of $\alpha \varepsilon \Phi$ and a finite number of P_α. Hence there exist α, β in Φ, $\alpha \neq \beta$, such that $P_\alpha = P_\beta$. Then $\eta = (\alpha - \beta)^{-1}(\xi + \alpha\eta - \xi - \beta\eta) \varepsilon P_\alpha$ and hence $\xi = \xi + \alpha\eta - \alpha\eta \varepsilon P_\alpha$. Thus $P_\alpha = \Phi(\xi, \eta)$ and this is generated by $\xi + \alpha\eta$.

EXERCISES

1. Let Φ_0 be of characteristic $p \neq 0$ and let $P = \Phi_0(\xi, \eta)$ the field of fractions of $\Phi[\xi, \eta]$, ξ, η indeterminates. Let $\Phi = \Phi_0(P^p)$ the subfield over Φ_0 generated by all the p-th powers. Show that $[P:\Phi] = p^2$ and that P does not have a primitive element over Φ.
2. Let P be the splitting field over the rationals of $(x^2 - 3)(x^2 - 2)$. Find a primitive element for P.
3. Do the same for $x^5 - 2$.

12. Normal bases. If P is finite dimensional Galois over Φ with Galois group $G = \{s_1, s_2, \cdots, s_n\}$, then $\rho \varepsilon P$ has a minimum polynomial of degree n over Φ if and only if the elements ρ^{s_i}, $i =$

$1, \cdots, n$, are distinct. This is clear since, if $\rho^{s_1}, \cdots, \rho^{s_r}$ are the distinct conjugates, then $f(x) = \prod_{j=1}^{r} (x - \rho^{s_j})$ is the minimum polynomial of ρ over Φ. It is clear also that ρ is a primitive element of P/Φ if and only if the degree of its minimum polynomial is $n = (G:1)$. Hence ρ is a primitive element if and only if the ρ^{s_i}, $i = 1, \cdots, n$, are distinct. A stronger condition than this is evidently that these elements are linearly independent. Then we have the basis $(\rho^{s_1}, \rho^{s_2}, \cdots, \rho^{s_n})$ of P over Φ. Such a basis, consisting of the conjugates of a single element, is called a *normal basis* for the Galois extension. We shall now show that such bases always exist if Φ is infinite. Our proof of this fact will be based on the notion of algebraic independence of isomorphisms which is of considerable interest on its own. We define this as follows:

Definition 2. *Let* E *be a field over* Φ *and* Ω *an extension field of* E. *Let* $s_1, \cdots, s_m$ *be isomorphisms of* E/Φ *into* Ω/Φ. *Then we shall call the* s_i algebraically independent *over* Ω *if the following is true: The only polynomial* $f(x_1, \cdots, x_m) \in \Omega[x_1, \cdots, x_m]$, x_i *indeterminates, such that* $f(\eta^{s_1}, \eta^{s_2}, \cdots, \eta^{s_m}) = 0$ *for all* $\eta \in E$ *is* $f = 0$.

We require the following

Lemma. *Let* Ω *be an extension field of an infinite field* Φ *and let* $f(x_1, \cdots, x_m) \in \Omega[x_1, \cdots, x_m]$ *satisfy* $f(\xi_1, \cdots, \xi_m) = 0$ *for all* $\xi_i \in \Phi$. *Then* $f = 0$.

Proof. Let (ω_α) be a basis for Ω over Φ. Then we can write $f(x_1, \cdots, x_m) = \sum_{1}^{r} f_j(x_1, \cdots, x_m)\omega_j$ where $\{\omega_j\}$ is a finite subset of (ω_α) and the $f_j \in \Phi[x_1, \cdots, x_m]$. Then $0 = f(\xi_1, \cdots, \xi_m) = \Sigma f_j(\xi_1, \cdots, \xi_m)\omega_j$ for all $\xi_i \in \Phi$. Since the ω_j are Φ-independent and the $f_j(\xi_1, \cdots, \xi_m) \in \Phi$, this implies that every $f_j(\xi_1, \cdots, \xi_m) = 0$. Hence, by a result proved in Vol. I, p. 112, $f_j(x_1, \cdots, x_m) = 0$ for all j and so $f(x_1, \cdots, x_m) = 0$.

We can now prove the following theorem on algebraic independence of isomorphisms.

Theorem 16. *Let* P *be finite dimensional Galois over an infinite field* Φ, E *a subfield of* P/Φ, *and* Ω *an arbitrary extension field of*

P. *Let* $s_1, s_2, \cdots, s_m$ *be the different isomorphisms of* E *over* Φ *into* P *over* Φ. *Then the* s_i *are algebraically independent over* Ω.

Proof. We recall that the number m of isomorphisms is $[E:\Phi]$ (§ 7). We note next that, if $(\epsilon_1, \epsilon_2, \cdots, \epsilon_m)$ is a basis for E/Φ, then the determinant $\det(\epsilon_i^{s_j})$ of the matrix whose (i, j) entry is $\epsilon_i^{s_j}$ is not 0. Otherwise, the rows of this matrix are P-dependent, so there exist ρ_j not all 0 in Φ such that $\Sigma\rho_j\epsilon_i^{s_j} = 0, i = 1, 2, \cdots, m$. If ϵ is any element in E, we can write $\epsilon = \Sigma\beta_i\epsilon_i$, $\beta_i \in \Phi$, and we obtain $\sum_{i,j} \beta_i\rho_j\epsilon_i^{s_j} = 0$. Since $\beta_i^{s_j} = \beta_i$, we have $\sum_j \rho_j\epsilon^{s_j} = 0$. This states that the operator $\Sigma s_j\rho_{jR} = 0$ contrary to Dedekind's independence theorem for isomorphisms. We have therefore established that $\det(\epsilon_i^{s_j}) \neq 0$. Now suppose $f \in \Omega[x_1, x_2, \cdots, x_m]$ and $f(\epsilon^{s_1}, \epsilon^{s_2}, \cdots, \epsilon^{s_m}) = 0$ for all $\epsilon \in E$. Then $f(\Sigma\beta_i\epsilon_i^{s_1}, \Sigma\beta_i\epsilon_i^{s_2}, \cdots, \Sigma\beta_i\epsilon_i^{s_m}) = 0$ for all β_i in the infinite field Φ. Now let $g(x_1, \cdots, x_m) = f(\Sigma x_i\epsilon_i^{s_1}, \Sigma x_i\epsilon_i^{s_2}, \cdots, \Sigma x_i\epsilon_i^{s_m}) \in \Omega[x_1, \cdots, x_m]$. This vanishes for all $x_i = \beta_i \in \Phi$, so by the lemma, $g(x_1, \cdots, x_m) = 0$. Now $\det(\epsilon_i^{s_j}) \neq 0$, so the matrix $(\epsilon_i^{s_j})$ has an inverse (μ_{ij}). Then $f(x_1, \cdots, x_m) = g(\Sigma x_i\mu_{i1}, \Sigma x_i\mu_{i2}, \cdots, \Sigma x_i\mu_{in}) = 0$. This proves that $s_1, \cdots, s_m$ are algebraically independent over Ω.

We can use the result just proved to establish

Theorem 17. *Let* P *be finite dimensional Galois over an infinite* Φ. *Then* P/Φ *has a normal basis.*

Proof. Let $G = \{s_1, \cdots, s_n\}$ be the Galois group of P/Φ. We have just seen that, if $(\rho_1, \cdots, \rho_n)$ is a basis of P over Φ, then $\det(\rho_i^{s_j}) \neq 0$. Conversely, this condition is sufficient for a basis; for, if $\Sigma\beta_i\rho_i = 0$ where the $\beta_i \in \Phi$, then $\Sigma\beta_i\rho_i^{s_j} = 0, j = 1, \cdots, m$. This implies that the columns of $(\rho_i^{s_j})$ are Φ-dependent unless every $\beta_i = 0$. Our criterion shows that, for a particular ρ, $(\rho^{s_1}, \rho^{s_2}, \cdots, \rho^{s_n})$ is a normal basis if and only if $\det(\rho^{s_i s_j}) \neq 0$. We now write $s_i s_j = s_{i_j}$ and we know that $(1_j, 2_j, \cdots, n_j)$ is a permutation of $(1, 2, \cdots, n)$. Consider the matrix whose (i, j) entry is the indeterminate x_{i_j} $(i_j = 1, 2, \cdots, n)$ in $P[x_1, \cdots, x_n]$. We assert that the polynomial $d(x_1, \cdots, x_n) = \det(x_{i_j}) \neq 0$. To see this we specialize $x_1 = 1$, $x_2 = \cdots = x_n = 0$. Since each row and column of (x_{i_j}) contains exactly one x_1, it follows that $\det(x_{i_j}) = \pm 1$ if the x_i are specialized as indicated. Therefore, $d(x_1, \cdots, x_n)$

$\neq 0$, and so by the algebraic independence of the x_i we can find a $\rho \in \mathrm{P}$ such that $\det (\rho^{s_i s_j}) = \det (\rho^{s_{i_j}}) \neq 0$. Then ρ determines a normal basis.

There is another, more sophisticated, formulation of the normal basis theorem which we shall now indicate. For this we introduce the group algebra $\Phi(G)$ of the group G: $\Phi(G)$ has the basis $G = \{s_1, \cdots, s_n\}$ and multiplication is defined by $(\Sigma\alpha_i s_i)(\Sigma\beta_j s_j) = \Sigma\alpha_i\beta_j s_i s_j = \Sigma\alpha_i\beta_j s_{i_j}$ (cf. Vol. I, ex. 2, p. 95). We consider two right modules for $\Phi(G)$: The first of these is $\Phi(G)$ itself considered in the usual way: xa, $x \in \Phi(G)$, $a \in \Phi(G)$ is the algebra product. Next we consider P as $\Phi(G)$-module by defining $\rho a = \Sigma\alpha_i\rho^{s_i}$ for $a = \Sigma\alpha_i s_i$ in $\Phi(G)$. It is immediate that the module axioms hold for this multiplication. The normal basis theorem is just the statement that these two modules are isomorphic. Thus, let (ρ^{s_i}) be a normal basis and consider the linear mapping of $\Phi(G)$ over Φ into P over Φ sending s_i into ρ^{s_i}. This is a Φ-linear isomorphism and, if $x = \Sigma\xi_i s_i$, then $xs_j = \Sigma\xi_i s_{i_j} \to \sum_i \xi_i\rho^{s_{i_j}} = \sum_i \xi_i\rho^{s_i s_j} = (\Sigma\xi_i\rho^{s_i})s_j$. Hence, if we denote the image of x by x', then $x's_j = (xs_j)'$. This implies that $x'a = (xa)'$ for all $a \in \Phi(G)$ so we have a $\Phi(G)$-isomorphism. It is easy to check that, conversely, if $x \to x'$ is a $\Phi(G)$-isomorphism of $\Phi(G)$ onto P, then the image of $(s_1 = 1, s_2, \cdots, s_n)$ is a normal basis for P/Φ.

EXERCISE

1. Prove the following generalization of Theorem 16: Let $f(x_1^{(1)}, \cdots, x_m^{(1)}, x_1^{(2)}, \cdots, x_m^{(2)}, \cdots, x_1^{(r)}, \cdots, x_m^{(r)})$ be a non-zero polynomial in indeterminates $x_i^{(j)}$. Then there exist $\eta_1, \cdots, \eta_r \in \mathrm{E}$ such that $f(\eta_1^{s_1}, \cdots, \eta_1^{s_m}, \cdots, \eta_r^{s_1}, \cdots, \eta_r^{s_m}) \neq 0$.

13. Finite fields. The main results on finite fields are readily obtained as applications of Galois theory. We proceed to derive these. At the same time we shall establish the validity of the theorem on primitive elements and normal basis for finite base fields.

We remark first that any finite field P is of characteristic $p \neq 0$, since otherwise P contains a subfield isomorphic to the field of rational numbers. Hence the prime field Φ_0 of P is isomorphic to

$I_p = I/(p)$. If Φ is any subfield, of course, $[P:\Phi] = n < \infty$. If $(\rho_1, \cdots, \rho_n)$ is a basis for P over Φ, every element $\rho \in P$ can be written in one and only one way as $\sum_1^n \alpha_i\rho_i$, $\alpha_i \in \Phi$. If the cardinal number $|\Phi| = q$, then it is clear from this that $|P| = q^n$. In particular, if $[P:\Phi_0] = N$, Φ_0 the prime field, then $|P| = p^N$. This shows that the number of elements in any finite field is a power of its characteristic.

We show next that for any prime power p^N there exists one and, in the sense of isomorphism, only one field with p^N elements. We consider the uniqueness first. Let P be a field with $|P| = p^N$. Then it is clear that the prime field Φ_0 of P is isomorphic to I_p. If ρ is a non-zero element of P, then $\rho^{p^N-1} = 1$ since the order of the multiplicative group P* of non-zero elements of P is $p^N - 1$. We have also that $\rho^{p^N} = \rho$, an equation which is valid for every $\rho \in P$. Thus every element of P is a root of $x^{p^N} - x = 0$ and $x^{p^N} - x \in \Phi_0[x]$, Φ_0 the prime field. Then

$$(13) \qquad x^{p^N} - x = \prod_1^{p^N} (x - \rho_i)$$

where the ρ_i are the elements of P. This shows that P/Φ_0 is a splitting field of the polynomial $x^{p^N} - x$. Now suppose P′ is a second field such that $|P'| = p^N$; then P′ has characteristic p so its prime field $\Phi_0' \cong \Phi_0$. Also P'/Φ_0 is a splitting field of $x^{p^N} - x$. Hence $P' \cong P$ by the uniqueness theorem on splitting fields.

The method just used also gives the existence of a field P with $|P| = p^N$, p a prime. For this we begin with $\Phi_0 = I_p$ which has p elements and we let P be a splitting field over Φ_0 of $x^{p^N} - x$. Let Σ be the set of roots of $x^{p^N} - x = 0$ contained in P. Since the derivative $(x^{p^N} - x)' = -1$, $x^{p^N} - x = 0$ has distinct roots so $|\Sigma| = p^N$. We note next that Σ is a subfield, since, if $\xi, \eta \in \Sigma$, then $\xi^{p^N} = \xi$, $\eta^{p^N} = \eta$ so $(\xi - \eta)^{p^N} = \xi^{p^N} - \eta^{p^N} = \xi - \eta$, $(\xi\eta)^{p^N} = \xi^{p^N}\eta^{p^N} = \xi\eta$ and $(\eta^{-1})^{p^N} = (\eta^{p^N})^{-1} = \eta^{-1}$ if $\eta \neq 0$. It now follows that $\Sigma \supseteq \Phi_0$ and, since P is a splitting field of $x^{p^N} - x$, $P = \Phi_0(\Sigma) = \Sigma$. Thus we have $|P| = |\Sigma| = p^N$.

We prove next the theorem on primitive elements: If Φ is a finite field and P is a finite dimensional extension of Φ, then $P = \Phi(\theta)$. Clearly, under the given conditions P is finite. We now show that the multiplicative group P* is cyclic. This is a consequence of the following useful general result.

Lemma 1. *Any finite subgroup A of the multiplicative group of a field is cyclic.*

Proof. Let m be the order of A and let m' be the highest order for the elements of A. It is known that, if a and b are two elements of a finite commutative group, then there exists a c in the group whose order is the least common multiple of the orders of a and b (Vol. II, ex. 1, p. 69). It follows that, if m' is the highest order, then $b^{m'} = 1$ for every b. On the other hand, we know that the equation $x^{m'} - 1 = 0$ has at most m' roots in a field. Since $m'|m$, we have $m' = m$. Moreover, if a is an element of order m, then the order of the cyclic group $[a]$ generated by a is m. Hence $A = [a]$.

Now if P is a finite field and Φ is a subfield, then surely $P = \Phi(\theta)$ if θ is chosen to be a generator of the cyclic group P*.

We consider next the automorphisms of a finite field P. If the characteristic is p, then we know that the mapping $\pi:\xi \rightarrow \xi^p$ is an isomorphism of P into P. Since P is finite, this is an automorphism. If $|P| = p^N$, then $\rho^{p^N} = \rho$ for every $\rho \in P$. Evidently π^i is the automorphism $\xi \rightarrow \xi^{p^i}$, so we have $\pi^N = 1$. On the other hand, if θ is a generator of the group P*, then $\theta^{p^m} \neq \theta$ if $m < N$. This implies that $\pi^m \neq 1$. Hence the cyclic group $G = [\pi]$ has the order N. Let Φ be the set of G-invariants of P. Then we know that Φ is a subfield and $[P:\Phi] = N$. On the other hand, we know that $[P:\Phi_0] = p^N$ if Φ_0 is the prime field. Hence $\Phi = \Phi_0$. We now see that the field P is Galois over its prime field Φ_0 and the Galois group is $G = [\pi]$. The Galois correspondence now gives a correspondence between the collection of subfields of P and the collection of subgroups of G. Since G is cyclic of order N, for each divisor n of N there exists one and only one subgroup H of index n. We have $H = [\tau]$ where $\tau = \pi^n$. The corresponding field Φ of H-invariants (or of τ-invariants) has dimensionality n over Φ_0. Hence $|\Phi| = p^n$ and we have shown that the subfields Φ of P have order p^n where $n|N$ and for each such order there is precisely one subfield of P of this order. The Galois group of P over Φ is the cyclic group $H = [\tau]$ as before. In general, we shall call an extension field P/Φ *cyclic*, *abelian*, or *solvable* if P/Φ is finite dimensional Galois and its Galois group is respectively cyclic, commutative, or solvable. Hence we can say that any finite field is a cyclic extension of any of its subfields. We shall therefore have the normal basis theorem also for finite base fields by proving

Lemma 2. *Any cyclic extension P/Φ has a normal basis over Φ.*

Proof. Let s be a generator of the Galois G group of P/Φ. We consider s as a linear transformation in P over Φ and let $\mu(x)$ ε Φ[x] be its minimum polynomial. Now Dedekind's independence theorem implies that the automorphisms $1, s, \cdots, s^{n-1}$ are P-independent if $(G:1) = n$. It follows that these are also Φ-independent and consequently $\deg \mu(x) \geq n$. On the other hand, [P:Φ] $= n$ so the degree of $\mu(x)$ cannot exceed n (Vol. II, p. 69). Hence $\deg \mu(x) = n$. Since $s^n = 1$, we see that $\mu(x) = x^n - 1$. Now we know that there exists a ρ ε P whose order polynomial relative to the linear transformation s is the minimum polynomial (Vol. II, p. 67). Then $\rho, \rho^s, \cdots, \rho^{s^{n-1}}$ are Φ-independent and so these elements form a normal basis for P/Φ.

EXERCISES

NOTE: A set of exercises on finite fields is given in Vol. I, pp. 112–113.

1. Let Φ be a finite field of order of $q(= p^N)$. Show that an irreducible polynomial $f(x)$ ε Φ[x] is a factor of $x^{q^n} - x$ if and only if $\deg f(x) \mid n$. (Hint: consider the field $\Phi[x]/(f(x))$.) Show that $x^{q^n} - x = \Pi f_i(x)$ where $f_i(x)$ runs over the irreducible polynomials with leading coefficients 1 of degrees divisors of n. Let $N(q, r)$ denote the number of these polynomials of degree r. Derive the formula

$$N(q, n) = \frac{1}{n} \sum_{r \mid n} \mu \left(\frac{n}{r}\right) q^r$$

where μ is the Möbius function (cf. Vol. I, ex. 5, p. 120).

2. Let $\mathfrak{M}$ be an n dimensional vector space over a finite field Φ of odd order q (characteristic $\neq 2$) and let $g(x, y)$ be a non-degenerate symmetric bilinear form on $\mathfrak{M}$ over Φ. Show that if $n \geq 2$, then there exists a vector u in $\mathfrak{M}$ such that $g(u, u) = 1$. Apply this and the reduction theory of Vol. II, pp. 152–154, to prove that $\mathfrak{M}$ has an orthogonal basis $(u_1, u_2, \cdots, u_n)$ such that $g(u_1, u_1) = \delta \neq 0, g(u_i, u_i) = 1$ if $i > 1$. Use this to prove that any two non-singular symmetric $n \times n$ matrices with entries in Φ are cogredient if and only if their determinants differ by a multiplicative factor which is a square ($\delta = \delta'\rho^2$) in Φ. Hence show that there are just two cogredience classes of non-singular symmetric matrices.

3. Let Φ, $\mathfrak{M}$, g be as in ex. 2. If $(e_1, e_2, \cdots, e_n)$ is a basis, then $\delta = \det(g(e_i, e_j))$ is called a *discriminant* of g. For $b \in \Phi$ let $N(g, b)$ be the number of vectors $u \in \mathfrak{M}$ satisfying $g(u, u) = b$. Show that

$$N(g,0) = \begin{cases} q^{2\nu-1} - q^{\nu} + q^{\nu-1}, \text{ if } n = 2\nu \text{ and} \\ \quad (-1)^{\nu}\delta \text{ is not a square} \\ q^{2\nu-1} + q^{\nu} - q^{\nu-1}, \text{ if } n = 2\nu \text{ and} \\ \quad (-1)^{\nu}\delta \text{ is a square} \\ q^{2\nu}, \text{ if } n = 2\nu + 1 \end{cases}$$

$$N(g,b) = \begin{cases} q^{2\nu-1} + q^{\nu-1}, \text{ if } b \neq 0,\ n = 2\nu, \text{ and} \\ \quad (-1)^{\nu}\delta \text{ is not a square} \\ q^{2\nu-1} - q^{\nu-1}, \text{ if } b \neq 0,\ n = 2\nu, \text{ and} \\ \quad (-1)^{\nu}\delta \text{ is a square} \\ q^{2\nu} - q^{\nu}, \text{ if } b \neq 0,\ n = 2\nu + 1, \text{ and} \\ \quad (-1)^{\nu}\delta b \text{ is not a square} \\ q^{2\nu} + q^{\nu}, \text{ if } b \neq 0,\ n = 2\nu + 1, \text{ and} \\ \quad (-1)^{\nu}\delta b \text{ is a square} \end{cases}$$

4. Let $O(n, g)$ denote the orthogonal group determined by g: $O(n, g)$ is the group of linear transformations A of M such that $g(xA, yA) = g(x, y)$ for all $x, y \in M$. If u is a non-isotropic vector, let O_u be the subgroup of $O(n, g)$ leaving u fixed. Show that O_u is isomorphic to $O(n-1, g')$ where g' is the restriction of g to $(\Phi u)^{\perp}$. Use Witt's theorem to show that the number of cosets O_uA of O_u in $O(n, g)$ is the number of vectors v satisfying $g(v, v) = g(u, u)$. Use this result and ex. 3 to establish the following formulas for the order $(O(n, g):1)$:

$$(O(n,g):1) = \begin{cases} 2 \cdot q^{(n-1)^2/4} \prod_{i=1}^{(n-1)/2} (q^{2i} - 1), \text{ if } n \text{ is odd} \\ 2 \cdot q^{n(n-2)/4}(q^{n/2} - \epsilon) \prod_{i=1}^{(n-2)/2} (q^{2i} - 1), \text{ if } n \text{ is even} \end{cases}$$

Here $\epsilon = 1$ if $(-1)^{\nu}\,\delta$ is a square; otherwise $\epsilon = -1$.

14. Regular representation, trace and norm. In this section we consider a finite dimensional extension field P/Φ and we shall define certain mappings of P into Φ called the trace and the norm. These functions can be defined just as easily for arbitrary finite dimensional algebras and are of importance for these also. We shall therefore begin by considering a finite dimensional algebra $\mathfrak{A}/\Phi$ with basis $(u_1, u_2, \cdots, u_n)$ over Φ. We define a (*finite dimensional*) *representation* of $\mathfrak{A}/\Phi$ to be a homomorphism of $\mathfrak{A}/\Phi$ into the algebra $\mathfrak{L}_{\Phi}(\mathfrak{M})$ of linear transformations of a finite dimensional vector space $\mathfrak{M}/\Phi$. If S is such a representation: $a \rightarrow$

a^S, then the defining conditions are

$$(a+b)^S = a^S + b^S, \quad (\alpha a)^S = \alpha a^S, \tag{14}$$
$$(ab)^S = a^S b^S, \quad 1^S = 1,$$

$a, b \in \mathfrak{A}$, $\alpha \in \Phi$. If $(x_1, x_2, \cdots, x_N)$ is a basis for $\mathfrak{M}$ over Φ, then we can determine the matrix a^S relative to this basis in the usual manner: We write

$$x_i a^S = \sum_{j=1}^{N} \alpha_{ij}(a) x_j, \quad i = 1, 2, \cdots, N. \tag{15}$$

This gives the matrix $\alpha(a) = (\alpha_{ij}(a))$ and the mapping $a \to \alpha(a)$ of $\mathfrak{A}/\Phi$ into the algebra Φ_N/Φ of $N \times N$ matrices with entries in Φ. Since the mapping $A \to (\alpha)$ of the linear transformation A into its matrix (α) relative to $(x_1, x_2, \cdots, x_N)$ is an isomorphism, the mapping $a \to \alpha(a)$ is a homomorphism of $\mathfrak{A}/\Phi$ into Φ_N/Φ. Such a homomorphism is called a *matrix representation*. We recall that, if we change the basis $(x_1, \cdots, x_N)$ to another basis $(y_1, \cdots, y_N)$ where $y_i = \Sigma \mu_{ij} x_j$, then the matrix representation defined by S and this basis is $a \to (\mu)\alpha(a)(\mu)^{-1}$ where $(\mu) = (\mu_{ij})$ (cf. Vol. II, p. 42).

The most important representation of $\mathfrak{A}/\Phi$ is the so-called *regular representation* R. Here $a^R = a_R$ the right multiplication $x \to xa$ defined by a. One checks directly that a_R is a linear transformation in $\mathfrak{A}$ over Φ and that $a \to a_R$ is an algebra homomorphism (cf. Vol. I, p. 82). Since $\mathfrak{A}$ has an identity, $a \to a_R$ is 1–1 and so is an isomorphism of $\mathfrak{A}/\Phi$ into $\mathfrak{L}_\Phi(\mathfrak{A})$. Since $xa_R = xa$, we obtain the matrix representation associated with the basis $(u_1, u_2, \cdots, u_n)$ of $\mathfrak{A}/\Phi$ by writing the products $u_i a$ as Φ-linear combinations of the u_i:

$$u_i a = \Sigma \rho_{ij}(a) u_j, \quad j = 1, 2, \cdots, n. \tag{16}$$

We write $\rho(a) = (\rho_{ij}(a))$ and we have the matrix representation $a \to \rho(a)$ which is 1–1. Also since $1_R = 1$, $\rho(1) = 1$, the identity matrix. As in the general case, a change to the basis $(v_1, v_2, \cdots, v_n)$, where $v_i = \Sigma \mu_{ij} u_j$, gives the new matrix representation $a \to \sigma(a)$, where

$$\sigma(a) = (\mu)\rho(a)(\mu)^{-1}, \quad (\mu) = (\mu_{ij}). \tag{17}$$

As an example of this we consider an algebra $\mathfrak{A} = \Phi[a]$ with a single generator. Since $[\mathfrak{A}:\Phi] < \infty$, $\Phi[a] \cong \Phi[x]/(f(x))$ where $f(x)$ is a non-zero polynomial with leading coefficient 1. We have $f(a) = 0$ and $f(x)$ is the non-zero polynomial of least degree (leading coefficient 1) having a as a root. Thus the polynomial $f(x)$ is the minimum polynomial of a (Introduction, p. 6). Also $\Phi[a]$ has the basis $(1, a, \cdots, a^{n-1})$ where $n = [\mathfrak{A}:\Phi] = \deg f(x)$. Suppose

$$f(x) = x^n - \alpha_1 x^{n-1} + \cdots + (-1)^n \alpha_n, \quad \alpha_i \in \Phi. \tag{18}$$

Then we have the relations

$$\begin{aligned} 1a = a,\ aa &= a^2, \cdots, a^{n-2}a = a^{n-1} \\ a^{n-1}a &= \alpha_1 a^{n-1} - \alpha_2 a^{n-2} + \cdots + (-1)^{n-1}\alpha_n. \end{aligned} \tag{19}$$

These show that, if $\rho(a)$ denotes the matrix of a_R relative to $(1, a, \cdots, a^{n-1})$, then we have

$$\rho(a) = \begin{bmatrix} 0 & 1 & 0 & \cdot & \cdot & \cdot \\ 0 & 0 & 1 & \cdot & \cdot & \cdot \\ \vdots & \vdots & \vdots & \vdots & \vdots & \vdots \\ \cdot & \cdot & \cdot & \cdot & \cdot & 1 \\ (-1)^{n-1}\alpha_n & & \cdot & \cdot & -\alpha_2 & \alpha_1 \end{bmatrix}, \tag{20}$$

which is called the companion matrix of the polynomial $f(x)$. In theory, once we know this matrix, we know $\rho(b)$ for any element b in $\Phi[a]$ since b is a polynomial in a.

We now consider the general case again and we define the *characteristic polynomial* of the element $a \in \mathfrak{A}$ to be the characteristic polynomial

$$f_a(x) = \det(x1 - \rho(a)) \tag{21}$$

of the linear transformation a_R in $\mathfrak{A}$ or of the corresponding matrix $\rho(a)$. By (17) we have

$$x1 - \sigma(a) = (\mu)(x1 - \rho(a))(\mu)^{-1}$$

which shows that

$$\begin{aligned} \det(x1 - \sigma(a)) &= \det(\mu)(x1 - \rho(a))(\mu)^{-1} \\ &= \det(\mu)\det(x1 - \rho(a))\det(\mu)^{-1} \\ &= \det(x1 - \rho(a)). \end{aligned}$$

Thus we see that $f_a(x)$ is independent of the choice of the basis for $\mathfrak{A}/\Phi$. We can write the characteristic polynomial as

$$f_a(x) = x^n - T(a)x^{n-1} + \cdots + (-1)^n N(a). \tag{22}$$

We have $T(a) = \text{trace}\, \rho(a) = \sum_1^n \rho_{ii}(a)$, $N(a) = \det(\rho(a))$ and we call these respectively the *trace* and *norm* of a in $\mathfrak{A}$ over Φ. We shall find it necessary at times to specify the base field of the algebra and also the algebra itself. In these cases we shall write $T_{\mathfrak{A}|\Phi}(a)$ for $T(a)$, $N_{\mathfrak{A}|\Phi}(a)$ for $N(a)$. Since the trace is a linear function of matrices and since $a \to \rho(a)$ is linear, it is clear that $a \to T_{\mathfrak{A}|\Phi}(a)$ is a linear mapping of $\mathfrak{A}$ into Φ. Also since $\rho(1) = 1$, we have $T_{\mathfrak{A}|\Phi}(1) = n1$ and $T_{\mathfrak{A}|\Phi}(\alpha a) = \alpha T_{\mathfrak{A}|\Phi}(a)$. Thus we have the following relations:

$$\begin{aligned} T_{\mathfrak{A}|\Phi}(a+b) &= T_{\mathfrak{A}|\Phi}(a) + T_{\mathfrak{A}|\Phi}(b) \\ T_{\mathfrak{A}|\Phi}(\alpha a) &= \alpha T_{\mathfrak{A}|\Phi}(a), \quad \alpha \in \Phi \\ T(1) &= n1. \end{aligned} \tag{23}$$

Since $a \to \rho(a)$ is multiplicative and $A \to \det A$ is a multiplicative mapping of the set of matrices, we have $N_{\mathfrak{A}|\Phi}(ab) = N_{\mathfrak{A}|\Phi}(a)N_{\mathfrak{A}|\Phi}(b)$. Also, it is clear that $N_{\mathfrak{A}|\Phi}(\alpha a) = \alpha^n N_{\mathfrak{A}|\Phi}(a)$ and $N_{\mathfrak{A}|\Phi}(1) = 1$. Thus we have:

$$\begin{aligned} N_{\mathfrak{A}|\Phi}(ab) &= N_{\mathfrak{A}|\Phi}(a)N_{\mathfrak{A}|\Phi}(b), \\ N_{\mathfrak{A}|\Phi}(\alpha a) &= \alpha^n N_{\mathfrak{A}|\Phi}(a), \quad \alpha \in \Phi, \\ N(1) &= 1. \end{aligned} \tag{24}$$

We recall that according to the Hamilton-Cayley theorem $\rho(a)$ is a root of $f_a(x) = 0$. If we apply the isomorphism $\rho(b) \to b$, we see that $f_a(a) = 0$. Thus we have

$$a^n - T(a)a^{n-1} + \cdots + (-1)^n N(a)1 = 0. \tag{25}$$

Let $m_a(x)$ be the minimum polynomial of $\rho(a)$ (or of a_R). Since $b \to \rho(b)$ is an isomorphism, it is clear that $m_a(x)$ is the minimum polynomial of a. We recall that the minimum polynomial of a matrix is a factor of its characteristic polynomial and these two have the same irreducible factors in $\Phi[x]$, differing only in the multiplicities of these factors (Vol. II, p. 99, or p. 102).

The trace function can be used to define an important bilinear form on the algebra $\mathfrak{A}/\Phi$. This is the *regular trace form*

$$(a, b) \equiv T_{\mathfrak{A}|\Phi}(ab). \tag{26}$$

Evidently, we have the following rules governing this function whose values are in Φ:

$$\begin{aligned} (a, b_1 + b_2) &= (a, b_1) + (a, b_2) \\ (a_1 + a_2, b) &= (a_1, b) + (a_2, b) \\ \alpha(a, b) &= (\alpha a, b) = (a, \alpha b) \\ (ab, c) &= (a, bc) \quad (= T_{\mathfrak{A}|\Phi}(abc)). \end{aligned} \tag{27}$$

Also we recall that, if M and N are matrices, then the tr $MN =$ tr NM (Vol. II, p. 104). This implies that

$$(a, b) = (b, a), \tag{28}$$

so (a, b) is a symmetric bilinear form. We shall define also the *discriminant* of $\mathfrak{A}$ over Φ relative to the basis $(u_1, u_2, \cdots, u_n)$ to be

$$\delta(u_i) = \det((u_i, u_j)) = \det(T_{\mathfrak{A}|\Phi}(u_iu_j)). \tag{29}$$

It is immediate that, if we replace $(u_1, \cdots, u_n)$ by the basis $(v_1, \cdots, v_n)$, $v_i = \Sigma\mu_{ij}u_j$, then the matrix $((u_i, u_j))$ is replaced by $((v_i, v_j)) = M((u_i, u_j))M'$, $M = (\mu_{ij})$ (Vol. II, p. 149). Hence the discriminant relative to (v_i) is $\delta(u_i)\mu^2$, $\mu = \det M$.

We now suppose that E is a subfield of Φ of finite co-dimension in Φ. Then $\mathfrak{A} \supseteq \Phi \supseteq \mathrm{E}$ and $[\mathfrak{A}:\mathrm{E}] = [\mathfrak{A}:\Phi][\Phi:\mathrm{E}]$ is finite so, if we consider $\mathfrak{A}$ as a vector space over E, this is finite dimensional. Hence $\mathfrak{A}$ is a finite dimensional algebra over E. We can therefore carry out all of the above considerations for the algebra $\mathfrak{A}/\mathrm{E}$. We can also consider Φ as an algebra over E and we can define $T_{\mathfrak{A}|\mathrm{E}}$, $T_{\Phi|\mathrm{E}}$, $N_{\mathfrak{A}|\mathrm{E}}$, $N_{\Phi|\mathrm{E}}$ as well as $T_{\mathfrak{A}|\Phi}$, $N_{\mathfrak{A}|\Phi}$. We shall now proceed to develop the following fundamental transitivity relations connecting these functions: If $\Phi \supseteq \mathrm{E}$,

$$T_{\mathfrak{A}|\mathrm{E}}(a) = T_{\Phi|\mathrm{E}}(T_{\mathfrak{A}|\Phi}(a)) \tag{30}$$

$$N_{\mathfrak{A}|\mathrm{E}}(a) = N_{\Phi|\mathrm{E}}(N_{\mathfrak{A}|\Phi}(a)). \tag{31}$$

As before, we let $(u_1, \cdots, u_n)$ be a basis for $\mathfrak{A}/\Phi$ and we suppose $(\gamma_1, \gamma_2, \cdots, \gamma_h)$ is a basis for Φ/E. Then

$$(32) \qquad (\gamma_1 u_1, \gamma_2 u_1, \cdots, \gamma_h u_1; \gamma_1 u_2, \cdots, \gamma_h u_2; \cdots \gamma_h u_n)$$

is a basis for $\mathfrak{A}/\mathrm{E}$. If $\rho \in \Phi$, we write

$$(33) \qquad \gamma_q \rho = \Sigma \lambda_{qt}(\rho)\gamma_t, \quad q, t = 1, \cdots, h$$

so we have the isomorphism $\rho \rightarrow (\lambda(\rho))$ where $\lambda(\rho)$ is the matrix $(\lambda_{qt}(\rho))$ with entries in E. Then $T_{\Phi|\mathrm{E}}(\rho) = \Sigma\lambda_{qq}(\rho)$ and $N_{\Phi|\mathrm{E}}(\rho) = \det(\lambda(\rho))$. We now combine the relations (33) and (16) to write

$$(34) \qquad (\gamma_q u_i)a = \Sigma\gamma_q \rho_{ij}(a)u_j = \Sigma\lambda_{qt}(\rho_{ij}(a))\gamma_t u_j,$$

$i, j = 1, \cdots, n$, $q, t = 1, \cdots, h$. This shows that if the basis $(\gamma_q u_i)$ is ordered as in (32), then the matrix of a_R in $\mathfrak{A}/\mathrm{E}$ is

$$(35) \qquad \Lambda(a) = \begin{bmatrix} \lambda(\rho_{11}) & \lambda(\rho_{12}) & \cdots & \lambda(\rho_{1n}) \\ \lambda(\rho_{21}) & \lambda(\rho_{22}) & \cdots & \lambda(\rho_{2n}) \\ \cdot & \cdot & \cdots & \cdot \\ \lambda(\rho_{n1}) & \lambda(\rho_{n2}) & \cdots & \lambda(\rho_{nn}) \end{bmatrix}$$

where $\lambda(\rho_{ij})$ is an $h \times h$ matrix with entries in E and we have abbreviated $\rho_{ij} = \rho_{ij}(a)$. It is clear from the form of (35) that

$$\begin{aligned} T_{\mathfrak{A}|\mathrm{E}}(a) &= \mathrm{tr}\,\Lambda(a) \quad (\mathrm{tr} \equiv \text{trace}) \\ &= \mathrm{tr}\,\lambda(\rho_{11}) + \mathrm{tr}\,\lambda(\rho_{22}) + \cdots + \mathrm{tr}\,\lambda(\rho_{nn}) \\ &= \mathrm{tr}\,\lambda(\rho_{11} + \rho_{22} + \cdots + \rho_{nn}) \\ &= \mathrm{tr}\,\lambda(T_{\mathfrak{A}|\Phi}(a)) \\ &= T_{\Phi|\mathrm{E}}(T_{\mathfrak{A}|\Phi}(a)). \end{aligned}$$

This establishes (30).

For the proof of (31) we require a general transitivity property of determinants (Vol. II, ex. 2, p. 135) which we proceed to derive. We suppose we have an $nh \times nh$ matrix with entries in a field E and we assume that, if we partition this as an $n \times n$ matrix $\Lambda = (\lambda_{ij})$ where each λ_{ij} is an $h \times h$ matrix, then the λ_{ij} all commute. This is equivalent to assuming that the λ_{ij} all belong to a commutative subalgebra $\mathfrak{B}$ of the matrix algebra Φ_h. This is precisely the situation for the matrix $\Lambda(a)$ and the blocks $\lambda(\rho_{ij})$ of (35). Since the $\lambda_{ij} \in \mathfrak{B}$ and $\mathfrak{B}$ is commutative, the usual definition and properties of determinants hold and we can consider

$$\det_n(\Lambda) = \sum_P \epsilon_P \lambda_{1i_1}\lambda_{2i_2}\cdots\lambda_{ni_n}, \tag{36}$$

where the summation is over the permutations $(i_1 i_2 \cdots i_n)$ of $(12 \cdots n)$ and $\epsilon_P = 1, -1$ according as P is even or odd. Now $\det_n(\Lambda)$ as defined above is an element of Φ_h. Hence we can take the usual determinant of this. We shall now establish the following formula:

$$\det(\det_n(\Lambda)) = \det \Lambda \tag{37}$$

where $\det \Lambda$ is the usual determinant of the $nh \times nh$ matrix.

To prove this result we extend the base field E to a splitting field over E of the product of the characteristic polynomials of all the matrices λ_{ij}. It suffices to prove the result in this field. Hence without loss of generality we may assume that E contains the characteristic roots of all the λ_{ij}. The theory of sets of commuting linear transformations (Vol. II, pp. 133–134) shows that there exists a matrix $\mu \in E_h$ such that every $\mu^{-1}\lambda_{ij}\mu$ is triangular:

$$\mu^{-1}\lambda_{ij}\mu = \eta_{ij} = \begin{bmatrix} \rho_{ij1} & & & * \\ & \rho_{ij2} & & \\ & & \ddots & \\ 0 & & & \rho_{ijh} \end{bmatrix}. \tag{38}$$

Hence if we set

$$M = \begin{bmatrix} \mu & & & \\ & \mu & & \\ & & \ddots & \\ & & & \mu \end{bmatrix}, \tag{39}$$

then

$$M^{-1}\Lambda M = \begin{bmatrix} \eta_{11} & \eta_{12} & \cdots & \eta_{1n} \\ \eta_{21} & \eta_{22} & \cdots & \eta_{2n} \\ \cdot & \cdot & \cdots & \cdot \\ \eta_{n1} & \eta_{n2} & \cdots & \eta_{nn} \end{bmatrix}. \tag{40}$$

We have $\det \Lambda = \det M^{-1}\Lambda M$ and, since $\eta_{ij} = \mu^{-1}\lambda_{ij}\mu$,

$$\det_n M^{-1}\Lambda M = \mu^{-1}(\det_n \Lambda)\mu.$$

Hence $\det(\det_n M^{-1}\Lambda M) = \det(\det_n \Lambda)$; so it suffices to verify that

$$\det(\det_n M^{-1}\Lambda M) = \det M^{-1}\Lambda M. \tag{41}$$

Now it follows directly from the definition of $\det_n$ and from the way triangular matrices are multiplied and added that

$$\det_n M^{-1}\Lambda M = \begin{bmatrix} \det \rho_1 & & & * \\ & \det \rho_2 & & \\ & & \ddots & \\ 0 & & & \det \rho_h \end{bmatrix} \tag{42}$$

where

$$\rho_k = \begin{bmatrix} \rho_{11k} & \rho_{12k} & \cdots & \rho_{1nk} \\ \rho_{21k} & \rho_{22k} & \cdots & \rho_{2nk} \\ \cdot & \cdot & \ddots & \cdot \\ \rho_{n1k} & \rho_{n2k} & \cdots & \rho_{nnk} \end{bmatrix}. \tag{43}$$

Hence

$$\det(\det_n M^{-1}\Lambda M) = \det \rho_1 \det \rho_2 \cdots \det \rho_h. \tag{44}$$

We need to calculate next $\det M^{-1}\Lambda M$. For this we make the following permutations of rows and columns:

$$\text{column } (i-1)h + j \rightarrow \text{column } (j-1)n + i$$

$$\text{row } (i-1)h + j \rightarrow \text{row } (j-1)n + i$$

for $i = 1, 2, \cdots, n$ and $j = 1, 2, \cdots, h$. This gives the matrix

$$\begin{bmatrix} \rho_1 & & & * \\ & \rho_2 & & \\ & & \ddots & \\ & & & \rho_h \end{bmatrix} \tag{45}$$

where ρ_k is as in (43). Hence, by Laplace's expansion, $\det M^{-1}\Lambda M = \det \rho_1 \det \rho_2 \cdots \det \rho_h = \det(\det_n M^{-1}\Lambda M)$. This proves (37), as required.

We now apply this to norms. Here we have $N_{\mathfrak{A}|\Phi}(a) = \det(\rho_{ij}(a)) \in \Phi$ and

$$N_{\mathrm{E}|\Phi}N_{\mathfrak{A}|\Phi}(a) = \det(\lambda(\det \rho_{ij})).$$

Since $\rho \rightarrow \lambda(\rho)$ is an isomorphism, we have

$$(\lambda(\det \rho_{ij})) = \det_n (\lambda(\rho_{ij}))$$

so

$$\det \lambda(\det \rho_{ij}) = \det \det_n (\lambda(\rho_{ij})) = \det \Lambda(a),$$

by (37). Since $\det \Lambda(a) = N_{\mathfrak{A}|\mathrm{E}}(a)$ we have the norm formula (31).

We shall now specialize all of this to the case: $\mathfrak{A} = \mathrm{P}$ a field.* We know that the minimum polynomial $m_a(x)$ of any $a \in \mathrm{P}$ is irreducible. Hence the characteristic polynomial $f_a(x) = m_a(x)^r$. We have $[\mathrm{P}:\Phi] = n = \deg f_a(x)$ and $[\Phi(a):\Phi] = \deg m_a(x)$; therefore $r = \deg f_a(x)/\deg m_a(x) = [\mathrm{P}:\Phi]/[\Phi(a):\Phi] = [\mathrm{P}:\Phi(a)]$. Hence we have

$$f_a(x) = m_a(x)^{[\mathrm{P}:\Phi(a)]}. \tag{46}$$

We shall now obtain some important formulas for the norm and trace of a field and we look first at the separable case. Thus let P/Φ be finite dimensional separable, Ω/Φ the normal closure of P/Φ. Then Ω/Φ is Galois and $[\Omega:\Phi] = (G:1)$ for the Galois group G of Ω/Φ. Let H be the subgroup of G corresponding to P/Φ (the Galois group of Ω/P). Since $[\mathrm{P}:\Phi] = n$, the index $(G:H) = n$ and we have n distinct cosets $Hs_1', Hs_2', \cdots, Hs_n'$. If s_i denotes the restriction of s_i' to P, then $s_1, s_2, \cdots, s_n$ are distinct isomorphisms of P/Φ into Ω/Φ and these are all the isomorphisms of P/Φ into Ω/Φ (§ 7). Next let $\rho \in \mathrm{P}$ and let K be the subgroup of G corresponding to $\Phi(\rho)$. Then $G \supseteq K \supseteq H$. Let $t_1', \cdots, t_m'$ be a complete set of representatives of the cosets Kt' in G and let $u_1', \cdots, u_r'$ be a complete set of representatives of the cosets Hu' in K. Then we have $G = \bigcup Kt_j'$, $K = \bigcup Hu_k'$ so $G =$

* A simplified version of the proof of the transitivity formula for norms in this case will be indicated in ex. 2 below.

$\bigcup Hu_k't_j'$ and the mr elements $u_k't_j'$ form a complete set of representatives of the cosets of H in G. We may assume that these are the s_i' which we indicated before. The restrictions of the t_j' to $\Phi(\rho)$ give all the isomorphisms of $\Phi(\rho)/\Phi$ into Ω/Φ and these are distinct. Since ρ generates $\Phi(\rho)$, it follows that the elements $\rho^{t_1'}, \rho^{t_2'}, \cdots, \rho^{t_m'}$ are distinct and these include all the conjugates $\rho^{s'}$, $s' \varepsilon G$. Hence the minimum polynomial of ρ over Φ is $m_\rho(x) = \prod_{j=1}^{m} (x - \rho^{t_j'})$. Also we have $\rho^{u_k't_j'} = \rho^{t_j'}$ for all k and j. Hence $\prod_{i=1}^{n} (x - \rho^{s_i}) = \prod_{k=1}^{r} \prod_{j=1}^{m} (x - \rho^{u_k't_j'}) = m_\rho(x)^r$. On the other hand, $r = [\mathrm{P}:\Phi(\rho)]$, so by (46) (for $a = \rho$), we see that the characteristic polynomial

$$f_\rho(x) = \prod_{1}^{n} (x - \rho^{s_i}) \tag{47}$$

where $s_1, s_2, \cdots, s_n$ are the different isomorphisms of P/Φ into its normal closure Ω/Φ. Comparison of this formula with (22) gives the following formulas for the trace and norm in the separable case:

$$T_{\mathrm{P}|\Phi}(\rho) = \sum_{1}^{n} \rho^{s_i}, \quad N_{\mathrm{P}|\Phi}(\rho) = \prod_{1}^{n} \rho^{s_i}. \tag{48}$$

Next let P/Φ be purely inseparable of characteristic $p \neq 0$. Then $[\mathrm{P}:\Phi] = p^f$. If $\rho \varepsilon \mathrm{P}$, the minimum polynomial $m_\rho(x)$ has the form $x^{p^e} - \alpha = (x - \rho)^{p^e}$. Since $\mathrm{P}/\Phi(\rho)$ is purely inseparable, $[\mathrm{P}:\Phi(\rho)] = p^g$ and $p^f = [\mathrm{P}:\Phi] = [\mathrm{P}:\Phi(\rho)][\Phi(\rho):\Phi] = p^g p^e$. Hence $f = g + e$. By (46), the characteristic polynomial is

$$f_\rho(x) = (x^{p^e} - \alpha)^{p^g} = (x - \rho)^{p^f}. \tag{49}$$

This shows that

$$T_{\mathrm{P}|\Phi}(\rho) = [\mathrm{P}:\Phi]\rho, \quad N_{\mathrm{P}|\Phi}(\rho) = \rho^{[\mathrm{P}:\Phi]}. \tag{50}$$

Now let P/Φ be arbitrary, Σ/Φ the maximal separable subfield, Ω/Φ the normal closure of P/Φ. Then Ω/Φ contains the normal closure Δ/Φ of Σ/Φ. Again we assume the characteristic is $p \neq 0$. Then $[\mathrm{P}:\Sigma] = p^f$, $f \geq 0$, and this is the degree of inseparability $[P:\Phi]_i$ (§ 10). If $\rho \in P$ we have, by (50) and (47),

$$N_{P|\Phi}(\rho) = N_{\Sigma|\Phi}(N_{P|\Sigma}(\rho)) = N_{\Sigma|\Phi}(\rho^{[P:\Phi]_i})$$
$$= (\rho^{[P:\Phi]_i})^{s_1}(\rho^{[P:\Phi]_i})^{s_2} \cdots (\rho^{[P:\Phi]_i})^{s_n}$$

where $s_1, s_2, \cdots, s_n$ are the different isomorphisms of Σ/Φ into Δ/Φ. Now it is easily seen that every s_i is the restriction of an isomorphism of P/Φ into Ω/Φ and distinct isomorphisms of P/Φ in Ω/Φ have distinct restrictions to Σ/Φ and map this field into Δ/Φ (ex. 1, § 10). It follows that the foregoing formula can be re-written as

$$N_{P|\Phi}(\rho) = (\rho^{s_1}\rho^{s_2} \cdots \rho^{s_n})^{[P:\Phi]_i} \tag{51}$$

where $s_1, s_2, \cdots, s_n$ are now considered as the different isomorphisms of P/Φ into Ω/Φ. In exactly the same way we obtain

$$T_{P|\Phi}(\rho) = [P:\Phi]_i(\rho^{s_1} + \rho^{s_2} + \cdots + \rho^{s_n}). \tag{52}$$

If P is not separable over Φ, then $f > 0$ and $[P:\Phi]_i = p^f$ is divisible by p. Hence we see that $T_{P|\Phi}(\rho) \equiv 0$ for inseparable P/Φ.

We obtain next some formulas for the discriminant of P/Φ relative to a basis $(\rho_1, \rho_2, \cdots, \rho_n)$. This is

$$\delta = \det (T_{P|\Phi}(\rho_i\rho_j)). \tag{53}$$

If P/Φ is inseparable, $T_{P|\Phi} = 0$ so $\delta = 0$. Now assume P/Φ separable and, as before, let $s_1, s_2, \cdots, s_n$ be the isomorphisms of P/Φ into Ω/Φ. Consider the matrix

$$A = (\rho_i^{s_j}), \quad i, j = 1, 2, \cdots, n. \tag{54}$$

We have shown in the proof of Theorem 16 that $\det A \neq 0$. We consider now the matrix AA', A' the transpose of A, whose (i, j)-entry is

$$\rho_i^{s_1}\rho_j^{s_1} + \rho_i^{s_2}\rho_j^{s_2} + \cdots + \rho_i^{s_n}\rho_j^{s_n} = T_{P|\Phi}(\rho_i\rho_j). \tag{55}$$

Hence $\delta = \det AA'$ and we have

$$\delta = (\det A)^2, \quad A = (\rho_i^{s_j}). \tag{56}$$

Since $\det A \neq 0$, this shows that $\delta \neq 0$. We recall that this implies that the trace bilinear form $(\rho, \sigma) = T_{P|\Phi}(\rho\sigma)$ is non-degenerate (Vol. II, p. 140). We therefore have the following

Theorem 18. *If* P/Φ *is finite dimensional separable, then the trace form* $(\rho, \sigma) = T_{P|\Phi}(\rho\sigma)$ *is non-degenerate and the discriminants* δ *of* P/Φ *are* $\neq 0$.

Let θ be a primitive element of the finite dimensional separable extension. Then it is clear from (46) that the characteristic polynomial $f(x)$ of θ is the same as the minimum polynomial. In $\Omega[x]$ we have $f(x) = (x - \theta_1)(x - \theta_2) \cdots (x - \theta_n)$, $\theta_1 = \theta$, and the θ_i are distinct. If $f'(x)$ is the derivative of $f(x)$, then

$$f'(\theta) = (\theta - \theta_2)(\theta - \theta_3) \cdots (\theta - \theta_n) \tag{57}$$

and this element is contained in $P = \Phi(\theta)$ since $f'(x) \in \Phi[x]$. The element $f'(\theta)$ is called the *different of* θ. We shall show that the discriminant δ determined by the basis $(1, \theta, \theta^2, \cdots, \theta^{n-1})$ is

$$\delta = (-1)^{\frac{n(n-1)}{2}} N_{P|\Phi}(f'(\theta)). \tag{58}$$

We have $\delta = \det T_{P|\Phi}(\theta^{i-1}\theta^{j-1})$. Now it is clear that we have an isomorphism of $\Phi(\theta)/\Phi$ into Ω/Φ sending θ into θ_i, $1 \leq i \leq n$. Hence the θ_i are the conjugates θ^{s_i} of θ and $(\theta^k)^{s_i} = \theta_i^k$. The matrix A of (54) for the basis $(1, \theta, \theta^2, \cdots, \theta^{n-1})$ now becomes

$$A = \begin{bmatrix} 1 & 1 & \cdots & 1 \\ \theta_1 & \theta_2 & \cdots & \theta_n \\ \cdot & \cdot & \cdots & \cdot \\ \theta_1^{n-1} & \theta_2^{n-1} & \cdots & \theta_n^{n-1} \end{bmatrix}.$$

It is well known that $\det A$, a so-called Vandermonde determinant, has the value $\prod_{i>j} (\theta_i - \theta_j)$. Consequently (56) gives the formula

$$\delta = \prod_{i<j} (\theta_i - \theta_j)^2, \quad i, j = 1, 2, \cdots, n, \tag{59}$$

for the discriminant. On the other hand, $f'(\theta) = \prod_{i\neq 1} (\theta_1 - \theta_i)$. Applying s_j which sends θ_1 into θ_j we obtain $f'(\theta)^{s_j} = \prod_{i\neq j} (\theta_j - \theta_i)$. It follows that

$$N_{P|\Phi}(f'(\theta)) = (-1)^{\frac{n(n-1)}{2}} \prod_{i<j} (\theta_i - \theta_j)^2. \tag{60}$$

Comparison of (59) and (60) proves (58).

EXERCISES

1. Let $\mathfrak{A}$ be the algebra $\Phi[x]/(x^n - 1)$, so that $\mathfrak{A}$ has the basis $(1, \theta, \cdots, \theta^{n-1})$ where θ is the coset $x + (x^n - 1)$. Show that, if $a = \alpha_0 + \alpha_1\theta + \cdots + \alpha_{n-1}\theta^{n-1}, \alpha_i \in \Phi$, then the matrix of a_R relative to the basis $(1, \theta, \cdots, \theta^{n-1})$ is the *circulant matrix*

$$A = \begin{bmatrix} \alpha_0 & \alpha_1 & \cdot & \cdot & \cdot & \alpha_{n-1} \\ \alpha_{n-1} & \alpha_0 & \alpha_1 & \cdot & \cdot & \alpha_{n-2} \\ \vdots & \vdots & \cdot & & & \vdots \\ \vdots & \vdots & & \cdot & \cdot & \vdots \\ \alpha_1 & \alpha_2 & \alpha_3 & \cdot & \cdot & \alpha_0 \end{bmatrix}.$$

Show that, if Φ contains n distinct n-th roots ζ_i of 1, then

$$N(a) = \det A = \prod_{j=1}^{n} \left(\sum_{0}^{n-1} \alpha_i \zeta_j^i \right).$$

2. Let $P \supseteq \Phi \supseteq E$ be finite dimensional extension fields of the field E. Let $a \in P$ and let $x^n - \alpha_1 x^{n-1} + \cdots + (-1)^n \alpha_n$ be the minimum polynomial of a over Φ so that (20) defines a matrix representation of $\Phi(a)/\Phi$. Show that one obtains a matrix representation of $\Phi(a)/E$ by replacing the entries 0, 1 which appear by the $h \times h$ zero and identity matrices respectively and the α_i by the representing matrices $\lambda(\alpha_i)$ for a matrix representation of Φ/E. Use Laplace's expansion to verify that the determinant of the resulting matrix is

$$N_{\Phi(a)/E}(a) = \det \lambda(\alpha_n).$$

Since $\alpha_n = N_{\Phi(a)/\Phi}(a)$, this gives

$$N_{\Phi(a)/E}(a) = N_{\Phi/E}(N_{\Phi(a)/\Phi}(a)).$$

Next show that $N_{P|E}(a) = N_{\Phi(a)/E}(a)^r, r = [P:\Phi(a)]$. Use these results to prove (31) for $\mathfrak{A} = P$.

3. Let $\mathfrak{A}/\Phi$ be an algebra with the basis $(u_1, u_2, \cdots, u_n)$ and let $\mathfrak{X} = \Phi(\xi_1, \xi_2, \cdots, \xi_n)$ the field of rational expressions in indeterminates ξ_i. Consider the algebra $(\mathfrak{A} \otimes_\Phi \mathfrak{X})$ over $\mathfrak{X}$ which has the basis $(u_1, u_2, \cdots, u_n)$ over $\mathfrak{X}$. Show that, if $X = \sum_1^n \xi_i u_i$, then the characteristic polynomial $f_X(x)$ of X is a homogeneous polynomial of degree n in $\Phi[x, \xi_1, \cdots, \xi_n]$, x, ξ_i, indeterminates. Use this and the arithmetic theory of polynomial rings of Vol. I, pp. 124–127, to show that the minimum polynomial $\mu_X(x)$ of X has the form $x^m - t(\xi_1, \cdots, \xi_n)x^{m-1} + \cdots + (-1)^m n(\xi_1, \cdots, \xi_n)$ where the coefficient of x^{m-i} is a homogeneous polynomial of degree i in the ξ's. If $a = \Sigma\alpha_i u_i \in \mathfrak{A}$, set $\mu_a(x) = x^m - t(\alpha_1, \cdots, \alpha_n)x^{m-1} + \cdots + (-1)^m n(\alpha_1, \cdots, \alpha_n)$. Prove that $\mu_a(a) = 0$.

4. Let the notations be as in 3 and assume $\mathfrak{A} = P$ is a field. Show that $\mu_X(x)$ is irreducible (Hint: Use ex. 4, § 9); hence show that $f_X(x)$ is a power of $\mu_X(x)$. Show that $n(\xi_1, \cdots, \xi_n)$ is irreducible and that the *norm form*

$$N(X) = \pm n(\xi_1, \cdots, \xi_n)^r.$$

15. Galois cohomology. One is often interested in studying mappings of the Galois group G of a finite dimensional Galois extension P/Φ into P or into the multiplicative group P^* of non-zero elements of P. More generally, one encounters mappings of the product sets $G \times G$, $G \times G \times G, \cdots$ (functions of several variables in G) into P or P^*. A particularly important type of mapping of G into P^*, $s \to \mu_s \in P^*$, is one which satisfies *Emmy Noether's equations*

$$(61) \qquad \mu_{st} = \mu_s{}^t \mu_t.$$

If the $\mu_s \in \Phi$, then $\mu_s{}^t = \mu_s$ and this reads: $\mu_{st} = \mu_s\mu_t$ which is just a *character* or multiplicative mapping of G into Φ. If G is cyclic with generator g: $G = \{1, g, \cdots, g^{n-1}\}$, $g^n = 1$, then any element $\mu \in P$ such that $N(\mu) = \mu\mu^g \cdots \mu^{g^{n-1}} = 1$ defines a mapping $s \to \mu_s$ satisfying (61) if we define

$$(62) \quad \mu_1 = 1, \quad \mu_g = \mu, \quad \mu_{g^2} = \mu\mu^g, \cdots, \mu_{g^{n-1}} = \mu\mu^g \cdots \mu^{g^{n-2}}.$$

Then $\mu_{g^{i+1}} = \mu\mu^g \cdots \mu^{g^i} = (\mu\mu^g \cdots \mu^{g^{i-1}})^g \, \mu = \mu_{g^i}{}^g \mu_g$ holds for $i = 1, \cdots, n-2$. Also (61) is clear for $t = 1$ since $\mu_1 = 1$ and $1 = \mu_{g^n} = \mu_{g^{n-1}}{}^g \mu_g = \mu^g \cdots \mu^{g^{n-1}}\mu = N(\mu)$. Hence (61) holds for all $s = g^i$ and $t = g$. It is easy to check by induction that it is valid for all s and all $t = g^j$.

In the general case, if γ is any element of P^*, we can set $\mu_s = \gamma(\gamma^s)^{-1}$ and we have

$$\begin{aligned}\mu_s{}^t\mu_t &= (\gamma(\gamma^s)^{-1})^t\gamma(\gamma^t)^{-1} \\ &= \gamma^t(\gamma^{st})^{-1}\gamma(\gamma^t)^{-1} \\ &= \gamma(\gamma^{st})^{-1} \\ &= \mu_{st}.\end{aligned}$$

Thus $\mu_s = \gamma(\gamma^s)^{-1}$ satisfies Noether's equations for any non-zero γ in P. We proceed to show that this "trivial" solution of Noether's equations is the only possible one, for we have

Theorem 19. *Let $s \to \mu_s$ be a mapping of G into P^* such that $\mu_{st} = \mu_s{}^t\mu_t$, $s, t \in G$. Then there exists a non-zero element γ in P such that $\mu_s = \gamma(\gamma^s)^{-1}$.*

Proof. Since the μ_s are $\neq 0$ and the automorphisms are right linearly independent over P, we see that the operator $\Sigma s\mu_s$ ($\equiv$

$\Sigma s\mu_{sR}$) is $\neq 0$. Thus we can find a $\beta \varepsilon$ P such that $\gamma \equiv \beta(\Sigma s\mu_s) = \Sigma\beta^s\mu_s \neq 0$. We now calculate

$$\gamma^t = \left(\sum_{s\varepsilon G} \beta^s\mu_s\right)^t = \sum_s \beta^{st}\mu_s{}^t$$

$$= \left(\sum_s \beta^{st}\mu_s{}^t\mu_t\right)\mu_t{}^{-1}$$

$$= \left(\sum_s \beta^{st}\mu_{st}\right)\mu_t{}^{-1}$$

$$= \left(\sum_s \beta^s\mu_s\right)\mu_t{}^{-1}$$

since st ranges over G if s ranges over G. Hence we have $\gamma^t = \gamma\mu_t{}^{-1}$ and $\mu_t = \gamma(\gamma^t)^{-1}$ as required.

We have seen that, if G is cyclic with generator g and μ is an element of norm one in P ($N_{P|\Phi}(\mu) = 1$), then Noether's equations hold for $\mu_{g^i} = \mu\mu^g \cdots \mu^{g^{i-1}}, 1 \leq i \leq n-1, \mu_1 = 1$. The theorem now states that there exists a $\gamma \varepsilon$ P* such that $\mu = \mu_g = \gamma(\gamma^g)^{-1}$. This gives the following corollary which is referred to in the literature as "Hilbert's Satz 90":

Corollary. *Let* P/Φ *be a finite dimensional cyclic extension field and let g be a generator of the Galois group of* P *over* Φ. *Then any element* $\mu \varepsilon$ P *such that* $N_{P|\Phi}(\mu) = 1$ *has the form* $\mu = \gamma(\gamma^g)^{-1}$ *for a suitable* $\gamma \varepsilon$ P.

The two results which we have just obtained have analogues for the additive group of the Galois extension P/Φ. We consider a mapping $s \to \delta_s$ of G into P. The additive analogue of Noether's equation is:

(63) $$\delta_{st} = \delta_s{}^t + \delta_t, \quad s, t \varepsilon G.$$

If $\gamma \varepsilon$ P and we set $\delta_s = \gamma - \gamma^s$, then $\delta_{st} = \gamma - \gamma^{st}$ and $\delta_s{}^t + \delta_t = \gamma^t - \gamma^{st} + \gamma - \gamma^t = \delta_{st}$; so (63) holds. The direct analogue of Theorem 19 is valid:

Theorem 20. *Let* δ_s, $s \varepsilon G$, *be elements of* P *satisfying* (63). *Then there exists a* $\gamma \varepsilon$ P *such that* $\delta_s = \gamma - \gamma^s$.

Proof. We choose an element $\rho \in P$ such that $T_{P|\Phi}(\rho) = \Sigma\rho^s \neq 0$. This can be done since $\sum_{s \neq G} s \neq 0$ by the Dedekind independence theorem. Set $\gamma = \sum_{s \in G} T(\rho)^{-1}\delta_s\rho^s$. Then

$$\begin{aligned}
\gamma - \gamma^t &= T(\rho)^{-1}\left(\sum_s (\delta_s\rho^s - \delta_s{}^t\rho^{st})\right)\\
&= T(\rho)^{-1}\left(\sum_s (\delta_{st}\rho^{st} - \delta_s{}^t\rho^{st})\right)\\
&= T(\rho)^{-1}\left(\sum_s \delta_t\rho^{st}\right)\\
&= \delta_t T(\rho)^{-1}\left(\sum_s \rho^{st}\right)\\
&= \delta_t T(\rho)^{-1}T(\rho) = \delta_t,
\end{aligned}$$

which is what we want.

In the cyclic case, G generated by g, the analogue of the condition $N_{P|\Phi}(\mu) = 1$ is $T_{P|\Phi}(\mu) = 0$. If μ is such an element, then we set $\delta_1 = 0$, $\delta_{g^i} = \mu + \mu^g + \cdots + \mu^{g^{i-1}}$ and it is easy to check that (63) holds. We therefore have the following additive analogue of Hilbert's Satz 90:

Corollary. *Let* P/Φ *be finite dimensional cyclic,* g *a generating automorphism of the Galois group of* P/Φ. *Then any element* $\mu \in P$ *such that* $T_{P|\Phi}(\mu) = 0$ *has the form* $\gamma - \gamma^g$ *for a suitable* $\gamma \in P$.

We recall that, if P/Φ is finite dimensional Galois with G as the Galois group, then the set $\mathfrak{L}_\Phi(P)$ of linear transformations of P as vector space over Φ coincides with the set $\mathfrak{A}$ of operators of the form $\sum_{s \in G} s\rho_s \equiv \sum_s s\rho_{sR}$ (Lemma of § 4). We know also that by the Dedekind independence theorem the group elements (s), $s \in G$, form a basis for $\mathfrak{L}_\Phi(P)$ as right vector space over P. We shall now show that Noether's equations arise in considering the following question: What are the automorphisms of the ring $\mathfrak{L}_\Phi(P)$ which leave fixed every element of the subring P_R $(= 1P)$? Let A be such an automorphism and set $s^A = u_s$, $s \in G$. Then if we apply A

to the basic relation $\rho_R s = s(\rho^s)_R$ (eq. (2)), we obtain

(64) $$\rho_R u_s = u_s(\rho^s)_R.$$

Hence,

$$\rho_R(s^{-1}u_s) = s^{-1}(\rho^{s^{-1}})_R u_s = s^{-1}u_s(\rho^{s^{-1}s})_R = (s^{-1}u_s)\rho_R.$$

Thus we see that $s^{-1}u_s$ is an endomorphism of the additive group of P which commutes with every right multiplication ρ_R. This implies that $s^{-1}u_s$ is itself a right multiplication (Vol. I, p. 83). Hence $s^{-1}u_s = \mu_{sR}$ and $u_s = s\mu_s$, $s \in G$. We now use the fact that $s \to s^A = u_s$ is a homomorphism of G. This implies that $u_{st} = u_s u_t$, $s, t \in G$ and so we have $st\mu_{st} = (s\mu_s)(t\mu_t) = st\mu_s{}^t\mu_t$. Hence the $\mu_s \in$ P* (since $u_s \neq 0$) satisfy Noether's equations. Conversely, it is easy to see by reversing the steps that, if the $\mu_s \neq 0$ satisfy Noether's equations, then the mapping

$$A: \sum_s s\rho_s \to \sum_s u_s\rho_s, \quad u_s = s\mu_s$$

is an automorphism of $\mathfrak{L}_\Phi(\mathrm{P})$ which is the identity on P_R. We now recall that any automorphism of $\mathfrak{L}_\Phi(\mathrm{P})$ which is the identity on Φ_R (acting in P) is an inner automorphism (Vol. II, ex. 5, p. 237). Hence there exists an element $C \in \mathfrak{L}_\Phi(\mathrm{P})$ such that $X^A = C^{-1}XC$ holds for all $X \in \mathfrak{L}_\Phi(\mathrm{P})$. In particular, we have $\rho_R = \rho_R{}^A = C^{-1}\rho_R C$ for all $\rho \in$ P, that is, C commutes with every ρ_R. This implies that $C = \gamma_R$, γ a non-zero element of P. Then

$$\begin{aligned} s\mu_s = u_s = s^A = C^{-1}sC &= \gamma_R{}^{-1}s\gamma_R \\ &= s(\gamma^{-1})^s{}_R\gamma_R = s((\gamma^{-1})^s\gamma)_R, \quad s \in G \end{aligned}$$

which implies that $\mu_s = \gamma(\gamma^s)^{-1}$. This gives another proof of Theorem 19. Of course, this is considerably less elementary than the first proof. However, the method is useful in related contexts in which the first method is not applicable.

The representation of $\mathfrak{L}_\Phi(\mathrm{P})$ as $\mathfrak{A} = \{\Sigma s\rho_s\}$ suggests a construction of a more general kind of ring, called a crossed product of the field P and its Galois group G. For this purpose we consider a right vector space $\mathfrak{B}$ over P with basis (u_s) in 1–1 correspondence $s \to u_s$ with the group G. Thus the elements of $\mathfrak{B}$ can be written in one and only one way in the form $\sum_s u_s\rho_s$, $\rho_s \in$ P, so that $[\mathfrak{B}:\mathrm{P}]_R$

$= (G:1)$. We now suppose we have a mapping of $G \times G$ into P^* so that for each ordered pair (s, t) of group elements we have a corresponding $\mu_{s,t} \varepsilon P^*$. We use these to define a multiplication in $\mathfrak{B}$ according to the formula

$$(65) \qquad \left(\sum_{s \varepsilon G} u_s \rho_s\right)\left(\sum_{t \varepsilon G} u_t \sigma_t\right) = \sum_{s,t} u_{st} \rho_s{}^t \sigma_t \mu_{s,t}.$$

It is easy to verify that the multiplication is both ways distributive relative to addition. Hence $\mathfrak{B}$ will be a ring if and only if the associative law of multiplication holds. Also because of the distributive laws it suffices to have $(ab)c = a(bc)$ for $a = u_s\rho$, $b = u_t\sigma$, $c = u_v\tau$, $s, t, v \varepsilon G$. Now

$$\begin{aligned} (ab)c &= (u_{st}\rho^t\sigma\mu_{s,t})(u_v\tau) \\ &= u_{stv}\rho^{tv}\sigma^v\tau\mu_{s,t}{}^v\mu_{st,v} \\ a(bc) &= (u_s\rho)(u_{tv}\sigma^v\tau\mu_{t,v}) \\ &= u_{stv}\rho^{tv}\sigma^v\tau\mu_{s,tv}\mu_{t,v}. \end{aligned}$$

Hence associativity holds if and only if

$$(66) \qquad \mu_{s,t}{}^v\mu_{st,v} = \mu_{s,tv}\mu_{t,v}, \quad s, t, v \varepsilon G.$$

A set of non-zero $\mu_{s,t}$, $s, t \varepsilon G$, satisfying these conditions is called a (G, P^*) *factor set*. Our argument shows that such a set defines a ring $\mathfrak{B}$ by means of (65), the associativity conditions corresponding precisely to the conditions (66). The ring $\mathfrak{B}$ is called the *crossed product* of G and P with respect to the factor set $\mu_{s,t}$. We shall write $\mathfrak{B} = (G, P, \mu)$ to indicate the ingredients G, P and the factor set $\mu = (\mu_{s,t})$.

If we consider again the representation of $\mathfrak{L}_\Phi(P)$ as $\mathfrak{A} = \{\Sigma s\rho_s\}$, we see that $\mathfrak{A}$ is isomorphic to the crossed product $(G, P, 1)$ where 1 is the factor set $\mu_{s,t} = 1$, $s, t \varepsilon G$, 1 the identity of P. This is clear if we compare (65) with the multiplication of elements of $\mathfrak{A}$. We now replace the right basis (s) of $\mathfrak{A}$ over P by (u_s) where $u_s = s\gamma_s$, γ_s a non-zero element of P. Then we have

$$\begin{aligned} u_s u_t &= (s\gamma_s)(t\gamma_t) = st\gamma_s{}^t\gamma_t \\ &= u_{st}\gamma_{st}{}^{-1}\gamma_s{}^t\gamma_t. \end{aligned}$$

Thus we see that $\mathfrak{A}$ is also isomorphic with the crossed product (G, P, μ) where

$$\mu_{s,t} = \gamma_{st}^{-1}\gamma_s^t\gamma_t. \tag{67}$$

It is easy to check that these satisfy the factor set conditions but this is unnecessary since these are equivalent to the associative law. A factor set μ which is obtained from a function $s \to \gamma_s \varepsilon P^*$ by means of (67) is said to be *equivalent to* 1 ($\mu \sim 1$). The result we have established is that, if $\mu \sim 1$, then (G, P, μ) is isomorphic to $\mathfrak{L}_\Phi(P)$. One might be tempted to guess that the analogue to Theorem 18 is valid for factor sets. However, this is not the case and we shall indicate this by considering the special case of a cyclic group.

Let G be cyclic with g as generator and let $(G:1) = n$. We set for $0 \leq i, j \leq n - 1$,

$$\mu_{g^i,g^j} = \begin{cases} 1 & \text{if } i + j < n \\ \alpha \neq 0 \text{ in } \Phi & \text{if } i + j \geq n. \end{cases} \tag{68}$$

We have to check the factor set conditions (66). Since $1, \alpha \varepsilon \Phi$ these simplify to

$$\mu_{g^i,g^j}\mu_{g^{i+j},g^k} = \mu_{g^i,g^{j+k}}\mu_{g^j,g^k}. \tag{69}$$

There are three cases: $i + j + k < n$, $n \leq i + j + k < 2n$, and $i + j + k \geq 2n$. In the first case, both sides reduce to 1. In the second, both are α; and in the third, both are α^2. A crossed product (G, P, μ) where G is cyclic and μ is of the type just defined is called a *cyclic algebra* or *cyclic crossed product.* The condition that $\mu \sim 1$ is that there exist non-zero elements γ_{g^i} such that $\mu_{g^i,g} = \gamma_{g^{i+1}}^{-1}\gamma_{g^i}^g\gamma_g$. This gives for $\gamma = \gamma_g$, $\gamma_{g^2} = \gamma\gamma^g, \cdots,$ $\gamma_{g^{n-1}} = \gamma\gamma^g \cdots \gamma^{g^{n-2}}$, $\alpha = \gamma_1^{-1}\gamma_{g^{n-1}}^g\gamma = \gamma_1^{-1}\gamma\gamma^g \cdots \gamma^{g^{n-1}} = \gamma_1^{-1}N_{P|\Phi}(\gamma)$. Also $1 = \mu_{1,1} = \gamma_1^{-1}\gamma_1^1\gamma_1$ gives $\gamma_1 = 1$ so we must have $\alpha = N_{P|\Phi}(\gamma)$. It is easily seen also that this condition implies that $\mu \sim 1$. Thus we see that we can get a factor set $\mu \not\sim 1$ simply by choosing an $\alpha \varepsilon \Phi$ which is not the norm of any element γ of P. For example, let Φ be the field of real numbers and P the field of complex numbers, $P = \Phi(i)$, $i^2 = -1$. Then if $\gamma = \gamma_1 + i\gamma_2$, γ_1, γ_2 in Φ, $\gamma^g = \bar{\gamma} = \gamma_1 - i\gamma_2$ and $N(\gamma) = \gamma_1^2 + \gamma_2^2 \geq 0$. Hence if $\alpha < 0$, then α is not a norm. We remark that it is easy to see that the cyclic crossed product constructed with such an α is isomorphic to Hamilton's quaternion algebra over Φ.

The notions with which we have been dealing are all special cases of notions in the cohomology theory of groups. We shall now indicate briefly the general situation. We begin with an arbitrary group G and the group ring $I(G)$ of G over the integers. The elements of $I(G)$ are the elements $\sum_{s \in G} m_s s$ where the m_s are integers and $m_s \neq 0$ for a finite subset of G (cf. Vol. I, ex. 2, p. 95). We consider $\Sigma m_s s = \Sigma n_s s$ if and only if $m_s = n_s$ for all s. Addition in $I(G)$ is by components: $\Sigma m_s s + \Sigma n_s s = \Sigma(m_s + n_s)s$. Multiplication is defined by $\left(\sum_{s \in G} m_s s\right)\left(\sum_{t \in G} n_t t\right) = \sum_{s, t \in G} m_s n_t st.$ Since G is associative, $I(G)$ is an associative ring. Let $\mathfrak{M}$ be a right $I(G)$-module so that $\mathfrak{M}$ is a commutative group under addition and a product xa, $x \in \mathfrak{M}$, $a \in I(G)$, $xa \in \mathfrak{M}$ is defined so that $(x + y)a = xa + ya$, $x(a + b) = xa + xb$, $x(ab) = (xa)b$, $x1 = 1$.

Let $C^r(G, \mathfrak{M})$ denote the set of mappings of the r-fold product $G \times G \times \cdots \times G$ into $\mathfrak{M}$. The elements of $C^r(G, \mathfrak{M})$ will be called *r-cochains* of G relative to the module $\mathfrak{M}$. These are the mappings $(s_1, s_2, \cdots, s_r) \to f(s_1, s_2, \cdots, s_r) \in \mathfrak{M}$, $s_i \in G$. We make a commutative group out of $C^r(G, \mathfrak{M})$ by defining $f + g$ in the usual way by $(f + g)(s_1, s_2, \cdots, s_r) = f(s_1, \cdots, s_r) + g(s_1, \cdots, s_r)$. We shall now define a homomorphism d, the *coboundary operator* of $C^r(G, \mathfrak{M})$ into $C^{r+1}(G, \mathfrak{M})$. We do this by defining df for f in $C^r = C^r(G, \mathfrak{M})$ by

$$\begin{aligned}(df)(s_1, s_2, \cdots, s_{r+1}) &= f(s_2, \cdots, s_{r+1}) \\ &+ \sum_{i=1}^{r} (-1)^i f(s_1, \cdots, s_{i-1}, s_i s_{i+1}, \cdots, s_{r+1}) \qquad (70) \\ &+ (-1)^{r+1} f(s_1, \cdots, s_r) s_{r+1}.\end{aligned}$$

It is clear that $d(f + g) = df + dg$, so d is a homomorphism of C^r into C^{r+1}. Strictly speaking we should denote the d we defined by (70) by d_r; however, it is convenient to use the same notation for all of these homomorphisms which are defined on C^r, $r = 1, 2, \cdots$. It is convenient to include also the group C^0 of 0-

cochains which we take to be the module $\mathfrak{M}$ itself. Then if $x \in C^0 = \mathfrak{M}$, dx is the element of C^1 such that $(dx)(s) = x - xs$.

The kernel of d (acting on C^r) is denoted as Z^r and its elements are called *r-cocycles* of G relative to the module $\mathfrak{M}$. The image in C^r of C^{r-1} under d is denoted as B^r and its elements are called *r-coboundaries*. Both Z^r and B^r are subgroups of C^r and it can be shown that $Z^r \supseteq B^r$. This is equivalent to showing that $d^2 = 0$ for the coboundary operator d. We shall leave the verification as an exercise (ex. 1 below). The factor group $H^r(G, \mathfrak{M}) = Z^r/B^r$ is called the *r-th cohomology group of G relative to the module* $\mathfrak{M}$. Here we take $r = 0, 1, 2, \cdots$, and we adopt the convention that $B^0 = 0$, so $H^0 = Z^0$, the group of 0-cocycles. The elements of this group are just the elements x of $\mathfrak{M}$ such that $xs - x = 0$ for all $s \in G$. Evidently these are just the set of invariants of $\mathfrak{M}$ relative to G.

We shall now show that the notions we have been considering in this section fit into this general picture. We take G to be the Galois group of the field P/Φ where P/Φ is finite dimensional Galois. For the module $\mathfrak{M}$ we take either the multiplicative group P^* of P or the additive group $(\mathrm{P}, +)$ of P. In the first case we make P^* a module for $I(G)$ by defining ρa, $\rho \in \mathrm{P}^*$, $a = \sum_{s \in G} m_s s$ to be the element $\prod_{s \in G} (\rho^s)^{m_s}$. Since G is a finite group there is no difficulty in defining this product. It is trivial to check the module axioms and we leave this to the reader. A 1-cochain $s \to \mu_s = \mu(s) \in \mathrm{P}^*$ is a cocycle if and only if $(d\mu)(s, t) = \mu_t \mu_{st}^{-1} \mu_s^t = 1$ (the 0 of P^*). This is equivalent to $\mu_{st} = \mu_t \mu_s^t$ which are Emmy Noether's equations (61). If $\gamma \in \mathrm{P}^*$, then γ is a 0-cochain and its coboundary is the 1-cochain $f(s) = \gamma(\gamma^s)^{-1}$. Theorem 19 can now be re-interpreted as the statement that every 1-cocycle of G relative to P^* is a coboundary. In other words, $Z^1/B^1 = 1$, or the *first cohomology group of G relative to* P^* is the identity.

If $(s, t) \to \mu_{s,t}$ is a 2-cochain, then the coboundary definition gives

$$(d\mu)(s, t, u) = \mu_{t,u} \mu_{st,u}^{-1} \mu_{s,tu} (\mu_{s,t}^u)^{-1}.$$

It follows that $\mu_{s,t}$ is a 2-cocycle if and only if $\mu_{s,tu} \mu_{t,u} = \mu_{st,u} \mu_{s,t}^u$, $s, t, u \in G$ and these are just the conditions (66) defining a factor set. Thus the 2-cocycles are just the factor sets. If $s \to \gamma_s$ is a 1-cochain, its coboundary $d\gamma$ is given by $(d\gamma)(s, t) = \gamma_t(\gamma_{st})^{-1}\gamma_s{}^t$. Hence 2-coboundaries are just the factor sets equivalent to 1. The general considerations imply that the set of factor sets form a group under multiplication: $(\mu\nu)_{s,t} = \mu_{s,t}\nu_{s,t}$. The factor sets equivalent to 1 form a subgroup and the second cohomology group $H^2(G, \mathrm{P}^*)$ is the factor group of the first of these groups relative to the second. As we have seen, in general the cohomology group $H^0(G, \mathrm{P}^*)$ is the set of G-invariants of P^*. Thus this is the multiplicative group Φ^* of the subfield Φ.

An entirely analogous discussion can be made for the additive group $(\mathrm{P}, +)$ considered as an $I(G)$-module by means of the definition $\rho a = \sum_{s \varepsilon G} m_s \rho^s$. It is easily seen that Theorem 20 states that *the first cohomology group of* G *relative to* $(\mathrm{P}, +)$ is 0. It can be shown that if the characteristic of P is either 0 or not a divisor of the order n of G, then all the cohomology groups $H^r(G, \mathrm{P}) = 0$, $r \geq 1$. This is an immediate consequence of ex. 2 below.

EXERCISES

1. Prove $d^2 = 0$.
2. Let G be a finite group of order n and let $\mathfrak{M}$ be a module for $I(G)$ which is *uniquely n-divisible* in the sense that for any $y \varepsilon \mathfrak{M}$ there exists a unique x $\left(\text{written as } \frac{1}{n}y\right)$ such that $nx = y$. Prove that the groups $H^r(G, \mathfrak{M}) = 0$ for $r \geq 1$.
3. Let P/Φ be a cyclic field extension, $[\mathrm{P}:\Phi] = n$, r a divisor of n, γ a non-zero element of Φ such that $\gamma^r = N_{\mathrm{P}|\Phi}(\rho)$, $\rho \varepsilon \mathrm{P}$. Prove that $\gamma = N_{\mathrm{E}|\Phi}(\eta)$ where E/Φ is the (unique) subfield of P/Φ such that $[\mathrm{P}:\mathrm{E}] = r$ and $\eta \varepsilon \mathrm{E}$. (Hint: Set $n = mr$ and consider the element $\beta = \rho\rho^g \cdots \rho^{g^{m-1}}$. Show that $N_{\mathrm{P}|\mathrm{E}}(\beta) = \gamma^r$ and apply Hilbert's Satz 90 to $\beta^{-1}\gamma$.)

16. Composites of fields. In this section we consider a problem which can be formulated roughly in the following manner. Given two extension fields E and P over Φ, to determine the ways these can be put together to form another extension field of Φ. More precisely we seek to determine the composite fields of E and P over Φ in the following precise sense.

Definition 3. *Let* E *and* P *be two fields over* Φ. *Then a* composite field *of* E/Φ *and* P/Φ *is a triple* (Γ, s, t) *where* Γ *is a field over* Φ *and* s *and* t *are isomorphisms of* E/Φ *and* P/Φ *respectively into* Γ/Φ *such* *that* Γ *is generated as a field by the images* E^s *and* P^t. *The composites* (Γ, s, t) *and* (Γ', s', t') *of* E/Φ *and* P/Φ *are* equivalent *if there exists an isomorphism* u *of* Γ/Φ *onto* Γ'/Φ *such that* $su = s'$ *and* $tu = t'$.

The problem is to determine the equivalence classes of composites. We shall consider this question now under the assumption that one of the fields, say P, is finite dimensional over Φ. In Chapter IV (§ 10) we shall investigate the problem for infinite dimensional extensions.

Suppose (Γ, s, t) is a field composite of E/Φ and P/Φ where $[P:\Phi] = n < \infty$. We consider the subset

$$E^sP^t = \left\{ \sum_i \epsilon_i^s \rho_i^t \,|\, \epsilon_i \in E, \rho_i \in P \right\}.$$

Clearly this is the subalgebra of Γ/Φ generated by the two subalgebras E^s/Φ and P^t/Φ. Also it is immediate that, if $(\rho_1, \rho_2, \cdots, \rho_n)$ is a basis for P/Φ, then $E^sP^t = E^s\rho_1{}^t + E^s\rho_2{}^t + \cdots + E^s\rho_n{}^t$, the set of E^s-linear combinations of the $\rho_i{}^t$, $1 \leq i \leq n$. Since Γ and hence E^sP^t is commutative, E^sP^t is an algebra over E^s and $[E^sP^t:E^s] \leq n < \infty$. Since E^sP^t is contained in a field, it has no zero divisors; hence by VII of the Introduction, E^sP^t is a field. Since Γ is the subfield of Γ generated by E^s and P^t, we see that $\Gamma = E^sP^t$. This important relation leads us to look at the tensor product algebra $E \otimes_\Phi P$ whose elements we indicate in the original notation: $\Sigma\epsilon_i \otimes \rho_i$. The basic property of the tensor product is that the mapping $\Sigma\epsilon_i \otimes \rho_i \to \Sigma\epsilon_i{}^s\rho_i{}^t$ is a homomorphism of $E \otimes_\Phi P$ onto $\Gamma = E^sP^t$. If $\mathfrak{J}$ is the kernel of this homomorphism, then $\Gamma \cong (E \otimes_\Phi P)/\mathfrak{J}$. Since Γ is a field, this implies that $\mathfrak{J}$ is a *maximal* ideal: $\mathfrak{J}$ is a proper subset of $E \otimes P$ and there exists no ideal $\mathfrak{J}'$ such that $E \otimes P \supset \mathfrak{J}' \supset \mathfrak{J}$. Conversely, if $\mathfrak{J}$ is a maximal ideal in $E \otimes P$, then $\Gamma = (E \otimes P)/\mathfrak{J} \neq 0$ and this has no ideals $\neq 0$, $E \otimes P$. Hence Γ is a field (Vol. I, p. 77). We can now state the following result.

Theorem 21. *Let* E/Φ *and* P/Φ *be fields such that* $[P:\Phi] < \infty$ *and let* $\mathfrak{J}$ *be a maximal ideal in* $E \otimes_\Phi P$. *Let* s *be the mapping* $\epsilon \to \epsilon \otimes 1 + \mathfrak{J}$ *of* E *into* $\Gamma = (E \otimes P)/\mathfrak{J}$ *and let* t *be the mapping* $\rho \to 1 \otimes \rho + \mathfrak{J}$ *of* P *into* Γ. *Then* (Γ, s, t) *is a field composite* *of* E/Φ *and* P/Φ. *Distinct maximal ideals* $\mathfrak{J}$, $\mathfrak{J}'$ *in* E ⊗ P *give rise in this way to inequivalent composites. Moreover, every field composite of* E/Φ *and* P/Φ *is equivalent to one of the* (Γ, *s*, *t*) *given by a maximal ideal* $\mathfrak{J}$ *in* E ⊗ P.

Proof. If $\mathfrak{J}$ is a maximal ideal in E ⊗ P, then $\epsilon \to \epsilon \otimes 1$ is a homomorphism into E ⊗ P so $s: \epsilon \to \epsilon \otimes 1 + \mathfrak{J}$ is a homomorphism into $\Gamma = (E \otimes P)/\mathfrak{J}$. Since $1 \to 1 + \mathfrak{J}$ and E is a field, s is an isomorphism. Similarly $t: \rho \to 1 \otimes \rho + \mathfrak{J}$ is an isomorphism of P/Φ into Γ. Any element of Γ has the form $\Sigma\epsilon_i \otimes \rho_i + \mathfrak{J}$ and $\epsilon_i \otimes \rho_i + \mathfrak{J} = (\epsilon_i \otimes 1 + \mathfrak{J})(1 \otimes \rho_i + \mathfrak{J}) = \epsilon_i^s\rho_i^t$; hence Γ is generated by E^s and P^t. Also Γ is a field since $\mathfrak{J}$ is maximal. Hence (Γ, *s*, *t*) is a composite. Next let $\mathfrak{J}$ and $\mathfrak{J}'$ be two maximal ideals, (Γ, *s*, *t*), (Γ, *s*′, *t*′) the associated composites and assume that there exists an isomorphism *u* of Γ/Φ onto Γ′/Φ such that $s' = su$, $t' = tu$. Let $\Sigma\epsilon_i \otimes \rho_i \in \mathfrak{J}$. Then the definitions of *s*, *t* give the relation $\Sigma\epsilon_i^s\rho_i^t = 0$ in Γ. Applying *u* we obtain $\Sigma\epsilon_i^{s'}\rho_i^{t'} = 0$ which means that $\Sigma\epsilon_i \otimes \rho_i \in \mathfrak{J}'$. Thus we see that $\mathfrak{J} \subseteq \mathfrak{J}'$. Since $\mathfrak{J}$ is maximal we have $\mathfrak{J} = \mathfrak{J}'$. We have therefore proved that, if the composites (Γ, *s*, *t*), (Γ, *s*′, *t*′) are equivalent, then $\mathfrak{J} = \mathfrak{J}'$. Finally, let (Γ′, *s*′, *t*′) be a composite of E/Φ and P/Φ constructed in any way. We have seen that the mapping $\Sigma\epsilon_i \otimes \rho_i \to \Sigma\epsilon_i^s\rho_i^t$ is a homomorphism of E ⊗ P onto Γ′ whose kernel $\mathfrak{J}$ is a maximal ideal in E ⊗ P. We have the induced isomorphism $u: \Sigma\epsilon_i \otimes \rho_i + \mathfrak{J} \to \Sigma\epsilon_i^s\rho_i^t$ of $\Gamma = (E \otimes P)/\mathfrak{J}$ onto Γ′. One checks that this is an equivalence of the composite (Γ, *s*, *t*) defined by $\mathfrak{J}$ with (Γ′, *s*′, *t*′). This completes the proof.

We have now established a bijection of the collection of equivalence classes of composites with the collection $\{\mathfrak{J}\}$ of maximal ideals in the tensor product $E \otimes_\Phi P$. Since $E \otimes_\Phi P$ can be considered as a finite dimensional algebra over E (Introduction), the following result implies that there are only a finite number of equivalence classes of composites of E/Φ and P/Φ.

Theorem 22. *A finite dimensional algebra with an identity element has only a finite number of distinct maximal ideals. If these are* $\mathfrak{J}_1, \mathfrak{J}_2, \cdots, \mathfrak{J}_h$ *and* $\mathfrak{R} = \bigcap_1^h \mathfrak{J}_j$, *then* $\bar{\mathfrak{A}} = \mathfrak{A}/\mathfrak{R} \cong \Gamma_1 \oplus \Gamma_2 \oplus \cdots \oplus \Gamma_h$ *where* $\Gamma_j = \mathfrak{A}/\mathfrak{J}_j$.

Proof. The direct sum $\Gamma_1 \oplus \Gamma_2 \oplus \cdots \oplus \Gamma_h$ is just the set of h-tuples $(\gamma_1, \gamma_2, \cdots, \gamma_h)$, $\gamma_i \in \Gamma_i$, where equality is defined by equality of components and addition and multiplication are also by components. Evidently the dimensionality of the direct sum is the sum of the dimensionalities of the Γ_j. Now let $\mathfrak{I}_1, \cdots, \mathfrak{I}_h$ be any distinct maximal ideals and let $\mathfrak{B} = \Gamma_1 \oplus \Gamma_2 \oplus \cdots \oplus \Gamma_h$, $\Gamma_j = \mathfrak{A}/\mathfrak{I}_j$. We define a homomorphism of $\mathfrak{A}$ into $\mathfrak{B}$ by mapping $a \rightarrow (a + \mathfrak{I}_1, a + \mathfrak{I}_2, \cdots, a + \mathfrak{I}_h)$. The fact that this mapping is a homomorphism is immediate. The kernel $\mathfrak{N}$ of this homomorphism is the set of elements a such that $a + \mathfrak{I}_j = \mathfrak{I}_j$ for every j. Hence $\mathfrak{N} = \bigcap_1^h \mathfrak{I}_j$. We shall now show that the homomorphism is surjective. We show first that $\mathfrak{I}_1 + \mathfrak{I}_2\mathfrak{I}_3 \cdots \mathfrak{I}_h = \mathfrak{A}$. Since the $\mathfrak{I}_j$ are distinct maximal ideals, $\mathfrak{A} = \mathfrak{I}_1 + \mathfrak{I}_2$ and $\mathfrak{A} = \mathfrak{I}_1 + \mathfrak{I}_3$. Multiplication gives $\mathfrak{A} = \mathfrak{A}^2 = \mathfrak{I}_1{}^2 + \mathfrak{I}_1\mathfrak{I}_3 + \mathfrak{I}_2\mathfrak{I}_1 + \mathfrak{I}_2\mathfrak{I}_3 = \mathfrak{I}_1 + \mathfrak{I}_2\mathfrak{I}_3$. Now suppose we already have $\mathfrak{A} = \mathfrak{I}_1 + \mathfrak{I}_2 \cdots \mathfrak{I}_k$. Since $\mathfrak{A} = \mathfrak{I}_1 + \mathfrak{I}_{k+1}$ a similar multiplication gives $\mathfrak{A} = \mathfrak{I}_1 + \mathfrak{I}_2\mathfrak{I}_3 \cdots \mathfrak{I}_{k+1}$. Hence we have $\mathfrak{A} = \mathfrak{I}_1 + \mathfrak{I}_2 \cdots \mathfrak{I}_h$ and this implies that $\mathfrak{A} = \mathfrak{I}_1 + (\mathfrak{I}_2 \cap \cdots \cap \mathfrak{I}_h)$ since $\mathfrak{I}_2 \cdots \mathfrak{I}_h \subseteq \mathfrak{I}_2 \cap \cdots \cap \mathfrak{I}_h$. If a is any element of $\mathfrak{A}$, then our relation shows that $a = b + c$ where $b \in \mathfrak{I}_1$, $c \in \mathfrak{I}_2 \cap \cdots \cap \mathfrak{I}_h$. Hence the image of c in our homomorphism is $(c + \mathfrak{I}_1, c + \mathfrak{I}_2, \cdots, c + \mathfrak{I}_h) = (a + \mathfrak{I}_1, \mathfrak{I}_2, \cdots, \mathfrak{I}_h)$, which shows that, if γ_1 is any element of Γ_1, then the element $(\gamma_1, 0, \cdots, 0)$ is in the image of the homomorphism. In a similar fashion, if γ_i is any element of Γ_i, then $(0, \cdots 0, \gamma_i, 0, \cdots 0)$ is in the image. Addition shows that any element $(\gamma_1, \gamma_2, \cdots, \gamma_h)$ is in the image so the homomorphism is surjective. It is now clear that, if $\mathfrak{I}_1, \mathfrak{I}_2, \cdots, \mathfrak{I}_h$ are distinct maximal ideals, then the dimensionality $[\mathfrak{A}:\Phi] \geq \sum_1^h [\Gamma_j:\Phi]$ where $\Gamma_j = \mathfrak{A}/\mathfrak{I}_j$. Since every $[\Gamma_j:\Phi] > 0$, this of course puts a bound on the number of $\mathfrak{I}_j$. Also we have seen that, if $\mathfrak{I}_1, \mathfrak{I}_2, \cdots, \mathfrak{I}_h$ are distinct maximal ideals and $\mathfrak{N} = \bigcap \mathfrak{I}_h$, then $\mathfrak{A}/\mathfrak{N} \cong \Gamma_1 \oplus \cdots \oplus \Gamma_h$.

We shall now obtain more precise information on composites under the assumption that $\mathrm{P} = \Phi(\theta)$ is a simple algebraic extension of Φ. Let $f(x)$ be the minimal polynomial of θ over Φ. Then $(1, \theta, \theta^2, \cdots, \theta^{n-1})$ is a basis for P/Φ and $\theta^n = \alpha_0 + \alpha_1\theta + \cdots +$

$\alpha_{n-1}\theta^{n-1}$, if $f(x) = x^n - \alpha_{n-1}x^{n-1} - \cdots - \alpha_0$. Now consider $E \otimes_\Phi P$. The elements of this algebra can be written in one and only one way as $\sum_0^{n-1} \epsilon_i \otimes \theta^i$, $\epsilon_i \in E$. We consider $E \otimes_\Phi P$ as an algebra over E by defining $\eta(\Sigma\epsilon_i \otimes \theta^i) = \Sigma\eta\epsilon_i \otimes \theta^i$, $\eta \in E$ (cf. Introd.). Then it is clear that $1 \otimes \theta$ is a generator of $E \otimes_\Phi P$ over E and the minimum polynomial over E of this element is $f(x)$. Thus we see that $E \otimes_\Phi P \cong E[x]/(f(x))$. The ideals of this algebra have the form $(p(x))/(f(x))$ where $p(x)$ is a divisor of $f(x)$ and such an ideal is maximal if and only if $p(x)$ is irreducible. Then the difference algebra $(E[x]/(f(x))/((p(x)/(f(x)))$ is isomorphic to the field $E[x]/(p(x))$.

Suppose finally that P is a finite dimensional separable extension of Φ. Then we know that $P = \Phi(\theta)$ where the minimum polynomial $f(x)$ of θ is irreducible and separable. The derivative criterion shows that in $E[x]$ we have the factorization $f(x) = p_1(x)p_2(x) \cdots p_h(x)$ where $p_j(x)$ is irreducible of positive degree and $p_i(x) \neq p_j(x)$ if $i \neq j$. Thus we see that we have h inequivalent field composites of P and E over Φ. These have the form (Γ_j, s_j, t_j) where $\Gamma_j \cong E[x]/(p_j(x))$. Also by ex. 1, § 2 of Introd., $E \otimes_\Phi P \cong E[x]/(f(x)) \cong \Gamma_1 \oplus \Gamma_2 \oplus \cdots \oplus \Gamma_h$ and $\sum_1^h [\Gamma_i : E] = [P:\Phi]$. We state this result as

Theorem 23. *Let* P/Φ *be finite dimensional separable and let* E/Φ *be an arbitrary extension field. If* θ *is a primitive element of* P *and* $f(x)$ *its minimum polynomial over* Φ, *then the field composites* (Γ, s, t) *are in* 1–1 *correspondence with the irreducible factors* $p(x)$ *of* $f(x)$ *in* $E[x]$. *If* (Γ_j, s_j, t_j), $j = 1, \cdots, h$ *are the inequivalent composites of* P/Φ *and* E/Φ, *then* $[P:\Phi] = \sum_{j=1}^h [\Gamma_j : E]$.

EXERCISES

1. Show that, if P/Φ is finite dimensional Galois, then there are $n = [P:\Phi]$ inequivalent composites of P with itself and, if (Γ, s, t) is one of these, then $\Gamma = P^s = P^t$. Use this to prove that $P \otimes_\Phi P \cong P^{(1)} \oplus P^{(2)} \oplus \cdots \oplus P^{(n)}$ where $P^{(i)} \cong P$.

2. Let P be finite dimensional Galois over Φ and let E be a subfield of P over Φ. Show that $P \otimes_\Phi E \cong P^{(1)} \oplus P^{(2)} \oplus \cdots \oplus P^{(m)}$ where $P^{(i)} \cong P$ and $m = [E:\Phi]$.

3. Let P/Φ be finite dimensional separable. Show that P/Φ and E/Φ have only one (in the sense of equivalence) composite if either 1) E/Φ is purely inseparable, or 2) $E = \Phi(\xi_1, \xi_2, \cdots, \xi_n)$ the field of rational expressions in indeterminates ξ_i.

4. Define a *composite* $(\Gamma, s_1, s_2, \cdots, s_r)$ of r extension fields P_i/Φ, $1 \leq i \leq r$, as a field Γ/Φ and isomorphisms s_i of P_i into Γ such that Γ is generated by the subfields $P_i^{s_i}$. Call two such composites $(\Gamma, s_1, \cdots, s_r)$, $(\Gamma', s_1', \cdots, s_r')$ of P_i *equivalent* if there exists an isomorphism u of Γ/Φ into Γ'/Φ such that $s_i' = s_i u$, $1 \leq i \leq r$. Assume every P_i/Φ is finite dimensional and prove that Theorem 21 generalizes to composites $(\Gamma, s_1, \cdots, s_r)$ and the tensor product $P_1 \otimes P_2 \otimes \cdots \otimes P_r$.

5. Let P/Φ be finite dimensional Galois and let $(s_1, s_2, \cdots, s_r)$ be an ordered r-tuple of automorphisms of P/Φ. Then $(P, s_1, s_2, \cdots, s_r)$ is an r-fold composite of P/Φ. Show that $(s_1, \cdots, s_r)$, $(s_1', \cdots, s_r')$ determine equivalent composites if and only if $s_i' = s_i u$ is an automorphism of P/Φ. Let $\mathfrak{J}(s_1, \cdots, s_r)$ be the ideal in $P \otimes P \otimes \cdots \otimes P$ (r factors) associated with $(P, s_1, \cdots, s_r)$. Use the fact that there are $[P:\Phi]^{r-1}$ distinct ideals $\mathfrak{J}(s_1, \cdots, s_r)$ and that

$$(P \otimes \cdots \otimes P)/\mathfrak{J}(s_1, \cdots, s_r) \cong P$$

to prove that the $\mathfrak{J}(s_1, \cdots, s_r)$ are the only maximal ideals in $P^{(r)} \equiv P \otimes \cdots \otimes P$ and that every r-fold composite of P/Φ is equivalent to one of the composites $(P, s_1, s_2, \cdots, s_r)$. Note that $\mathfrak{J}(s_1, \cdots, s_r)$ is the kernel of the homomorphism of $P^{(r)}/\Phi$ into P/Φ such that

$$\rho_1 \otimes \rho_2 \otimes \cdots \otimes \rho_r \rightarrow \rho_1^{s_1} \rho_2^{s_2} \cdots \rho_r^{s_r}.$$

Chapter II

GALOIS THEORY OF EQUATIONS

In this chapter we shall consider the classical application of Galois theory: Galois' criterion for solvability by radicals of a polynomial equation $f(x) = 0$. To say that an equation is solvable by radicals means roughly that its roots can be obtained from the coefficients by rational operations and root extractions. A criterion for this was given by Galois after Abel and Ruffini had proved that the general equation of the fifth degree is not solvable by radicals. Galois was led to the development of his theory in order to give the criterion for solvability by radicals. Besides the fundamental group-field correspondence which we gave in the last chapter and whose scope goes far beyond the theory of equations, some results of a more special nature are needed. These concern cyclotomic fields, that is, fields of the roots of 1, and "pure" extensions $P = \Phi(\theta)$, $\theta^n = \alpha$ in Φ. The study of these fields is interesting also beyond the theory of equations and we shall undertake a detailed study of such fields in the next chapter. In the present one we confine ourselves to the minimum which is needed for the theory of equations.

1. The Galois group of an equation. Let Φ be a field, $f(x)$ a polynomial of positive degree in $\Phi[x]$ having leading coefficient 1. Let P/Φ be a splitting field of $f(x)$, so $P = \Phi(\rho_1, \cdots, \rho_m)$ and $f(x) = (x - \rho_1)^{e_1}(x - \rho_2)^{e_2} \cdots (x - \rho_m)^{e_m}$ in $P[x]$ where the ρ_i are distinct and the e_i are positive integers. Since the ρ_i are generators of P/Φ, any automorphism s of P/Φ is completely determined by its action on the finite set of roots $R = \{\rho_1, \rho_2, \cdots, \rho_m\}$. Also each ρ_i^s is again a root of $f(x)$, so $\rho_i^s \in R$ and the restriction s_f of s to R is a permutation of this finite set. Thus we see that every s

of the Galois group G of P/Φ defines a permutation s_f of R. The mapping $s \to s_f$ is a homomorphism of G into the symmetric group $S(R)$ of 1–1 mappings of R. Moreover, if $s \in G$ has the property that $\rho_i^s = \rho_i$, $1 \leq i \leq m$, then $s = 1$ in $P = \Phi(\rho_1, \cdots, \rho_m)$. Consequently, $s \to s_f$ is an isomorphism of G with a subgroup $G_f = \{s_f\}$ of the symmetric group $S(R)$. In view of this isomorphism we are led, in studying the equation $f(x) = 0$, to shift our attention from the group G to the permutation group G_f. Accordingly, we give the following

Definition 1. *If Φ is a field and $f(x)$ is a non-zero polynomial in $\Phi[x]$, then* the Galois group of the equation $f(x) = 0$ over Φ *is the group G_f induced by the Galois group G of a splitting field* P/Φ *in the set of roots of $f(x) = 0$ in* P.

Since any two splitting fields are isomorphic, G_f is essentially uniquely determined by Φ and $f(x)$.

It is convenient to identify the permutation $\rho_i \to \rho_i'$ of R with the permutation $i \to i'$ of $\{1, 2, \cdots, m\}$. In this way we can consider G_f as a subgroup of the symmetric group S_m of permutations of $\{1, 2, \cdots, m\}$. We shall do this from now on. Moreover, we assume in the sequel that $f(x)$ has simple roots, that is, the $e_i = 1$. This implies that P/Φ is Galois. Hence we have the fundamental correspondence between the collection of subgroups of the Galois group G and the collection of subfields E/Φ of P/Φ. Combining this with the isomorphism of G onto G_f we obtain a 1–1 correspondence between the collection of subgroups of G_f with the collection of subfields E/Φ. We shall refer to the subfield E/Φ corresponding to a subgroup H_f of G_f as the "field of invariants of H_f." In reality, of course, E/Φ is the field of invariants of the subgroup H of G corresponding to H_f. In the other direction, H_f is the Galois group of $f(x) = 0$ over the subfield E.

We recall that the symmetric group S_m contains the alternating group A_m as an invariant subgroup of index 2. A_m is the set of even permutations, that is, the set of permutations which can be written as products of an even number of transpositions (ij) (Vol. I, pp. 35–36). If G_f is the Galois group of the equation $f(x) = 0$ over Φ, then $G_f \cap A_m$ is a subgroup of index 1 or 2 in G_f. We shall now give an identification of the corresponding subfield of P/Φ, assuming the characteristic is not two (see ex. 1 below for the characteristic 2 case). The result is the following

Theorem 1. *Let* Φ *be a field of characteristic* $\neq 2$ *and* $f(x)$ *a non-zero polynomial* $\varepsilon\ \Phi[x]$ *without multiple roots. Let* P/Φ *be a splitting field of* $f(x)$, $\rho_1, \rho_2, \cdots, \rho_m$ *its roots,* G_f *the Galois group of the equation* $f(x) = 0$ *considered as a permutation group of* $\{1, 2, \cdots, m\}$. *Then the subfield of invariants of* $G_f \cap A_m$ *is* $\Phi(\Delta)$, *where*

$$(1) \qquad \Delta = \prod_{i<j=1}^{m} (\rho_i - \rho_j).$$

Proof. We recall a standard characterization of the alternating group. For this one considers the ring $\Phi[x_1, x_2, \cdots, x_m]$, x_i indeterminates. If $i \to i^\sigma$ is a permutation of $1, 2, \cdots, m$, then we have the automorphism $A(\sigma)$ of $\Phi[x_1, \cdots, x_m]$ over Φ such that $x_i^{A(\sigma)} = x_{i^\sigma}$ (Vol. I, p. 107, and Introd.). Let $X = \prod_{i<j} (x_i - x_j)$. Then $X^{A(\sigma)} = \chi(\sigma)X$ where $\chi(\sigma) = 1$ or -1 according as σ is even or odd (cf. Vol. I, ex. 2, p. 110). Now let π be the homomorphism of $\Phi[x_1, x_2, \cdots, x_m]$ over Φ into P/Φ such that $x_i^\pi = \rho_i$, $1 \leq i \leq m$. Let s be in the Galois group of P/Φ, s_f the corresponding permutation of the ρ_i. Then if we apply π to the relation $X^{A(s_f)} = \chi(s_f)X$ we obtain $\Delta^s = \chi(s_f)\Delta$ where Δ is given by (1). Since $\Delta \neq 0$ we see that $\Delta^s = \Delta$ if and only if $s_f \,\varepsilon\, G_f \cap A_m$. Thus the Galois group of the equation $f(x) = 0$ over $\Phi(\Delta)$ is $G_f \cap A_m$. Consequently, by the Galois correspondence, $\Phi(\Delta)$ is the set of invariants of $G_f \cap A_m$.

We have seen that for any $s \,\varepsilon\, G$, $\Delta^s = \pm\Delta$; hence if $\delta = \Delta^2$, then $\delta^s = \delta$ for all s and so $\delta \,\varepsilon\, \Phi$. We have seen also that the statement of the theorem is equivalent to the assertion that the Galois group of the equation $f(x) = 0$ over $\Phi(\Delta)$ is $A_m \cap G$. This implies the

Corollary. *The Galois group of* $f(x) = 0$ *over* Φ *is a subgroup of the alternating group if and only if* δ *is the square of an element of* Φ.

Proof. Clearly the condition $G_f \subseteq A_m$ or $G_f = G_f \cap A_m$ is that $\Phi(\Delta) = \Phi$. If this holds, then $\Delta \,\varepsilon\, \Phi$ and $\delta = \Delta^2$ is a square of an element of Φ. Conversely, if $\delta = \alpha^2$, $\alpha \,\varepsilon\, \Phi$, then $\delta = \Delta^2$ gives $\Delta = \pm\alpha \,\varepsilon\, \Phi$. Hence $\Phi(\Delta) = \Phi$.

We recall that, if θ is a separable algebraic element over Φ and $\theta_1 = \theta, \theta_2, \cdots, \theta_n$ are the distinct images of θ under the isomorphisms of $\Phi(\theta)$ into its normal closure, then $\delta = \prod_{i<j} (\theta_i - \theta_j)^2$ is a discriminant of the field $\Phi(\theta)/\Phi$ (cf. § 1.14). If $f(x) = (x - \rho_1)(x - \rho_2) \cdots (x - \rho_m)$, as before, then we shall call $\delta = \Delta^2 = \prod_{i<j} (\rho_i - \rho_j)^2$ the *discriminant of the polynomial* $f(x)$ *or of the equation* $f(x) = 0$. It follows from the main theorem on symmetric polynomials (Vol. I, p. 109) that δ can be expressed as a polynomial with coefficients in the prime field in the coefficients of

$$f(x) \equiv x^m - \alpha_1 x^{m-1} + \alpha_2 x^{m-2} - \cdots + (-1)^m \alpha_m. \tag{2}$$

We shall now indicate how this can be done. We begin with the Vandermonde formula:

$$\begin{vmatrix} 1 & 1 & \cdots & 1 \\ \rho_1 & \rho_2 & \cdots & \rho_m \\ \cdot & \cdot & \cdots & \cdot \\ \rho_1^{m-1} & \rho_2^{m-1} & \cdots & \rho_m^{m-1} \end{vmatrix} = \prod_{i>j} (\rho_i - \rho_j). \tag{3}$$

Squaring we get

$$\delta = \begin{vmatrix} m & \sigma_1 & \sigma_2 & \cdots & \sigma_{m-1} \\ \sigma_1 & \sigma_2 & \sigma_3 & \cdots & \sigma_m \\ \cdot & \cdot & \cdot & \cdots & \cdot \\ \sigma_{m-1} & \sigma_m & \sigma_{m+1} & \cdots & \sigma_{2m-2} \end{vmatrix} \tag{4}$$

where $\sigma_i = \rho_1^i + \rho_2^i + \cdots + \rho_m^i$. Since the power-sums can be expressed as polynomials in the α_i with coefficients in the prime field, (4) will give the same kind of expression for δ.*

We shall now carry this out for the cases $m = 2, 3$.

$m = 2$. We have $f(x) = x^2 - \alpha_1 x + \alpha_2 = (x - \rho_1)(x - \rho_2)$ so $\sigma_1 = \rho_1 + \rho_2 = \alpha_1$ and $\rho_1\rho_2 = \alpha_2$. Then $\sigma_2 = \rho_1^2 + \rho_2^2 = (\rho_1 + \rho_2)^2 - 2\rho_1\rho_2 = \alpha_1^2 - 2\alpha_2$. The formula (4) gives

$$\delta = 2\sigma_2 - \sigma_1^2 = \alpha_1^2 - 4\alpha_2. \tag{5}$$

* That this can be done follows from the fundamental theorem on symmetric polynomials (Vol. I, p. 109). Explicit recursion formulas, the so-called Newton's identities can be used to express the σ_i in terms of α_j (Vol. I, ex. 4, p. 110).

$m = 3$. Here $f(x) = x^3 - \alpha_1 x^2 + \alpha_2 x - \alpha_3 = (x - \rho_1)(x - \rho_2)(x - \rho_3)$ so $\sigma_1 = \rho_1 + \rho_2 + \rho_3 = \alpha_1$, $\rho_1\rho_2 + \rho_2\rho_3 + \rho_1\rho_3 = \alpha_2$, $\rho_1\rho_2\rho_3 = \alpha_3$. Then $\sigma_2 = \rho_1^2 + \rho_2^2 + \rho_3^2 = (\rho_1 + \rho_2 + \rho_3)^2 - 2(\rho_1\rho_2 + \rho_1\rho_3 + \rho_2\rho_3) = \alpha_1^2 - 2\alpha_2$. To calculate σ_3 and σ_4 we use the relations $\rho_k^3 = \alpha_1\rho_k^2 - \alpha_2\rho_k + \alpha_3$, $\rho_k^4 = \alpha_1\rho_k^3 - \alpha_2\rho_k^2 + \alpha_3\rho_k$. Then

$$\begin{aligned}
\sigma_3 &= \rho_1^3 + \rho_2^3 + \rho_3^3 \\
&= \alpha_1(\rho_1^2 + \rho_2^2 + \rho_3^2) - \alpha_2(\rho_1 + \rho_2 + \rho_3) + 3\alpha_3 \\
&= \alpha_1(\alpha_1^2 - 2\alpha_2) - \alpha_2\alpha_1 + 3\alpha_3 \\
&= \alpha_1^3 - 3\alpha_1\alpha_2 + 3\alpha_3 \\
\sigma_4 &= \alpha_1\sigma_3 - \alpha_2\sigma_2 + \alpha_3\sigma_1 \\
&= \alpha_1(\alpha_1^3 - 3\alpha_1\alpha_2 + 3\alpha_3) - \alpha_2(\alpha_1^2 - 2\alpha_2) + \alpha_3\alpha_1 \\
&= \alpha_1^4 - 4\alpha_1^2\alpha_2 + 4\alpha_1\alpha_3 + 2\alpha_2^2.
\end{aligned}$$

Using (4) and these formulas we obtain

$$\begin{aligned}
(6) \qquad \delta &= 3\sigma_2\sigma_4 + 2\sigma_1\sigma_2\sigma_3 - \sigma_2^3 - 3\sigma_3^2 - \sigma_1^2\sigma_4 \\
&= -4\alpha_1^3\alpha_3 + \alpha_1^2\alpha_2^2 + 18\alpha_1\alpha_2\alpha_3 - 4\alpha_2^3 - 27\alpha_3^2.
\end{aligned}$$

We obtain next a criterion on G_f as permutation group of $\{1, 2, \cdots, m\}$ that $f(x)$ be irreducible in $\Phi[x]$. This is the following

Theorem 2. *Let $f(x) \in \Phi[x]$ have no multiple roots in its splitting field* P. *Then $f(x)$ is irreducible in $\Phi[x]$ if and only if the Galois group G_f of $f(x) = 0$ over Φ is a transitive permutation group.*

Proof. We recall that a transformation group of a set M is called transitive if given any pair (x, y), $x, y \in M$ there exists a σ in the group such that $x^\sigma = y$. Suppose first that $f(x)$ is irreducible in $\Phi[x]$ and let ρ_1, ρ_2 be two of its roots in P. Since $f(x)$ is irreducible and $f(\rho_1) = 0 = f(\rho_2)$, there exists an isomorphism of $\Phi(\rho_1)/\Phi$ onto $\Phi(\rho_2)/\Phi$ mapping ρ_1 on ρ_2. This isomorphism can be extended to an automorphism s of P/Φ. Then $s \in G$ and $\rho_1^s = \rho_2$. This implies that G_f is transitive. Conversely, suppose G_f is transitive. Let $f_1(x)$ be an irreducible factor of $f(x)$ of positive degree and let ρ_1 be one of its roots. Let ρ_2 be any root of $f(x)$. Then there exists an $s \in G$ such that $\rho_1^s = \rho_2$. Then

$f_1(\rho_2) = f_1(\rho_1^s) = f_1(\rho_1)^s = 0$. This shows that every root of $f(x)$ is a root of $f_1(x)$. Hence $f(x) = f_1(x)$ is irreducible.

The two results which we have derived make it trivial to calculate the Galois groups of quadratic and cubic equations. Similar ideas can be applied to quartics. We shall look at the first two cases now and will indicate how quartics can be handled in the exercises which follow. We assume that the characteristic of Φ is not 2 and that $f(x)$ has distinct roots. If $f(x)$ is a quadratic: $f(x) = x^2 - \alpha_1 x + \alpha_2$, then the group is the symmetric group S_2 or $A_2 = 1$ according as $\delta = \alpha_1{}^2 - 4\alpha_2$ is not or is a square in Φ. Next let $f(x) = x^3 - \alpha_1 x + \alpha_2 x - \alpha_3$. If $f(x) = (x - \rho)g(x)$ in $\Phi[x]$, then the Galois group of $f(x) = 0$ is the same as that of the quadratic $g(x)$. Hence we may assume $f(x)$ irreducible in $\Phi[x]$. Since the only transitive subgroups of S_3 are S_3 and A_3, the Galois group G_f is one of these. The corollary to Theorem 1 shows that $G_f = A_3$ if $\delta = -4\alpha_1{}^3\alpha_3 + \alpha_1{}^2\alpha_2{}^2 + 18\alpha_1\alpha_2\alpha_3 - 4\alpha_2{}^3 - 27\alpha_3{}^2$ is a square in Φ. Otherwise, $G_f = S_3$.

EXERCISES

1. Assume Φ has characteristic 2 and $f(x) \varepsilon \Phi[x]$ has distinct roots $\rho_1, \rho_2, \cdots, \rho_m$ in its splitting field P. Let $\Delta' = \sum_{\sigma \varepsilon A_m} \rho_{1\sigma}{}^{n-1}\rho_{2\sigma}{}^{n-2} \cdots \rho_{(n-2)\sigma}$. Show that the subfield of invariants of $G_f \cap A_m$ is $\Phi(\Delta')$.

2. Let Φ be a finite field, $f(x)$ an irreducible polynomial of n-th degree with coefficients in Φ. Show that G_f consists of the powers of an n-cycle which may be taken to be $(123 \cdots n)$.

In the remainder of the exercises we assume the characteristic of the base field Φ is $\neq 2$, $f(x) = x^4 - \alpha_1 x^3 + \alpha_2 x^2 - \alpha_3 x + \alpha_4$ has distinct roots $\rho_1, \rho_2, \rho_3, \rho_4$ in the splitting field P/Φ, G the Galois group of P/Φ.

3. Show that the subgroup V (Klein's Vierergruppe) $= \{1, (12)(34), (13)(24), (14)(23)\}$ is invariant in S_4.

4. Show that the subfield of invariants relative to $G_f \cap V$ is $\Phi(\tau_1, \tau_2, \tau_3)$ where $\tau_1 = \rho_1\rho_2 + \rho_3\rho_4$, $\tau_2 = \rho_1\rho_3 + \rho_2\rho_4$, $\tau_3 = \rho_1\rho_4 + \rho_2\rho_3$.

5. Let $g(x) = (x - \tau_1)(x - \tau_2)(x - \tau_3)$ (the *resolvent cubic* of $f(x)$). Verify that

$$g(x) = x^3 - \beta_1 x^2 + \beta_2 x - \beta_3 \tag{7}$$

where

$$\begin{aligned} \beta_1 &= \alpha_2 \\ \beta_2 &= \alpha_1\alpha_3 - 4\alpha_4 \\ \beta_3 &= \alpha_1{}^2\alpha_4 + \alpha_3{}^2 - 4\alpha_2\alpha_4 \end{aligned} \tag{8}$$

and that $g(x)$ and $f(x)$ have the same discriminant.

6. Prove that the transitive subgroups of S_4 are (i) S_4, (ii) A_4, (iii) V, (iv) $C = \{1, (1234), (13)(24), (1432)\}$ and its conjugates, (v) $D = V \cup \{(12), (34), (1423), (1324)\}$ a Sylow 2-group (subgroup of order 8) and its conjugates.

7. Show that the Galois group G_g of $g(x) = 0$ is isomorphic to $G_f/(G_f \cap V)$. Assume $f(x)$ is irreducible and verify that, if (i) $G_f = S_4$, then G_g is of order 6, (ii) $G_f = A_4$, G_g is of order 3, (iii) $G_f = V$, $G_g = 1$, (iv) $G_f = C$ or one of the conjugates (that is, any cyclic subgroup of order 4 of S_4), then G_g is of order 2, (v) $G_f = D$ or one of its conjugates (any Sylow subgroup of order 8 in S_4), then G_g is of order 2. Note that these results identify G_f if we know G_g unless G_f is either as in (iv) or (v).

8. Prove that, if G_g is of order 2, then $G_f \cong D$ or $G_f \cong C$ according as $f(x)$ is or is not irreducible in $\Phi(\sqrt{\delta})$ where δ is the discriminant of $f(x)$.

9. Determine the Galois group of $x^4 + 3x^3 - 3x - 2 = 0$ over the field of rational numbers.

2. Pure equations. In this section we shall derive the special results which are needed for Galois' criterion. We shall formulate these in the invariant fashion in terms of splitting fields rather than, as in the last section, of the groups of equations. The results we need concern equations of the form $x^n - \alpha = 0$ (or $x^n = \alpha$) which are called *pure* (or *binomial*) equations. Occasionally, we use the notation $\rho = \sqrt[n]{\alpha}$ or $\rho = \alpha^{1/n}$ to indicate that ρ is a root of $x^n = \alpha$. We consider first the case $\alpha = 1$. The roots of $x^n = 1$ are called the n-th roots of 1 and a splitting field P of this equation is called a *cyclotomic field of order n* over Φ. The derivative $(x^n - 1)' = nx^{n-1}$ is not relatively prime to $x^n - 1$ if and only if the characteristic is $p \neq 0$ and $p \mid n$. Then we can write $n = p^e n'$, $(n', p) = 1$ and we have $x^n - 1 = (x^{n'} - 1)^{p^e}$. Hence the cyclotomic field of order n coincides with that of order n'. We shall therefore assume from now on that $p \nmid n$ in the characteristic $p \neq 0$ case.

Let P/Φ be a cyclotomic field of order n over Φ. Because of our assumption on the characteristic, the set $Z(n) = \{\zeta_i\}$ of n-th roots of 1 contains n elements. If $m \mid n$, then $\eta^m = 1$ implies $\eta^n = 1$; hence the cyclotomic field of order n contains that of order m for every divisor m of n. We observe next that $Z(n)$ is a subgroup of the multiplicative semigroup of P. This is clear since $\zeta_1^n = 1 = \zeta_2^n$ imply $(\zeta_1\zeta_2)^n = 1$, $(\zeta_1^{-1})^n = 1$. Since $Z(n)$ is finite, this is a cyclic group (Lemma 1, § 1.13). Hence there exists a $\zeta \in Z(n)$ such that $Z(n) = \{\zeta^i \mid i = 0, 1, \cdots, n-1\}$. Such a ζ is called a *primitive n-th root of* 1. Since P/Φ is generated by the ζ_i, we have $P = \Phi(\zeta)$ so ζ is a primitive element of the field P/Φ. We shall now prove

Theorem 3. *If the characteristic of Φ is not a divisor of n (0 included), then the Galois group G of the cyclotomic field P/Φ of order n is isomorphic to a subgroup of the multiplicative group $U(n)$ of units in $I/(n)$, I the ring of integers.*

Proof. As in § 1 let G_f denote the group of permutations of the set $Z(n)$ of roots induced by G. Since the elements of G_f are restrictions of automorphisms, it is clear that they are automorphisms of the multiplicative group of $Z(n)$. Hence G_f ($\cong G$) is isomorphic to a subgroup of the group of automorphisms of $Z(n)$. Now $Z(n)$ is a cyclic group of order n and it is well known that the group of automorphisms of such a group is isomorphic to $U(n)$ (Vol. I, ex. 3, p. 47, and ex. 1, p. 82). Hence the Galois group G is isomorphic to a subgroup of $U(n)$.

It is important to note that G is commutative since $U(n)$ is commutative. Moreover, we observe that, if l is a prime, then $U(l)$ is just the multiplicative group (of order $l - 1$) of the field $I/(l)$ and this is a cyclic group. Hence G which is isomorphic to a subgroup of $U(l)$ is cyclic. We therefore have the

Corollary. *If the notation is as in Theorem 3, then G is a commutative group and G is cyclic if n is a prime.*

Next we consider the Galois group of any pure equation $x^n = \alpha$ under the assumption that the base field Φ contains n distinct n-th roots of 1. We have seen that this implies that the characteristic is not a divisor of n. Then $x^n - \alpha$ is prime to its derivative nx^{n-1} if $\alpha \neq 0$ (the case $\alpha = 0$ is trivial) and so $x^n - \alpha$ has n distinct roots. We have the following

Theorem 4. *If Φ contains n distinct n-th roots of 1 then the Galois group of the equation $x^n = \alpha$ over Φ is cyclic of order a divisor of n.*

Proof. Let P/Φ be a splitting field over Φ of $x^n - \alpha$, G its Galois group. We have to show that G is cyclic. If $\alpha = 0$, we have $P = \Phi$, $G = 1$. Hence we assume $\alpha \neq 0$. Let ρ be one of the roots of $x^n - \alpha$ in P. If $Z(n) = \{\zeta_1 = 1, \zeta_2, \cdots, \zeta_n\}$ is the set of n-th roots of 1 contained in Φ, then we know that $Z(n)$ is a cyclic group under multiplication and, since $\rho \neq 0$, it is clear that $\{\rho\zeta_1, \rho\zeta_2, \cdots, \rho\zeta_n\}$ is the set of roots of $x^n - \alpha$. Evidently $P = \Phi(\rho)$; so an automorphism $s \in G$ is completely determined by its effect on ρ. We have $\rho^s = \zeta_{i(s)}\rho$ where $\zeta_{i(s)}$ is one of the ζ's $\in Z(n)$ and is uniquely determined by s. If $t \in G$ and $\rho^t = \zeta_{i(t)}\rho$, then

$$\rho^{st} = \zeta_{i(s)}\rho^t = \zeta_{i(s)}\zeta_{i(t)}\rho.$$

This shows that the mapping $s \to \zeta_{i(s)}$ is a homomorphism of G into $Z(n)$. If $\zeta_{i(s)} = 1$, we have $\rho^s = \rho$ so $s = 1$. Hence $s \to \zeta_i(s)$ is an isomorphism. Thus G is isomorphic to a subgroup of the cyclic group $Z(n)$ and the result is clear.

We shall need one more special result for the proof of Galois' criterion. This is the following converse of Theorem 4.

Theorem 5. *Assume Φ has n distinct n-th roots of* 1 *and let* P/Φ *be a cyclic n dimensional extension field. Then* $P = \Phi(\xi)$ *where* $\xi^n = \alpha \in \Phi$.

Proof. The hypothesis on P is that P/Φ is Galois with Galois group G which is cyclic of order n. Since P is separable over Φ it has a primitive element so $P = \Phi(\theta)$. Let s be a generator of G and let ξ be the "Lagrange resolvent":

$$\xi = \theta + \theta^s\zeta^{-1} + \theta^{s^2}\zeta^{-2} + \cdots + \theta^{s^{n-1}}\zeta^{-(n-1)} \tag{9}$$

where ζ is a primitive n-th root of 1. Then

$$\begin{aligned}\xi^s &= \theta^s + \theta^{s^2}\zeta^{-1} + \cdots + \theta\zeta^{-(n-1)}\\ &= \zeta(\theta + \theta^s\zeta^{-1} + \cdots + \theta^{s^{n-1}}\zeta^{-(n-1)}) = \zeta\xi.\end{aligned}$$

Then $\xi^{s^k} = \zeta^k\xi$ so ξ has n distinct conjugates and hence its minimum polynomial is of degree n. Consequently $P = \Phi(\xi)$. Set $\xi^n = \alpha$. Then $(\xi^n)^s = (\zeta\xi)^n = \xi^n$ implies that $\alpha^s = \alpha$ so $\alpha \in \Phi$ and the proof is complete.

EXERCISES

1. Let p be a prime not equal to the characteristic of the field Φ. Show that, if $\alpha \in \Phi$, then $x^p - \alpha$ is irreducible in $\Phi[x]$ or it has a root in this field.
2. Let Φ be the field of rational numbers, p a prime, and let $x^p - \alpha$ be irreducible in $\Phi[x]$. Show that the Galois group of the equation $x^p = \alpha$ over Φ is isomorphic to the group of transformations in $I/(p)$ which have the form $y \to \gamma y + \delta, \gamma \neq 0$.

3. Let Φ be a field of characteristic $p \neq 0$. Show that $x^p - x - \alpha$ is irreducible in $\Phi[x]$ unless $\alpha = \beta^p - \beta$, β in Φ. Show also that, if $x^p - x - \alpha$ is irreducible, then the group of the equation $x^p = x + \alpha$ is cyclic of order p. (Hint: Show that, if ρ is a root of $x^p - x - \alpha = 0$, then $\rho, \rho + 1, \rho + 2, \cdots, \rho + (p - 1)$are roots. Hence show that the Galois group of the equation is isomorphic to a subgroup of the additive group of $I/(p)$.)

4. Let Φ be of characteristic $p \neq 0$, P/Φ cyclic and p dimensional. Show that P can be generated over Φ by an element ξ such that $\xi^p - \xi = \alpha \varepsilon \Phi$.

3. Galois' criterion for solvability by radicals. It is essential first to have a precise formulation of the statement that an equation $f(x) = 0$ is solvable by radicals over a field Φ. We give this in the following

Definition 2. *Let Φ be a field and let $f(x) \varepsilon \Phi[x]$ be of positive degree. Then the equation $f(x) = 0$ is said to be* solvable by radicals *over Φ if the splitting field P/Φ can be imbedded in a field Σ which possesses a tower of subfields:*

$$\Phi = \Phi_1 \subseteq \Phi_2 \subseteq \Phi_3 \subseteq \cdots \subseteq \Phi_{r+1} = \Sigma \tag{10}$$

where each $\Phi_{i+1} = \Phi_i(\xi_i)$ and $\xi_i^{n_i} = \alpha_i \varepsilon \Phi_i$. A chain of fields such as (2) *is called a* root tower *for Σ/Φ.*

For the sake of simplicity we restrict our attention to fields of characteristic 0. This will avoid the complications of inseparability and some difficulties with roots of 1 in the characteristic $p \neq 0$ case. Our objective is to establish the following criterion of Galois:

An equation $f(x) = 0$ is solvable by radicals over a field Φ of characteristic 0 *if and only if its Galois group is solvable.*

We recall that a group G is defined to be solvable if it has a chain of subgroups $G = G_1 \supseteq G_2 \supseteq G_3 \supseteq \cdots \supseteq G_{r+1} = 1$ such that each G_{i+1} is invariant in G_i and G_i/G_{i+1} is commutative. Every subgroup and homomorphic image of a solvable group is solvable. Moreover, if G contains an invariant subgroup H such that H and G/H are solvable, then G is solvable. A finite group G is solvable if and only if it has a composition series $G = G_1 \supset G_2 \supset \cdots \supset G_{s+1} = 1$ whose composition factors G_i/G_{i+1} are cyclic of prime order. We recall also that the alternating group A_n, $n \geq 5$, is simple and this implies that the symmetric group S_n on n letters is not solvable if $n \geq 5$. A proof of the statement about A_n is given in Vol. I, p. 139. All of the other results which we have stated are easy consequences of the theory of normal series and most of these have been given as exercises in Vol. I, pp. 139, 143. At any rate we shall assume all of these results.

In order to prove the necessity of Galois' criterion we shall need the following

Lemma. *If Σ has a root tower over Φ of characteristic* 0, *then Σ has an extension field Ω which is finite dimensional Galois over Φ and also has a root tower over Φ.*

Proof. We are given $\Phi = \Phi_1 \subseteq \Phi_2 \subseteq \cdots \subseteq \Phi_{r+1} = \Sigma$ where $\Phi_{i+1} = \Phi_i(\xi_i)$, $\xi_i^{n_i} = \alpha_i \in \Phi_i$. We shall show that there exists a field $\Delta_i \supseteq \Sigma$ which also contains a subfield Ω_i such that 1) $\Omega_i \supseteq \Phi_i$, 2) Ω_i is Galois over Φ, 3) Ω_i has a root tower over Φ:

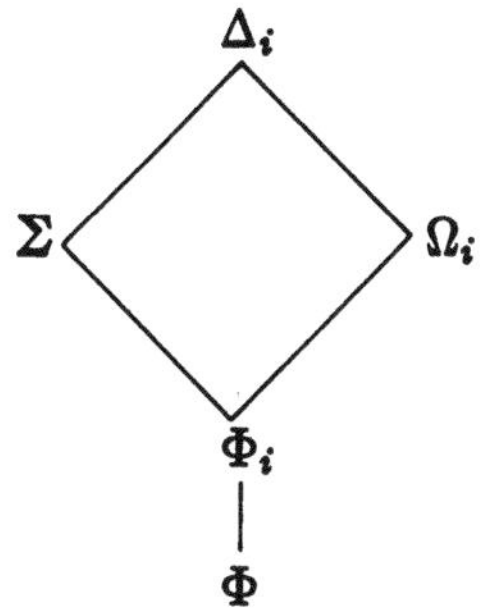

Now for $i = 1$ we take $\Delta_1 = \Sigma$, $\Omega_1 = \Phi_1$ and we suppose we are given Δ_i and Ω_i for a certain i. Let G_i be the Galois group of Ω_i over Φ and let $\alpha_i^{s_1}, \cdots, \alpha_i^{s_{k_i}}$ be the conjugates of the element α_i under the automorphisms $s_j \in G_i$. Set $g_i(x) = \prod_{j=1}^{k_i} (x^{n_i} - \alpha_i^{s_j})$. Then $g_i(x) \in \Phi[x]$. Let Δ_{i+1} be a splitting field over Δ_i of $g_i(x)$ and let $\xi_i, \xi_i', \xi_i'', \cdots$ be the roots of $g_i(x)$ in Δ_{i+1}. Note that one of these is the ξ_i such that $\Phi_{i+1} = \Phi_i(\xi_i)$ since $g_i(\xi_i) = 0$ and $\Delta_{i+1} \supseteq \Delta_i \supseteq \Sigma$. Set $\Omega_{i+1} = \Omega_i(\xi_i, \xi_i', \xi_i'', \cdots)$. Since Ω_i/Φ is a splitting field of a polynomial $f_i(x) \in \Phi[x]$, Ω_{i+1}/Φ is a splitting field of $f_i(x)g_i(x)$ and (since the characteristic is 0) Ω_{i+1} is Galois over Φ. Since $\Omega_{i+1} \supseteq \Omega_i$ and $\xi_i \in \Omega_i$, $\Omega_{i+1} \supseteq \Phi_{i+1} = \Phi_i(\xi_i)$. Let $\xi_i^{(h)}$ be any one of the elements $\xi_i, \xi_i', \xi_i'', \cdots$; then $g_i(\xi_i^{(h)}) = 0$ and $g_i(x) = \Pi(x^{n_i} - \alpha_i^{s_j})$ show that $(\xi_i^{(h)})^{n_i}$ is one of the $\alpha_i^{s_j}$.

Hence $\Omega_{i+1} = \Omega_i(\xi_i, \xi_i', \xi_i'', \cdots)$ has a root tower over Φ. This shows that Δ_{i+1} and Ω_{i+1} satisfies the conditions 1), 2), 3). We now take $\Omega = \Omega_{r+1}$ and this satisfies the conditions stated in the lemma.

Remark. Note that the integers n_i for the root tower for Ω/Φ are the same as those for the given tower for Σ/Φ.

We can now prove the necessity of Galois' condition. Thus let $f(x) = 0$ be solvable by radicals over Φ (of characteristic 0) so the splitting field P/Φ of $f(x)$ can be imbedded in a field Σ which has a root tower over Φ. By the lemma we may assume that Σ/Φ is Galois. Let n be the least common multiple of the exponents n_i occurring in a root tower for Σ and let Δ be a splitting field over Σ of $x^n - 1$ so $\Delta = \Sigma(\zeta)$ where ζ is a primitive n-th root of 1 and Δ is Galois over Φ and has a root tower over Φ. Moreover, it is clear that we can obtain a root tower for Δ which has the form:

$$(11) \qquad \begin{aligned}\Phi = \Phi_1 \subseteq \Phi_2 = \Phi_1(\zeta) \subseteq \Phi_3 \\ = \Phi_2(\xi_1) \subseteq \cdots \subseteq \Phi_{r+1}(\xi_r) = \Delta\end{aligned}$$

where $\xi_i^{n_i} \in \Phi_{i+1}$. If H is the Galois group of Δ over Φ, then the chain of subfields (11) gives rise to a decreasing chain of subgroups

$$(12) \qquad H = H_1 \supseteq H_2 \supseteq \cdots \supseteq H_{r+2} = 1$$

where H_i is the Galois group of Δ over Φ_i. By Theorem 3, Φ_2 is Galois over Φ_1 with commutative Galois group and since Φ_2 contains the necessary roots of 1, Φ_{i+1} is cyclic over Φ_i if $i \geq 2$. This implies that H_{j+1} is an invariant subgroup of H_j for $j \geq 1$. The factor group H_1/H_2 is isomorphic to the Galois group of Φ_2 over Φ_1 and so is commutative while the factor group H_i/H_{i+1}, $i \geq 2$, is isomorphic to the Galois group of Φ_{i+1} over Φ_i and so is cyclic. Thus the sequence of groups (12) shows that H is solvable. Now we have $\Delta \supseteq \text{P} \supseteq \Phi$ where P/Φ is the splitting field of $f(x)$. Hence if K is the subgroup corresponding to P, then K is invariant in H and $H/K \cong G$, the Galois group of P/Φ. Since H is solvable, this shows that G is solvable; hence the Galois group G_f of the equation $f(x) = 0$ is a solvable group.

In order to prove the sufficiency of Galois' condition we require the following result which is of independent interest.

Theorem 6. *Let* P/Φ *be finite dimensional Galois over* Φ *and let* P' *be an extension field of* P *such that* P' *is generated by* P *and a second subfield* $\Phi' \supseteq \Phi$. *Then* P'/Φ' *is finite dimensional Galois and its Galois group* G' *is isomorphic to a subgroup of the Galois group* G *of* P/Φ.

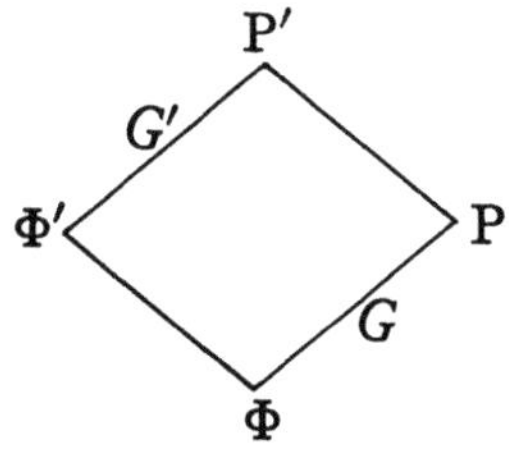

Proof. We know that $P = \Phi(\xi_1, \cdots, \xi_n)$ where the ξ_i are the roots of a separable polynomial $f(x) \in \Phi[x]$. Since P' is generated by $\Phi' \supseteq \Phi$ and P, we have $P' = \Phi'(\xi_1, \cdots, \xi_n)$. Hence P' is a splitting field over Φ' of $f(x)$. Since separability is invariant under extension of the base field, $f(x)$ is separable over Φ' and consequently P' is Galois over Φ'. Let s' belong to the Galois group G' of P'/Φ'. Then s' is the identity mapping in $\Phi \subseteq \Phi'$ and s' maps the set $R = \{\xi_1, \xi_2, \cdots, \xi_n\}$ into itself. Hence s' maps $P = \Phi(R)$ into itself and so the restriction of s' to P is an element s of the Galois group of P over Φ. The mapping $s' \to s$ is a homomorphism of G' into G. Since $s = 1$ implies that $\xi_i^s = \xi_i$, $1 \leq i \leq n$, and this implies that $s' = 1$, we see that $s' \to s$ is an isomorphism, so G' is isomorphic to a subgroup of G.

We can now give the proof of the sufficiency of Galois' condition. We assume that $f(x) = 0$ has a solvable Galois group G_f; hence the Galois group G of the splitting field P/Φ of $f(x)$ is solvable. We are assuming also that Φ is of characteristic 0. Let $n = (G:1)$ and let $P' = P(\zeta)$ where ζ is a primitive n-th root of 1. Then P' is generated by P and the subfield $\Phi' = \Phi(\zeta)$. Hence, by Theorem 6, P' is Galois over Φ' and its Galois group G' over Φ' is isomorphic to a subgroup of G. Hence G' is solvable and has a composition series $G' = G_1' \supset G_2' \supset \cdots \supset G_{s+1}' = 1$ whose composition factors G_i'/G_{i+1}' are cyclic of prime order. Evidently these orders are divisors of $n = (G:1)$. Let $\Phi' = \Phi_1' \subset \Phi_2' \subset \cdots \subset \Phi_{s+1}' = P'$ be the chain of subfields corresponding to the composition series for G'. Since G_{i+1}' is invariant in G_i' and G_i'/G_{i+1}'

is cyclic, Φ_{i+1}' is Galois over Φ_i' with cyclic Galois group whose order n_i is a divisor of n. Now Φ_i' contains a primitive n_i-th root of 1 since $\Phi_i' \supseteq \Phi' = \Phi(\zeta)$. Hence, by Theorem 5, $\Phi_{i+1}' = \Phi_i(\xi_i)$ where $\xi_i^{n_i} = \alpha_i \varepsilon \Phi_i$. Thus $\Phi_1' \subset \Phi_2' \subset \cdots \subset \Phi_{s+1}' = \mathrm{P}'$ is a root tower for P' over Φ'. Since $\Phi' = \Phi(\zeta)$, $\zeta^n = 1$, $\Phi \subseteq \Phi_1' \subset \Phi_2' \subset \cdots \subset \Phi_{s+1}' = \mathrm{P}'$ is a root tower for P' over Φ. Since $\mathrm{P}' \supseteq \mathrm{P}$, this shows that $f(x) = 0$ is solvable by radicals over Φ.

EXERCISES

1. Let P/Φ be a splitting field over Φ of characteristic 0 for $x^p - 1$, p a prime. Prove that P/Φ can be imbedded in a field Σ/Φ which has a root tower (10) for which the n_i are primes and $[\Phi_{i+1}:\Phi_i] = n_i$. Call such a root tower *normalized*. (Hint: Use induction on p and ex. 1 of § 2.)

2. Obtain normalized root tower fields over the cyclotomic fields of 5th and 7th roots of 1 over the field R_0 of rational numbers.

3. Prove that, if $f(x) = 0$ has a solvable Galois group over a field of characteristic 0, then its splitting field can be imbedded in an extension which has a normalized root tower.

4. Let Φ be of characteristic $p \neq 0$. Call an equation $f(x) = 0$, $f(x) \varepsilon \Phi[x]$, *solvable by equations* $x^p - x = \alpha$ if its splitting field P/Φ can be imbedded in a field Σ which has a tower of fields $\Phi_1 = \Phi \subseteq \Phi_2 \subseteq \cdots \subseteq \Phi_{r+1} = \Sigma$ where $\Phi_{i+1} = \Phi_i(\xi_i)$, $\xi_i^p - \xi_i = \alpha_i \varepsilon \Phi_i$. Show that, if $f(x)$ has distinct roots, then $f(x) = 0$ is solvable by equations $x^p - x = \alpha$ if and only if its Galois group is of order p^e. (Hint: Use ex. 3, 4 of § 2 and the fact that a finite group of prime power order is solvable.)

4. The general equation of n-th degree. The formula $x = (a \pm \sqrt{a^2 - 4b})/2$ for the solutions of the quadratic equation $x^2 - ax + b = 0$ (characteristic $\neq 2$) is valid if a, b are considered as indeterminates. When this is done one has a "general quadratic equation." Particular quadratic equations are obtained by specializing the coefficients. The corresponding specialization for the solutions gives the solutions of the particular equations. Similar solutions for general cubic and quartic equations by radicals are known (ex. 3, 4 below). We shall now consider the question of solvability by radicals of the general equation of n-th degree for any n.

Let Φ be a field and let $\Sigma = \Phi(t_1, t_2, \cdots, t_n)$ be the field of rational expressions in indeterminates t_i over Φ. Then the equation

$$(13) \qquad f(x) = x^n - t_1x^{n-1} + t_2x^{n-2} - \cdots + (-1)^n t_n = 0$$

is called the *general equation of the n-th degree over* Φ. We wish to determine the Galois group G_f over Σ of this equation. Let $P = \Sigma(x_1, x_2, \cdots, x_n)$ be a splitting field over Σ of $f(x)$ such that $f(x) = (x - x_1)(x - x_2) \cdots (x - x_n)$ in $P[x]$. Then

$$(14) \qquad t_1 = \Sigma x_i, t_2 = \sum_{i<j} x_i x_j, \cdots, t_n = x_1 x_2 \cdots x_n;$$

hence

$$(15) \quad P = \Sigma(x_1, x_2, \cdots, x_n) = \Phi(t_1, t_2, \cdots, t_n; x_1, \cdots, x_n) = \Phi(x_1, x_2, \cdots, x_n).$$

In order to determine G_f we consider first a simpler problem. We introduce new indeterminates $\xi_1, \xi_2, \cdots, \xi_n$ over Φ and the field $\bar{P} = \Phi(\xi_1, \xi_2, \cdots, \xi_n)$ of rational expressions in the ξ_i. Consider the polynomial

$$(16) \qquad \bar{f}(x) = (x - \xi_1)(x - \xi_2) \cdots (x - \xi_n)$$

in $\bar{P}[x]$. We have

$$(17) \qquad \bar{f}(x) = x^n - \tau_1 x^{n-1} + \tau_2 x^{n-2} - \cdots + (-1)^n \tau_n$$

where

$$(18) \qquad \tau_1 = \Sigma \xi_i, \quad \tau_2 = \sum_{i<j} \xi_i \xi_j, \cdots, \tau_n = \xi_1 \xi_2 \cdots \xi_n.$$

We now consider the subfield $\bar{\Sigma} = \Phi(\tau_1, \tau_2, \cdots, \tau_n)$ of $\bar{P}/\Phi$ and we note that the relation $\bar{P} = \bar{\Sigma}(\xi_1, \xi_2, \cdots, \xi_n)$ and (16) show that $\bar{P}$ is a splitting field over $\bar{\Sigma}$ of $\bar{f}(x)$. We assert that the Galois group $G_{\bar{f}}$ of the equation $\bar{f}(x) = 0$ over $\bar{\Sigma}$ is the symmetric group. Thus we have to show that, if $\xi_i \to \xi_{i'}$ is any permutation of the ξ_i, then there exists an $\bar{s}_{\bar{f}} \in G_{\bar{f}}$ such that $\xi_i^{\bar{s}_{\bar{f}}} = \xi_{i'}$. Now we know that we have an automorphism $\bar{s}$ of the polynomial algebra $\Phi[\xi_1, \xi_2, \cdots, \xi_n]$ over Φ such that $\xi_i^{\bar{s}} = \xi_{i'}$, $1 \leq i \leq n$. We know also that $\bar{s}$ has an extension to an automorphism $\bar{s}$ of the field $\bar{P} = \Phi(\xi_1, \xi_2, \cdots, \xi_n)$ over Φ. Finally $\bar{s}$ can be extended to an automorphism $\bar{s}$ of $\bar{P}[x]$ so that $x^{\bar{s}} = x$. Then we have $\bar{f}^{\bar{s}}(x) = (x - \xi_{1'})(x - \xi_{2'}) \cdots (x - \xi_{n'}) = \bar{f}(x)$ which, by (17), implies $\tau_i^{\bar{s}} = \tau_i$, $1 \leq i \leq n$. (This can be seen also by using the expression (18) for the τ_i.) Now $\tau_i^{\bar{s}} = \tau_i$ implies that the elements of $\bar{\Sigma} = \Phi(\tau_1, \tau_2, \cdots, \tau_n)$ are fixed under $\bar{s}$. Hence $\bar{s}$ is in the Galois group of $\bar{P}/\bar{\Sigma}$ and the induced mapping $\bar{s}_{\bar{f}}$ satisfies $\xi_i^{\bar{s}_{\bar{f}}} = \xi_{i'}$, $1 \leq i \leq n$, as required.

We shall now carry over the result we have just obtained on the pair of fields $\bar{P}$, $\bar{\Sigma}$ to the pair P, Σ by establishing an isomorphism of P onto $\bar{P}$ which maps Σ onto $\bar{\Sigma}$. We consider first the algebra homomorphism η over Φ of

$$\Phi[t_1, t_2, \cdots, t_n] \underset{\eta}{\rightarrow} \Phi[\tau_1, \tau_2, \cdots, \tau_n]$$

such that $t_i^\eta = \tau_i$, $1 \leq i \leq n$. The existence of η is clear since the t_i are indeterminates. We assert that η is an isomorphism. To see this we note that we have the homomorphism ζ over Φ of

$$\Phi[\xi_1, \xi_2, \cdots, \xi_n] \underset{\zeta}{\rightarrow} \Phi[x_1, x_2, \cdots, x_n]$$

so that $\xi_i^\zeta = x_i$. Again this is clear since the ξ_i are indeterminates. Note also that $\Phi[\xi_1, \xi_2, \cdots, \xi_n] \supseteq \Phi[\tau_1, \tau_2, \cdots, \tau_n]$ so $\eta\zeta$ is defined. Now the formulas (18) and (14) show that $\tau_i^\zeta = t_i$. Hence $t_i^{\eta\zeta} = \tau_i^\zeta = t_i$ and consequently $g^{\eta\zeta} = g$ for every g in $\Phi[t_1, t_2, \cdots, t_n]$. This implies that our first mapping η is an isomorphism since $g^\eta = 0$ gives $g = g^{\eta\zeta} = 0$ for g in $\Phi[t_1, \cdots, t_n]$.

We are now in a position to extend η to an isomorphism η of $\Sigma = \Phi(t_1, t_2, \cdots, t_n)$ onto $\bar{\Sigma} = \Phi(\tau_1, \tau_2, \cdots, \tau_n)$ and this extends to an isomorphism η of $\Sigma[x]$ onto $\bar{\Sigma}[x]$ so that $x^\eta = x$. Then

$$f(x)^\eta = (x^n - t_1x^{n-1} + \cdots)^\eta = x^n - \tau_1x^{n-1} + \cdots = \bar{f}(x).$$

On the other hand, P is a splitting field over Σ of $f(x)$ and $\bar{P}$ is a splitting field over $\bar{\Sigma}$ of $\bar{f}(x)$. Hence the general uniqueness theorem for splitting fields (Th. 1.7) provides an isomorphism η of P onto $\bar{P}$ which coincides with the given η on Σ. It is immediate from the existence of such an isomorphism that the Galois group G of P/Σ is isomorphic to the Galois group $\bar{G}$ of $\bar{P}/\bar{\Sigma}$. In fact, it is clear that the mapping $s \rightarrow \eta^{-1}s\eta$ is an isomorphism of G onto $\bar{G}$. The fact that $G_{\bar{f}} = S_n$ now implies that the Galois group G_f of $f(x) = 0$ over Σ is S_n. It is clear also that the roots of $f(x)$ are distinct and Theorem 2 shows that $f(x)$ is irreducible in $\Sigma[x]$. The results we have obtained can be stated as

Theorem 7. *The general equation of the n-th degree* (13) *is irreducible in* $\Sigma = \Phi(t_1, t_2, \cdots, t_n)$ *and has distinct roots. The Galois group of* $f(x) = 0$ *is the symmetric group* S_n.

Since S_n is not solvable if $n > 4$ this implies the

Theorem of Abel-Ruffini. *The general equation of the n-th degree is not solvable by radicals if* $n > 4$ (*characteristic* 0).

EXERCISES

1. Use the fact that every finite group is isomorphic to a subgroup of S_n to construct a field P whose Galois group over a suitable field Φ is isomorphic to a given finite group G. (The construction of P for a given Φ and G is an open problem. In fact, for Φ the field of rational numbers this is a classical problem which is still unsolved.)

2. Use the Galois theory to prove that, if $r(x_1, x_2, \cdots, x_n) \in \Phi(x_1, x_2, \cdots, x_n)$ the field of rational expressions in indeterminates x_i over the field Φ and r is symmetric in the sense that $r(x_{1'}, x_{2'}, \cdots, x_{n'}) = r(x_1, x_2, \cdots, x_n)$ for every permutation $x_i \to x_{i'}$ of the x's, then r is a rational expression with coefficients in Φ in the elementary symmetric polynomials $t_1 = \Sigma x_i$, $t_2 = \sum_{i<j} x_i x_j, \cdots, t_n = x_1 x_2 \cdots x_n$. (Compare the fundamental theorem on symmetric polynomials, Vol. I, p. 109.)

3. Assume the characteristic of Φ is not two or three and consider the general cubic $x^3 - t_1 x^2 + t_2 x - t_3 = (x - x_1)(x - x_2)(x - x_3)$. Here the t_i are indeterminates and the x_i are in a splitting field P over $\Sigma = \Phi(t_1, t_2, t_3)$. Nothing is lost in replacing x_i by $y_i = x_i - \frac{1}{3}(x_1 + x_2 + x_3) = x_i - \frac{1}{3}t_1$. Then the given equation is replaced by $y^3 + py + q = 0$ whose roots are y_1, y_2, y_3 where $y_1 + y_2 + y_3 = 0$. Then the formula (6) for the discriminant gives $\delta = -4p^3 - 27q^2$. The group of P over $\Sigma(\sqrt{\delta})$ is the alternating group A_3 which is cyclic of order 3. Let ζ be a primitive cube root of 1 (e.g., $\zeta = -\frac{1}{2} + \frac{1}{2}\sqrt{-3}$) and set $z_1 = y_1 + \zeta^{-1}y_2 + \zeta^{-2}y_3 = y_1 + \zeta^2 y_2 + \zeta y_3$, $z_2 = y_1 + \zeta^{-2}y_2 + \zeta^{-4}y_3 = y_1 + \zeta y_2 + \zeta^2 y_3$, $z_3 = y_1 + y_2 + y_3 = 0$. Verify that $z_1^3 = -\frac{27}{2}q - \frac{3}{2}\sqrt{-3}\sqrt{\delta}$ if $\zeta = -\frac{1}{2} + \frac{1}{2}\sqrt{-3}$ and $z_2^3 = -\frac{27}{2}q + \frac{3}{2}\sqrt{-3}\sqrt{\delta}$, $z_1 z_2 = -3p$. Hence

$$
\begin{aligned}
z_1 &= \sqrt[3]{-\tfrac{27}{2}q - \tfrac{3}{2}\sqrt{-3\delta}} \\
z_2 &= \sqrt[3]{-\tfrac{27}{2}q + \tfrac{3}{2}\sqrt{-3\delta}}
\end{aligned}
\tag{19}
$$

where the determination of $\sqrt{-3\delta}$ is the same in both formulas and that of $\sqrt[3]{\ }$ is such that $z_1 z_2 = -3p$. Solve the equations $z_1 = y_1 + \zeta^2 y_2 + \zeta y_3$, $z_2 = y_1 + \zeta y_2 + \zeta^2 y_3$, $z_3 = y_1 + y_2 + y_3$ for y_1, y_2, y_3 to obtain Cardan's solution of the equation $y^3 + py + q = 0$.

4. Assume the characteristic is not two or three and consider the general quartic $x^4 - t_1 x^3 + t_2 x^2 - t_3 x + t_4 = (x - x_1)(x - x_2)(x - x_3)(x - x_4)$. Replacing x_i by $y_i = x_i - \frac{1}{4}t_1$ gives an equation $f(y) \equiv y^4 + py^2 + qy + r = 0$ whose roots are y_1, y_2, y_3, y_4. Show that the resolvent cubic of $f(y) = 0$ is $g(z) = z^3 - 2pz^2 + (p^2 - 4r)z + q^2 = 0$ (cf. the exercises in § 1). Show that the Galois group of $P = \Phi(x_1, x_2, x_3, x_4) = \Phi(y_1, y_2, y_3, y_4)$ over $\Phi(z_1, z_2, z_3)$, z_i the roots of $g(z) = 0$ is the Vierergruppe. Obtain formulas for y_1, y_2, y_3, y_4 in terms of z_1, z_2, z_3 and square roots of elements of $\Phi(z_1, z_2, z_3)$.

5. Consider a splitting field P over $\Sigma = \Phi(t_1, \cdots, t_n)$, t_i indeterminates, of the general equation (13) and let $x_1, x_2, \cdots, x_n$ be the roots. Assume Φ contains n distinct elements $c_1, c_2, \cdots, c_n$. Prove that $\theta = c_1 x_1 + c_2 x_2 + \cdots + c_n x_n$ is a primitive element of P/Σ.

5. Equations with rational coefficients and symmetric group as Galois group. The theorem of Abel-Ruffini shows that equations of degree ≥ 5 with indeterminate coefficients are not solvable by radicals. On the other hand, it is clear that for certain fields Φ, e.g., the field of real numbers or the field of complex numbers, every equation with coefficients in Φ is solvable by radicals. We shall now show that there exist equations with rational coefficients which are not solvable by radicals. We shall do this by showing that there exist rational equations of any prime degree p with Galois group the symmetric group S_p. We prove first the following result on permutation groups.

Lemma. *If G is a permutation group on p elements where p is a prime and G contains an element of order p and a transposition, then $G = S_p$.*

Proof. We recall that any permutation can be written as a product of disjoint cycles (Vol. I, p. 35). Moreover, the order of a cycle is the number of letters it contains. This implies that, if $\sigma \varepsilon G$ has order p, then σ is a cycle containing all the letters 1, 2, $\cdots$, p. By re-ordering the elements 1, 2, $\cdots$ suitably, we may assume that G contains the transposition (12). Since a suitable power of the p-cycle σ has the form $(12 \cdots)$ further re-ordering of the elements 12, $\cdots$, p, if necessary, permits us to assume that G contains (12) and $\sigma = (123 \cdots p)$. We recall that, if τ is any element of S_p (or S_n), then $\tau^{-1}(ij)\tau = (i^\tau j^\tau)$ where i^τ, j^τ are the images of i, j respectively under τ. This shows that $\sigma^{-1}(12)\sigma = (23)$, $\sigma^{-2}(12)\sigma^2 = (34)$, $\cdots$, $(p-1, p)$ and $(p1)$ are contained in G. Since

$$(13) = (12)(23)(12)$$

$$(14) = (13)(34)(13)$$

$$\cdot \qquad \cdot$$

$$\cdot \qquad \cdot$$

$$(1p) = (1\ p-1)(p-1\ p)(1\ p-1)$$

all of these elements are contained in G. Since $(ij) = (1i)(1j)(1i)$ if $1, i, j$ are different, this shows that every transposition is contained in G. Since every element of S_p is a product of transpositions, we have $G = S_p$.

We shall now prove the following

Theorem 8. *Let $f(x)$ be a polynomial of prime degree with rational coefficients which is irreducible in the rational field. Suppose $f(x) = 0$ has exactly two non-real roots in the field C of complex numbers. Then the group G_f of $f(x) = 0$ over the rationals is the symmetric group.*

Proof. The fundamental theorem of algebra asserts that $f(x) = (x - \rho_1)(x - \rho_2) \cdots (x - \rho_p)$ in $C[x]$. Then the subfield $P = R_0(\rho_1, \rho_2, \cdots, \rho_p)$, R_0 the rationals, of C is a splitting field of $f(x)$ over R_0. Since $P \supseteq R_0(\rho_1)$ and $[R_0(\rho_1):R_0] = \deg f(x) = p$, $[P:R_0]$ is divisible by p. Hence p is a divisor of $(G:1)$, G the Galois group of P over R_0. It follows from Sylow's theorem that G contains an element of order p. Now consider the automorphism $a = \alpha + \beta\sqrt{-1} \to \alpha - \beta\sqrt{-1} = \bar{a}$, α, β real, of C over the field of real numbers. This maps $f(x)$ into itself since the coefficients of $f(x)$ are real. Hence it maps the set $\{\rho_1, \rho_2, \cdots, \rho_p\}$ of the roots of $f(x)$ belonging to C into itself. Let ρ_1, ρ_2 be the non-real roots of $f(x)$. Then $a \to \bar{a}$ interchanges ρ_1 and ρ_2 and leaves fixed all the ρ_i, $i > 2$. Thus the restriction of the automorphism $a \to \bar{a}$ of C to the set of roots is an element of G_f which is a transposition. Hence G_f contains an element of order p and a transposition and $G_f = S_p$ by the lemma.

We shall now indicate how one can construct polynomials satisfying the conditions of the theorem.* Let m be a positive integer, $n_1 < n_2 < \cdots < n_{r-2}$ be $r - 2$ even integers where r is odd and >3. Consider the polynomial

$$g(x) = (x^2 + m)(x - n_1)(x - n_2) \cdots (x - n_{r-2}). \tag{20}$$

The real roots of $g(x)$ are $n_1, n_2, \cdots, n_{r-2}$ and the graph of $y = g(x)$ has the form:

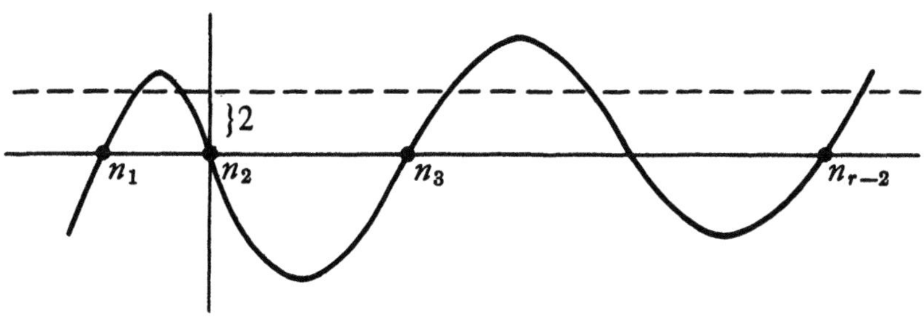

* The construction we give is due to R. Brauer.

This has $(r - 3)/2$ relative maxima and, since $|g(k)| > 2$ for any odd integer k, it is clear that the values of these relative maxima are >2. This implies that $f(x) = g(x) - 2$ has $(r - 3)/2$ positive relative maxima between n_1 and n_{r-2}. It follows that $f(x)$ has $r - 3$ real roots in the interval (n_1, n_{r-2}). Since $f(n_{r-2}) = -2$ and $f(\infty) = \infty$, there is also a real root $>n_{r-2}$. This gives $r - 2$ real roots for $f(x)$. Let $\alpha_1, \alpha_2, \cdots, \alpha_r$ be the complex roots of $f(x)$. Then $f(x) = \Pi(x - \alpha_i) = (x^2 + m)(x - n_1) \cdots (x - n_{r-2}) - 2$, and equating coefficients of x^{r-1} and x^{r-2}, we obtain

$$(21) \qquad \sum_1^r \alpha_i = \sum_1^{r-2} n_k, \quad \sum_{i<j} \alpha_i\alpha_j = \sum_{k<l} n_k n_l + m.$$

Hence

$$(22) \qquad \begin{aligned} \Sigma\alpha_i^2 &= (\Sigma\alpha_i)^2 - 2\sum_{i<j}\alpha_i\alpha_j \\ &= \Sigma n_k^2 - 2m. \end{aligned}$$

If we choose m sufficiently large, (22) shows that $\Sigma\alpha_i^2 < 0$ and this implies that not every α_i is real. If α_1 is a non-real root, then $\bar{\alpha}_1 \neq \alpha_1$ is another such root so we have at least two non-real roots. Since in any case we have $r - 2$ real roots, we see that $f(x)$ has exactly $r - 2$ real roots. We now write $f(x) = x^r + a_1x^{r-1} + \cdots + a_r$. Clearly the a_i are even integers. Moreover, since the constant term of $g(x)$ is divisible by 4, that of $f(x) = g(x) - 2$ is not divisible by 4. It follows by Eisenstein's criterion applied to the prime $q = 2$ that $f(x)$ is irreducible in the rational field. We therefore see that we can satisfy the conditions of the theorem for every prime $p = r \geq 5$. It is easy to see that this holds also for $p = 2, 3$. Hence the conditions hold for every prime, so we see that there exist rational equations of every prime degree p with Galois group the symmetric group S_p.

EXERCISES

1. Let $f(x) \in \Phi[x]$ have distinct roots $\rho_1, \rho_2, \cdots, \rho_n$ in a splitting field P/Φ and let $G_f \subseteq S_n$ be the Galois group of the equation f. Let $y_1, y_2, \cdots, y_n$ be indeterminates and set

$$\begin{aligned} F(x) &= \prod_{t \in S_n} (x - (\rho_{1^t}y_1 + \rho_{2^t}y_2 + \cdots + \rho_{n^t}y_n)) \\ &= \prod_{t \in S_n} (x - (\rho_1 y_{1^t} + \rho_2 y_{2^t} + \cdots + \rho_n y_{n^t})). \end{aligned}$$

Show that $F(x) \varepsilon \Phi[y_1, y_2, \cdots, y_n, x]$. Let $F(x) = F_1(x)F_2(x) \cdots F_r(x)$ be the factorization of $F(x)$ into irreducible factors with leading coefficient 1 in $\Phi(y_1, y_2, \cdots, y_n)[x]$. Show that, if $x - \sum_i \rho_{i^t} y_i$ is a factor of $F_1(x)$, then

$$F_1(x) = \prod_{s \varepsilon G_f} (x - \sum_i \rho_{i^{ts}})$$

Hence show that $\deg F_i(x) = (G_f:1)$.

2. Same notations as 1. Assume, moreover, that $\Phi = R_0$ the field of rational numbers and that $f(x)$ has integer coefficients and leading coefficient 1. Assume p is a prime such that the polynomial $\bar{f}(x)$ obtained by replacing the coefficients of $f(x)$ by their residues modulo p has distinct roots in a splitting field $\bar{P}/I_p$. Show that $\bar{F}(x) = \prod_{t \varepsilon S_n} (x - \bar{\rho}_i y_{i^t})$ in $\bar{P}[x, y_1, \cdots, y_n]$ where $\bar{\rho}_1, \bar{\rho}_2, \cdots, \bar{\rho}_n$ are the roots of $\bar{f}(x)$ in $\bar{P}$. Use this and ex. 1 to prove that, if the $\bar{\rho}_i$ are suitably ordered, then $G_{\bar{f}}$ is a subgroup of G_f.

3. Show that any transitive subgroup of S_n which contains an $(n-1)$-cycle and a transposition coincides with S_n.

4. Show that the equation

$$x^6 + 22x^5 - 9x^4 + 12x^3 - 37x^2 - 29x - 15$$

over R_0 has the group S_6. (Hint: Apply ex. 2 using the primes $p = 2, 3, 5$.)

Chapter III

ABELIAN EXTENSIONS

In this chapter we shall investigate several types of abelian extension fields. First, we shall consider cyclotomic fields over the field of rational numbers and we shall determine their dimensionalities and Galois groups. Next we shall consider Kummer extensions, which are obtained by adjoining the roots of a finite number of pure equations $x^m = \alpha$ to a field containing m distinct m-th roots of 1. Finally, we shall study the so-called abelian p-extensions, which are defined to be abelian extensions of p^f dimensions of a field of characteristic $p \neq 0$. The theory of characters of finite commutative groups is a basic tool for the investigation of Kummer extensions and abelian p-extensions. Besides this, our study of abelian p-extensions will be based on a certain type of ring, a ring of Witt vectors which can be constructed from any commutative algebra $\mathfrak{A}$ over a field of characteristic $p \neq 0$. For any such $\mathfrak{A}$ and integer $m = 1, 2, \cdots$, we have a ring of Witt vectors $\mathfrak{W}_m(\mathfrak{A})$ of characteristic p^m. In the theory of valuations it is useful to pass to the limit as $m \to \infty$ and to consider also rings $\mathfrak{W}(\mathfrak{A})$ of infinite Witt vectors. This will be considered in Chapter V. A number of the results of this chapter will be needed for an application to the theory of formally real fields which we shall take up in Chapter VI.

1. Cyclotomic fields over the rationals. We have defined the cyclotomic field of order m over a field Φ to be the splitting field over Φ of the polynomial $x^m - 1$ (§ 2.2). We have shown that, if the characteristic of Φ is not a divisor of m, then the Galois group of the cyclotomic field is isomorphic to a subgroup of the group $U(m)$ of units in the ring $I/(m)$ (Th. 2.3). We now assume that the base field $\Phi = R_0$, the field of rational numbers, and we let

$P^{(m)}$ denote the cyclotomic field of the m-th roots of 1 over R_0. Let $Z(m)$ be the multiplicative group of the m-th roots of 1. We recall that $Z(m)$ is cyclic and its generators are called primitive m-th roots of 1. Also $P^{(m)} = R_0(\zeta)$, where ζ is any primitive m-th root of 1; hence the dimensionality $[P^{(m)}:R_0]$ is the degree of the minimum polynomial of ζ over R_0. If ζ is a primitive m-th root of 1, then any other primitive m-th root of 1 has the form ζ^k where $(k, m) = 1$. Hence the number of primitive m-th roots of 1 is $\varphi(m)$ the number of positive integers not exceeding m which are relatively prime to m. This is also the order of the group $U(m)$.

Now let

$$\lambda_m(x) = \prod_{\zeta \text{ primitive}} (x - \zeta). \tag{1}$$

This is a polynomial of degree $\varphi(m)$ with coefficients in $P^{(m)}$. If s is in the Galois group G of $P^{(m)}$ over R_0, then clearly s maps the set of primitive m-th roots of 1 into itself. Hence we have the relation $\lambda_m{}^s(x) = \lambda_m(x)$ for every $s \in G$. Since $P^{(m)}$ is Galois over R_0, we see that $\lambda_m(x) \in R_0[x]$, that is, $\lambda_m(x)$ has rational coefficients. We can see this also in a more elementary way which, at the same time, gives an inductive procedure for calculating $\lambda_m(x)$. Since the order of any m-th root of 1 is a divisor of m and since every d-th root of 1 for $d|m$ is an m-th root of 1, we clearly have the formula

$$x^m - 1 = \prod_{\substack{d|m \\ 1 \leq d \leq m}} \lambda_d(x). \tag{2}$$

Evidently we have $\lambda_1(x) = x - 1$ and, assuming that $\lambda_d(x) \in R_0[x]$ for all d such that $1 \leq d < m$, then the formula (2) gives

$$\lambda_m(x) = (x^m - 1)/ \prod_{d|m} \lambda_d(x) \tag{3}$$

which shows that $\lambda_m(x) \in R_0[x]$. This gives a practical way of calculating $\lambda_m(x)$. For example, we have $\lambda_1(x) = x - 1$,

$$\lambda_2(x) = (x^2 - 1)/\lambda_1(x) = x + 1,$$

$$\lambda_3(x) = (x^3 - 1)/\lambda_1(x) = x^2 + x + 1,$$

$$\lambda_4(x) = (x^4 - 1)/\lambda_1(x)\lambda_2(x) = x^2 + 1,$$

$$\lambda_6(x) = (x^6 - 1)/\lambda_1(x)\lambda_2(x)\lambda_3(x) = x^2 - x + 1$$

and

$$\lambda_{12}(x) = (x^{12} - 1)/\lambda_1(x)\lambda_2(x)\lambda_3(x)\lambda_4(x)\lambda_6(x) = x^4 - x^2 + 1.$$

If p is a prime, we have

$$(4) \qquad \lambda_p(x) = (x^p - 1)/(x - 1) = x^{p-1} + x^{p-2} + \cdots + 1$$

and it is easy to see, using Eisenstein's criterion, that $\lambda_p(x)$ is irreducible in $R_0[x]$ (Vol. I, ex. 2, p. 127). We shall now prove the following general result

Theorem 1. *$\lambda_m(x)$ is irreducible in the rational field.*

Proof. We observe first that $\lambda_m(x)$ has integer coefficients. For, assuming this holds for every $\lambda_d(x)$, $d < m$, and setting $p(x) = \prod_{\substack{d|m \\ 1 \leq d < m}} \lambda_d(x)$, we obtain by the usual division algorithm that $x^m - 1 = p(x)q(x) + r(x)$ where $q(x)$ and $r(x) \in I[x]$ and $\deg r(x) < \deg p(x)$. On the other hand, we have $x^m - 1 = p(x)\lambda_m(x)$, so by the uniqueness of the quotient and remainder, $\lambda_m(x) = q(x)$ has integer coefficients. Now suppose that $\lambda_m(x) = h(x)k(x)$ where $h(x)$ is irreducible in $R_0[x]$ and $\deg h(x) \geq 1$. By Gauss' lemma (Vol. I, p. 125) we may assume that $h(x)$ and $k(x)$ have integer coefficients and leading coefficients 1. Let p be a prime integer such that $p \nmid m$ and let ζ be a root of $h(x)$. We shall show that ζ^p is a root of $h(x)$. Since $(p, m) = 1$, ζ^p is a primitive m-th root of 1 and, if ζ^p is not a root of $h(x)$, ζ^p is a root of $k(x)$; consequently ζ is a root of $k(x^p)$. Since $h(x)$ is irreducible in $R_0[x]$ and has ζ as a root, $h(x) | k(x^p)$. It follows (as above) that $k(x^p) = h(x)l(x)$, where $l(x)$ has integer coefficients and leading coefficient 1. Also we have $x^m - 1 = \lambda_m(x)p(x) = h(x)k(x)p(x)$ and all of these polynomials have integer coefficients and leading coefficients 1. We now pass to congruences modulo p or, what is the same thing, to relations in the polynomial ring $I_p[x]$. Then we obtain

$$(5) \qquad x^m - \bar{1} = \bar{h}(x)\bar{k}(x)\bar{p}(x)$$

where in general, if $f(x) = a_0x^n + a_1x^{n-1} + \cdots + a_n \in I[x]$, then $\bar{f}(x) = \bar{a}_0x^n + \bar{a}_1x^{n-1} + \cdots + \bar{a}_n$, $\bar{a}_i = a_i + (p)$ in I_p. Similarly, we have $\bar{k}(x^p) = \bar{h}(x)\bar{l}(x)$. On the other hand, using $\bar{a}^p = \bar{a}$ for every integer a, we see that

$$\bar{f}(x)^p = (\bar{a}_0x^n + \cdots + \bar{a}_n)^p = \bar{a}_0{}^px^{pn} + \cdots + \bar{a}_n{}^p$$
$$= \bar{a}_0x^{pn} + \cdots + \bar{a}_n = \bar{f}(x^p)$$

for any polynomial $f(x)$. Hence $\bar{k}(x)^p = \bar{k}(x^p) = \bar{h}(x)\bar{l}(x)$ which implies that $(\bar{h}(x), \bar{k}(x)) \neq 1$. Then (5) shows that $x^m - \bar{1}$ has multiple roots in its splitting field over I_p. Since $p \nmid m$ this is impossible and so we have proved that ζ^p is a root of $h(x)$ for every prime p satisfying $p \nmid m$. A repetition of this process shows that ζ^r is a root of $h(x)$ for every integer r prime to m. Since any primitive m-th root of 1 has the form ζ^r, $(r, m) = 1$ we see that every primitive m-th root of 1 is a root of $h(x)$. Hence $h(x) = \lambda_m(x)$ and $\lambda_m(x)$ is irreducible in $R_0[x]$.

We now see that $\lambda_m(x)$ is the minimum polynomial over R_0 of any primitive m-th root of 1. Since $\mathrm{P}^{(m)} = R_0(\zeta)$, ζ primitive we have established the formula

$$[\mathrm{P}^{(m)}:R_0] = \varphi(m). \tag{6}$$

This implies that $(G:1) = \varphi(m)$ for the Galois group G of $\mathrm{P}^{(m)}/R_0$. Since $(U(m):1) = \varphi(m)$ and G is isomorphic to a subgroup of $U(m)$, this proves

Theorem 2. *Let $\mathrm{P}^{(m)}$ be the cyclotomic field of order m over the rationals R_0. Then the Galois group of $\mathrm{P}^{(m)}/R_0$ is isomorphic to $U(m)$, the multiplicative group of units in the ring $I/(m)$.*

We shall now proceed to determine the structure of the Galois group G or, what is the same thing, that of $U(m)$. It is easy to see that, if $m = p_1{}^{e_1}p_2{}^{e_2} \cdots p_r{}^{e_r}$ where the p_i are distinct primes, then $U(m)$ is isomorphic to the direct product of the $U(p_i{}^{e_i})$. For this reason we shall confine our attention to the case $m = p^e$ a prime power. Then $U(p^e)$ is a commutative group of order $\varphi(p^e) = p^e - p^{e-1} = p^{e-1}(p - 1)$. We prove first

Theorem 3. *If p is an odd prime, then the multiplicative group $U(p^e)$ of units in $I/(p^e)$ is cyclic.*

Proof. Since the order of this group is $p^{e-1}(p - 1)$, $U(p^e)$ is a direct product of its subgroup H of order p^{e-1} consisting of the elements which satisfy $x^{p^{e-1}} = 1$ and the subgroup K of order $p - 1$ of the elements satisfying $x^{p-1} = 1$. It suffices to show that both H and K are cyclic since the direct product of cyclic groups having relatively prime orders is cyclic. If $e = 1$, then $U(p) = K$ is the multiplicative group of the field $I/(p)$ and this is cyclic. Hence we can choose an integer a such that $a + (p)$, $a^2 + (p)$, $\cdots$, $a^{p-1} + (p)$ are distinct in $I/(p)$. Set $b = a^{p^{e-1}}$. Since $(a, p) = 1$, $(b, p^e) = 1$ and $b + (p^e)$ and $a + (p^e) \varepsilon U(p^e)$. Also $b^{p-1} = (a^{p^{e-1}})^{p-1} = a^{\varphi(p^e)} \equiv 1 \pmod{p^e}$ so $b + (p^e) \varepsilon K$. Since $b = a^{p^{e-1}} \equiv a \pmod{p}$, $b + (p)$, $b^2 + (p)$, $\cdots$, $b^{p-1} + (p)$ are distinct. Hence also $b + (p^e)$, $b^2 + (p^e)$, $\cdots$, $b^{p-1} + (p^e)$ are distinct. This implies that the order of $b + (p^e)$ is precisely $p - 1$. Since $(K:1) = p - 1$, it follows that K is cyclic with generator $b + (p^e)$. It remains to prove that H is cyclic, and we may assume that $e \geq 2$, since, otherwise, $H = (1)$ and the result is clear. Assuming $e \geq 2$, we can conclude that H is a direct product of $k \geq 1$ cyclic groups of order p^{e_i}, $e_i \geq 1$. Then the number of solutions of the equation $x^p = 1$, $x \varepsilon H$ is p^k. Hence it will be enough to show that the number of integers n, $0 < n < p^e$, satisfying $n^p \equiv 1 \pmod{p^e}$ does not exceed p. Now if n satisfies these conditions, then, since $n^p \equiv n \pmod{p}$, we have $n \equiv 1 \pmod{p}$. Then if $n \not\equiv 1$, we may write $n = 1 + yp^f + zp^{f+1}$ where $1 \leq f \leq e - 1$, $0 < y < p$, and z is a non-negative integer. Then

$$
\begin{aligned}
n^p &= 1 + \binom{p}{1}(y + zp)p^f + \binom{p}{2}(y + zp)^2 p^{2f} \\
&\quad + \cdots + (y + zp)^p p^{pf} \\
&\equiv 1 + yp^{f+1} \pmod{p^{f+2}}.
\end{aligned}
$$

If $n^p \equiv 1 \pmod{p^e}$ and $f < e - 1$, this gives $yp^{f+1} \equiv 0 \pmod{p^{f+2}}$ so $y \equiv 0 \pmod{p}$ contrary to $0 < y < p$. Hence we see that, if $1 < n < p^e$ satisfies $n^p \equiv 1 \pmod{p^e}$, then $n = 1 + yp^{e-1}$, $0 < y < p$. This gives altogether at most p solutions including 1 and completes the proof of the theorem.

We consider next the case of the prime 2 in the following

Theorem 4. *$U(2)$ and $U(4)$ are cyclic and, if $e \geq 3$, then $U(2^e)$ is a direct product of a cyclic group of order 2 and one of order 2^{e-2}.*

Proof. The order of $U(2^e)$ is $\varphi(2^e) = 2^{e-1}$. If $e = 1$, $(U(2):1) = 1$ and if $e = 2$, $U(2^e) = U(4)$ has only two elements and so is cyclic. Suppose $e \geq 3$. We show first that there are four distinct elements $x \in U(2^e)$ satisfying $x^2 = 1$. This will imply that $U(2^e)$ is a direct product of at least two distinct cyclic groups $\neq 1$. Set $a_1 = 1$, $a_2 = -1$, $a_3 = 1 + 2^{e-1}$, $a_4 = -1 + 2^{e-1}$, $x_i = a_i + (2^e)$. Then the x_i are distinct and satisfy $x_i^2 = 1$, which proves our assertion. Also since $U(2^e)$ is a direct product of at least two cyclic groups $\neq 1$ and the order of $U(2^e)$ is 2^{e-1}, we see that, if $x \in U(2^e)$, then $x^{2^{e-2}} = 1$ or, what is the same thing, if a is an odd integer, then $a^{2^{e-2}} \equiv 1 \pmod{2^e}$. The proof will be completed by displaying an x such that $x^{2^{e-3}} \neq 1$. Then we shall have a cyclic subgroup of order 2^{e-2} and this can happen only if $U(2^e)$ is a direct product of a cyclic group of this order and one of order 2. We proceed to show that we may take $x = 5 + (2^e)$. Note first that, if $e = 3$, then $5^{2^{e-3}} = 5 \not\equiv 1 \pmod{2^e}$ but $5^{2^{e-3}} \equiv 1 \pmod{2^{e-1}}$. Now let $f \geq 3$ and let $k(f)$ be the largest integer k such that $5^{2^{f-3}} \equiv 1 \pmod{2^k}$. Then we have $k(3) = 2$. Also for any $f \geq 3$ we have $5^{2^{f-3}} = 1 + y2^{k(f)}$ where y is odd. This gives

$$5^{2^{(f+1)-3}} = (5^{2^{f-3}})^2 = 1 + y2^{k(f)+1} + y^2 2^{2k(f)}$$

which shows first that $k(f+1) \geq k(f)$, so $k(f) \geq 2$ if $f \geq 3$. Then the relation shows that $5^{2^{(f+1)-3}} = 1 + z2^{k(f)+1}$ where $z = y + 2^{k(f)-1}y^2$ is odd. Hence $k(f+1) = k(f) + 1$. This and $k(3) = 2$ imply that $k(f) = f - 1$ for all $f \geq 3$. Thus $5^{2^{e-3}} \not\equiv 1 \pmod{2^e}$ if $e \geq 3$ which is what we needed. This completes the proof.

Theorems 2, 3, and 4 give a description of the Galois group of the field of the p^e-th roots of 1 over the rationals. The result is the following

Theorem 5. *Let $m = p^e$, p a prime, and let $P^{(m)}$ be the field of the m-th roots of 1 over the field R_0 of rational numbers. Then the Galois group G of $P^{(m)}/R_0$ is cyclic unless $p = 2$ and $e \geq 3$, in which case G is a direct product of a cyclic group of order 2 and one of order 2^{e-2}.*

EXERCISES

1. Use the Möbius inversion formula (Vol. I, ex. 5, p. 120) to prove that

$$\lambda_m(x) = \prod_{d|m} (x^d - 1)^{\mu\left(\frac{m}{d}\right)}.$$

2. Let p be a prime and let $P^{(p)}$ be the cyclotomic field of p-th roots of 1 over the field R_0 of rational numbers. Let $g + (p)$ be a generator of the cyclic group $U(p)$ and let s be the automorphism of $P^{(p)}/R_0$ such that $\zeta^s = \zeta^g$, ζ a fixed primitive p-th root of 1. Show that $(\zeta, \zeta^s, \zeta^{s^2}, \cdots, \zeta^{s^{p-2}})$ form a basis (normal basis) for $P^{(p)}/R_0$. Suppose $p - 1 = ef$, e, f positive integers and let E/R_0 be the subfield of e dimensions of $P^{(p)}/R_0$. Show that, if $t = s^e$ and $\eta = \zeta + \zeta^t + \zeta^{t^2} + \cdots + \zeta^{t^{f-1}}$, then $(\eta, \eta^s, \cdots, \eta^{s^{e-1}})$ is a basis for E/R_0. Show that the multiplication table for this basis has integer coefficients.

3. Let P be the field of the 17-th roots of 1 over R_0. Determine the subfields R_i, $i = 1, 2, 3$ such that $R_0 \subset R_1 \subset R_2 \subset R_3 \subset R_4 = P$ and $[R_i:R_{i-1}] = 2$. Find an element ω_i in R_i so that $R_i = R_{i-1}(\omega_i)$, $\omega_i^2 \in R_{i-1}$, $1 \leq i \leq 4$.

4. (O. Todd) Let P be the field of p-th roots of 1 over R_0 where p is a prime of the form $4n + 3$. Show that P is a tensor product of a quadratic subfield and a subfield of odd dimensionality. Show that the quadratic subfield is not real (if P is considered as a subfield of the field of complex numbers).

5. Let $E^{(m)}$ be the cyclotomic field of degree m over $\Phi_0 = I_p$. Write $m = m'p^e$, $(m', p) = 1$. Show that $[E^{(m)}:\Phi_0]$ is the order of the element $p + (m')$ in the group $U(m')$.

2. Characters of finite commutative groups. In the remainder of this chapter we shall study two classes of abelian extension fields: Kummer extensions and abelian extensions of p^e dimensions over a field of characteristic p. For both of these the theory of characters of finite commutative groups is basic, so we shall develop this first.

Let A and B be two commutative groups (written multiplicatively) and let χ and ψ be homomorphisms of A into B. We define the product $\chi\psi$ by $a^{\chi\psi} = a^\chi a^\psi$. One checks that this is again a homomorphism and that the set Hom (A, B) of all the homomorphisms of A into B is a commutative group under the product $\chi\psi$ (cf. Vol. I, p. 78). We shall be interested particularly in the case A finite and $B = Z$ a finite cyclic group whose order is divisible by the orders of all the elements of A. We shall call the maximum order of the elements of A the *exponent* of A. We recall that the order of every element is a divisor of the exponent (Vol. II, ex. 1, p. 69), so the condition we have imposed on Z is equivalent to: the order of Z is divisible by the exponent of A.

We wish to determine Hom (A, Z) from a particular decomposition of A as $A = A_1 \times A_2 \times \cdots \times A_r$ where the A_i are cyclic subgroups. Thus we are assuming that $A = A_1 \cdots A_r$ and $A_i \cap A_1 \cdots A_{i-1}A_{i+1} \cdots A_r = 1$. Let $n_i = (A_i:1)$ and let C_i be the subgroup of Z of elements z satisfying $z^{n_i} = 1$. Since n_i

is a divisor of the order of Z, C_i is the subgroup of order n_i of Z. Let C be the group of r-tuples $(c_1, c_2, \cdots, c_r)$ where $c_i \varepsilon C_i$ and multiplication is defined componentwise. Hence C is the (external) direct product of the groups $C_1, C_2, \cdots, C_r$ and $C \cong A$ (cf. Vol. I, p. 144). We shall now obtain an isomorphism of Hom (A, Z) onto C. For this purpose we choose a generator a_i of A_i, $i = 1, 2, \cdots, r$. If $\chi \varepsilon$ Hom (A, Z), then $a_i^{\chi} = c_i$ satisfies $c_i^{n_i} = 1$, since $a_i^{n_i} = 1$. Thus $c_i \varepsilon C_i$. We now map χ into the element $(c_1, c_2, \cdots, c_r) \equiv (a_1^{\chi}, a_2^{\chi}, \cdots, a_r^{\chi}) \varepsilon C$. If $\chi, \psi \varepsilon$ Hom (A, Z), then

$$\begin{aligned}(a_1^{\chi\psi}, \cdots, a_r^{\chi\psi}) &= (a_1^{\chi}a_1^{\psi}, \cdots, a_r^{\chi}a_r^{\psi}) \\ &= (a_1^{\chi}, \cdots, a_r^{\chi})(a_1^{\psi}, \cdots, a_r^{\psi}),\end{aligned}$$

so $\chi \rightarrow (a_1^{\chi}, \cdots, a_r^{\chi})$ is a homomorphism of Hom (A, Z) into C. If $a_i^{\chi} = 1$, $i = 1, \cdots, r$, then $a^{\chi} = 1$ for every $a \varepsilon A$, since the a_i are generators of A. Thus $a_i^{\chi} = 1$ for all i implies $\chi = 1$, which shows that $\chi \rightarrow (a_1^{\chi}, \cdots, a_r^{\chi})$ is an isomorphism into C. It remains to show that this mapping is surjective. Let c_i be any element of C_i. Then $c_i^{n_i} = 1$ and it is clear that we have a homomorphism χ_i of A_i onto C_i such that $a_i^{\chi_i} = c_i$. Since $A = A_1 \times A_2 \times \cdots \times A_r$ the mapping $x_1x_2 \cdots x_r \rightarrow x_1^{\chi_1}x_2^{\chi_2} \cdots x_r^{\chi_r}$, $x_i \varepsilon A_i$, is a homomorphism χ of A into Z. Clearly $\chi \rightarrow (a_1^{\chi}, \cdots, a_r^{\chi}) = (c_1, c_2, \cdots, c_r)$. This shows that the mapping of Hom (A, Z) into C is surjective. Thus we have shown that $A \cong C \cong$ Hom (A, Z).

Theorem 6. *Let A be a finite commutative group and let Z be a finite cyclic group whose order is divisible by the exponent of A. Then the group* Hom (A, Z) *is isomorphic to A.*

If Z satisfies the condition of the theorem, then we shall call the group Hom (A, Z) a *character group* of the group A and we shall call the elements of this group *characters* of A.

We are now in a position to derive in quick succession the results on characters which we need. We note first the following

Corollary 1. *If $a \neq 1$ in A, then there exists a character $\chi \varepsilon$* Hom (A, Z) *such that $a^{\chi} \neq 1$.*

Proof. Let B be the subgroup of A of elements b such that $b^{\chi} = 1$ for all $\chi \varepsilon$ Hom (A, Z). Then we see immediately that our assertion will follow if we can show that $B = 1$. Now let $\chi \in$ Hom (A, Z). Since $b^\chi = 1$, $b \in B$, B is in the kernel of χ and so we have an induced homomorphism $\bar{\chi}$ of A/B into Z defined by $(aB)^{\bar{\chi}} = a^\chi$. If $\chi, \psi \in$ Hom (A, Z) and $\bar{\chi} = \bar{\psi}$, then the definition shows that $\chi = \psi$. Hence the mapping $\chi \to \bar{\chi}$ of Hom (A, Z) into Hom $(A/B, Z)$ is 1–1. Since $(A:1) =$ (Hom $(A, Z):1)$ and $(A/B:1) =$ (Hom $(A/B, Z):1)$, by Theorem 6, we must have equality of all of these numbers. This implies that $B = 1$, which is what we needed.

If a is a fixed element of A, then we can define a mapping η_a of Hom (A, Z) into Z by $\chi^{\eta_a} = a^\chi$. If $\chi, \psi \in$ Hom (A, Z) we have $(\chi\psi)^{\eta_a} = a^{\chi\psi} = a^\chi a^\psi = \chi^{\eta_a}\psi^{\eta_a}$, which shows that η_a is a homomorphism of Hom (A, Z) into Z. Thus η_a is a character of the group Hom (A, Z). Then we have the basic

Corollary 2. *For $a \in A$ define a mapping η_a of* Hom (A, Z) *into Z by $\chi^{\eta_a} = a^\chi$. Then $\eta_a \in$* Hom (Hom $(A, Z), Z)$ *and the mapping $a \to \eta_a$ is an isomorphism of A onto* Hom (Hom $(A, Z), Z)$.

Proof. Observe first that $a \to \eta_a$ is a homomorphism since $\chi^{\eta_{ab}} = (ab)^\chi = a^\chi b^\chi = \chi^{\eta_a}\chi^{\eta_b} = \chi^{\eta_a\eta_b}$ (the last equation by the definition of the product in a character group). Next suppose $\eta_a = 1$. Then $a^\chi = 1$ for all χ so, by Cor. 1, $a = 1$. This shows that the kernel of the homomorphism $a \to \eta_a$ is the identity. Hence the mapping is an isomorphism. Since $(A:1) =$ (Hom $(A, Z):1) =$ (Hom (Hom $(A, Z), Z):1)$, by Theorem 6, $a \to \eta_a$ is surjective and the proof is complete.

Corollary 2 permits us to identify A with the character group (relative to Z) of Hom (A, Z). By virtue of this result we have a perfect duality between A and Hom (A, Z). We use this in the proof of

Corollary 3. *A set $\{\chi_1, \chi_2, \cdots, \chi_r\}$ of characters generate the character group* Hom (A, Z) *if and only if the only $a \in A$ satisfying $a^{\chi_i} = 1$, $i = 1, 2, \cdots, r$ is $a = 1$.*

Proof. This is equivalent to the dual statement $\{a_1, a_2, \cdots, a_r\}$ generate A if and only if $a_i^\chi = 1$, for $i = 1, 2, \cdots, r$ holds only for the character 1. This is easy; for, if $a_1, \cdots, a_r$ generate A and $a_i^\chi = 1$ holds for the character χ, then $a^\chi = 1$ holds since χ is a homomorphism. This implies that $\chi = 1$. On the other hand, if the subgroup B generated by $a_1, \cdots, a_r$ is a proper subgroup, then there exists a character $\bar{\chi} \neq 1$ for A/B. If $a \in A$ the mapping defined by $a \to aB \to (aB)^{\bar{\chi}}$ is an element $\neq 1$ of Hom (A, Z) satisfying $a_i^{\chi} = 1, i = 1, \cdots, r$.

3. Kummer extensions. It is generally a difficult problem to obtain a survey of the abelian extensions of a given field Φ. For example, if Φ is the field of rational numbers, this requires deep arithmetic considerations. However, there are two types of abelian extensions which can be quite exhaustively studied by comparatively elementary algebraic means. One of these, which we shall call abelian p-extensions, are the abelian extensions of p^e dimensions over a field of characteristic $p \neq 0$. We shall consider these in § 5. In the present section we shall develop the theory of Kummer extensions, which are defined as follows.

Definition 1. *Let* P *be an abelian extension of a field* Φ. *Then* P/Φ *is called a* Kummer m-extension *if the Galois group of* P/Φ *is of exponent* m *and* Φ *contains* m *distinct* m*-th roots of* 1.

We shall now suppose that Φ is a given field which contains m distinct m-th roots of 1. *The field* Φ *and the integer* m *will be fixed throughout our discussion.* We are interested in obtaining a survey of the Kummer m'-extensions P/Φ where $m'|m$. We recall that the condition that Φ contain m distinct m-th roots of 1 implies that the characteristic is not a divisor of m (§ 2.2). If P/Φ is a Kummer m'-extension where $m'|m$, then $[\mathrm{P}:\Phi] = (G:1)$ and, since the exponent and order of a finite commutative group are divisible by the same primes, we see that the characteristic is not a divisor of $[\mathrm{P}:\Phi]$.

Let Φ and m be as indicated and let P/Φ be a Kummer m'-extension, $m'|m$. Let P^* and Φ^* be the multiplicative groups of non-zero elements of P and Φ respectively. For $\rho \in \mathrm{P}^*$, the mapping $\rho \to \rho^m$ is an endomorphism of P^* which maps Φ^* into itself. The kernel of $\rho \to \rho^m$ is $Z(m)$ the group of order m of m-th roots of 1 and $Z(m) \subseteq \Phi^*$. Let

$$M(\mathrm{P}) = \{\rho \in \mathrm{P}^* | \rho^m \in \Phi^*\} \tag{7}$$

$$N(\mathrm{P}) = \{\rho^m | \rho \in M(\mathrm{P})\}. \tag{8}$$

Thus $M(P)$ consists of the m-th roots in P of the elements of Φ^* and $N(P)$ is the set of elements of Φ^* which are m-th powers of elements of P. It is clear that $M(P)$ is a subgroup of P^* containing Φ^* and $N(P)$ is a subgroup of Φ^* containing $\Phi^{*m} = \{\alpha^m | \alpha \in \Phi^*\}$.

Let $\rho \in M(P)$ and set $\chi_\rho(s) = \rho^s\rho^{-1}$, $s \in G$. Since $\rho^m = \alpha \in \Phi$, $(\rho^s)^m = \alpha$ so $\rho^s\rho^{-1} \in Z(m)$. Moreover, since $Z(m) \subseteq \Phi$,

$$\chi_\rho(st) = \rho^{st}\rho^{-1} = (\rho^s\rho^{-1})^t(\rho^t\rho^{-1}) = \chi_\rho(s)\chi_\rho(t).$$

Thus we see that $\chi_\rho \in \text{Hom}(G, Z)$, $Z = Z(m)$, which is a character group of the finite commutative group G since the exponent of G is a divisor of m. Conversely, let χ be any element of Hom (G, Z). Then we have $\chi(st) = \chi(s)\chi(t) = \chi(s)^t\chi(t)$, so Noether's equations are satisfied. Consequently, by Noether's theorem (Th. 1.19), there exists a non-zero element $\rho \in P$ such that $\chi(s) = \rho^s\rho^{-1}$. Since $\rho^s\rho^{-1} \in Z$ we have $(\rho^s)^m = \rho^m$ or $(\rho^m)^s = \rho^m$ for every $s \in G$. This implies that $\rho^m \in \Phi$ and so $\rho \in M(P)$. We have therefore shown that every element of the character group Hom (G, Z) is of the form $\chi(s) = \rho^s\rho^{-1}$, ρ in $M(P)$. If $\rho_1, \rho_2 \in M(P)$ and $\chi_{\rho_1}, \chi_{\rho_2}$ are the corresponding characters of G, then $\chi_{\rho_1\rho_2}(s) = (\rho_1\rho_2)^s(\rho_1\rho_2)^{-1} = \rho_1{}^s\rho_1{}^{-1}\rho_2{}^s\rho_2{}^{-1} = \chi_{\rho_1}(s)\chi_{\rho_2}(s)$. Hence the mapping $\rho \to \chi_\rho(s)$ is a homomorphism of $M(P)$ onto Hom (G, Z). The kernel of this homomorphism is the set of elements $\rho \in M(P)$ such that $\rho^s\rho^{-1} = 1$, $s \in G$. This is just the set of elements satisfying $\rho^s = \rho$, $s \in G$, $\rho \neq 0$ and so it is Φ^*.

It is convenient to state the result which we have just obtained on the homomorphism of $M(P)$ onto Hom (G, Z) as a result on exact sequences of group homomorphisms. If $G_1, G_2, \cdots, G_k$ are groups and η_i is a homomorphism of G_i into G_{i+1}, then we say that the sequence

$$G_1 \xrightarrow[\eta_1]{} G_2 \xrightarrow[\eta_2]{} \cdots \to G_{k-1} \xrightarrow[\eta_{k-1}]{} G_k$$

is *exact* if for each $i = 1, 2, \cdots, k-2$ the image of G_i under η_i coincides with the kernel of η_{i+1}. If 1 denotes the group consisting of 1 alone then the only homomorphism of 1 into any group G is $1 \to 1$. It follows from this and the definition of exactness that $1 \to G_1 \xrightarrow[\eta]{} G_2$ is exact if and only if η is 1–1 and $G_1 \xrightarrow[\eta]{} G_2 \to 1$ is exact if and only if η is surjective.

Using this terminology we can state the following theorem.

Theorem 7. *Let* Φ *be a field containing* m *distinct* m*-th roots of* 1 *and let* P/Φ *be a Kummer* m'*-extension where* $m'|m$. *Let* $M(P)$ *be defined by* (7) *where* P^* *is the multiplicative group of* P *and* Φ^* *is the multiplicative group of* Φ. *Then we have the exact sequence of multiplicative groups*

$$1 \to \Phi^* \to M(P) \to \text{Hom}\,(G, Z) \to 1$$

where the homomorphism of Φ^* *is the inclusion mapping and that of* $M(P)$ *is* $\rho \to \chi_\rho$, $\chi_\rho(s) = \rho^s\rho^{-1}$. *The factor group* $M(P)/P^*$ *is finite and isomorphic to* G. *We have* $P = \Phi(M(P))$ *and* $P = \Phi(\rho_1, \rho_2, \cdots, \rho_r)$, ρ_i *in* $M(P)$, *if and only if the cosets* $\rho_i\Phi^*$ *generate* $M(P)/\Phi^*$.

Proof. The first statement on the exactness of the displayed sequence means that Φ^* is the kernel of the mapping $\rho \to \chi_\rho$ and this mapping is surjective on Hom (G, Z). Both of these facts were established above. Consequently, we have Hom $(G, Z) \cong M(P)/\Phi^*$. Since Hom $(G, Z) \cong G$, by Theorem 6, we have $M(P)/\Phi^* \cong G$. This proves the second statement. Now let $\rho_1, \cdots, \rho_r$ be elements of $M(P)$ such that the cosets $\rho_i\Phi^*$ generate the finite group $M(P)/\Phi^*$. Clearly the homomorphism $\rho \to \chi_\rho$ of $M(P)$ gives the isomorphism $\rho\Phi^* \to \chi_\rho$ of $M(P)/\Phi^*$ onto Hom (G, Z). Hence we see that the characters χ_{ρ_i} generate Hom (G, Z). Now let $P' = \Phi(\rho_1, \rho_2, \cdots, \rho_r)$ and let H be the subgroup of G corresponding to P' (the Galois group of P/P'). If $t \in H$, we have $\rho_i^t = \rho_i$, $1 \leq i \leq r$, so $\chi_{\rho_i}(t) = 1$. This implies that $\chi(t) = 1$ for every $\chi \in$ Hom (G, Z). It follows from Corollary 1 to Theorem 6 that $t = 1$. Thus $H = 1$ which implies that $P' = \Phi(\rho_1, \cdots, \rho_r) = P$ and $P = \Phi(M(P))$. Conversely, let $\rho_1, \cdots, \rho_r \in M(P)$ satisfy $\Phi(\rho_1, \cdots, \rho_r) = P$ and let $s \in G$. Then $\rho_i^s = \rho_i$, $1 \leq i \leq r$, will imply that $\rho^s = \rho$, $\rho \in P$. Hence we see that $\chi_{\rho_i}(s) = 1$, $1 \leq i \leq r$, implies $s = 1$. Then Corollary 3 to Theorem 6 implies that the χ_{ρ_i} generate Hom (G, Z). In particular, if $\rho \in M(P)$, then $\chi_\rho = \chi_{\rho_1}{}^{k_1}\chi_{\rho_2}{}^{k_2}\cdots\chi_{\rho_r}{}^{k_r}$. Thus for every $s \in G$ we have

$$\rho^s\rho^{-1} = (\rho_1{}^s\rho_1{}^{-1})^{k_1}(\rho_2{}^s\rho_2{}^{-1})^{k_2}\cdots(\rho_r{}^s\rho_r{}^{-1})^{k_r}.$$

Hence $(\rho\rho_1{}^{-k_1}\rho_2{}^{-k_2}\cdots\rho_r{}^{-k_r})^s = \rho\rho_1{}^{-k_1}\rho_2{}^{-k_2}\cdots\rho_r{}^{-k_r}$, $s \in G$. It follows that $\rho = \beta\rho_1^{k_1} \cdots \rho_r^{k_r}$, $\beta \in \Phi^*$. Since ρ was any element of $M(\mathrm{P})$, this shows that the cosets $\rho_i\Phi^*$ generate $M(\mathrm{P})/\Phi^*$. This completes the proof.

Next we consider the mapping $\rho \to \rho^m\Phi^{*m}$ of $M(\mathrm{P})$ onto the factor group $N(\mathrm{P})/\Phi^{*m}$. This is a homomorphism whose kernel is the set of elements of $M(\mathrm{P})$ such that $\rho^m = \alpha^m$ where $\alpha \in \Phi^*$. Then $\rho = \zeta\alpha$ where $\zeta^m = 1$. Since $Z \subseteq \Phi^*$, these are just the elements of Φ^*. Hence we have the isomorphism $\rho\Phi^* \to \rho^m\Phi^{*m}$ of $M(\mathrm{P})/\Phi^*$ onto $N(\mathrm{P})/\Phi^{*m}$. Since $M(\mathrm{P})/\Phi^*$ is isomorphic to the Galois group of P/Φ, it is clear that $N(\mathrm{P})/\Phi^{*m}$ is a finite subgroup of Φ^*/Φ^{*m} and we have

$$N(\mathrm{P})/\Phi^{*m} \cong M(\mathrm{P})/\Phi^* \cong G. \tag{9}$$

We shall now shift our attention to the subgroup $N(\mathrm{P})$ of Φ^*. This satisfies the two conditions: $N(\mathrm{P}) \supseteq \Phi^{*m}$ and $N(\mathrm{P})/\Phi^{*m}$ is finite. We shall see that these subgroups, which are defined by Φ and m, can be used to give a survey of the Kummer extensions P/Φ. We observe first that, if $\alpha_1, \alpha_2, \cdots, \alpha_r$ are elements of $N(\mathrm{P})$ such that the cosets $\alpha_i\Phi^{*m}$ generate $N(\mathrm{P})/\Phi^{*m}$, then P/Φ is the splitting field of

$$f(x) = (x^m - \alpha_1)(x^m - \alpha_2) \cdots (x^m - \alpha_r). \tag{10}$$

For, we have $\rho_i^m = \alpha_i$ where $\rho_i \in M(\mathrm{P})$ and the isomorphism $\rho\Phi^* \to \rho^m\Phi^{*m}$ of $M(\mathrm{P})/\Phi^*$ with $N(\mathrm{P})/\Phi^{*m}$ implies that the cosets $\rho_i\Phi^*$ generate $M(\mathrm{P})/\Phi^*$. Hence, by Theorem 7, $\mathrm{P} = \Phi(\rho_1, \cdots, \rho_r)$. If $Z = \{\zeta_j\}$, then the roots of $f(x)$ are $\rho_i\zeta_j$, so we see that $\mathrm{P} = \Phi(\rho_i\zeta_j)$ is a splitting field over Φ of $f(x)$.

We proceed to show next that any subgroup N of Φ^* satisfying the stated conditions arises from a Kummer extension. The precise result is the following

Theorem 8. *Let Φ be a field containing m distinct m-th roots of* 1 *and let N be a subgroup of Φ^* containing Φ^{*m} such that N/Φ^{*m} is finite. Then there exists a Kummer m'-extension P/Φ with $m' | m$ such that $N(\mathrm{P}) = N$ where $N(\mathrm{P})$ and $M(\mathrm{P})$ are defined by* (8) *and* (7).

Proof. The foregoing analysis of Kummer extensions gives the clue to the definition of P/Φ. In view of this, we are led to choose

$\alpha_1, \alpha_2, \cdots, \alpha_r$ in the given group N so that the cosets $\alpha_i\Phi^{*m}$ generate N/Φ^{*m}. Let P/Φ be a splitting field of the polynomial $f(x)$ given in (10). Since $x^m - \alpha_i$ has m distinct roots, $f(x)$ is separable and P/Φ is finite dimensional Galois. Let G be the Galois group. If ρ_i is a root of $x^m - \alpha_i$, then all the roots of this polynomial are the elements $\rho_i\zeta_j$, ζ_j in the group Z of m-th roots of 1. Hence if $s \varepsilon G$, then $\rho_i^s = \zeta_i(s)\rho_i$, $\zeta_i(s) \varepsilon Z$. If $s, t \varepsilon G$, we have $\rho_i^{st} = (\zeta_i(s)\rho_i)^t = \zeta_i(s)\rho_i^t = \zeta_i(s)\zeta_i(t)\rho_i$. Hence $\rho_i^{st} = \rho_i^{ts}$, $i = 1, 2, \cdots, r$, and since it is clear that $P = \Phi(\rho_1, \rho_2, \cdots, \rho_r)$, $st = ts$ for all $s, t \varepsilon G$. This shows that G is a commutative group. Also we have $\rho_i^{s^k} = \zeta_i(s)^k\rho_i$, $k = 1, 2, \cdots$, and consequently $\rho_i^{s^m} = \rho_i$ which implies that $s^m = 1$, $s \varepsilon G$. Thus the exponent of G is a divisor of m and P/Φ is a Kummer m'-extension since $Z \subseteq \Phi$. It remains to show that, if $N(P)$ is defined by (8), then $N(P) = N$. Since $\rho_i^m = \alpha_i \varepsilon \Phi$, $\rho_i \varepsilon M(P)$ defined by (7). Since $P = \Phi(\rho_1, \cdots, \rho_r)$, Theorem 7 shows that the cosets $\rho_i\Phi^*$ generate $M(P)/\Phi^*$. Applying the isomorphism of $M(P)/\Phi^*$ with $N(P)/\Phi^{*m}$ we see that the cosets $\alpha_i\Phi^{*m}$ generate $N(P)/\Phi^{*m}$. On the other hand, we know that the cosets $\alpha_i\Phi^{*m}$ generate N/Φ^{*m}. This implies that $N(P) = N$

We now consider two Kummer m_i-extensions P_i/Φ, $i = 1, 2$, where $m_i | m$. It is clear from the definitions of $M(P_i)$, $N(P_i)$ that, if $P_1/\Phi \cong P_2/\Phi$, then the subgroups $N(P_1)$ and $N(P_2)$ of Φ^* coincide. Conversely, suppose we have $N(P_1) = N(P_2)$. We have seen that, if $\alpha_1, \alpha_2, \cdots, \alpha_r$ are elements of $N(P_i)$ such that the cosets $\alpha_i\Phi^{*m}$ generate $N(P_i)/\Phi^{*m}$, then P_i is a splitting field over Φ of $f(x) = (x^m - \alpha_1)(x^m - \alpha_2) \cdots (x^m - \alpha_r)$. The uniqueness of splitting fields implies that $P_1/\Phi \cong P_2/\Phi$ if $N(P_1) = N(P_2)$. Next we look at the Kummer m'-extensions, P/Φ, $m' | m$, which are contained in one extension field Ω/Φ (e.g., the algebraic closure of Φ in the sense of § 4.1). We have seen that, if P_i/Φ is one of our extensions, then $P_i = \Phi(M(P_i))$. Hence it is clear that $P_1 \supseteq P_2$ if and only if $M(P_1) \supseteq M(P_2)$. Also it is clear that $M(P_1) \supseteq M(P_2)$ if and only if $N(P_1) \supseteq N(P_2)$. Hence $P_1 \supseteq P_2$ if and only if the $N(P_1) \supseteq N(P_2)$. It is apparent that our results give a completely satisfactory internal description of the Kummer extensions P/Φ by means of the subgroups N of Φ^* satisfying the two conditions: $N \supseteq \Phi^{*m}$, N/Φ^{*m} is finite.

EXERCISES

1. Show that there exist an infinite number of non-isomorphic quadratic extensions of the field of rational numbers.

2. Assume Φ contains m distinct m-th roots of 1 and let P/Φ be cyclic of m dimensions over Φ, s a generator of the Galois group of P/Φ. Show that $P = \Phi(\rho)$ where $\rho^m = \alpha \varepsilon \Phi$ and $\rho^s = \zeta\rho$ where ζ is a primitive m-th root of 1. Show that, if $\sigma \varepsilon P$ satisfies $\sigma^m \varepsilon \Phi$, then $\sigma = \beta\rho^k$ where $\beta \varepsilon \Phi$ and $1 \leq k \leq m$.

3. (Albert). Let P be cyclic of $n = l^e$ dimensions over Φ where l is a prime and Φ contains l distinct l-th roots of 1. Let s be a generator of the Galois group G of P/Φ, H the subgroup of order l of G generated by $t = s^m$, $m = l^{e-1}$, E the subfield of H-invariants, so E/Φ is the unique subfield of m dimensions in P/Φ. By 2, $P = E(\rho)$ where $\rho^l = \alpha \varepsilon E$ and $\rho^t = \zeta\rho$, ζ a primitive l-th root of 1. Show that $\rho^s = \beta\rho^k$ where $\beta \varepsilon E$ and $1 \leq k < l$. Show that $\rho^t = \gamma\rho^{k^m} = \zeta\rho$ where $\gamma \varepsilon E$ and hence that $k^m \equiv 1 \pmod{l}$ and $k = 1$. Show that $N_{E|\Phi}(\beta) = \zeta$ and $\alpha^s\alpha^{-1} = \beta^l$.

4. (Albert). Assume Φ has l distinct l-th roots of 1, l a prime, and that E/Φ is cyclic of $m = l^{e-1}$ dimensions over Φ, $e > 1$. Suppose E contains an element β such that $N_{E|\Phi}(\beta) = \zeta$ a primitive l-th root of 1. Show that there exists an $\alpha \varepsilon E$ such that $\alpha^s\alpha^{-1} = \beta^l$ where s is a generator of the Galois group of E/Φ. Show that α is not an l-th power in E so that, if $P = E(\rho)$ with $\rho^l = \alpha$, then $[P:E] = l$. Show that P is cyclic of l^e dimensions over Φ.

5. Note that ex. 3 and 4 imply the following: If Φ contains l distinct l-th roots of 1, l a prime, and E/Φ is cyclic of $l^f > 1$ dimensions, then E/Φ can be imbedded in an extension P/Φ which is cyclic of l^{f+1} dimensions if and only if the primitive l-th root of 1, ζ is a norm of an element of E. Use this to prove that, if Φ is of characteristic $\neq 2$, the quadratic extension $E = \Phi(\epsilon)$, $\epsilon^2 = \gamma \varepsilon \Phi$, can be imbedded in a quartic cyclic extension of Φ if and only if γ is a sum of two squares of elements of Φ. In particular, show that, if R_0 is the field of rational numbers, then an imaginary quadratic extension $R_0(\epsilon)$, $\epsilon^2 = \gamma < 0$ in R_0 cannot be imbedded in a cyclic quartic extension.

6. (O. Todd). Let P be the field of p-th roots of 1 over R_0 where p is a prime of the form $4n + 1$. Show that P contains a real quadratic subfield.

7. Assume Φ contains four distinct fourth roots of 1. Show that any quadratic extension E/Φ can be imbedded in a cyclic quartic extension P/Φ.

4. Witt vectors. We have defined abelian p-extensions of a field Φ of characteristic $p \neq 0$ to be abelian extensions of p^e dimensions of Φ. Cyclic p-extensions of dimensionality p and p^2 were encountered first by Artin and Schreier in connection with a problem on real fields (see § 6.9). Their construction was generalized by Albert to give an inductive construction of cyclic p-extensions of p^e dimensions. Slightly later Witt gave a direct construction and survey of abelian p-extensions along the lines of the theory of Kummer extensions which we have just considered. Witt's method is based on an ingenious definition of a ring of vectors defined by a given field of characteristic p. This construction has important application in other connections (e.g., valuation theory) and we shall consider it now in its general form.

We shall begin first with the polynomial ring $\mathfrak{X} = R_0[x_i, y_j, z_k]$ in indeterminates x_i, y_j, z_k, $i, j, k = 0, 1, \cdots, m-1$, over the field R_0 of rational numbers. Let $\mathfrak{X}^{(m)}$ be set of m-tuples $(a_0, a_1, \cdots, a_{m-1})$, $a_i \in \mathfrak{X}$, with the usual definition of equality and with addition and multiplication by components. If $a = (a_0, \cdots, a_{m-1})$, $b = (b_0, \cdots, b_{m-1})$, then we denote the sum and product by $a \oplus b$, $a \odot b$, so that $a \oplus b = (a_0 + b_0, \cdots, a_{m-1} + b_{m-1})$, $a \odot b = (a_0 b_0, \cdots, a_{m-1} b_{m-1})$. Let p be a fixed prime number. We use this to define a mapping φ in $\mathfrak{X}^{(m)}$ by the rule that, if $a = (a_0, a_1, \cdots, a_{m-1})$, then $a^\varphi = (a^{(0)}, a^{(1)}, \cdots, a^{(m-1)})$ where

$$(11) \quad a^{(\nu)} = a_0^{p^\nu} + p a_1^{p^{\nu-1}} + \cdots + p^\nu a_\nu, \; \nu = 0, 1, \cdots, m-1.$$

Thus $a^{(0)} = a_0$, $a^{(1)} = a_0^p + p a_1, \cdots$. We introduce also the mapping $P: a \to a^P = (a_0^p, a_1^p, \cdots, a_{m-1}^p)$. Then the definition (11) gives

$$(12) \qquad a^{(0)} = a_0, \quad a^{(\nu)} = (a^P)^{(\nu-1)} + p^\nu a_\nu, \quad \nu \geq 1.$$

Next let $A = (a^{(0)}, a^{(1)}, \cdots, a^{(m-1)})$ be arbitrary and define a mapping ψ by $A^\psi = (a_0, a_1, \cdots, a_{m-1})$ where

$$(13) \quad \begin{aligned} a_0 &= a^{(0)}, \\ a_\nu &= \frac{1}{p^\nu}\left(a^{(\nu)} - a_0^{p^\nu} - p a_1^{p^{\nu-1}} - \cdots - p^{\nu-1} a_{\nu-1}\right), \quad \nu \geq 1. \end{aligned}$$

One checks directly that $a^{\varphi\psi} = a$, $A^{\psi\varphi} = A$, which shows that φ is 1–1 surjective with ψ as its inverse.

We shall now use the mapping φ and $\psi = \varphi^{-1}$ to define a new addition and multiplication composition in $\mathfrak{X}^{(m)}$. These are respectively

$$(14) \quad \begin{aligned} a + b &= (a^\varphi \oplus b^\varphi)^{\varphi^{-1}} \\ ab &= (a^\varphi \odot b^\varphi)^{\varphi^{-1}}. \end{aligned}$$

These provide another ring structure in $\mathfrak{X}^{(m)}$ (Vol. I, ex. 6, p. 71). We denote the new ring as $\mathfrak{X}_m$ so $\mathfrak{X}_m$ and $\mathfrak{X}^{(m)}$ coincide as sets and $a \to a^\varphi$ is an isomorphism of $\mathfrak{X}_m$ onto $\mathfrak{X}^{(m)}$. Hence $\mathfrak{X}_m$, like $\mathfrak{X}^{(m)}$, is commutative.

We now examine the formulas for $x + y$, xy, and $x - y$ for the "generic" vectors $x = (x_0, x_1, \cdots, x_{m-1})$, $y = (y_0, y_1, \cdots, y_{m-1})$, x_i, y_j the given indeterminates. For example, we have

$$(x + y)_0 = x_0 + y_0, \quad (x + y)_1 = x_1 + y_1 - \frac{1}{p}\sum_{i=1}^{p-1}\binom{p}{i} x_0^{\,i} y_0^{\,p-i},$$

$$(xy)_0 = x_0 y_0, \quad (xy)_1 = x_0^{\,p} y_1 + x_1 y_0^{\,p} + p x_1 y_1.$$

In general, if $\circ$ denotes any one of the compositions $+, \cdot, -$, then it is clear from the definitions that the ν-th component $(x \circ y)_\nu$ of $x \circ y$ is a polynomial with rational coefficients in $x_0, y_0, x_1, y_1, \cdots, x_\nu, y_\nu$. Also one sees easily that

$$(15) \qquad (x + y)_\nu = x_\nu + y_\nu + f_\nu(x_0, y_0, \cdots, x_{\nu-1}, y_{\nu-1})$$

where f_ν is a polynomial in the indicated indeterminates. The basic result which we shall now establish is that $(x \circ y)_\nu$ is a polynomial in $x_0, y_0, \cdots, x_\nu, y_\nu$ with *integer* coefficients.

Throughout our discussion we write $a^\varphi = (a^{(0)}, a^{(1)}, \cdots, a^{(m-1)})$ if $a = (a_0, a_1, \cdots, a_{m-1})$ etc. Let $I[x_i, y_j]$ be the ring of polynomials in $x_0, y_0, \cdots, x_{m-1}, y_{m-1}$ with coefficients in the ring of integers I. If μ is a non-negative integer we denote the ideal $p^\mu I[x_i, y_j]$ by (p^μ) and we write $c \equiv d\ (p^\mu)$ for $c - d \in (p^\mu)$. Then we have

Lemma 1. *Let* $\mu \geq 1$, $0 \leq k \leq m - 1$, $a = (a_\nu)$, $b = (b_\nu)$, $0 \leq \nu \leq m - 1$, a_ν, $b_\nu \in I[x_i, y_j]$. *Write* $a^\varphi = (a^{(\nu)})$, $b^\varphi = (b^{(\nu)})$. *Then the system of congruences*

$$(16) \qquad a_\nu \equiv b_\nu(p^\mu), \quad 0 \leq \nu \leq k$$

is equivalent to

$$(17) \qquad a^{(\nu)} \equiv b^{(\nu)}(p^{\mu+\nu}), \quad 0 \leq \nu \leq k.$$

Proof. We have $a^{(0)} = a_0$, $b^{(0)} = b_0$, so the result is clear for $k = 0$. To prove the result by induction on k we may assume that both sets (16) and (17) hold for $0 \leq \nu \leq k - 1$ and prove that under these conditions $a_k \equiv b_k(p^\mu)$ if and only if $a^{(k)} \equiv b^{(k)}$ $(p^{\mu+k})$. It is clear that $a_k \equiv b_k(p^\mu)$ if and only if $p^k a_k \equiv p^k b_k$ $(p^{\mu+k})$. Hence, using (12), it is enough to show that $(a^P)^{(k-1)} \equiv (b^P)^{(k-1)}(p^{\mu+k})$ holds under the induction hypothesis. We have

$a_\nu \equiv b_\nu(p^\mu)$, $0 \leq \nu \leq k-1$. Using $\binom{p}{i} \equiv 0\ (p)$, $1 \leq i \leq p-1$, this gives $a_\nu{}^p \equiv b_\nu{}^p(p^{\mu+1})$, $0 \leq \nu \leq k-1$. Hence the induction on k applies to a^P and b^P to give $(a^P)^{(k-1)} \equiv (b^P)^{(k-1)}(p^{\mu+1+k-1})$ which is what is required.

We can now prove the basic

Theorem 9. *If $x \circ y$ denotes $x + y$, xy or $x - y$, then $(x \circ y)_\nu$ is a polynomial in $x_0, y_0, x_1, y_1, \cdots, x_\nu, y_\nu$ with integer coefficients.*

Proof. Since $(x \circ y)_\nu$ is a polynomial in $x_0, y_0, \cdots, x_\nu, y_\nu$ with rational coefficients, it suffices to prove that $(x \circ y)_\nu \, \varepsilon \, I[x_i, y_j]$. This is clear for $(x \circ y)_0$ and we assume it for $(x \circ y)_k$, $0 \leq k \leq \nu - 1$. We have

$$p^\nu(x \circ y)_\nu = (x \circ y)^{(\nu)} - ((x \circ y)^P)^{(\nu-1)}, \tag{18}$$

by (12) and $(x \circ y)^{(\nu)} = x^{(\nu)} \dot{\pm} y^{(\nu)} \, \varepsilon \, I[x_i, y_j]$. The induction hypothesis implies that $((x \circ y)^P)_{(\nu-1)} \, \varepsilon \, I[x_i, y_j]$. Hence, by (18), it suffices to show that $(x \circ y)^{(\nu)} \equiv ((x \circ y)^P)^{(\nu-1)}\ (p^\nu)$. We have $x^{(\nu)} \equiv (x^P)^{(\nu-1)}\ (p^\nu)$ and $y^{(\nu)} \equiv (y^P)^{(\nu-1)}\ (p^\nu)$, by (12). Hence

$$\begin{aligned}(x \circ y)^{(\nu)} &= x^{(\nu)} \dot{\pm} y^{(\nu)} \equiv (x^P)^{(\nu-1)} \dot{\pm} (y^P)^{(\nu-1)} \\ &= (x^P \circ y^P)^{(\nu-1)}\ (p^\nu).\end{aligned} \tag{19}$$

We are assuming that $(x \circ y)_k \, \varepsilon \, I[x_i, y_j]$, $0 \leq k \leq \nu - 1$. For any polynomial with integer coefficients one has $f(x_0, y_0, \cdots)^p \equiv f(x_0{}^p, y_0{}^p, \cdots)\ (p)$. It follows that $(x \circ y)_k{}^P \equiv (x^P \circ y^P)_k\ (p)$, $0 \leq k \leq \nu - 1$. Hence, by Lemma 1, we have

$$((x \circ y)^P)^{(\nu-1)} \equiv (x^P \circ y^P)^{(\nu-1)}\ (p^\nu). \tag{20}$$

By (19) and (20), $(x \circ y)^{(\nu)} \equiv ((x \circ y)^P)^{(\nu-1)}\ (p^\nu)$, which is what was needed.

It is convenient to write the result we have proved as follows:

$$\begin{aligned}(x + y)_\nu &= s_\nu(x_0, y_0, \cdots, x_\nu, y_\nu) \, \varepsilon \, I[x_i, y_j] \\ (xy)_\nu &= m_\nu(x_0, y_0, \cdots, x_\nu, y_\nu) \, \varepsilon \, I[x_i, y_j] \\ (x - y)_\nu &= d_\nu(x_0, y_0, \cdots, x_\nu, y_\nu) \, \varepsilon \, I[x_i, y_j].\end{aligned} \tag{21}$$

We note also that, since $(0, \cdots, 0)$ and $(1, \cdots, 1)$ are the zero and identity elements of $\mathfrak{X}^{(m)}$ and $(0, \cdots, 0)^\varphi = (0, \cdots, 0)$, $(1, 0, \cdots,$

$0)^\varphi = (1, \cdots, 1)$, then $(0, \cdots, 0)$ and $(1, 0, \cdots, 0)$ are the zero and identity of $\mathfrak{X}_m$. Let η be an algebra homomorphism of $\mathfrak{X}$ over R_0 into itself and assume that $x_\nu^\eta = a_\nu$, $y_\nu^\eta = b_\nu$, $0 \leq \nu \leq m - 1$. Then we have $(x^{(\nu)})^\eta = a^{(\nu)}$, $(y^{(\nu)})^\eta = b^{(\nu)}$, $((x + y)^{(\nu)})^\eta = a^{(\nu)} + b^{(\nu)}$ and $((x + y)_\nu)^\eta = (a + b)_\nu$. Hence, by (21), $(a + b)_\nu = s_\nu(a_0, b_0, \cdots, a_\nu, b_\nu)$ and similar formulas hold for $(ab)_\nu$ and $(a - b)_\nu$. Since there exists a homomorphism η of $\mathfrak{X}$ over R_0 such that x_ν^η and y_ν^η are arbitrary elements of $\mathfrak{X}$, these formulas hold for all $a, b \in \mathfrak{X}_m$. Evidently they imply that, if $\mathfrak{B}$ is any subring of $R_0[x_i, y_j, z_k]$, then the set $\mathfrak{B}_m$ of vectors $(b_0, b_1, \cdots, b_{m-1})$ with the $b_\nu \in \mathfrak{B}$ is a subring of $\mathfrak{X}_m$. In particular, this holds for $\mathfrak{B} = \mathfrak{Y} \equiv I[x_i, y_j, z_k]$ and for $\mathfrak{B} = \mathfrak{Y}' \equiv I[x_i, y_j]$.

We are now ready to define the ring $\mathfrak{W}_m(\mathfrak{A})$ of Witt vectors. Here $\mathfrak{A}$ is any commutative algebra over the field I_p of p elements, where p is the prime used above. The elements of $\mathfrak{W}_m(\mathfrak{A})$ are the "vectors" $(a_0, a_1, \cdots, a_{m-1})$, $a_\nu \in \mathfrak{A}$, with equality defined as usual. If $a = (a_0, \cdots, a_{m-1})$, $b = (b_0, \cdots, b_{m-1})$, then we define addition and multiplication in $\mathfrak{W}_m(\mathfrak{A})$ by

$$(22) \qquad \begin{aligned} (a + b)_\nu &= \bar{s}_\nu(a_0, b_0, \cdots, a_\nu, b_\nu) \\ (ab)_\nu &= \bar{m}_\nu(a_0, b_0, \cdots, a_\nu, b_\nu). \end{aligned}$$

Here we understand that $a + b = ((a + b)_0, \cdots, (a + b)_{m-1})$, $ab = ((ab)_0, \cdots, (ab)_{m-1})$ and, if $f(x_0, y_0, \cdots)$ is a polynomial with integer coefficients, then $\bar{f}(a_0, a_1, \cdots)$ is the element of $\mathfrak{A}$ obtained by replacing the integer coefficients of $f(x_0, y_0, \cdots)$ by their cosets in I_p, x_ν by a_ν, y_ν by b_ν, $0 \leq \nu \leq m - 1$. These replacements amount to applying the homomorphism of $I[x_i, y_j]$ into $\mathfrak{A}$ such that $n \to \bar{n} = n + (p)$, $n \in I$, $x_\nu \to a_\nu$, $y_\nu \to b_\nu$.

Now suppose $a = (a_\nu)$, $b = (b_\nu)$, $c = (c_\nu)$ are any three elements of $\mathfrak{W}_m(\mathfrak{A})$. We have a homomorphism of $I[x_i, y_j, z_k]$ into $\mathfrak{A}$ such that $n \to \bar{n}$, $n \in I$, $x_\nu \to a_\nu$, $y_\nu \to b_\nu$, $z_\nu \to c_\nu$. Consider the subring $\mathfrak{Z}_m$ of $\mathfrak{X}_m$ of vectors $(w_0, w_1, \cdots, w_{m-1})$ where $w_\nu \in I[x_i, y_j, z_k]$. We have seen that, if $t = (t_0, \cdots, t_{m-1}) \in \mathfrak{Z}_m$, then $(w + t)_\nu = s_\nu(w_0, t_0, \cdots, w_\nu, t_\nu)$, $(wt)_\nu = m_\nu(w_0, t_0, \cdots, w_\nu, t_\nu)$. It follows that the mapping $(w_0, \cdots, w_{m-1}) \to (w_0^\eta, \cdots, w_{m-1}^\eta)$ is a homomorphism of $\mathfrak{Z}_m$ into the system $(\mathfrak{W}_m(\mathfrak{A}), +, \cdot)$ where $+$ and $\cdot$ are defined by (22). Note that our homomorphism maps x into a, y into b, z into c. We remark also that any element $w =$

(w_ν) such that the $w_\nu \in (p)$ is in the kernel of the homomorphism of $\mathfrak{J}_m$ into $\mathfrak{A}$.

We can now prove

Theorem 10. *$\mathfrak{W}_m(\mathfrak{A})$ is a commutative ring.*

Proof. Let $a = (a_\nu)$, $b = (b_\nu)$, $c = (c_\nu)$ be any three elements of $\mathfrak{W}_m(\mathfrak{A})$. Then we have just seen that we have a homomorphism of $\mathfrak{J}_m$ into $\mathfrak{W}_m(\mathfrak{A})$ such that $x = (x_\nu) \to a$, $y = (y_\nu) \to b$, $z = (z_\nu) \to c$. Then the associative, commutative, and distributive laws of addition and multiplication in $\mathfrak{J}_m$ give the same rules for the elements a, b, c (e.g., $(ab)c = a(bc)$). The image of $0 = (0, \cdots, 0)$ and $1 = (1, 0, \cdots, 0)$ under our homomorphism are $0 \equiv (0, \cdots, 0)$ and $1 \equiv (\bar{1}, 0, \cdots, 0)$ and the relations $x + 0 = x$, $x1 = x$ give $a + 0 = a$, $a1 = a$ in $\mathfrak{W}_m(\mathfrak{A})$. Also if we write a' for the image of $-x$ under the homomorphism, then we have $a + a' = 0$. Since a, b, c are arbitrary in $\mathfrak{W}_m(\mathfrak{A})$, these remarks show that $\mathfrak{W}_m(\mathfrak{A})$ is a commutative ring with $0 = (0, \cdots, 0)$, $1 = (\bar{1}, 0, \cdots, 0)$ as 0 and identity elements.

We shall call $\mathfrak{W}_m(\mathfrak{A})$ the *ring of Witt vectors of length m* over $\mathfrak{A}$. We remark that $\mathfrak{W}_1(\mathfrak{A})$ can be identified with $\mathfrak{A}$ itself since we have the isomorphism $a \to (a)$ of $\mathfrak{A}$ onto $\mathfrak{W}_1(\mathfrak{A})$.

Now let $\mathfrak{B}$ be a subalgebra of $\mathfrak{A}$ over I_p and form the ring $\mathfrak{W}_m(\mathfrak{B})$ of Witt vectors over $\mathfrak{B}$. Then it is clear that $b = (b_\nu) \to b$ is an isomorphism of $\mathfrak{W}_m(\mathfrak{B})$ into $\mathfrak{W}_m(\mathfrak{A})$. In this way we can identify $\mathfrak{W}_m(\mathfrak{B})$ with the subring of $\mathfrak{W}_m(\mathfrak{A})$ of the Witt vectors b with $b_\nu \in \mathfrak{B}$. In particular, if we take $\mathfrak{B} = I_p$ we obtain the subring $\mathfrak{W}_m(I_p)$ of vectors with components in I_p. This subring evidently consists of p^m elements.

We define the mapping P of $\mathfrak{W}_m(\mathfrak{A})$ into itself by $a^P = (a_0{}^p, a_1{}^p, \cdots, a_{m-1}{}^p)$ for $a = (a_0, a_1, \cdots, a_{m-1})$. We have noted that, if $f(x_0, y_0, \cdots) \in I[x_i, y_j]$, then $\bar{f}(a_0, b_0, \cdots)^p = \bar{f}(a_0{}^p, b_0{}^p, \cdots)$. This and the definitions of addition and multiplication in $\mathfrak{W}_m(\mathfrak{A})$ imply that

$$(a + b)^P = a^P + b^P; \quad (ab)^P = a^P b^P. \tag{23}$$

We shall call P the *Frobenius endomorphism* in $\mathfrak{W}_m(\mathfrak{A})$. We introduce the *restriction mapping* R of $\mathfrak{W}_m(\mathfrak{A})$ into $\mathfrak{W}_{m-1}(\mathfrak{A})$ by $(a_0, \cdots, a_{m-1})^R = (a_0, \cdots, a_{m-2})$ and the *shift mapping* V of $\mathfrak{W}_{m-1}(\mathfrak{A})$

into $\mathfrak{W}_m(\mathfrak{A})$ by $(a_0, \cdots, a_{m-2})^V = (0, a_0, \cdots, a_{m-2})$. It is immediate that R is a ring homomorphism and we shall see that V is a homomorphism of the additive group of $\mathfrak{W}_{m-1}(\mathfrak{A})$ into that of $\mathfrak{W}_m(\mathfrak{A})$. We have

$$(a_0, \cdots, a_{m-1})^{VR} = (0, a_0, \cdots, a_{m-2}) = (a_0, \cdots, a_{m-1})^{RV}.$$

Also it is clear that $PV = VP$, $RP = PR$, and $(VR)^m = 0$ hold in $\mathfrak{W}_m(\mathfrak{A})$.

We prove next the important

Lemma 2. *The following relations hold in Witt rings:*

$$p1 = \overbrace{1 + 1 + 1 \cdots + 1}^{p} = 1^{VR} \tag{24}$$

$$(a + b)^V = a^V + b^V \tag{25}$$

$$a^V b = (ab^{PR})^V, \quad a \in \mathfrak{W}_m(\mathfrak{A}), \quad b \in \mathfrak{W}_{m+1}(\mathfrak{A}) \tag{26}$$

$$pa = a^{PVR}. \tag{27}$$

Proof. Consider the subrings $\mathfrak{J}_{m-1}, \mathfrak{J}_m, \mathfrak{J}_{m+1}$ of $\mathfrak{X}_{m-1}, \mathfrak{X}_m, \mathfrak{X}_{m+1}$ of elements with components in $I[x_i, y_j, z_k]$ and define the mappings R and V for these in the same way as for the Witt rings. Also we have the mapping P defined before. Consider the element $1 = (1, 0, \cdots, 0)$ of I. Set $\mathbf{p} = p1$. We have $1^\varphi = (1, 1, \cdots, 1)$ and

$$\overbrace{1^\varphi \oplus 1^\varphi \oplus \cdots \oplus 1^\varphi}^{p} = (p, \cdots, p).$$

Hence $\mathbf{p}^{(\nu)} = p$, $0 \leq \nu \leq m - 1$. On the other hand, $1^{VR} = (0, 1, 0, \cdots, 0)$, so the definition of φ gives $(1^{VR})^{(0)} = 0$, $(1^{VR})^{(\nu)} = p$, $1 \leq \nu \leq m - 1$. Then we have $(1^{VR})^{(\nu)} \equiv \mathbf{p}^{(\nu)}\ (p^{\nu+1})$, $0 \leq \nu \leq m - 1$. By Lemma 1, this implies that $(1^{VR})_\nu \equiv \mathbf{p}_\nu\ (p)$. We have seen that there is a homomorphism of $\mathfrak{J}_m$ into $\mathfrak{W}_m(\mathfrak{A})$ such that every $w = (w_\nu)$, $w_\nu \in (p)$, is in the kernel. If we apply this to 1^{VR} and to $\mathbf{p}$ and use the foregoing relations on components, we obtain (24) in $\mathfrak{W}_m(\mathfrak{A})$. Next we note that $x^V = (0, x_0, \cdots, x_{m-1})$, $y^V = (0, y_0, \cdots, y_{m-1})$ for $x = (x_0, \cdots, x_{m-1})$, $y = (y_0, \cdots, y_{m-1})$ in $\mathfrak{J}_m$. Then

$$(x^V)^{(\nu)} = px_0^{p^{\nu-1}} + p^2 x_1^{p^{\nu-2}} + \cdots + p^\nu x_{\nu-1}, \quad 1 \leq \nu \leq m,$$

by (11); hence

$$(28) \qquad (x^V)^{(\nu)} = px^{(\nu-1)}, \quad 1 \leq \nu \leq m.$$

Since $(x+y)^{(\nu)} = x^{(\nu)} + y^{(\nu)}$, this and $(x^V)^{(0)} = (y^V)^{(0)} = ((x+y)^V)^{(0)} = 0$ give $((x+y)^V)^{(\nu)} = (x^V)^{(\nu)} + (y^V)^{(\nu)}$, $0 \leq \nu \leq m$. Hence $(x+y)^V = x^V + y^V$ holds in $\mathfrak{J}_{m+1}$. If we apply the homomorphism of $I[x_i, y_j, z_k]$ into $\mathfrak{A}$ such that $n \to \bar{n} = n + (p)$, $x_\nu \to a_\nu$, $y_\nu \to b_\nu$, $z_\nu \to c_\nu$ to the components of $(x+y)^V$ and $x^V + y^V$, we obtain (25) for $a, b \in \mathfrak{W}_m(\mathfrak{A})$. To prove (26) we shall show that

$$(29) \qquad (x^V y)_\nu \equiv ((xy^{PR})^V)_\nu \ (p), \quad 0 \leq \nu \leq m$$

if $x = (x_0, x_1, \cdots, x_{m-1}) \in \mathfrak{J}_m$ and $y = (y_0, y_1, \cdots, y_m) \in \mathfrak{J}_{m+1}$, x_i, y_j indeterminates. Set $x^V y = (w_0, w_1, \cdots, w_m)$, $(xy^{PR})^V = (t_0, t_1, \cdots, t_m)$. Then we have to show that $w_\nu \equiv t_\nu \ (p)$, $0 \leq \nu \leq m$. By Lemma 1, this is equivalent to $w^{(\nu)} \equiv t^{(\nu)} \ (p^{\nu+1})$. This holds for $\nu = 0$ since $w^{(0)} = 0 = t^{(0)}$. For $\nu \geq 1$, we have, by (28), that $w^{(\nu)} = px^{(\nu-1)}y^{(\nu)}$ and $t^{(\nu)} = px^{(\nu-1)}(y^{PR})^{(\nu-1)}$. Since $y^{(\nu)} = (y^P)^{(\nu-1)} + p^\nu y_\nu$, this gives the congruences

$$\begin{aligned} w^{(\nu)} &= px^{(\nu-1)}y^{(\nu)} \equiv px^{(\nu-1)}(y^P)^{(\nu-1)} \\ &\equiv px^{(\nu-1)}(y^{PR})^{(\nu-1)} \equiv t^{(\nu)} \ (p^{\nu+1}). \end{aligned}$$

Hence (29) holds. Applying a suitable homomorphism into $\mathfrak{A}$, we obtain (26). If we apply R to both sides of (26), we obtain $a^{VR}b^R = (ab^{PR})^{VR}$. Setting $a = 1$ and $b^R = c \in \mathfrak{W}_m(\mathfrak{A})$, we obtain $1^{VR}c = c^{PVR}$. Since $1^{VR} = p1$, by (24), this gives $pc = c^{PVR}$. Since $c = b^R$ can be taken to be any element of $\mathfrak{W}_m(\mathfrak{A})$, this is equivalent to (27).

We can now derive the basic properties of $\mathfrak{W}_m(\mathfrak{A})$ which we shall need. We prove first

Theorem 11. *$\mathfrak{W}_m(\mathfrak{A})$ is a ring of characteristic p^m.*

Proof. It suffices to show that the order of 1 in the additive group of $\mathfrak{W}_m(\mathfrak{A})$ is p^m. We have seen that $p1 = 1^{VR} = (0, 1, 0, \cdots, 0)$ and by iterating (27) we obtain $p^2 1 = (0, 0, 1, 0, \cdots)$ etc. This shows that $p^{m-1}1 = (0, \cdots, 0, 1) \neq 0$ but $p^m 1 = 0$, as required.

We have seen that, if $\mathfrak{B}$ is a subalgebra of $\mathfrak{A}$, then we can consider $\mathfrak{W}_m(\mathfrak{B})$ as a subring of $\mathfrak{W}_m(\mathfrak{A})$. In particular this holds if we take $\mathfrak{B} = I_p$. Then $Z \equiv \mathfrak{W}_m(I_p)$ is the set of vectors with components in the field I_p and so the number of elements in Z is p^m. On the other hand, Theorem 11 shows that there are p^m distinct elements of the form $k1$, k an integer, in $\mathfrak{W}_m(\mathfrak{A})$ and these belong to Z. Hence it is clear that Z is just the set of integral multiples of the identity of $\mathfrak{W}_m(\mathfrak{A})$. Evidently Z is isomorphic to the ring $I/(p^m)$ of residues modulo p^m. The following result gives an insight into the structure of $\mathfrak{W}_m(\mathfrak{A})$.

Theorem 12. *The mapping $a = (a_0, a_1, \cdots, a_{m-1}) \rightarrow a_0$ is a homomorphism of $\mathfrak{W}_m(\mathfrak{A})$ onto $\mathfrak{A}$ whose kernel $\mathfrak{N}$ is a nilpotent ideal.*

Proof. We have seen that R is a homomorphism of $\mathfrak{W}_m(\mathfrak{A})$ onto $\mathfrak{W}_{m-1}(\mathfrak{A})$. Iteration of this shows that R^{m-1} is a homomorphism of $\mathfrak{W}_m(\mathfrak{A})$ onto $\mathfrak{W}_1(\mathfrak{A}) = \mathfrak{A}$. Evidently R^{m-1} is the mapping we have indicated. The kernel of our homomorphism is the ideal $\mathfrak{N}$ of elements of the form $(0, a_0, a_1, \cdots, a_{m-2})$. Hence $\mathfrak{N} = \mathfrak{W}_m(\mathfrak{A})^{VR}$. If we apply R to (26) we obtain $a^{VR}b^R = (ab^{PR})^{VR}$. Since b^R can be taken to be any element c in $\mathfrak{W}_m(\mathfrak{A})$, this gives the relation $a^{VR}c = (ac^P)^{VR}$ in $\mathfrak{W}_m(\mathfrak{A})$. Then $a^{VR}c^{VR} = (ac^{PVR})^{VR} = (a^Pc^P)^{(VR)^2} \varepsilon \mathfrak{W}_m(\mathfrak{A})^{(VR)^2}$. Thus $\mathfrak{N}^2 = (\mathfrak{W}_n(\mathfrak{A})^{VR})^2 \subseteq \mathfrak{N}^{VR}$. Now assume that for some $k \geq 2$, $\mathfrak{N}^k \subseteq \mathfrak{N}\mathfrak{N}^{(VR)^{k-2}} \subseteq \mathfrak{N}^{(VR)^{k-1}}$. Then if $d = a^{VR} \varepsilon \mathfrak{N}$ and $b \varepsilon \mathfrak{N}^k$, we have $b = c^{(VR)^k}$, $c \varepsilon \mathfrak{W}_m(\mathfrak{A})$, since $b \varepsilon \mathfrak{N}^{(VR)^{k-1}} = \mathfrak{W}_m(\mathfrak{A})^{(VR)^k}$. Hence $db = a^{VR}c^{(VR)^k} \varepsilon \mathfrak{N}\mathfrak{N}^{(VR)^{k-1}}$ and so $\mathfrak{N}^{k+1} \subseteq \mathfrak{N}\mathfrak{N}^{(VR)^{k-1}}$. Moreover, if $a, c \varepsilon \mathfrak{W}_m(\mathfrak{A})$, then $a^{VR}c^{(VR)^k} = (a^{PVR}c^{(VR)^{k-1}})^{VR} \varepsilon (\mathfrak{N}\mathfrak{N}^{(VR)^{k-2}})^{VR} \subseteq (\mathfrak{N}^{(VR)^{k-1}})^{VR} = \mathfrak{N}^{(VR)^k}$. Hence $\mathfrak{N}\mathfrak{N}^{(VR)^{k-1}} \subseteq \mathfrak{N}^{(VR)^k}$ and so $\mathfrak{N}^{k+1} \subseteq \mathfrak{N}^{(VR)^k}$. This shows that $\mathfrak{N}^k \subseteq \mathfrak{N}^{(VR)^{k-1}}$ holds for all $k \geq 2$. Since $\mathfrak{N} = \mathfrak{W}_m(\mathfrak{A})^{VR}$ and $\mathfrak{W}_m(\mathfrak{A})^{(VR)^m} = 0$ this gives $\mathfrak{N}^m = 0$.

Corollary. *An element $a = (a_0, a_1, \cdots, a_{m-1})$ is a unit in $\mathfrak{W}_m(\mathfrak{A})$ if and only if a_0 is a unit in $\mathfrak{A}$.*

Proof. This follows from Theorem 12 and the remark that, if $\mathfrak{N}$ is a nilpotent ideal in a ring $\mathfrak{W}$, then an element $a \varepsilon \mathfrak{W}$ is a unit in $\mathfrak{W}$ if and only if the coset $a + \mathfrak{N}$ is a unit in $\mathfrak{W}/\mathfrak{N}$. We leave the proof as an exercise.

5. Abelian p-extensions. It will be instructive to consider first briefly the abelian extensions of a field Φ of characteristic $p \neq 0$

whose Galois groups G have exponent p (cf. ex. 3, 4, p. 98). In this case we let Z be the additive cyclic group generated by the element 1 of Φ and we consider the character group Hom (G, Z) where G is the Galois group of an extension P of the type specified. The elements $\chi \varepsilon$ Hom (G, Z) are the mappings of G into Z satisfying $\chi(st) = \chi(s) + \chi(t)$. Since $\chi(s) \varepsilon Z \subseteq \Phi$, this can be written also in the form $\chi(st) = \chi(s)^t + \chi(t)$ so that we have an instance of the additive analogue of Noether's equations. Hence by Theorem 1.20, there exists a $\rho \varepsilon P$ such that $\chi(s) = \rho^s - \rho$. Since $\chi(s) \varepsilon Z$, $\chi(s)^p = \chi(s)$ so $(\rho^s - \rho)^p = \rho^s - \rho$. This gives the equation $(\rho^p - \rho)^s = \rho^p - \rho$, $s \varepsilon G$; hence $\rho^p - \rho = \alpha \varepsilon \Phi$. Conversely, let ρ be any element of P such that $\rho^p - \rho = \alpha \varepsilon \Phi$ and define $\chi(s) = \rho^s - \rho$. Then $\chi(s)^p - \chi(s) = (\rho^p - \rho)^s - (\rho^p - \rho) = \alpha^s - \alpha = 0$. Hence $\chi(s)^p = \chi(s)$ and this implies that $\chi(s)$ is in the prime field, so $\chi(s) \varepsilon Z$. Also we have $\chi(st) = \rho^{st} - \rho = (\rho^s - \rho)^t + (\rho^t - \rho) = (\rho^s - \rho) + (\rho^t - \rho) = \chi(s) + \chi(t)$; hence $\chi \varepsilon$ Hom (G, Z). Following the pattern of the Kummer theory this leads us to consider the subset $S(\mathrm{P})$ of P of elements ρ such that $\rho^p - \rho \varepsilon \Phi$. This is a subgroup of the additive group (P, +) containing $(\Phi, +)$ and we have the mapping $\rho \to \chi_\rho$, where $\chi_\rho(s) = \rho^s - \rho$, of $S(\mathrm{P})$ onto Hom (G, Z). Since Z is an additive group the composition in Hom (G, Z) is $(\chi + \psi)(s) = \chi(s) + \psi(s)$. Moreover, if $\rho, \sigma \varepsilon S(\mathrm{P})$, then $\chi_{\rho+\sigma}(s) = (\rho + \sigma)^s - (\rho + \sigma) = \chi_\rho(s) + \chi_\sigma(s)$; hence $\rho \to \chi_\rho$ is a homomorphism of $S(\mathrm{P})$ onto Hom (G, Z). It is clear that the kernel of this homomorphism is Φ. Hence $S(\mathrm{P})/\Phi \cong$ Hom $(G, Z) \cong G$.

The next step in the discussion is to consider the subset $Q(\mathrm{P})$ of Φ of elements of the form $\rho^p - \rho$, $\rho \varepsilon S(\mathrm{P})$. This is a subgroup of the additive group $(\Phi, +)$ containing the subgroup of elements of the form $\alpha^p - \alpha$, $\alpha \varepsilon \Phi$. One sees easily that the factor group of $Q(\mathrm{P})$ relative to the last subgroup is isomorphic to $S(\mathrm{P})/\Phi$, hence to Hom (G, Z) and to G. It can be shown that any subgroup of $(\Phi, +)$ containing the subgroup of elements $\alpha^p - \alpha$, $\alpha \varepsilon \Phi$, and having a finite factor group relative to this subgroup is a group $Q(\mathrm{P})$ for an abelian p-extension with Galois group of exponent $\leq p$. The groups Q give a survey of these extensions in the same manner that the group $N(\mathrm{P})$ gave a survey of the Kummer extensions. We shall not work out the details here but instead we shall proceed to the general case of arbitrary p-extensions. The idea here is to work in the ring $\mathfrak{W}_m(P)$ of Witt vectors over the given extension P where $m \geq e$, p^e the exponent of G. Then the subgroup Z of the additive group of $\mathfrak{W}_m(P)$ generated by 1 is cyclic of order p^m; hence, Hom (G, Z) is a character group of G. We shall need first of all the generalization of Theorem 1.20 to the ring of Witt vectors and we proceed to derive this result.

Suppose first that P is a finite dimensional Galois extension field of the field Φ of characteristic $p \neq 0$ with Galois group G. Let $\mathfrak{W}_m(P)$ be the ring of Witt vectors of length $m \geq 1$ over P. We have seen that we can identify $\mathfrak{W}_m(\Phi)$ with the subset of $\mathfrak{W}_m(P)$ of vectors $\beta = (\beta_0, \beta_1, \cdots, \beta_{m-1})$ with the $\beta_\nu \in \Phi$. If $\rho = (\rho_0, \cdots, \rho_{m-1}) \in \mathfrak{W}_m(P)$ and $s \in G$, we define $\rho^s = (\rho_0{}^s, \cdots, \rho_{m-1}{}^s)$. It is clear that $\rho \rightarrow \rho^s$ is an automorphism of $\mathfrak{W}_m(P)$ and that the set of these automorphisms is a group isomorphic to G. We denote this group again as G. Evidently $\rho^s = \rho$ if and only if $\rho_\nu{}^s = \rho_\nu$, $0 \leq \nu \leq m - 1$. Hence $\mathfrak{W}_m(\Phi)$ can be characterized as the subring of G-invariants of the ring $\mathfrak{W}_m(P)$.

If $\rho \in \mathfrak{W}_m(P)$ we define its *trace* $T(\rho) = \sum_{s \in G} \rho^s$. Evidently $T(\rho)^s = T(\rho)$, $s \in G$, so $T(\rho) \in \mathfrak{W}_m(\Phi)$. If $\rho = (\rho_0, \rho_1, \cdots, \rho_{m-1})$, then the first component of $T(\rho)$ is $T(\rho_0)$ (trace in P over Φ), since first components are added in forming a sum in $\mathfrak{W}_m(P)$. We recall that the automorphisms $s \in G$ in P are P-independent and this implies that there exists a $\rho_0 \in P$ such that $T(\rho_0) \neq 0$. If ρ_0 is chosen in this way and $\rho = (\rho_0, \cdots)$, then $T(\rho) = (T(\rho_0), \cdots)$ has non-zero first component. It follows from the corollary to Theorem 12 that $T(\rho)$ is a unit in $\mathfrak{W}_m(\Phi)$. Hence we have proved the following

Lemma 1. *There exist* $\rho \in \mathfrak{W}_m(P)$ *such that* $T(\rho)^{-1}$ *exists in* $\mathfrak{W}_m(\Phi)$.

We use this to prove the following key cohomology result.

Theorem 13. *Let* $s \rightarrow \mu_s$ *be a mapping of* G *into* $\mathfrak{W}_m(P)$ *such that* $\mu_{st} = \mu_s{}^t + \mu_t$, $s, t \in G$. *Then there exists an element* $\sigma \in \mathfrak{W}_m(P)$ *such that* $\mu_s = \sigma^s - \sigma$. *Conversely, if* $\sigma \in \mathfrak{W}_m(P)$, *then* $\mu_s = \sigma^s - \sigma$ *satisfies the given equations.*

Proof. The proof is identical with that of the special case of Galois extension fields treated in Theorem 1.20. We choose ρ in $\mathfrak{W}_m(\mathrm{P})$ so that $T(\rho)^{-1}$ exists in $\mathfrak{W}_m(\Phi)$ and we let $\tau = T(\rho)^{-1}\left(\sum_{s \varepsilon G} \mu_s \rho^s\right)$. Then

$$\begin{aligned}\tau - \tau^t &= T(\rho)^{-1}\left(\sum_s (\mu_{st}\rho^{st} - \mu_s{}^t\rho^{st})\right)\\ &= T(\rho)^{-1}\left(\sum_s \mu_t\rho^{st}\right)\\ &= T(\rho)^{-1}\mu_t T(\rho)\\ &= \mu_t.\end{aligned}$$

Hence if we take $\sigma = -\tau$, then we have $\mu_s = \sigma^s - \sigma$ as required. Conversely, if we take $\mu_s = \sigma^s - \sigma$ where σ is any element of P, then we have $\mu_s{}^t + \mu_t = \sigma^{st} - \sigma^t + \sigma^t - \sigma = \sigma^{st} - \sigma = \mu_{st}$.

We recall that the Frobenius mapping $\rho \to \rho^P = (\rho_0{}^p, \rho_1{}^p, \cdots, \rho_m{}^p)$ is an endomorphism of the ring $\mathfrak{W}_m(\mathrm{P})$. We shall now introduce the mapping $\mathfrak{P}$ in $\mathfrak{W}_m(\mathrm{P})$ defined by

$$\mathfrak{P}(\rho) = \rho^P - \rho. \tag{30}$$

It is clear that $\mathfrak{P}$ is an endomorphism of the additive group of $\mathfrak{W}_m(\mathrm{P})$ (but not of the ring $\mathfrak{W}_m(\mathrm{P})$). The kernel of $\mathfrak{P}$ is the set of vectors $(\rho_0, \rho_1, \cdots, \rho_{m-1})$ such that $\rho_\nu{}^p = \rho_\nu$, $0 \le \nu \le m-1$. Evidently this is just the set of vectors with components ρ_i in the prime field Φ_0 ($\cong I_p$). Hence the kernel of $\mathfrak{P}$ is the set of Witt vectors $(\rho_0, \rho_1, \cdots, \rho_{m-1})$ with the $\rho_i \,\varepsilon\, \Phi_0$. We have seen (after Th. 11) that this is just the set Z of integral multiples of the identity 1, and Z is a cyclic group of order p^m under addition.

We now assume that the Galois group G is an abelian group of order p^f and that $m \ge e$ where p^e is the exponent of G. Let

$$S(\mathfrak{W}_m(\mathrm{P})) = \{\rho \,\varepsilon\, \mathfrak{W}_m(\mathrm{P}) \mid \mathfrak{P}(\rho) \,\varepsilon\, \mathfrak{W}_m(\Phi)\}. \tag{31}$$

Then $S(\mathfrak{W}_m(\mathrm{P}))$ is a subgroup of the additive group $(\mathfrak{W}_m(\mathrm{P}), +)$ containing $\mathfrak{W}_m(\Phi)$. If $\rho \,\varepsilon\, S(\mathfrak{W}_m(\mathrm{P}))$, then we define the mapping χ_ρ of G by $\chi_\rho(s) = \rho^s - \rho$. Then $\chi_\rho(s)^P = \rho^{sP} - \rho^P = \rho^{Ps} - \rho^P = (\rho^s + \alpha) - (\rho + \alpha)$ if $\mathfrak{P}(\rho) = \alpha$. Hence $\chi_\rho(s)^P = \rho^s - \rho =$

$\chi_\rho(s)$. We have seen that this implies that $\chi_\rho(s) \in Z$. Also we have $\chi_\rho(st) = \rho^{st} - \rho = \rho^{st} - \rho^t + \rho^t - \rho = (\rho^s - \rho) + (\rho^t - \rho) = \chi_\rho(s) + \chi_\rho(t)$. Hence $\chi_\rho \in \mathrm{Hom}(G, Z)$. Next let $\rho, \sigma \in S(\mathfrak{W}_m(\mathrm{P}))$. Then $\rho + \sigma \in S(\mathfrak{W}_m(\mathrm{P}))$ and $\chi_{\rho+\sigma}(s) = (\rho + \sigma)^s - (\rho + \sigma) = (\rho^s - \rho) + (\sigma^s - \sigma) = \chi_\rho(s) + \chi_\sigma(s)$. This shows that the mapping $\rho \to \chi_\rho$ is a homomorphism of $S(\mathfrak{W}_m(\mathrm{P}))$ into $\mathrm{Hom}(G, Z)$. If $\chi_\rho(s) = 0$ for all $s \in G$, then we have $\rho^s = \rho$, $s \in G$, and this implies that $\rho \in \mathfrak{W}_m(\Phi)$. Hence the kernel of $\rho \to \chi_\rho$ is $\mathfrak{W}_m(\Phi)$. Finally, we note that our homomorphism is surjective. For, let $\chi \in \mathrm{Hom}(G, Z)$. Then $\chi(st) = \chi(s) + \chi(t)$ and, since the $\chi(s) \in Z$, we have also $\chi(st) = \chi(s)^t + \chi(t)$. Hence, by Th. 13, there exists a $\rho \in \mathfrak{W}_m(\mathrm{P})$ such that $\chi(s) = \rho^s - \rho$. Since $\chi(s) \in Z$, $\chi(s)^P = \chi(s)$ and this gives $(\rho^P - \rho)^s = \rho^P - \rho$. Hence $\mathfrak{P}(\rho) = \rho^P - \rho \in \mathfrak{W}_m(\Phi)$ and so $\rho \in S(\mathfrak{W}_m(\mathrm{P}))$. We now see that the mapping $\rho \to \chi_\rho$ of $S(\mathfrak{W}_m(\mathrm{P}))$ into $\mathrm{Hom}(G, Z)$ is surjective and since the kernel is $\mathfrak{W}_m(\Phi)$ we have $S(\mathfrak{W}_m(\mathrm{P}))/\mathfrak{W}_m(\Phi) \cong \mathrm{Hom}(G, Z) \cong G$. We have therefore proved the first two statements of the following theorem which is a perfect analogue of Theorem 7:

Theorem 14. *Let Φ be a field of characteristic $p \neq 0$, P/Φ an abelian p-extension whose Galois group G is of exponent p^e and let $\mathfrak{W}_m(\mathrm{P})$ be the ring of Witt vectors of length m over P where $m \geq e$. Let $S(\mathfrak{W}_m(\mathrm{P}))$ be defined by* (31). *Then we have the exact sequence of additive groups*

$$0 \to \mathfrak{W}_m(\Phi) \to S(\mathfrak{W}_m(\mathrm{P})) \to \mathrm{Hom}(G, Z) \to 0$$

where the homomorphism of $\mathfrak{W}_m(\Phi)$ is the inclusion mapping and that of $S(\mathfrak{W}_m(\mathrm{P}))$ into $\mathrm{Hom}(G, Z)$ *is $\rho \to \chi_\rho$, $\chi_\rho(s) = \rho^s - \rho$. The factor group $S(\mathfrak{W}_m(\mathrm{P}))/\mathfrak{W}_m(\Phi)$ is finite and is isomorphic to G. The field P/Φ is generated by the components of the vectors $\rho \in S(\mathfrak{W}_m(\mathrm{P}))$ and*

$$\mathrm{P} = \Phi(\rho_0^{(1)}, \cdots, \rho_{m-1}^{(1)}; \rho_0^{(2)}, \cdots, \rho_{m-1}^{(2)}; \cdots; \rho_0^{(r)}, \cdots, \rho_{m-1}^{(r)})$$

if and only if the cosets $\rho^{(i)} + \mathfrak{W}_m(\Phi)$, $\rho^{(i)} = (\rho_0^{(i)}, \cdots, \rho_{m-1}^{(i)})$, generate $S(\mathfrak{W}_m(\mathrm{P}))/\mathfrak{W}_m(\Phi)$.

The proof of the last statement is exactly like that of the corresponding statement of Theorem 7. We leave it to the reader to check the details.

Following the pattern of our treatment of the Kummer theory we introduce next the set

$$(32)\quad Q(\mathfrak{W}_m(\mathrm{P})) = \{\mathfrak{P}(\rho) \mid \rho \in S(\mathfrak{W}_m(\mathrm{P}))\} = \mathfrak{W}_m(\Phi) \cap \mathfrak{P}(\mathfrak{W}_m(\mathrm{P})).$$

This is a subgroup of the additive group $(\mathfrak{W}_m(\Phi), +)$ containing $\mathfrak{P}(\mathfrak{W}_m(\Phi))$ the subgroup of vectors $\mathfrak{P}(\alpha)$, $\alpha \in \mathfrak{W}_m(\Phi)$. Consider the homomorphism

$$\rho \to \mathfrak{P}(\rho) + \mathfrak{P}(\mathfrak{W}_m(\Phi))$$

of $S(\mathfrak{W}_m(\mathrm{P}))$ onto $Q(\mathfrak{W}_m(\mathrm{P}))/\mathfrak{P}(\mathfrak{W}_m(\Phi))$. An element ρ is in the kernel of this homomorphism if and only if $\mathfrak{P}(\rho) = \mathfrak{P}(\alpha)$, $\alpha \in \mathfrak{W}_m(\Phi)$. This is equivalent to $\mathfrak{P}(\rho - \alpha) = 0$ which means that $\rho - \alpha \in Z$. Hence it is clear that the kernel of the homomorphism is $\mathfrak{W}_m(\Phi)$ and we have the isomorphism

$$(33)\qquad Q(\mathfrak{W}_m(\mathrm{P}))/\mathfrak{P}(\mathfrak{W}_m(\Phi)) \cong S(\mathfrak{W}_m(\mathrm{P}))/\mathfrak{W}_m(\Phi).$$

This implies that $Q(\mathfrak{W}_m(\mathrm{P}))/\mathfrak{P}(\mathfrak{W}_m(\Phi))$ is a finite group isomorphic to Hom (G, Z) and to G. We wish to show next that, if Q is any subgroup of $\mathfrak{W}_m(\Phi)$ containing $\mathfrak{P}(\mathfrak{W}_m(\Phi))$ as a subgroup of finite index, then $Q = Q(\mathfrak{W}_m(\mathrm{P}))$ for an abelian p-extension P over Φ. For this we need

Lemma 2. *Let* $\beta = (\beta_0, \beta_1, \cdots, \beta_{m-1}) \in \mathfrak{W}_m(\Phi)$. *Then there exists a finite dimensional separable extension field* P *of* Φ *such that* $\mathrm{P} = \Phi(\rho) \equiv \Phi(\rho_0, \rho_1, \cdots, \rho_{m-1})$ *and the element* $\rho = (\rho_0, \rho_1, \cdots, \rho_{m-1})$ *of* $\mathfrak{W}_m(\mathrm{P})$ *satisfies* $\mathfrak{P}(\rho) = \beta$.

Proof. If $m = 1$ we just have to construct a separable extension $\mathrm{P} = \Phi(\rho)$ generated by a root ρ of an equation $x^p - x = \beta$, β a given element in Φ. Since the derivative $(x^p - x - \beta)' = -1$ the given equation has distinct roots so any field generated by a root of this equation will satisfy the condition. Now suppose we have already constructed a separable extension $\mathrm{E} = \Phi(\rho_0, \cdots, \rho_{m-2})$ so that the vector $\sigma = (\rho_0, \cdots, \rho_{m-2})$ of $\mathfrak{W}_{m-2}$ (E) satisfies $\mathfrak{P}(\sigma) = (\beta_0, \cdots, \beta_{m-2})$. Consider the polynomial ring E$[x]$ and the Witt ring $\mathfrak{W}_m(\mathrm{E}[x])$. We take the vector $y = (\rho_0, \cdots, \rho_{m-2}, x)$ in this ring and we form

$$\mathfrak{P}(y) = (\rho_0^{\,p}, \cdots, \rho_{m-2}^{\,p}, x^p) - (\rho_0, \cdots, \rho_{m-2}, x).$$

Then $\mathfrak{P}(y) = (\beta_0, \beta_1, \cdots, \beta_{m-2}, f(x))$, $f(x) \varepsilon \mathrm{E}[x]$. Hence $(\beta_0, \beta_1, \cdots, \beta_{m-2}, f(x)) + (\rho_0, \cdots, \rho_{m-2}, x) = (\rho_0{}^p, \cdots, \rho_{m-2}{}^p, x^p)$. Using the formula (15) we see that

$$x^p = f(x) + x + \gamma \tag{34}$$

where $\gamma \varepsilon \mathrm{E}$. Hence $f(x) = x^p - x - \gamma$. The derivative argument shows that $f(x) = \beta_{m-1}$ has distinct roots. If $\mathrm{P} = \mathrm{E}(\rho_{m-1})$ where $f(\rho_{m-1}) = \beta_{m-1}$, then P is separable over E so $\mathrm{P} = \Phi(\rho_0, \cdots, \rho_{m-1})$ is separable over Φ. Moreover, it is clear from the formulas given above that $\rho = (\rho_0, \cdots, \rho_{m-1})$ is an element of $\mathfrak{W}_m(\mathrm{P})$ such that $\mathfrak{P}(\rho) = \beta$.

We can now prove

Theorem 15. *Let Q be a subgroup of $(\mathfrak{W}_m(\Phi), +)$ containing $\mathfrak{P}(\mathfrak{W}_m(\Phi))$ and having the property that $Q/\mathfrak{P}(\mathfrak{W}_m(\Phi))$ is finite. Then there exists an abelian p-extension* P *of Φ such that the exponent of the Galois group is p^e, $e \leq m$, and $Q(\mathfrak{W}_m(\mathrm{P})) = Q$.*

Proof. Let $\beta^{(1)}, \beta^{(2)}, \cdots, \beta^{(r)}$ be elements of Q such that the cosets $\beta^{(i)} + \mathfrak{P}(\mathfrak{W}_m(\Phi))$ generate $Q/\mathfrak{P}(\mathfrak{W}_m(\Phi))$. By Lemma 2, we can construct a field P which is finite dimensional separable over Φ and is generated by elements $\rho_\nu{}^{(i)}$, $1 \leq i \leq r$, $0 \leq \nu \leq m-1$, such that $\mathfrak{P}(\rho_0{}^{(i)}, \cdots, \rho_{m-1}{}^{(i)}) = (\beta_0{}^{(i)}, \cdots, \beta_{m-1}{}^{(i)})$ in $\mathfrak{W}_m(\mathrm{P})$. Let Ω be a finite dimensional Galois extension field of Φ containing P. We form $\mathfrak{W}_m(\Omega)$ and let the Galois group G of Ω/Φ act in $\mathfrak{W}_m(\Omega)$ as before. If $s \varepsilon G$ and $\rho^{(i)} = (\rho_0{}^{(i)}, \cdots, \rho_{m-1}{}^{(i)})$, then $\mathfrak{P}(\rho^{(i)}) = \beta^{(i)}$ gives $\mathfrak{P}(\rho^{(i)s}) = \beta^{(i)}$. Hence $\mathfrak{P}(\rho^{(i)s} - \rho^{(i)}) = 0$ so $\rho^{(i)s} - \rho^{(i)} \varepsilon Z \subseteq \mathfrak{W}_m(\Phi)$. This implies that $\mathrm{P}^s \subseteq \mathrm{P}$, $s \varepsilon G$. It follows that P is Galois over Φ and so we may take $\Omega = \mathrm{P}$. If s, t are in the Galois group G of P over Φ, then $\rho^{(i)s} = \rho^{(i)} + \gamma^{(i)}$ and $\rho^{(i)t} = \rho^{(i)} + \delta^{(i)}$ where $\gamma^{(i)}, \delta^{(i)} \varepsilon \mathfrak{W}_m(\Phi)$. Hence $\rho^{(i)st} = \rho^{(i)} + \gamma^{(i)} + \delta^{(i)} = \rho^{(i)ts}$ which implies that G is commutative. Also $\rho^{(i)s^k} = \rho^{(i)} + k\gamma^{(i)}$, so $\rho^{(i)s^{p^m}} = \rho^{(i)}$ since $\mathfrak{W}_m(\mathrm{P})$ has characteristic p^m. This shows that $s^{p^m} = 1$ and so G is of order p^f and of exponent p^e, $e \leq m$. Let χ_i be the character of G determined by $\rho^{(i)}$: $\chi_i(s) \equiv \rho^{(i)s} - \rho^{(i)}$. Then it is clear that $\chi_i(s) = 1$, $1 \leq i \leq r$, implies that $s = 1$. It follows that the χ_i generate the character group Hom (G, Z). Hence if ρ is any element of $\mathfrak{W}_m(\mathrm{P})$ such that $\mathfrak{P}(\rho) \varepsilon \mathfrak{W}_m(\Phi)$, then we have $\chi_\rho = \Pi\chi_i{}^{m_i}$. This implies that $\rho =$

$\Sigma m_i \rho^{(i)} + \beta$, $\beta \in \mathfrak{W}_m(\Phi)$, m_i integers. Then $\mathfrak{P}(\rho) = \Sigma m_i \beta^{(i)} + \mathfrak{P}(\beta) \in Q$. Since ρ is any element of $S(\mathfrak{W}_m(\mathrm{P}))$ this shows that $Q(\mathfrak{W}_m(\mathrm{P})) \subseteq Q$. The converse is clear so the proof is complete.

The results which we have now obtained correspond to the main results on Kummer extensions. They have the consequence that two abelian p-extensions P_1/Φ, P_2/Φ with Galois groups of exponent p^e, $e \leq m$, are isomorphic if and only if $Q(\mathfrak{W}_m(\mathrm{P}_1)) = Q(\mathfrak{W}_m(\mathrm{P}_2))$ (ex. 2 below). We have also the order preserving correspondence between the subfields P/Φ of a particular Ω/Φ and the subgroups $Q(\mathfrak{W}_m(\mathrm{P}))$ of the additive group $(\mathfrak{W}_m(\Phi), +)$ (ex. 1 below). We shall now consider the special case of cyclic p-extensions. We note first that it is an immediate consequence of our results that the cyclic extensions of p dimensions of Φ have the form $\Phi(\rho)$ where $\rho^p - \rho = \beta \in \Phi$ and $\beta \notin \mathfrak{P}(\Phi)$, that is, $\beta \neq \alpha^p - \alpha$, $\alpha \in \Phi$. We shall now show that, if such an extension exists over Φ, which is equivalent to the condition $\Phi \neq \mathfrak{P}(\Phi)$, then there exist cyclic extensions of p^m dimensions over Φ for any $m = 1, 2, \cdots$. This will follow from

Lemma 3. *If* $\beta_0, \cdots, \beta_{m-1} \in \Phi$, *then* $\beta_0 \in \mathfrak{P}(\Phi)$ *if and only if* $\beta = (\beta_0, \beta_1, \cdots, \beta_{m-1})$ *satisfies* $p^{m-1}\beta \in \mathfrak{P}(\mathfrak{W}_m(\Phi))$.

Proof. By (27), $p^{m-1}\beta = (0, \cdots, 0, \beta_0^{p^{m-1}})$. We have $(0, \cdots, 0, \beta_0) - (0, \cdots, 0, \beta_0^{p^{m-1}}) = (0, \cdots, 0, \beta_0) - (0, \cdots, 0, \beta_0^p) + (0, \cdots, 0, \beta_0^p) - (0, \cdots, 0, \beta_0^{p^2}) + \cdots + (0, \cdots, 0, \beta_0^{p^{m-2}}) - (0, \cdots, 0, \beta_0^{p^{m-1}}) \in \mathfrak{P}(\mathfrak{W}_m(\Phi))$. Hence $p^{m-1}\beta = (0, \cdots, 0, \beta_0^{p^{m-1}}) \in \mathfrak{P}(\mathfrak{W}_m(\Phi))$ if and only if $(0, \cdots, 0, \beta_0) \in \mathfrak{P}(\mathfrak{W}_m(\Phi))$. Suppose this holds, say, $(0, \cdots, 0, \beta_0) = \alpha^P - \alpha$ where $\alpha = (\alpha_0, \alpha_1, \cdots, \alpha_{m-1})$. Then $\alpha^{PR} - \alpha^R = (0, \cdots, 0, \beta_0)^R = 0$ so $\alpha^{RP} = \alpha^R$ and, if $\gamma = (\alpha_0, \alpha_1, \cdots, \alpha_{m-2}, 0)$, then $\gamma^P = \gamma$ so $\delta = \alpha - \gamma$ satisfies $\delta^P - \delta = (0, \cdots, 0, \beta_0)$. Moreover, $\delta = (0, \cdots, 0, \delta_{m-1})$. This implies that $\delta_{m-1}{}^p - \delta_{m-1} = \beta_0$ so $\beta_0 \in \mathfrak{P}(\Phi)$. Conversely, if this condition holds so that $\beta_0 = \alpha_{m-1}{}^p - \alpha_{m-1}$, then $\alpha^P - \alpha = (0, \cdots, 0, \beta_0)$ for $\alpha = (0, \cdots, 0, \alpha_{m-1})$.

We can now prove

Theorem 16. *Let* Φ *be a field of characteristic* $p \neq 0$. *Then there exist cyclic extensions of* p^m *dimensions,* $m = 1, 2, 3, \cdots$ *over* Φ *if and only if there exist such extensions of* p *dimensions. The condition for this is* $\Phi \neq \mathfrak{P}(\Phi)$.

Proof. We have seen that there exists a cyclic extension of p dimensions over Φ if and only if $\Phi \neq \mathfrak{P}(\Phi)$. Suppose this condition holds and choose $\beta_0 \in \Phi, \notin \mathfrak{P}(\Phi)$. Let $\beta = (\beta_0, \beta_1, \cdots, \beta_{m-1})$ where the β_i, $i > 0$, are any elements of Φ. We have shown that $p^{m-1}\beta \notin \mathfrak{P}(\Phi)$ and this implies that the subgroup Q of $\mathfrak{W}_m(\Phi)$ generated by β and $\mathfrak{P}(\mathfrak{W}_m(\Phi))$ has the property that $Q/\mathfrak{P}(\mathfrak{W}_m(\Phi))$ is cyclic of order p^m. By Theorem 15, $Q = Q(\mathrm{P})$ for an abelian p-extension P. Moreover, we have seen that the Galois group G of P/Φ is isomorphic to $Q/\mathfrak{P}(\mathfrak{W}_m(\Phi))$ and so this is cyclic of p^m dimensions over Φ.

EXERCISES

1. Let P_1 and P_2 be two abelian p-extensions of Φ contained in the same field Ω. Show that $\mathrm{P}_1 \supseteq \mathrm{P}_2$ if and only if $Q(\mathfrak{W}_m(\mathrm{P}_1)) \supseteq Q(\mathfrak{W}_m(\mathrm{P}_2))$ where $m \geq e_i$, p^{e_i}, the exponent of the Galois group of P_i/Φ.

2. Let P_i, m be as in 1, but do not assume that the P_i are contained in the same Ω. Show that P_1 and P_2 are isomorphic over Φ if and only if $Q(\mathfrak{W}_m(\mathrm{P}_1)) = Q(\mathfrak{W}_m(\mathrm{P}_2))$.

3. Prove that if β is an element of $\mathfrak{W}_m(\Phi)$ such that $p^{m-1}\beta \in \mathfrak{P}(\mathfrak{W}_m(\Phi))$, then there exists a γ in $\mathfrak{W}_m(\Phi)$ such that $p\gamma = \beta$. Use this to prove that any cyclic extension of p^{m-1} dimensions over Φ of characteristic p can be imbedded in a cyclic extension of p^m dimensions over Φ.

Chapter IV

STRUCTURE THEORY OF FIELDS

In this chapter we shall analyze arbitrary extension fields of a field Φ. A study of finite dimensional extension fields and a partial study of algebraic extensions has been made in Chapter I. In this chapter our primary concern will be with infinite dimensional extensions and we shall begin again with the algebraic ones. We define algebraically closed fields and prove the existence of an algebraic closure of any field. We shall extend the classical Galois theory to apply to infinite dimensional normal and separable extensions. After this we shall consider arbitrary extension fields and we shall show that these can be built up in two stages: first a purely transcendental one and then on top of this an algebraic extension. The invariant of this mode of generating a field is the transcendency degree which is the cardinal number of a transcendency basis. We shall obtain conditions for the existence of a transcendency basis such that the extension is separable algebraic over the purely transcendental extension determined by the basis. We shall also give a definition of separability of an extension field that generalizes the notion of algebraic separability. The notion of a derivation plays an important role in these considerations. Moreover, this notion can be used to develop a Galois theory for finite dimensional purely inseparable extensions of exponent one. We shall consider also briefly the notion of a higher derivation that is useful for purely inseparable extensions of exponent greater than one. At the end of the chapter we consider the tensor product of extension fields, neither of which is algebraic, and we apply this to the study of free composites of fields.

1. Algebraically closed fields. The "fundamental theorem of algebra" states that every algebraic equation $f(x) = 0$ with coefficients in the field of complex numbers has a root in this field. Any field that has this property is called *algebraically closed.* If Φ is an algebraically closed field, then every polynomial $f(x) \in \Phi[x]$ of positive degree has a linear factor $x - \rho$ in $\Phi[x]$ and, consequently, every $f(x)$ can be written as a product of linear factors in $\Phi[x]$. Clearly, the converse holds also: If every polynomial of positive degree in $\Phi[x]$ is a product of linear factors in $\Phi[x]$, then Φ is algebraically closed. We recall that a field Φ is called algebraically closed in an extension field P if the only elements of P which are algebraic over Φ are the elements belonging to Φ (§ 1.9). We now note that a field Φ is algebraically closed if and only if it is algebraically closed in every extension field. Thus let Φ be algebraically closed and let P be an extension field. Let $\rho \in$ P be algebraic over Φ and suppose $f(x)$ is its minimum polynomial. Since $f(x)$ is irreducible and Φ is algebraically closed, $f(x)$ is of first degree. Hence $\rho \in \Phi$. Conversely, suppose Φ is algebraically closed in every extension field and let $f(x)$ be an irreducible polynomial of positive degree belonging to $\Phi[x]$. We can form the extension field $\mathrm{P} = \Phi[x]/(f(x))$ whose dimensionality is the degree of $f(x)$. Since P is algebraic over Φ and Φ is algebraically closed in P, $\mathrm{P} = \Phi$. Hence $\deg f(x) = 1$, which shows that the only irreducible polynomials of positive degree in $\Phi[x]$ are the linear ones. This means that Φ is algebraically closed.

Let Φ be an arbitrary field and let P be an algebraically closed extension field of Φ. Let A/Φ be the subfield of elements of P/Φ of algebraic elements. If $f(x) \in \mathrm{A}[x]$, we have $f(x) = \Pi(x - \rho_i)$ in $\mathrm{P}[x]$ and the ρ_i are evidently algebraic over A. Since A is algebraically closed in P (§ 1.9), the $\rho_i \in$ A. Hence we see that every polynomial of positive degree in $\mathrm{A}[x]$ is a product of linear factors in $\mathrm{A}[x]$. This implies that A is algebraically closed. It is therefore clear that, if there exists an algebraically closed field containing a given field Φ, then there exists such a field which is, moreover, algebraic over Φ. This leads to the definition: An extension field A/Φ is called an *algebraic closure* of Φ if: 1) A is algebraic over Φ and 2) A is algebraically closed. We proceed to prove the existence and uniqueness in the sense of isomorphism of an algebraic closure for any field Φ.

If Φ is countable there is a rather straightforward way of constructing an algebraic closure of Φ. Thus, in this case, it is easy to enumerate the polynomials of positive degree with leading coefficients 1. Let $f_1(x), f_2(x), f_3(x), \cdots$ be such an enumeration. Then we begin with $\Phi_0 = \Phi$ and we construct Φ_i inductively as a splitting field over Φ_{i-1} of $f_i(x)$. There is a simple way of making precise the notion of the union $A = \bigcup \Phi_i$ of all the Φ_i. Once this has been done, one can prove that A is an algebraic closure of Φ in the following way. First, it is clear that A/Φ is algebraic. Let P be an algebraic extension of A and let $\rho \in P$. Since ρ is algebraic over A and A is algebraic over Φ, ρ is algebraic over Φ. Hence the minimum polynomial $f(x)$ over Φ is one of the polynomials $f_i(x)$, say $f(x) = f_n(x)$. Since Φ_n contains all the roots of $f(x)$, $\rho \in \Phi_n \subseteq A$. This shows that A is algebraically closed.

The procedure just sketched can be used also in the general case by invoking transfinite induction. However, we prefer to give another construction which will be based on Zorn's lemma.* We shall need also the following

Lemma. *If* A *is an algebraic extension of an infinite field* Φ, *then the cardinal number* $|A| = |\Phi|$.

Proof. Let Σ be the subset of $\Phi[x]$ of polynomials of positive degree with leading coefficients 1 and let $\Sigma^{(n)}$ be the subset of Σ of polynomials of degree $n + 1 = 1, 2, 3, \cdots$. The elements of $\Sigma^{(n)}$ have the form $x^{n+1} + \alpha_1 x^n + \alpha_2 x^{n-1} + \cdots + \alpha_n$, $\alpha_i \in \Phi$, so $\Sigma^{(n)}$ has the same cardinal number as the n-fold product set $\Phi \times \Phi \times \cdots \times \Phi$. Since Φ is infinite, $|\Sigma^{(n)}| = |\Phi \times \cdots \times \Phi| = |\Phi|$. Also $|\Sigma| = |\bigcup \Sigma^{(n)}| = |\Phi|$.† We now map each $f(x) \in \Sigma$ into the finite set R_f (possibly vacuous) of its roots in A. Since every element of A is algebraic, $\bigcup_{f \in \Sigma} R_f = A$. Since each R_f is finite the cardinal number of the collection $\{R_f\}$ of these subsets of A is the same as $|A|$. Hence $|A| = |\{R_f\}| \leq |\Sigma| = |\Phi|$. This implies that $|A| = |\Phi|$.

* This has been used in several places in Vol. II. An adequate account of this lemma or "maximum principle" can be found in Kelley's *General Topology*, D. Van Nostrand Co., Inc., Princeton, N. J., 1955, p. 33.

† For properties of cardinal numbers, see Sierpinski's *Leçons sur les Nombres Transfinis*, Paris, 1928.

We can now prove

Theorem 1. *Any field has an algebraic closure.*

Proof. If Φ is a given field, then we can imbed Φ in a set Ω which is very large compared to Φ in the following sense: if Φ is finite, then Ω is not countable and, if Φ is infinite, then $|\Omega| > |\Phi|$. We now make extension fields out of subsets E of Ω containing Φ. More precisely, we consider the collection Γ of all triples $(E, +, \cdot)$ where E is a subset of Ω containing Φ, and $+$ and $\cdot$ are binary compositions in E such that E with these compositions as addition and multiplication is an algebraic extension field of Φ. We partially order Γ by defining $(E_1, +_1, \cdot_1) < (E_2, +_2, \cdot_2)$ if E_2 is an extension field of E_1. Any linearly ordered subcollection $(E_\alpha, +_\alpha, \cdot_\alpha)$ of Γ has an upper bound whose underlying set is the union of the E_α and whose addition and multiplication are defined in the obvious way. Thus Zorn's lemma is applicable and it gives a maximal element $(A, +, \cdot)$ in the collection Γ. We assert that A is algebraically closed. Otherwise, A has a proper algebraic extension B. By the lemma, $|B| = |A| = |\Phi|$ if Φ is infinite. For finite Φ, $|A|$ and hence $|B|$ is countable. Hence, in both cases, $|B| < |\Omega|$. This implies that there exists a 1–1 mapping of B into Ω which is the identity on A. We can use this mapping to convert the image B' in Ω into a field over A isomorphic to E over A. Then $(B', +, \cdot)$, where the $+$ and $\cdot$ are the addition and multiplication obtained by carrying over the $+$ and $\cdot$ of B, is in the collection Γ. Moreover $B' \supset A$ and this contradicts the maximality of $(A, +, \cdot)$. Hence A is algebraically closed. Since A is algebraic over Φ, A is an algebraic closure of Φ.

We shall generalize next the notion of a splitting field and we shall prove a result for these which will give the uniqueness of algebraic closures as a special case. For this we consider a collection Ω of polynomials of positive degree with coefficients in Φ. We shall say that an extension field P/Φ is a *splitting field* of Ω if (1) every polynomial in Ω is a product of linear factors in $P[x]$ and (2) no proper subfield of P/Φ satisfies (1). Let A be an algebraic closure of Φ and let P be the subfield of A/Φ generated by all the roots of the polynomials $f \in \Omega$. Evidently P is a splitting field over Φ of the set Ω. It is clear also that A itself is a splitting field over Φ of the complete set of polynomials of positive degree belonging to $\Phi[x]$. The isomorphism over Φ of any two algebraic closures of Φ is clearly a consequence of the following extension of the theorem on splitting fields of a polynomial (Th. 1.8).

Theorem 2. *Let $\alpha \to \bar{\alpha}$ be an isomorphism of a field Φ onto a field $\bar{\Phi}$ and let Ω be a set of polynomials of positive degree contained in $\Phi[x]$, $\bar{\Omega}$ the set of images of the $f \varepsilon \Omega$ under the isomorphism $g(x) \to \bar{g}(x)$ of $\Phi[x]$ onto $\bar{\Phi}[x]$. Let P/Φ be a splitting field of Ω and $\bar{P}/\bar{\Phi}$ a splitting field of $\bar{\Omega}$. Then the isomorphism of Φ onto $\bar{\Phi}$ can be extended to an isomorphism of P onto $\bar{P}$.*

Proof. We consider the collection Δ of isomorphisms s of subfields of P/Φ onto subfields of $\bar{P}/\bar{\Phi}$ which coincide with the given isomorphism $\alpha \to \bar{\alpha}$ of Φ onto $\bar{\Phi}$. We can partially order $\Delta = \{s\}$ by defining $s_1 < s_2$ if s_2 is an extension of s_1. Then it is clear that Δ is inductive, that is, every linearly ordered subset of Δ has an upper bound. We may therefore invoke Zorn's lemma to obtain a maximal element $t \varepsilon \Delta$. We assert that t is an isomorphism of P onto $\bar{P}$ extending $\alpha \to \bar{\alpha}$. Otherwise, the domain of definition of t is a proper subfield E of P/Φ. Since P/Φ is a splitting field of Ω and $E \subset P$, there exists a polynomial $f(x) \varepsilon \Omega$ that does not have all of its roots in E. Hence if $\rho_1, \rho_2, \cdots, \rho_n$ are these roots in P, then $E(\rho_1, \rho_2, \cdots, \rho_n) \supset E$ and evidently $E(\rho_1, \cdots, \rho_n)$ is a splitting field over E of $f(x)$. On the other hand, $\bar{E} = E^t$ can be imbedded in a subfield of $\bar{P}$ which is a splitting field over $\bar{E}$ of $\bar{f}(x) \varepsilon \bar{\Omega}$. The theorem on a single polynomial can now be applied to give an extension of t to an isomorphism of $E(\rho_1, \cdots, \rho_n)$ onto the splitting field over $\bar{E}$ of $\bar{f}(x)$. This contradicts the maximality of t, so we see that $E = P$. Evidently the image P^t is a splitting field over $\bar{\Phi}$ of $\bar{\Omega}$. Hence $P^t = \bar{P}$ and the theorem is proved.

If we take $\bar{\Phi} = \Phi$ and $\bar{\alpha} \equiv \alpha$ in this result we see that any two splitting fields over Φ of a set of polynomials are isomorphic over Φ. In particular, we have the

Corollary. *Any two algebraic closures of a field Φ are isomorphic over Φ.*

Let A be an algebraic closure of a field Φ. There are two subfields of A/Φ which are of particular interest. The first of these is the subfield Σ of separable elements over Φ. This can be defined also as a splitting field over Φ of the set of separable polynomials belonging to $\Phi[x]$. We shall call Σ a *separable algebraic closure* of Φ. Next let Φ be of characteristic $p \neq 0$ and let $\Phi^{p^{-\infty}}$ be the subfield of elements of A which are purely inseparable over Φ. By Lemma 2 of § 1.9, these are the elements of A which are roots of equations of the form $x^{p^e} - \alpha = 0$, α in Φ. If Φ is of characteristic $p \neq 0$, we call $\Phi^{p^{-\infty}}$ a *perfect closure* of Φ, and if Φ is of characteristic 0, then the perfect closure of Φ is taken to be Φ itself. It is immediate that for $p \neq 0$ every element of $\Phi^{p^{-\infty}}$ is a p-th power so the mapping $\alpha \rightarrow \alpha^p$ is an automorphism of $\Phi^{p^{-\infty}}$. Moreover, $\Phi^{p^{-\infty}}$ is the smallest subfield of A over Φ which has the property that all of its elements are p-th powers in this subfield.

A field Φ is called *perfect* if every algebraic extension of Φ is separable. The perfect closure of any field which we have just defined is a perfect field; for we have the following

Theorem 3. *Any field of characteristic* 0 *is perfect and a field* Φ *of characteristic* $p \neq 0$ *is perfect if and only if* $\Phi = \Phi^p$, *that is, every element of* Φ *is a* p*-th power in* Φ.

Proof. The first statement is clear since inseparable polynomials exist only for characteristic $p \neq 0$. Now let Φ be of characteristic $p \neq 0$ and suppose $\Phi^p \subset \Phi$. Let α be an element of Φ which is not a p-th power in Φ. Then we know that $x^p - \alpha$ is irreducible and inseparable in $\Phi[x]$ (Lemma in § 1.6). Then $P = \Phi[x]/(x^p - \alpha)$ is an inseparable extension of Φ different from Φ so Φ is not perfect. Conversely, assume that $\Phi^p = \Phi$ and let $f(x)$ be a polynomial in $\Phi[x]$ such that $f'(x) = 0$. Then we can write $f(x) = g(x^p)$ where $g(x) = x^m + \beta_1 x^{m-1} + \cdots + \beta_m$. Let $\gamma_i^p = \beta_i$, $i = 1, \cdots, m$, and set $h(x) = x^m + \gamma_1 x^{m-1} + \cdots + \gamma_m$. Then we have $f(x) = g(x^p) = h(x)^p$. Thus every polynomial in $\Phi[x]$ having zero derivative is a p-th power and so there exist no irreducible inseparable polynomials of positive degree in $\Phi[x]$. Hence Φ has no proper inseparable algebraic extension field.

If Φ is a finite field of characteristic p, then the isomorphism $\alpha \rightarrow \alpha^p$ of Φ into Φ^p is necessarily an automorphism. It follows from Theorem 3 that every finite field is a perfect field.

EXERCISES

1. Let E be an algebraic extension of a field Φ and let A be an algebraic closure of Φ. Show that E/Φ is isomorphic to a subfield of A/Φ. (Hint: Consider the algebraic closure of E and note that this is an algebraic closure of Φ.)
2. Show that, if Φ is of characteristic $p \neq 0$ and ξ is transcendental over Φ, then $\Phi(\xi)$ is not perfect.
3. Prove that any algebraic extension of a perfect field is perfect.
4. Let Φ be a field, Φ^* its perfect closure. Prove that either $\Phi^* = \Phi$ or $[\Phi^*:\Phi]$ is infinite.
5. Prove that any algebraically closed field is infinite.

A field is called ***absolutely algebraic*** if it is algebraic over its prime field. Examples are finite fields. We recall that for every prime p and every integer n there exists one and in the sense of isomorphism only one field of cardinality p^n (§ 1.13). This result was generalized by Steinitz to arbitrary absolutely algebraic fields of characteristic $p \neq 0$. We indicate this in the following exercise.

6. A *Steinitz number* is a formal product $N = \Pi p_i^{k_i}$ over all primes p_i where $k_i = 0, 1, 2, \cdots$, or ∞. If $M = \Pi p_i^{l_i}$ is a second Steinitz number, we say that M is a divisor of N $(M|N)$ if $l_i \leq k_i$ for all i. This leads in an obvious way to a definition of the least common multiple (L.C.M.) of any collection of Steinitz numbers. Let Φ be absolutely algebraic of characteristic p. Define $\deg \Phi$ to be the Steinitz number L.C.M. of the degrees of the minimum polynomials over the prime field $(\cong I_p)$ of the elements of Φ. Note that if Φ is finite, then $|\Phi| = p^{\deg \Phi}$. Prove that for any given prime p and Steinitz number N there exists an absolutely algebraic field $\Phi_{p,N}$ of characteristic p and $\deg \Phi_{p,N} = N$. (Hint: Let r_n be the highest common factor of N and $n!$, so that $r_n | r_{n+1}$, $n = 1, 2, \cdots$. Let Φ_n be a field of cardinality p^{r_n} and suppose $\Phi_n \subseteq \Phi_{n+1} \subseteq \cdots$. Then $\Phi_{p,N} = \cup \Phi_n$.) Show that any two absolutely algebraic fields having the same prime characteristic and Steinitz degree are isomorphic. Prove that $\Phi_{p,M}$ is isomorphic to a subfield of $\Phi_{p,N}$ if and only if $M | N$.

2. Infinite Galois theory. In this section we shall give a generalization of the fundamental theorem of Galois theory to certain infinite dimensional algebraic extensions P/Φ. We assume that P is a splitting field over Φ of a set Ω of separable polynomials and we prove first the following

Lemma 1. *Any finite subset of* P *is contained in a subfield* E/Φ *which is finite dimensional Galois.*

Proof. Let f be a polynomial which is a product of a finite number of polynomials contained in the set Ω. Then it is clear that P contains a splitting field P_f/Φ of f. Moreover, we know that P_f is finite dimensional Galois over Φ (Th. 1.10). It is clear also that, if f and g are both products of polynomials belonging to Ω, then P_{fg} is the subfield of P generated by P_f and P_g. It follows that $\cup P_f$, the union of all of these subfields, is a subfield of P over Φ. Since $\cup P_f$ contains a splitting field of every $g \in \Omega$, it is clear that $\cup P_f = P$. This implies that any $\rho_i \in P$ is contained in a subfield P_{f_i} and consequently any finite subset $\{\rho_1, \rho_2, \cdots, \rho_m\}$ is contained in the subfield $P_{\Pi f_i}$ which is finite dimensional Galois over Φ.

Let G be the Galois group of P/Φ. The following result gives essentially the first half of the Galois correspondence.

Lemma 2. $\Phi = I(G)$, *that is, the only elements of* P *which are* G-*invariant are the elements of* Φ.

Proof. We have to show that, if $\rho \in P, \notin \Phi$, then there exists an automorphism s of P over Φ such that $\rho^s \neq \rho$. By Lemma 1, ρ is contained in a subfield E/Φ which is finite dimensional Galois over Φ. Since $\rho \notin \Phi$ there exists an element $\bar{s}$ of the Galois group of E/Φ such that $\rho^{\bar{s}} \neq \rho$. On the other hand, it is clear that P is a splitting field over E of the set of polynomials Ω and consequently, by Theorem 2, the automorphism $\bar{s}$ of E can be extended to an automorphism s of P/Φ. Evidently $s \in G$ and $\rho^s = \rho^{\bar{s}} \neq \rho$.

The full intermediate subfield-subgroup correspondence which holds in the finite dimensional case fails if P is of infinite dimensionality. As an example of this we consider the algebraic closure P of the field $\Phi = I_p$ of p elements. Since Φ is perfect, all polynomials of $\Phi[x]$ are separable and so P is a splitting field over Φ of a set Ω of separable polynomials contained in $\Phi[x]$. Let G be the Galois group of P/Φ and let H be the subgroup generated by the automorphism $\pi: \rho \to \rho^p$. (It is clear that this is an automorphism.) The subfield $I(H)$ of H-invariants is Φ since the only elements ρ such that $\rho^p = \rho$ are the p elements of Φ. We shall now show that H is a proper subgroup of G; then we shall have two subgroups of G, namely, G and H which have the same subfield of invariants. To do this we note that, if p^e is any power of p, $e \geq 1$, then P contains a subfield Φ_e of order p^e. We recall also that $\Phi_e \subseteq \Phi_f$ if and only if $e \mid f$ (§ 1.13). Now let l be a prime and let Φ_{l^∞} denote the union of the fields in the sequence $\Phi_l \subset \Phi_{l^2} \subset \Phi_{l^3} \subset \cdots$. It is immediate that Φ_{l^∞} is a proper subfield of P and P is a splitting field over Φ_{l^∞} of the set Ω. Hence Lemma 2 shows that there exists an automorphism s of P over Φ_{l^∞} such that

$s \neq 1$. Now s is not a power of the automorphism π; for if $s = \pi^k$, then the subfield of s-invariants is the finite set of elements satisfying $\rho^{p^k} = \rho$. This set must include Φ_{l^∞} and this is impossible since Φ_{l^∞} is an infinite subfield of P.

This type of difficulty in the infinite Galois theory was first observed (in the field of algebraic numbers of the rationals) by Dedekind. The way out of the difficulty was found by Krull who saw that it was necessary to restrict the correspondence to subgroups of the Galois group which are closed in a certain topology which we shall now define.

The topology one needs is essentially the same as the finite topology which we introduced in Vol. II, p. 248, for the set of linear transformations of one vector space into a second one. We consider the set P^P of (single-valued) mappings of the field P into itself. If $(\xi_1, \xi_2, \cdots, \xi_m)$ and $(\eta_1, \eta_2, \cdots, \eta_m)$ are finite sequences of elements of P, then we let $O(\xi_i, \eta_i)$ be the subset of P^P of all s such that $\xi_i^s = \eta_i$, $i = 1, \cdots, m$. The sets $O(\xi_i, \eta_i)$ can be used as a basis for a set of open sets which make P^P a topological space (cf. Vol. II, p. 248). This topology of P^P is called the *finite topology*.

If G is any subset of P^P, that is, any set of mappings of P into itself, then we topologize G as a subspace of P^P. In particular, we shall do this for the Galois group G of P/Φ. We now prove that the fact that P is algebraic over Φ implies that G is a closed subset of P^P. Thus let $\bar{s}$ belong to the closure of G and let $\xi, \eta \in P$, $\alpha \in \Phi$. Then there exists an $s \in G$ such that $\alpha^s = \alpha^{\bar{s}}$, $\xi^s = \xi^{\bar{s}}$, $\eta^s = \eta^{\bar{s}}$, $(\xi + \eta)^s = (\xi + \eta)^{\bar{s}}$, $(\xi\eta)^s = (\xi\eta)^{\bar{s}}$. Since $\alpha^s = \alpha$, $(\xi + \eta)^s = \xi^s + \eta^s$, $(\xi\eta)^s = \xi^s\eta^s$ we have the same relations for $\bar{s}$ and these show that $\bar{s}$ is an isomorphism of P/Φ into itself. To see that $\bar{s}$ is surjective we let ξ be any element of P and we let E be the subfield of P/Φ generated by all the roots ξ' in P of the minimum polynomial $f(x)$ of ξ over Φ. Clearly $[E:\Phi] < \infty$. Since $\bar{s}$ is an isomorphism of P/Φ into itself, $E^{\bar{s}} \subseteq E$. Hence the restriction of $\bar{s}$ is a linear isomorphism of E/Φ into itself and so this mapping is surjective. Thus there exists an $\eta \in E$ such that $\eta^{\bar{s}} = \xi$. Hence $\bar{s}$ is an automorphism of P/Φ, so $\bar{s} \in G$ and G is closed.

We can prove the following fundamental theorem of the infinite Galois theory.

Theorem 4. *Let* P/Φ *be a splitting field of a set* Ω *of separable polynomials with coefficients in* Φ *and let* G *be the Galois group of* P/Φ. *With each closed subgroup* H *of* G *we associate the subfield* $E = I(H)$ *of* H*-invariants and with each subfield* E *of* P *over* Φ *we associate the Galois group* $A(E)$ *of* P *over* E. *Then these two correspondences are inverses of each other. Moreover, a closed subgroup* H *is invariant in* G *if and only if the corresponding field* $E = I(H)$ *is Galois over* Φ *and, in this case, the Galois group of* E/Φ *is isomorphic to* G/H.

Proof. If E is a subfield of P/Φ, then P is a splitting field over E of Ω. Hence if $H = A(E)$ the Galois group of P/E, then H is closed and Lemma 2 shows that $I(A(E)) = E$. Next let H be a closed subgroup of G and let $E = I(H)$. We have to show that, if s is an automorphism of P/E, then $s \in H$. Since H is closed it is enough to show that s is in the closure of H, that is, if $\rho_1, \cdots, \rho_n \in P$, then there exists a $t \in H$ such that $\rho_i^t = \rho_i^s$, $1 \leq i \leq n$. Let Λ/E be a subfield of P/E which is finite dimensional Galois and contains $\{\rho_i\}$ (Lemma 1). Then s and the $t \in H$ map Λ into itself and so their restrictions are elements of the Galois group of Λ/E. If the restriction s' of s to Λ coincides with no restriction t' of $t \in H$ to Λ, then the group H' of the restrictions of $t \in H$ is a proper subgroup of the Galois group of Λ over E. Consequently, there exists an element $\xi \in \Lambda, \notin E$ such that $\xi^t = \xi$ for every $t \in H$. This contradicts the definition of E as $I(H)$. This proves the first statement. If H is a closed subgroup and $s \in G$, then $s^{-1}Hs$ is closed, and if $E = I(H)$, then $E^s = I(s^{-1}Hs)$. It follows that H is invariant in G if and only if $E^s = E$ for every $s \in G$. If this condition holds, then the set of restrictions to E of the $s \in G$ is a group of automorphisms $\bar{G}$ in E whose set of invariants is Φ. Hence E is Galois over Φ. Conversely, assume E is Galois over Φ and let $\bar{G}$ be the Galois group of E/Φ. If $\epsilon \in E$ and $\bar{s} \in \bar{G}$, then ϵ and $\epsilon^{\bar{s}}$ have the same minimum polynomial over Φ. Hence ϵ has only a finite number of conjugates $\epsilon^{\bar{s}}$, $\bar{s} \in \bar{G}$. If these are $\epsilon_1 = \epsilon$, $\epsilon_2, \cdots, \epsilon_r$, then the polynomial $f(x) = \Pi(x - \epsilon_i)$ has coefficients in Φ and ϵ is a root of $f(x) = 0$. If $s \in G$, then ϵ^s is also a root of $f(x) = 0$ so $\epsilon^s = \epsilon_i \in E$. Since ϵ is arbitrary, this shows that $E^s \subseteq E$ for $s \in G$. This implies that H is invariant in G. Since

P is a splitting field over E of a set of polynomials belonging to $\Phi[x]$, any automorphism of E/Φ can be extended to an automorphism of P/Φ. It follows easily from this that, if E is Galois over Φ, then the mapping $s \to \bar{s}$ the restriction of s to E is a homomorphism of G onto $\bar{G}$. The kernel is $H = A(\mathrm{E})$ so $\bar{G} \cong G/H$.

EXERCISES

1. Show that the hypothesis of Theorem 4 can be replaced by: P/Φ is a separable and normal algebraic extension, where we define normality by the condition that, if $f(x)$ is irreducible in $\Phi[x]$ and has a root in P, then $f(x)$ is a product of linear factors in P[x], (cf. § 1.8).
2. A topological space is called *discrete* if every subset is open. Let P be a splitting field over Φ of a set of separable polynomials and let G be the Galois group of P over Φ. Show that G is discrete if and only if $[\mathrm{P}:\Phi] < \infty$.
3. Let P, Φ, and G be as in ex. 2. Use the fact that every $\rho \in \mathrm{P}$ has only a finite number of conjugates and the Tychonoff theorem to prove that G is a compact group.
4. Let P be the algebraic closure of the field $\Phi = I_p$ and let G be the Galois group of P over Φ. Show that G is a commutative group. Let Φ_{l^∞} be the subfield of P defined for the prime l as in the text. Let π be the automorphism $\rho \to \rho^p$ restricted to Φ_{l^∞}. Show that $\pi^{l^k} \to 1$ in the sense that, if S is any finite subset of Φ_{l^∞}, then there exists a positive integer N such that $\xi^{\pi^{l^k}} = \xi$ for all $\xi \in S$ provided $k \geq N$. Let $m_1, m_2, \cdots$ be a sequence of integers such that for any positive integer k there exists an N such that $m_r \equiv m_s \pmod{l^k}$ if $r, s \geq N$. Show that the sequence of automorphisms $\pi^{m_1}, \pi^{m_2}, \cdots$ converges to an automorphism σ of Φ_{l^∞} over Φ in the sense that $\pi^{m_k}\sigma^{-1} \to 1$.
5. Let G be a group of automorphisms in a field P and let $\Phi = I(G)$. Assume G is a compact subset of P^{P}. Show that this implies that for every $\xi \in \mathrm{P}$ the set $\{\xi^s \mid s \in G\}$ is finite. Hence prove that P is the splitting field over Φ of a set of separable polynomials and that G is the Galois group of P/Φ.
6. Let G be the Galois group of P/Φ where P is algebraic over Φ and let $\{G_\alpha\}$ be the collection of invariant subgroups of finite index in G. Show that $\bigcap G_\alpha = 1$.
7. Let Φ be a finite field, A its algebraic closure, and G the Galois group of A/Φ. Show that G has no elements of finite order $\neq 1$.
8. Let A_p be the algebraic closure of the field I_p of p elements, G_p the Galois group of A_p/I_p. Show that $G_p \cong G_q$ for any two primes p, q.
9. Let $\mathrm{P} = \Phi(\xi_1, \xi_2, \cdots)$ the field of rational expressions in an infinite number of indeterminates. Show that the Galois group of P/Φ is not closed in the finite topology.

3. Transcendency basis. We have defined the property of algebraic independence over Φ for a finite subset $\{\xi_1, \xi_2, \cdots, \xi_n\}$ of a field P over Φ in Vol. I and this definition has been repeated in Introduction, p. 4. We now extend this notion to arbitrary subsets by stating that such a set S is algebraically independent if every finite subset of S is algebraically independent. A set which is not algebraically independent will be called algebraically dependent; hence a set is algebraically dependent if and only if it contains a non-vacuous algebraically dependent finite subset. We shall now introduce another notion, which we shall see in our first theorem is intimately related to those just given.

Definition 1. *Let S be a subset of* P *over* Φ *and let* ρ *be an element of* P; *then* ρ *is said to be* algebraically dependent over Φ on S *if* ρ *is algebraic over* $\Phi(S)$.

We note first that, if ρ is algebraically dependent over Φ on S and $f(x) \in \Phi(S)[x]$ is the minimum polynomial of ρ over $\Phi(S)$, then the coefficients of $f(x)$ are contained in a subfield $\Phi(F)$ where F is a finite subset of S. Hence it is clear that ρ is algebraically dependent over Φ on a set S if and only if ρ has this property for a finite subset F of S.

Theorem 5. *A non-vacuous subset S of a field* P/Φ *is algebraically dependent over* Φ *if and only if there exists an element* $\xi \in S$ *which is algebraically dependent over* Φ *on the complementary set* $S - \{\xi\}$.

Proof. The remarks we have made show that it is sufficient to assume S is finite, say, $S = \{\xi_1, \xi_2, \cdots, \xi_n\}$. Assume the condition stated holds. Then we may suppose that ξ_n is algebraic over $\Phi(\xi_1, \cdots, \xi_{n-1})$. Let $f(x) \in \Phi(\xi_1, \cdots, \xi_{n-1})[x]$ be the minimum polynomial of ξ_n over $\Phi(\xi_1, \cdots, \xi_{n-1})$ and let $\beta_1, \beta_2, \cdots, \beta_m$ be its coefficients. Now every element of $\Phi(\xi_1, \cdots, \xi_{n-1})$ has the form $g(\xi_1, \cdots, \xi_{n-1})h(\xi_1, \cdots, \xi_{n-1})^{-1}$ where $g, h \in \Phi[x_1, \cdots, x_{n-1}]$, x_i indeterminates, and $h(\xi_1, \cdots, \xi_{n-1}) \neq 0$. In particular, $\beta_j = g_j(\xi_1, \cdots, \xi_{n-1})h_j(\xi_1, \cdots, \xi_{n-1})^{-1}$, $h_j(\xi_1, \cdots, \xi_{n-1}) \neq 0$. Set $h(x_1, \cdots, x_{n-1}) = \Pi h_j(x_1, \cdots, x_{n-1})$ and

$$\begin{aligned} F(x_1, \cdots, x_n) = h(x_1, \cdots, x_{n-1})\{x_n{}^m & \\ + g_1(x_1, \cdots, x_{n-1})h_1(x_1, \cdots, x_{n-1})^{-1}x_n{}^{m-1} & \\ + \cdots + g_m(x_1, \cdots, x_{n-1})h_m(x_1, \cdots, x_{n-1})^{-1}\}. & \end{aligned}$$

Then F is a non-zero element in $\Phi[x_1, \cdots, x_n]$, x_i indeterminates, and we have $F(\xi_1, \cdots, \xi_n) = 0$. This means that the ξ_i are algebraically dependent. Conversely, assume there exists a non-zero polynomial $F(x_1, \cdots, x_n) \in \Phi[x_1, \cdots, x_n]$ such that $F(\xi_1, \cdots, \xi_n) = 0$, which amounts to saying that the ξ_i are algebraically dependent over Φ. We may assume that n is minimal and we may write $F(x_1, \cdots, x_n) = f_0(x_1, \cdots, x_{n-1})x_n^m + f_1(x_1, \cdots, x_{n-1})x_n^{m-1} + \cdots + f_m(x_1, \cdots, x_{n-1})$ where $f_0 \neq 0$ and $m \geq 1$. Since n is minimal, $f_0(\xi_1, \cdots, \xi_{n-1}) \neq 0$. Then

$$f(x) = x^m + \sum_{1}^{m} f_i(\xi_1, \cdots, \xi_{n-1})f_0(\xi_1, \cdots, \xi_{n-1})^{-1}x^{m-i}$$

is a non-zero element of $\Phi(\xi_1, \cdots, \xi_{n-1})[x]$ such that $f(\xi_n) = 0$. Hence ξ_n is algebraically dependent on $\xi_1, \cdots, \xi_{n-1}$.

The relation of algebraic dependence in a field P/Φ is a special kind of relation between elements of P and subsets of P. Another relation of a similar type is that of linear dependence of a vector in a vector space on a subset of the space, and we shall encounter still others. It is therefore worthwhile to treat such relations axiomatically and we shall do this by considering an arbitrary set P. A relation $<$ between elements of P and subsets S of P ($\xi < S$) is called a *dependence relation* if the following conditions hold.

I. If $\xi \in S$, then $\xi < S$.
II. If $\xi < S$, then $\xi < F$ for some finite subset F of S.
III. If $\xi < S$ and every η in S satisfies $\eta < T$, then $\xi < T$.
IV. If $\xi < S$ and $\xi \nless S - \{\eta\}$ where $\eta \in S$, then $\eta < (S - \{\eta\}) \cup \{\xi\}$ (Exchange axiom).

Now let P be a field over Φ and let $\xi < S$ for ξ in P, S a subset of P, mean that ξ is algebraically dependent on S over Φ. Then we have

Theorem 6. *Algebraic dependence in* P/Φ *is a dependence relation in the sense of I–IV.*

Proof. I. This is evident. II. This was proved before. III. Let ξ be algebraic over $\Phi(S)$ and suppose every $\eta \in S$ is algebraic over $\Phi(T)$. Consider the subset A of P of elements which are algebraic over $\Phi(T)$. Then we know that A is a subfield of $P/\Phi(T)$ and A is algebraically closed in P. Now $S \subseteq A$ and ξ is algebraic over $\Phi(S)$, so over A. Hence $\xi \in A$ which means that $\xi < T$.

IV. Suppose $\xi < S$ and $\xi \nless T = S - \{\eta\}$ where $\eta \in S$. Let $E = \Phi(T)$. Then ξ is transcendental over E and algebraic over $E(\eta)$. Hence there exists a polynomial $f(x, y) \in E[x, y]$, x, y indeterminates over E, such that $f(x, y) \neq 0$ and $f(\xi, \eta) = 0$. We write $f(x, y) = a_0(x)y^m + a_1(x)y^{m-1} + \cdots + a_m(x)$ where the $a_i(x) \in E[x]$ and $a_0(x) \neq 0$. Then $a_0(\xi) \neq 0$ and $m > 0$, since ξ is transcendental over E. The polynomial $f(\xi, y)$ is a non-zero polynomial belonging to $E(\xi)[y]$ and η is a root of $f(\xi, y) = 0$. Hence η is algebraic over $E(\xi)$, which implies that η is algebraic over $\Phi(T \cup \{\xi\})$. Thus $\eta < T \cup \{\xi\}$.

We now return to the general theory of dependence relations. As before, P is an arbitrary set. We define a subset S of P to be *independent* (relative to $<$) if no $\xi \in S$ is dependent on $S - \{\xi\}$. Then we have the following

Lemma. *If B is independent and ξ is not dependent on B, then $B \cup \{\xi\}$ is independent.*

Proof. Otherwise, we have an $\eta \in B$ such that $\eta < (B \cup \{\xi\}) - \{\eta\}$. Since $\eta \nless B - \{\eta\}$ the exchange axiom implies that $\xi < B = (B - \{\eta\}) \cup \{\eta\}$ contrary to hypothesis.

A subset B of P will be called a *basis* for P (relative to $<$) if (1) B is independent and (2) every ξ in P is dependent on S. The main result on dependence relations is the following

Basis theorem. *The set* P *has a basis. Moreover, any two bases have the same cardinal number.*

Proof. To prove the existence of a basis we consider the collection I of subsets of P which are independent. (It may happen that the vacuous set is the only member of I.) We order I by the inclusion relation. If $\{S\}$ is a linearly ordered subset of I, then $\cup S$ is contained in I. Otherwise, there is a $\xi \in \cup S$ which is dependent on $\cup S - \{\xi\}$. Then $\xi < F$ where F is a finite subset of $\cup S - \{\xi\}$ and $F \cup \{\xi\}$ is a finite subset which is not independent. Since $\{S\}$ is linearly ordered, $F \cup \{\xi\} \subseteq T$ for some $T \in \{S\}$ and this contradicts the assumption that T is an independent set. We now see that I is inductive and so, by Zorn's lemma, there exists a maximal element B in I. Now let ξ be any element of P. Then ξ is dependent on B since otherwise $B \cup \{\xi\}$ is independent, by the lemma. This would contradict the maximality of B.

Hence every ξ in P is dependent on B. This means that B is a basis.

Now let B and C be two bases for P. We have to show that the cardinal numbers $|B| = |C|$. Assume first that B is finite, say, $B = \{\beta_1, \beta_2, \cdots, \beta_n\}$. We assert that there is a $\gamma = \gamma_1 \varepsilon C$ such that γ is not dependent on $\{\beta_2, \cdots, \beta_n\}$. Otherwise, by III, every element of P is dependent on $\{\beta_2, \cdots, \beta_n\}$. In particular, β_1 has this property, contrary to the independence of B. Now if γ_1 is not dependent on $\{\beta_2, \cdots, \beta_n\}$, then $\{\gamma_1, \beta_2, \cdots, \beta_n\}$ is independent. Moreover, the exchange axiom shows that $\beta_1 < \{\gamma_1, \beta_2, \cdots, \beta_n\}$ so every $\beta_i < \{\gamma_1, \beta_2, \cdots, \beta_n\}$. Thus $\{\gamma_1, \beta_2, \cdots, \beta_n\}$ is a basis. We can repeat this process and obtain γ_2 in C so that $\{\gamma_1, \gamma_2, \beta_3, \cdots, \beta_n\}$ is a basis. Continuing in this way we obtain a basis $\{\gamma_1, \cdots, \gamma_n\}$ which is a subset of C and has the same cardinal number as B. Since C is independent, this is all of C and we have $|C| = |B|$. Next assume $|C|$ and $|B|$ are infinite. In this case we use a counting argument which is due to Löwig (cf. Vol. II, p. 241). Let $\gamma \varepsilon C$. Then γ is dependent on a finite subset B_γ of B. Consequently $|\{B_\gamma\}| \leq |C|$ and

$$\Big|\bigcup_{\gamma \varepsilon C} B_\gamma\Big| \leq \aleph_0 |C| = |C|.$$

Next we note that $\cup B_\gamma = B$. Otherwise, we have a $\beta \varepsilon B$, $\notin \cup B_\gamma$. Since $\beta < C$ and every $\gamma \varepsilon C$ satisfies $\gamma < \cup B_\gamma$, we have $\beta < \cup B_\gamma$ which does not contain β. This contradicts the independence of B. Thus $\cup B_\gamma = B$ and the above relation on cardinals gives $|B| \leq |C|$. By symmetry $|C| \leq |B|$; hence $|B| = |C|$.

This result is applicable in particular to algebraic dependence in P/Φ. In this case a basis B is a set of algebraically independent elements of P/Φ such that every ξ in P is algebraically dependent on B. Such a set B is called a *transcendency basis* for P/Φ and its cardinal number, which is the same for all bases, is called the *transcendency degree* (tr. d.) of P/Φ. An extension P/Φ is algebraic if and only if P has a vacuous transcendency basis over Φ, hence if and only if the transcendency degree is 0. If a field P has a transcendency basis B over Φ such that P $= \Phi(B)$, then P is called a *purely transcendental extension* of Φ. The theorem on the existence of a transcendency basis can be interpreted in the following manner: Every field can be obtained as an algebraic extension of a purely transcendental extension $\Phi(B)$ of the base field Φ. If $x_1, x_2, \cdots, x_r$ are indeterminates, then the field of fractions of the algebra $\Phi[x_1, \cdots, x_r]$ is a purely transcendental extension $\Phi(x_1, \cdots, x_r)$ of Φ with the transcendency basis $\{x_i\}$. Moreover, it is clear that any purely transcendental extension of degree $r < \infty$ is essentially identical with $\Phi(x_1, \cdots, x_r)$.

Of particular interest in algebraic geometry are the fields $P = \Phi(\xi_1, \xi_2, \cdots, \xi_n)$ which are generated over the base field Φ by a finite set of elements ξ_i. If B is a maximal algebraically independent subset of the set $\{\xi_1, \xi_2, \cdots, \xi_n\}$, then B is a transcendency basis. We may assume $B = \{\xi_1, \xi_2, \cdots, \xi_r\}$. A field of the form $P = \Phi(\xi_1, \xi_2, \cdots, \xi_n)$ is called a *field of algebraic functions* over Φ and the transcendency degree r ($\leq n$) is called the *number of variables* of P. If $\{\xi_1, \cdots, \xi_r\}$ is a transcendency basis, then P is a finite dimensional extension of $\Phi(\xi_1, \xi_2, \cdots, \xi_r)$. If this is separable over $\Phi(\xi_1, \cdots, \xi_r)$, then one of the theorems on primitive elements shows that $P = \Phi(\xi_1, \cdots, \xi_r, \eta)$ for a suitable η in P. This is always the case for characteristic 0 and we shall see in § 5 that simple conditions can be given to insure the existence of a basis $\{\xi_i\}$ for a field of algebraic functions such that P is separable algebraic over $\Phi(\xi_1, \xi_2, \cdots, \xi_r)$.

EXERCISES

1. Show that, if C is a subset of P/Φ such that every element of P is algebraically dependent on C, then C contains a transcendency basis. Show also that, if D is an algebraically independent subset of P/Φ, then D can be imbedded in a transcendency basis.

2. Let E/Φ be a subfield of P/Φ. Show that the transcendency degree tr. d. $P/E \leq$ tr. d. P/Φ and that tr. d. $E/\Phi \leq$ tr. d. P/Φ.

3. Let E/Φ be a subfield of P/Φ and let B and C be transcendency bases for E/Φ and P/E respectively. Show that $B \cup C$ is a transcendency basis for P/Φ. Hence prove the formula

$$\text{tr. d. } P/\Phi = \text{tr. d. } P/E + \text{tr. d. } E/\Phi. \tag{1}$$

Note that ex. 2 is a consequence of this. (Hint: Since E is algebraic over $\Phi(B)$ the subalgebra generated by E/Φ and $\Phi(C)$ is a field which is algebraic over $\Phi(B, C)$ (p. 45). Hence $E(C)$ is algebraic over $\Phi(B, C)$.)

4. Prove that, if Φ is a field of characteristic $\neq 3$ and $P = \Phi(\xi, \eta)$ where ξ is transcendental and $\eta^3 + \xi^3 = 1$, then P is not purely transcendental over Φ.

5. Let P be the field of complex numbers, Φ the subfield of rationals. Show that tr. d. $P/\Phi = c = |P|$. Show that, if B is a transcendency basis of P/Φ, then any 1–1 surjective mapping of B can be extended to an automorphism of P/Φ. Hence show that P has as many automorphisms as 1–1 surjective mappings.

6. Prove that, if P is finitely generated over Φ, then this holds for any subfield E/Φ.

4. Lüroth's theorem. The purely transcendental extensions $P = \Phi(\xi_1, \xi_2, \cdots, \xi_r)$ appear to be the simplest types of extension fields. Nevertheless, it is easy to ask difficult questions about such extensions, particularly about subfields of P/Φ if $r > 1$. If $r = 1$ the situation is comparatively simple and we shall look at this in this section.

Let $P = \Phi(\xi)$, ξ transcendental, and let η be an element of P which is not contained in Φ. We can write $\eta = f(\xi)g(\xi)^{-1}$ where $f(\xi)$ and $g(\xi)$ are polynomials in ξ which we may assume have no common factor of positive degree in ξ. We may write $f(\xi) = \alpha_0 + \alpha_1\xi + \cdots + \alpha_n\xi^n$, $g(\xi) = \beta_0 + \beta_1\xi + \cdots + \beta_n\xi^n$ where either $\alpha_n \neq 0$ or $\beta_n \neq 0$, so n is the larger of the degrees of f and g. The relation $\eta = f(\xi)g(\xi)^{-1}$ gives $f(\xi) - \eta g(\xi) = 0$ and

$$0 = (\alpha_n - \eta\beta_n)\xi^n + (\alpha_{n-1} - \eta\beta_{n-1})\xi^{n-1} + \cdots + (\alpha_0 - \eta\beta_0).$$

Moreover, $\alpha_n - \eta\beta_n \neq 0$ since α_n or $\beta_n \neq 0$ and $\eta \notin \Phi$. Thus we see that ξ is a root of the equation of degree n: $\sum_0^n (\alpha_i - \eta\beta_i)x^i = 0$ with coefficients in $\Phi(\eta)$. We proceed to show that $\sum_0^n (\alpha_i - \eta\beta_i)x^i$ is irreducible in $\Phi(\eta)[x]$. First, it is clear that η is transcendental over Φ, since ξ is algebraic over $\Phi(\eta)$; hence η algebraic over Φ implies ξ algebraic over Φ, contrary to assumption. The ring $\Phi[\eta, x] = \Phi[\eta][x]$ is the polynomial ring in two indeterminates η, x and we know that this ring is Gaussian, that is, the theorem on unique factorization into irreducible elements holds in $\Phi[\eta, x]$ (Vol. I, p. 126). We recall also that a polynomial in $\Phi[\eta, x]$ of positive degree in x is irreducible in $\Phi(\eta)[x]$ if it is irreducible in $\Phi[\eta, x]$ Now $f(\eta, x) = \Sigma(\alpha_i - \eta\beta_i)x^i = f(x) - \eta g(x)$ is of degree 1 in η. Hence if $f(\eta, x)$ is reducible in $\Phi(\eta)[x]$, then it has a factor $h(x)$ of positive degree in x. This implies that $f(x)$ and $g(x)$ are divisible by $h(x)$ contrary to assumption. We have therefore shown that

$f(\eta, x)$ is irreducible in $\Phi(\eta)[x]$. Thus ξ is algebraic of degree n over $\Phi(\eta)$. This proves

Theorem 7. *Let* $P = \Phi(\xi)$, ξ *transcendental over* Φ *and let* η *be an element of* P *not in* Φ. *Write* $\eta = f(\xi)g(\xi)^{-1}$ *where* $f(\xi)$ *and* $g(\xi)$ *are polynomials in* ξ *with no common factor of positive degree in* ξ. *Let* $n = \max(\deg f, \deg g)$. *Then* ξ *is algebraic over* $\Phi(\eta)$ *and* $[\Phi(\xi):\Phi(\eta)] = n$. *Moreover,* $f(x, \eta) = f(x) - \eta g(x)$ *is irreducible in* $\Phi(\eta)[x]$.

This result enables us to determine the automorphisms of $\Phi(\xi)$ over Φ. Such an automorphism is completely specified by the image η of the generator ξ. For, if $\xi \to \eta$, then $u(\xi)v(\xi)^{-1} \to u(\eta)v(\eta)^{-1}$ for u, v polynomials in ξ. It is clear also that, if η is the image of ξ under an automorphism, then $\Phi(\eta) = \Phi(\xi)$. If $\eta = f(\xi)g(\xi)^{-1}$ as above, then $[\Phi(\xi):\Phi(\eta)] = n = \max(\deg f, \deg g)$. This shows that $\Phi(\eta) = \Phi(\xi)$ if and only if $\max(\deg f, \deg g) = 1$. Then we have

$$\eta = \frac{\alpha\xi + \beta}{\gamma\xi + \delta}, \tag{2}$$

where $\alpha \neq 0$ or $\gamma \neq 0$ and $\alpha\xi + \beta$, $\gamma\xi + \delta$ have no common factor of positive degree. It is easy to see that these conditions are equivalent to the single condition:

$$\alpha\delta - \beta\gamma \neq 0. \tag{3}$$

If this condition holds, then $\Phi(\eta) = \Phi(\xi)$ and the mapping $u(\xi)v(\xi)^{-1} \to u(\eta)v(\eta)^{-1}$ is an automorphism of P/Φ.

The condition (3) is equivalent to the requirement that the matrix

$$\begin{bmatrix} \alpha & \beta \\ \gamma & \delta \end{bmatrix} \tag{4}$$

is non-singular. With each such matrix we associate the automorphism of $\Phi(\xi)$ over Φ such that $\xi \to \eta$ given by (2). One verifies directly that the mapping of the non-singular matrix into the corresponding automorphism is a group homomorphism. The kernel is the set of matrices $\begin{bmatrix} \alpha & \beta \\ \gamma & \delta \end{bmatrix}$ such that $(\alpha\xi + \beta)(\gamma\xi + \delta)^{-1} = \xi$ or $\alpha\xi + \beta = \xi(\gamma\xi + \delta)$. This implies $\gamma = 0$, $\beta = 0$, $\alpha = \delta$.

Hence the kernel is the set of scalar matrices $\begin{bmatrix} \alpha & 0 \\ 0 & \alpha \end{bmatrix} \neq 0$. It is now clear that the group of automorphisms of $\Phi(\xi)$ is isomorphic to the factor group of the group $L(\Phi, 2)$ of 2×2 non-singular matrices relative to the subgroup of scalar matrices. This factor group is called the *projective group* $PL(\Phi, 2)$.

We now consider an arbitrary subfield E of $\Phi(\xi)/\Phi$. We may assume $E \neq \Phi$. Then E contains an element η not in Φ so $P = \Phi(\xi)$ is algebraic over $\Phi(\eta)$ and hence is algebraic over $E \supseteq \Phi(\eta)$. Let the minimum polynomial of ξ over E be $f(x) = x^n + \gamma_1 x^{n-1} + \cdots + \gamma_n$. The γ_i have the form $\mu_i(\xi)\nu_i(\xi)^{-1}$ where μ_i, ν_i are polynomials in the transcendental element ξ. Multiplication of $f(x)$ by a suitable polynomial in ξ will give a polynomial

$$f(\xi, x) = c_0(\xi)x^n + c_1(\xi)x^{n-1} + \cdots + c_n(\xi) \tag{5}$$

in $\Phi[\xi, x]$, ξ, x indeterminates, which is a primitive polynomial in x in the sense that the highest common factor of the $c_i(\xi)$ is 1. Also we have $\gamma_i = c_i(\xi)c_0(\xi)^{-1} \in E$ and not all of these are in Φ since ξ is transcendental over Φ. Thus one of the γ's has the form $\gamma = g(\xi)h(\xi)^{-1}$ where $g(\xi)$, $h(\xi)$ have no common factor of positive degree in ξ and $\max(\deg g, \deg h) = m > 0$. We have seen before that $g(x) - \gamma h(x)$ is irreducible in $\Phi(\gamma)[x]$ and $[P:\Phi(\gamma)] = m$. Since $E \supseteq \Phi(\gamma)$ and $[P:E] = n$, clearly $m \geq n$. We shall show that $m = n$ and this will prove that $E = \Phi(\gamma)$.

Since ξ is a root of $g(x) - \gamma h(x) = 0$ and the coefficients of this polynomial are contained in E, we have $g(x) - \gamma h(x) = f(x)q(x)$ in $E[x]$. We have $\gamma = g(\xi)h(\xi)^{-1}$ and we can replace the coefficients of f and q by their rational expressions in ξ and then multiply by a suitable polynomial in ξ to obtain a relation in $\Phi[\xi, x]$ of the form

$$k(\xi)[g(x)h(\xi) - g(\xi)h(x)] = f(\xi, x)q(\xi, x), \tag{6}$$

where $f(\xi, x)$ is the primitive polynomial given in (5). It now follows that $k(\xi)$ is a factor of $q(\xi, x)$ and so cancelling this we may assume the relation is

$$g(x)h(\xi) - g(\xi)h(x) = f(\xi, x)q(\xi, x). \tag{7}$$

Now the degree in ξ of the left-hand side is at most m. Since

$\gamma = g(\xi)h(\xi)^{-1}$ with $(g(\xi), h(\xi)) = 1$ and $\max(\deg g, \deg h) = m$, the ξ-degree of $f(\xi, x)$ is at least m. It follows that it is exactly m and $q(\xi, x) = q(x) \in \Phi[x]$. Then the right hand side of (7) is primitive as a polynomial in x. This holds also for the left hand side. By symmetry, the left hand side is primitive as a polynomial in ξ also, and this implies that $q(x) = q$ is a non-zero element of Φ. Then (7) implies that the x-degree and ξ-degree of $f(\xi, x)$ are the same. Thus $m = n$ and $\mathrm{E} = \Phi(\gamma)$. As we saw before, $\mathrm{E} \supset \Phi$ implies that γ is transcendental. We have proved the following

Theorem 8 (Lüroth). *If* $\mathrm{P} = \Phi(\xi)$, ξ *transcendental over* Φ, *then any subfield* $\mathrm{E} \supset \Phi$ *is also a simple transcendental extension:* $\mathrm{E} = \Phi(\gamma)$, γ *transcendental.*

The theorem of Lüroth is not valid for purely transcendental extensions P/Φ of transcendency degree $r > 1$. The best positive result in this direction is a theorem of Castelnuovo-Zariski which states that, if Φ is algebraically closed and $r = 2$, then a subfield E/Φ of tr. d. 2 such that P/E is separable is a purely transcendental extension.*

EXERCISES

1. Show that, if $\mathrm{P} = \Phi(\xi, \eta)$ where ξ is transcendental and $\eta^2 + \xi^2 = 1$, then P is purely transcendental.
2. Let Φ be a finite field, $|\Phi| = q = p^m$. Determine the order of the Galois group of $\Phi(\xi)/\Phi$, ξ transcendental.
3. Give an example of a subalgebra of $\Phi[\xi]$, ξ transcendental, which does not have a single generator.

5. Linear disjointness and separating transcendency bases. Let Φ be of characteristic $p \neq 0$ and let $\mathrm{P} = \Phi(\xi, \eta)$ where ξ is transcendental and $\eta^p = \xi$. Then $\{\xi\}$ is a transcendency basis for P/Φ and P is inseparable over $\Phi(\xi)$. On the other hand, $\mathrm{P} = \Phi(\eta)$ is separable over P. This simple example shows that certain transcendency bases B for an extension may be preferable to others in that $\mathrm{P}/\Phi(B)$ is separable algebraic. We remark also that such bases may not always exist, as is shown by the example

* See O. Zariski, *On Castelnuovo's criterion of rationality* $p_a = P_2 = 0$, *Illinois Jour. of Math.*, Vol. 2 (1958), pp. 303–315.

of any P/Φ which is algebraic and not separable over Φ. A transcendency basis B for P/Φ such that P is separable algebraic over $\Phi(B)$ is called a *separating transcendency basis*. If P/Φ has such a basis, then we shall say that P/Φ is *separably generated*. In this section we shall derive a criterion that P/Φ is separably generated, based on the following important notion.

Definition 2. *Let* E_1 *and* E_2 *be subalgebras of an arbitrary field* P/Φ. *Then* E_1 *and* E_2 *are said to be* linearly disjoint *over* Φ *if the subalgebra* E_1E_2 *generated by* E_1 *and* E_2 *is the tensor product* $E_1 \otimes_\Phi E_2$. *More precisely, what is meant here is that the canonical homomorphism of* $E_1 \otimes_\Phi E_2$ *into* E_1E_2 *sending* $\epsilon_1 \otimes \epsilon_2$ *into* $\epsilon_1\epsilon_2$ *is an isomorphism* (see Introd., § 3).

It is well to recall the conditions that $E_1E_2 = E_1 \otimes_\Phi E_2$ which we obtained in the Introduction. We recall first that a sufficient condition is that there exist bases (u_α), (v_β) of E_1 and E_2 over Φ respectively such that $(u_\alpha v_\beta)$ is a basis for E_1E_2 over Φ. Since every element of the subalgebra E_1E_2 is anyhow a linear combination of the elements $u_\alpha v_\beta$, we see that a sufficient condition for $E_1E_2 = E_1 \otimes_\Phi E_2$ is that there exist bases (u_α), (v_β) for E_1/Φ and E_2/Φ such that $\{u_\alpha v_\beta\}$ is Φ-independent. If E_1 is a subfield of P/Φ, then $E_1E_2 = E_1 \otimes_\Phi E_2$ if there exists a basis (v_β) for E_2 over Φ such that the set (v_β) is E_1-independent.

Conversely, assume E_1 and E_2 are linearly disjoint subalgebras. Then $(u_\alpha v_\beta)$ is a basis for E_1E_2/Φ for any basis (u_α) of E_1/Φ and any basis (v_β) of E_2/Φ. Also if E_1 is a subfield, then (v_β) is a basis for E_1E_2/E_1. We note also that linear disjointness of the subalgebras E_1 and E_2 implies that, if $\{u_\alpha\}$ is any linearly independent subset of E_1/Φ and $\{v_\beta\}$ is any linearly independent subset of E_2/Φ, then $\{u_\alpha v_\beta\}$ is Φ-independent.

We note next that E_1 and E_2 are linearly disjoint subalgebras over Φ if and only if the subfields Q_1 and Q_2 generated by E_1 and E_2 respectively are linearly disjoint over Φ. For, if Q_1 and Q_2 have the property, then so do E_1 and E_2, since subalgebras of linearly disjoint algebras are clearly linearly disjoint by our criteria. Conversely, suppose E_1 and E_2 are linearly disjoint. Let $\xi_1, \xi_2, \cdots, \xi_m$ be elements of Q_1 which are Φ-independent and $\eta_1, \eta_2, \cdots, \eta_n$ elements of Q_2 which are Φ-independent. We can write $\xi_i = \xi_i'\xi^{-1}$, $\eta_j = \eta_j'\eta^{-1}$ where ξ_i', $\xi \in E_1$, η_j', $\eta \in E_2$. Then $\{\xi_1', \xi_2', \cdots, \xi_m'\}$ is a Φ-independent subset of E_1 and $\{\eta_1', \eta_2', \cdots, \eta_n'\}$ is a Φ-independent subset of E_2. Hence the mn elements $\xi_i'\eta_j'$ are Φ-independent and so also the elements $\xi_i\eta_j$ are Φ-independent. This implies that Q_1 and Q_2 are linearly disjoint over Φ.

We are interested primarily in subfields of P and we shall require the following lemma which will enable us to prove linear disjointness by steps.

Lemma. *Let* E_1 *and* E_2 *be subfields of* P/Φ, Δ_1 *a subfield of* E_1/Φ. *Then* E_1 *and* E_2 *are linearly disjoint over* Φ *if and only if the following two conditions hold:* (1) Δ_1 *and* E_2 *are linearly disjoint over* Φ *and* (2) *the field* $\Delta_1(E_2)$ *and* E_1 *are linearly disjoint over* Δ_1.

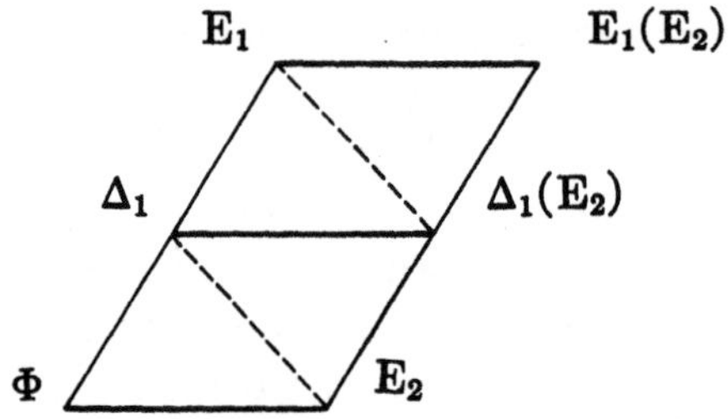

Proof. Assume (1) and (2). Let (u_α) be a basis for E_2/Φ. By (1), the u_α are linearly independent over Δ_1. Since these elements are contained in $\Delta_1(E_2)$, and $\Delta_1(E_2)$ and E_1/Δ_1 are linearly disjoint over Δ_1, by (2), the u_α are linearly independent over E_1. Hence E_1 and E_2 are linearly disjoint over Φ. Conversely, assume this holds. Then it is clear that Δ_1 and E_2 are linearly disjoint over Φ, that is, (1) holds. Also the hypothesis implies that, if (u_α) is a basis for E_1 over Δ_1, (v_β) a basis for Δ_1 over Φ, (w_γ) a basis for E_2 over Φ, then $(u_\alpha v_\beta w_\gamma)$ is a basis for E_1E_2 over Φ. This implies that, if we have a relation $\Sigma c_i u_{\alpha_i} = 0$, $c_i \in E_2\Delta_1$, then every $c_i = 0$. Now suppose we have $\Sigma \delta_i u_{\alpha_i} = 0$, $\delta_i \in \Delta_1(E_2)$. Then we can write $\delta_i = c_i d^{-1}$, c_i, $d \in E_2\Delta_1$ and we obtain $\Sigma c_i u_{\alpha_i} = 0$, $c_i = 0$ and $d_i = 0$. Thus we have shown that the basis (u_α) of E_1/Δ_1 is $\Delta_1(E_2)$-independent. This implies that E_1 and $\Delta_1(E_2)$ are linearly disjoint over Δ_1.

We now embark on the study of linear disjointness and separability. We assume that P is an extension field of a field Φ of characteristic $p \neq 0$ and we shall operate in an algebraic closure A of P (which contains an algebraic closure of Φ). We consider the subset $\Phi^{p^{-1}}$ of A of elements γ such that $\gamma^p \in \Phi$. This is a subfield over Φ. We shall be interested in linear disjointness of P and $\Phi^{p^{-1}}$ over Φ. In studying this question it is useful to note the following simple criterion: A set $\{\rho_1, \rho_2, \cdots, \rho_n\}$, $\rho_i \in$ P is $\Phi^{p^{-1}}$-independent if and only if $\{\rho_1{}^p, \rho_2{}^p, \cdots, \rho_n{}^p\}$ is Φ-independent; for, suppose we have $\Sigma\beta_i\rho_i = 0$, $\beta_i \in \Phi^{p^{-1}}$. Then $\Sigma\alpha_i\rho_i{}^p = 0$ for $\alpha_i = \beta_i{}^p$ in Φ. On the other hand, if $\Sigma\alpha_i\rho_i{}^p = 0$, α_i in Φ, then, since A contains an algebraic closure of Φ, $\alpha_i = \beta_i{}^p$, β_i in $\Phi^{p^{-1}}$. Then $\Sigma\beta_i{}^p\rho_i{}^p = 0$. Hence $(\Sigma\beta_i\rho_i)^p = 0$ and $\Sigma\beta_i\rho_i = 0$. Also it is clear in both situations that $\alpha_i = 0$ if and only if $\beta_i = 0$. We shall now establish the following criterion.

Theorem 9. *If* P *is an algebraic extension of* Φ *(possibly infinite dimensional), then* P *is separable over* Φ *if and only if* P *is linearly disjoint to* $\Phi^{p^{-1}}$ *over* Φ.

Proof. We recall that an algebraic element ρ of P over Φ is separable if and only if $\rho \in \Phi(\rho^p)$ (Lemma 2 of § 1.9). Suppose first that P and $\Phi^{p^{-1}}$ are linearly disjoint over Φ and let $\rho \in$ P. Let $[\Phi(\rho):\Phi] = n$. Then $(1, \rho, \rho^2, \cdots, \rho^{n-1})$ is a basis for $\Phi(\rho) = \Phi[\rho]$ and hence these elements are $\Phi^{p^{-1}}$-independent. This implies that the elements $1, \rho^p, \rho^{2p}, \cdots, \rho^{(n-1)p}$ are Φ-independent. Since there are n of these and they are contained in $\Phi(\rho)$, they form a basis for $\Phi(\rho)$. Evidently this implies that $\rho \in \Phi[\rho^p]$ so ρ is separable over Φ. Conversely, assume P separable over Φ and let $\{\rho_1, \cdots, \rho_n\}$ be a finite Φ-independent subset of P. We may imbed this set in a subfield E/Φ which is finite dimensional, and we can choose a basis $(\rho_1, \cdots, \rho_n, \rho_{n+1}, \cdots, \rho_q)$ for E/Φ. Any element ϵ of E is a Φ-linear combination of the ρ_j, $1 \leq j \leq q$. Then ϵ^p is a Φ-combination of the $\rho_j{}^p$. The same holds for $\epsilon^{2p} = (\epsilon^2)^p$, $\epsilon^{3p} = (\epsilon^3)^p, \cdots$. On the other hand, since ϵ is separable, $\epsilon \in \Phi(\epsilon^p) = \Phi[\epsilon^p]$. Consequently, ϵ itself is a Φ-linear combination of the $\rho_j{}^p$. Since $[E:\Phi] = q$ this implies that $(\rho_1{}^p, \rho_2{}^p, \cdots, \rho_q{}^p)$ is a basis for E/Φ. Hence $\{\rho_1{}^p, \cdots, \rho_n{}^p\}$ is Φ-independent and $\{\rho_1, \cdots, \rho_n\}$ is $\Phi^{p^{-1}}$-independent. Since $\{\rho_1, \cdots, \rho_n\}$ was an arbitrary finite Φ-independent subset of P, this proves that P and $\Phi^{p^{-1}}$ are linearly disjoint over Φ.

We prove next the following

Theorem 10. *If* P *is purely transcendental over* Φ, *then* P *is linearly disjoint to* $\Phi^{p^{-1}}$ *over* Φ.

Proof. Our assumption is that $P = \Phi(B)$ where B is an algebraically independent set. We have seen also that P is linearly disjoint to $\Phi^{p^{-1}}$ over Φ if and only if the subalgebra $\Phi[B]$ of polynomials in the elements of B is linearly disjoint to $\Phi^{p^{-1}}$. To prove that the latter holds, it suffices to give a basis for $\Phi[B]/\Phi$ which is $\Phi^{p^{-1}}$-independent. For this we take the basis M consisting of the monomials in $\xi_\alpha \in B$. Now it is clear that, if m_1 and m_2 are distinct monomials, then $m_1{}^p$ and $m_2{}^p$ are distinct monomials. Hence it is clear that the set M^p of p-th powers of the elements of M is a Φ-independent set. We have seen that this implies that M is $\Phi^{p^{-1}}$-independent. Hence $\Phi[B]$ is linearly disjoint to $\Phi^{p^{-1}}$ and the proof is complete.

We can now prove our main result which is

MacLane's Criterion. *If* P/Φ *is separably generated (of characteristic* p*), then* P *and* $\Phi^{p^{-1}}$ *are linearly disjoint over* Φ. *On the other hand, if* P *is finitely generated over* Φ *and* P *and* $\Phi^{p^{-1}}$ *are linearly disjoint over* Φ, *then* P *is separably generated over* Φ.

Proof. Suppose first that P is separably generated over Φ, which means that P has a transcendency basis B over Φ such that P is separable algebraic over $\Sigma = \Phi(B)$. Then, by Theorem 10, Σ and $\Phi^{p^{-1}}$ are linearly disjoint over Φ. Also, by Theorem 9, P and $\Sigma^{p^{-1}}$ are linearly disjoint over Σ. Hence P and $\Sigma(\Phi^{p^{-1}})$ which is a subfield of $\Sigma^{p^{-1}}$ over Σ are linearly disjoint over Σ. The lemma now shows that P and $\Phi^{p^{-1}}$ are linearly disjoint over Φ.

Next we assume $P = \Phi(\xi_1, \xi_2, \cdots, \xi_m)$ and P and $\Phi^{p^{-1}}$ are linearly disjoint over Φ. We may assume also that $\{\xi_1, \xi_2, \cdots, \xi_r\}$ is a transcendency basis. Suppose we know already that $\xi_{r+1}, \cdots, \xi_s$ are separable algebraic over $\Phi(\xi_1, \cdots, \xi_r)$. If $s = m$, then $\{\xi_1, \cdots, \xi_r\}$ is a separating transcendency basis. Hence we suppose that ξ_{s+1} is inseparable algebraic over $\Sigma = \Phi(\xi_1, \cdots, \xi_r)$. Let $f(x)$ be the minimum polynomial of ξ_{s+1} over Σ. If we multiply f by a suitable polynomial in $\xi_1, \cdots, \xi_r$ we obtain a polynomial $F(\xi_1, \cdots, \xi_r, x) \in \Phi[\xi_1, \cdots, \xi_r, x]$ which is irreducible in $\Phi[\xi_1, \cdots, \xi_r, x]$ and satisfies $F(\xi_1, \cdots, \xi_r, \xi_{s+1}) = 0$. Since f is inseparable it is a polynomial in x^p; hence $F(\xi_1, \cdots, \xi_r, x)$ is a polynomial in x^p. We assert that there exists a ξ_i, $1 \leq i \leq r$, such that F is not a polynomial in ξ_i^p. Otherwise, the monomials in $\xi_1, \xi_2, \cdots, \xi_r, x$ which actually occur in F are all p-th powers and this implies that $F(\xi_1, \cdots, \xi_r, x) = H(\xi_1, \cdots, \xi_r, x)^p$ in $\Phi^{p^{-1}}[\xi_1, \cdots, \xi_r, x]$. Then $H(\xi_1, \cdots, \xi_r, \xi_{s+1}) = 0$, so the monomials in $\xi_1, \cdots, \xi_r, \xi_{s+1}$ occurring in H are linearly dependent over $\Phi^{p^{-1}}$. The assumption of linear disjointness of P and $\Phi^{p^{-1}}$ therefore implies that the same monomials are Φ-dependent. This implies that ξ_{s+1} is a root of a polynomial $h(x) \varepsilon \Sigma[x]$, $h(x) \neq 0$ of lower degree than f, contradicting the fact that f is the minimum polynomial of ξ_{s+1} over Σ. This shows that we may suppose that $F(\xi_1, \cdots, \xi_r, x)$ is not a polynomial in ξ_1^p. The relation $F(\xi_1, \cdots, \xi_r, \xi_{s+1}) = 0$ shows that ξ_1 is algebraic over $\Phi(\xi_2, \cdots, \xi_r, \xi_{s+1})$. Since $\xi_2, \cdots, \xi_r$ are also algebraic over this subfield, it is clear that $\{\xi_2, \cdots, \xi_r, \xi_{s+1}\}$ is a transcendency basis.

We shall show that ξ_1 is separable over $\Sigma' = \Phi(\xi_2, \cdots, \xi_r, \xi_{s+1})$. We recall that $F(\xi_1, \cdots, \xi_r, y)$ is irreducible in $\Phi[\xi_1, \cdots, \xi_r, y]$. Hence $F(x, x_2, \cdots, x_r, y)$ is irreducible in $\Phi[x, x_2, \cdots, x_r, y]$, x, x_i, y, indeterminates, and consequently this polynomial is irreducible in $\Phi(x_2, \cdots, x_r, y)[x]$ where $\Phi(x_2, \cdots, x_r, y)$ is the field of fractions of $\Phi[x_2, \cdots, x_r, y]$. Since $\xi_2, \cdots, \xi_r, \xi_{s+1}$ are algebraically independent over Φ, $\Phi(\xi_2, \cdots, \xi_r, \xi_{s+1}) \cong \Phi(x_2, \cdots, x_r, y)$ under a Φ-isomorphism such that $x_i \to \xi_i, 2 \leq i \leq r, y \to \xi_{s+1}$. It follows that $F(x, \xi_2, \cdots, \xi_r, \xi_{s+1})$ is irreducible in $\Phi(\xi_2, \cdots, \xi_r, \xi_{s+1})[x]$ and so this is a multiple of the minimum polynomial of ξ_1 over Σ'. Since this polynomial is not a polynomial in x^p, ξ_1 is separable algebraic over Σ'. Also ξ_i, $1 \leq i \leq s$, are separable algebraic over $\Phi(\xi_1, \cdots, \xi_r, \xi_{s+1})$ and, since ξ_1 is separable algebraic over Σ', ξ_i is separable algebraic over Σ'. If we re-number the ξ's we now have a transcendency basis $\xi_1, \cdots, \xi_r$ such that every ξ_j, $1 \leq j \leq s + 1$, is separable algebraic over $\Phi(\xi_1, \cdots, \xi_r)$. This establishes the inductive step to show that we can choose $\xi_1, \cdots, \xi_r$ among the generators $\xi_1, \cdots, \xi_m$ so that every ξ_i is separable algebraic over $\Sigma = \Phi(\xi_1, \cdots, \xi_r)$. Then $\{\xi_1, \cdots, \xi_r\}$ is a separating transcendency basis for P/Φ. This completes the proof of the second assertion of the theorem.

It is important to note for future use that, if $P = \Phi(\xi_1, \cdots, \xi_m)$ is linearly disjoint to $\Phi^{p^{-1}}$ over Φ, then we have shown that a separating transcendency basis can be extracted from the set $\{\xi_1, \xi_2, \cdots, \xi_m\}$. We note also the following

Corollary (F. K. Schmidt). *If Φ is perfect, then any field of algebraic functions $\Phi(\xi_1, \cdots, \xi_m)$ has a separating transcendency basis over Φ.*

This is an immediate consequence of MacLane's criterion since P is certainly linearly disjoint to $\Phi^{p^{-1}} = \Phi$.

The results which we have proved, particularly Theorem 9, make it natural to extend the notion of separability to arbitrary (not necessarily algebraic) field extensions in the following way.

Definition 3. *A field* P *is* separable *over Φ if it is either of characteristic* 0 *or if it is of characteristic $p \neq 0$ and* P *is linearly disjoint to $\Phi^{p^{-1}}$ over Φ.*

Theorem 9 shows that this is equivalent to the usual notion of separability if P is algebraic over Φ. Also MacLane's criterion shows that, if P is finitely generated over Φ, then it is separable over Φ if and only if P is separably generated over Φ. The following theorem gives two other properties of separability which are familiar in the algebraic case.

Theorem 11. (1) *If* P *is separable over Φ and* E *is a subfield of* P *over Φ, then* E *is separable over Φ.* (2) *If* P *is separable over* E *and* E *is separable over Φ, then* P *is separable over Φ.*

Proof. We may assume the characteristic is $p \neq 0$. (1) This is clear since the linear disjointness of P and $\Phi^{p^{-1}}$ implies the linear disjointness of E and $\Phi^{p^{-1}}$. (2) We are assuming that $\Phi^{p^{-1}}$ is linearly disjoint to E over Φ and $E^{p^{-1}}$ is linearly disjoint to P over E. Then $E(\Phi^{p^{-1}})$ which is a subfield of $E^{p^{-1}}$ is linearly disjoint over E to P. The lemma now applies to show that $\Phi^{p^{-1}}$ and P are linearly disjoint over Φ. Hence P is separable over Φ.

We close the present discussion with two negative results. First, we recall that, if P is separable algebraic over Φ, then P is separable algebraic over any intermediate field. This fails in the general case since, for ξ transcendental, $\Phi(\xi)$ is separable over Φ of characteristic $p \neq 0$, but $\Phi(\xi)$ is not separable over $\Phi(\xi^p)$.

Next we note that a field may be separable over Φ and not separably generated. An example of this is given in ex. 1 below.

EXERCISES

1. Let Φ be of characteristic $p \neq 0$ and let $P = \Phi(\xi, \xi^{p^{-1}}, \xi^{p^{-2}}, \cdots)$ where ξ is transcendental over Φ. Show that P is separable over Φ but not separably generated over Φ.
2. Let $\Phi^{p^{-m}}$ be the subfield of the algebraic closure of $P \supseteq \Phi$ of elements ξ such that $\xi^{p^m} \in \Phi$ and let $\Phi^{p^{-\infty}} = \bigcup \Phi^{p^{-m}}$. Show that P is separable over Φ if and only if P and $\Phi^{p^{-\infty}}$ are linearly disjoint over Φ.
3. Let E/Φ and Δ/Φ be subfields of P/Φ such that E/Φ is purely transcendental and Δ/Φ is algebraic. Show that E and Δ are linearly disjoint over Φ.
4. Let $P = \Phi(\xi, \eta, \zeta, \tau)$ where Φ is of characteristic $p \neq 0$, ξ, η, ζ are algebraically independent and $\tau^p = \xi\zeta^p + \eta$. Show that P is not separably generated over $E = \Phi(\xi, \eta)$.
5. (MacLane). Let Φ be a perfect field of characteristic $p \neq 0$, P an imperfect extension field of Φ such that tr. d. $P/\Phi = 1$. Show that P is separably generated over Φ.

6. Derivations. We have found it useful to introduce the usual formal derivative of a polynomial in considering multiple roots (§ 1.6). The mapping of the polynomial algebra $\Phi[x]$ into itself defined by: $f(x) \rightarrow f'(x)$ the formal derivative of $f(x)$, is an example of a derivation in the algebra $\Phi[x]$. More generally it is convenient to consider derivations from a subalgebra into an algebra. This general notion, which is of great importance in algebra, is given in the following

Definition 4. *If* $\mathfrak{A}$ *is a subalgebra of an algebra* $\mathfrak{B}$, *a* derivation D of $\mathfrak{A}$ into $\mathfrak{B}$ *is a linear mapping of* $\mathfrak{A}$ *into* $\mathfrak{B}$ *such that*

$$(ab)D = (aD)b + a(bD), \quad a, b \in \mathfrak{A}. \tag{8}$$

If $\mathfrak{A} = \mathfrak{B}$, *then we speak of a* derivation in $\mathfrak{A}$.

We shall be interested mainly in derivations in fields of algebraic functions. In this section we consider some general results on extension of derivations and on the algebraic system consisting of all the derivations of an algebra into itself. We begin our considerations by noting first that the study of derivations is equivalent to the study of a certain type of algebra isomorphisms. This will enable us to derive the main facts about derivations as consequences of corresponding results on homomorphisms. For this purpose we introduce the algebra $\mathfrak{T}$ with basis $(1, t)$ over the base field Φ and multiplication rule $t^2 = 0$. Thus $\mathfrak{T} = \Phi[x]/(x^2)$, x an indeterminate, and t is the coset $x + (x^2)$. If $\mathfrak{B}$ is an arbitrary algebra, then we form the algebra $\mathfrak{B} \otimes \mathfrak{T}$. If we identify $\mathfrak{B}$ in the usual way with the subalgebra of elements $b \otimes 1$ and $\mathfrak{T}$ with the subalgebra of elements $1 \otimes u$, $u \in \mathfrak{T}$, then we see that the elements of $\mathfrak{B} \otimes \mathfrak{T}$ can be written in one and only one way in the form $b_0 + b_1 t$, $b_i \in \mathfrak{B}$, the generator t of $\mathfrak{T}$. We have $bt = tb$ and in general the multiplication rule in $\mathfrak{B} \otimes \mathfrak{T}$ is

$$(9) \qquad (b_0 + b_1 t)(c_0 + c_1 t) = b_0 c_0 + (b_0 c_1 + b_1 c_0)t,$$

$b_i, c_i \in \mathfrak{B}$. The algebra $\mathfrak{B} \otimes \mathfrak{T}$ is called the *algebra of dual numbers over* $\mathfrak{B}$.

Now let D be a derivation of $\mathfrak{A}$ into $\mathfrak{B}$. Then we can use this to define a mapping $s = s(D)$ of $\mathfrak{A}$ into $\mathfrak{B} \otimes \mathfrak{T}$ by

$$(10) \qquad a \to a^s \equiv a + (aD)t.$$

Evidently s is linear. Furthermore, if $a, b \in \mathfrak{A}$, then

$$\begin{aligned} a^s b^s &= (a + (aD)t)(b + (bD)t) \\ &= ab + (a(bD) + (aD)b)t \\ &= ab + ((ab)D)t \\ &= (ab)^s. \end{aligned}$$

Hence s is a homomorphism of the algebra $\mathfrak{A}$ into the algebra of dual numbers $\mathfrak{B} \otimes \mathfrak{T}$. The homomorphism s has a simple characterization. For this we introduce the mapping π: $a + bt \to a$, $a, b \in \mathfrak{B}$, of $\mathfrak{B} \otimes \mathfrak{T}$ into $\mathfrak{B}$. It is clear that this is a homomorphism of $\mathfrak{B} \otimes \mathfrak{T}$ into $\mathfrak{B}$ which is the identity mapping on the subalgebra $\mathfrak{B}$. Now we see that, if $a \in \mathfrak{A}$ and s is defined by the derivation D of $\mathfrak{A}$ into $\mathfrak{B}$ as before, then $a^{s\pi} = (a + (aD)t)^\pi = a$. Evidently, this requirement guarantees that s is an isomorphism.

Conversely, let s be any homomorphism of $\mathfrak{A}$ into $\mathfrak{B} \otimes \mathfrak{T}$ such that $a^{s\pi} = a$, $a \in \mathfrak{A}$. Then $a^s = a + bt$, $a, b \in \mathfrak{B}$ and b is uniquely determined by a. Hence we have the mapping D: $a \to b$, and we may write $a^s = a + (aD)t$. It is clear that the linearity of s implies the linearity of D. Also since $(ab)^s = a^s b^s$ for any a, b in $\mathfrak{A}$,

$$\begin{aligned} (a + (aD)t)(b + (bD)t) &= ab + (a(bD) + (aD)b)t \\ &= ab + ((ab)D)t. \end{aligned}$$

Hence we have $(ab)D = (aD)b + a(bD)$ so D is a derivation. We can therefore state the following

Theorem 12. *If $\mathfrak{A}$ is a subalgebra of $\mathfrak{B}$ and D is a derivation of $\mathfrak{A}$ into $\mathfrak{B}$, then $s: a \rightarrow a + (aD)t$ is an isomorphism of $\mathfrak{A}$ into the algebra of dual numbers $\mathfrak{B} \otimes \mathfrak{T}$ over $\mathfrak{B}$ such that $a^{s\pi} = a$. Conversely, any homomorphism of $\mathfrak{A}$ into $\mathfrak{B} \otimes \mathfrak{T}$ satisfying this condition has the form $a \rightarrow a + (aD)t$ where D is a derivation of $\mathfrak{A}$ into $\mathfrak{B}$.*

We shall now obtain some simple consequences of this connection between derivations and isomorphisms. First, let $\mathfrak{X}$ be a set of generators of the subalgebra $\mathfrak{A}$ of the algebra $\mathfrak{B}$ and let D_1 and D_2 be derivations of $\mathfrak{A}$ into $\mathfrak{B}$. Suppose $xD_1 = xD_2$ for every $x \in \mathfrak{X}$. Then $x^{s_1} = x^{s_2}$ for the associated isomorphisms $s_1 = s(D_1)$, $s_2 = s(D_2)$ of $\mathfrak{A}$ into $\mathfrak{B} \otimes \mathfrak{T}$. It follows that $a^{s_1} = a^{s_2}$ for every $a \in \mathfrak{A}$ so $s_1 = s_2$ and $D_1 = D_2$. This shows that, if two derivations coincide on a set of generators of $\mathfrak{A}$, then they are identical on $\mathfrak{A}$. We remark next that, if s is a homomorphism of $\mathfrak{A}$ into $\mathfrak{B} \otimes \mathfrak{T}$ such that $x^{s\pi} = x$ for $x \in \mathfrak{X}$ a set of generators, then $a^{s\pi} = a$ for all $a \in \mathfrak{A}$. Hence s defines a derivation in the manner indicated.

An element c of $\mathfrak{A}$ such that $cD = 0$ is called a *D-constant.* Evidently c is a D-constant if and only if $c^s = c$ for the isomorphism $s = s(D)$. It follows from this—or directly—that the set of D-constants is a subalgebra of $\mathfrak{A}$. In particular, 1 is a D-constant for every derivation D. If $\mathfrak{A}$ is commutative and Φ is of characteristic p, then every p-th power in $\mathfrak{A}$ is a D-constant. For, in any commutative algebra the basic property (8) for D implies that $(a^k)D = ka^{k-1}(aD)$. Hence if $k = p$, then $(a^p)D = 0$. We note also that, if $\mathfrak{A} = \mathrm{P}$ is a field, then the set of D-constants of P forms a subfield Γ of P. This is clear from the consideration of $s = s(D)$ or it follows directly, by noting the rule for the derivative of $\gamma^{-1}: \gamma^{-1}D = -(\gamma D)\gamma^{-2}$, which follows by taking the derivative of the relation $\gamma\gamma^{-1} = 1$. If $\rho \in \mathrm{P}$ and $\gamma \in \Gamma$, then $(\gamma\rho)D = \gamma(\rho D)$ for the derivation D. This shows that $D \in \mathfrak{D}_\Gamma(\mathrm{P}, \mathfrak{B})$ the set of derivations of P/Γ into $\mathfrak{B}/\Gamma$. In considering a particular derivation D of a field, it is often convenient to shift from the original base field to the field of constants Γ of D or to some subfield E/Φ of Γ/Φ.

We shall now carry over to derivations the two basic results I and IV′ on extensions of homomorphisms of commutative rings which we derived in the Introduction. We remark that these results are valid for algebras over a field Φ and we shall use them in this form. Our first result on extension of derivations is

Theorem 13. *Let* P *be a field over* Φ, $\mathfrak{A}$ *a subalgebra of* P/Φ *(containing* 1), *M a multiplicatively closed subset of non-zero elements of* $\mathfrak{A}$ *containing* 1, *and let* $\mathfrak{A}_M$ *be the subalgebra of* P *of elements of the form* ab^{-1}, $a \in \mathfrak{A}$, $b \in M$. *Let D be a derivation of* $\mathfrak{A}$ *into* P. *Then D can be extended in one and only one way to a derivation of* $\mathfrak{A}_M$ *into* P.

Proof. Let s be the isomorphism $a \to a + (aD)t$ of $\mathfrak{A}$ into $P \otimes \mathfrak{T}$. If $a \neq 0$, then $a^s = a + (aD)t$ has the inverse $a^{-1} - a^{-2}(aD)t$ since

$$(a + (aD)t)(a^{-1} - a^{-2}(aD)t) = 1 + (aD)a^{-1}t - a^{-1}(aD)t = 1.$$

By I of the Introduction, s can be extended to an isomorphism of $\mathfrak{A}_M$ into P. The extension is unique and maps $ab^{-1} \to a^s(b^s)^{-1}$. We have

$$\begin{aligned} a^s(b^s)^{-1} &= (a + (aD)t)(b^{-1} - b^{-2}(bD)t) \\ &= ab^{-1} + ((aD)b^{-1} - ab^{-2}(bD))t. \end{aligned}$$

This formula shows that, if s denotes the extension of s to $\mathfrak{A}_M$, then $(ab^{-1})^{s\pi} = (a^s(b^s)^{-1})^\pi = ab^{-1}$. It follows that s defines the derivation:

$$ab^{-1} \to (aD)b^{-1} - ab^{-2}(bD) \tag{11}$$

of $\mathfrak{A}_M$ into P. The argument shows also that this extension of D is unique.

Next we consider a subalgebra $\mathfrak{A}$ of P/Φ and an extension of this which has the form $\mathfrak{A}[\xi_1, \xi_2, \cdots, \xi_m]$, ξ_i elements of the field P. We suppose we are given a derivation D of $\mathfrak{A}$ into P and elements $\eta_1, \eta_2, \cdots, \eta_m$ of P. We seek conditions on D and the η_i which insure that D can be extended to a derivation D of $\mathfrak{A}[\xi_1, \xi_2, \cdots, \xi_m]$ such that $\xi_i D = \eta_i$, $i = 1, 2, \cdots, m$. Since $\mathfrak{A}$ and the ξ_i generate $\mathfrak{A}[\xi_1, \cdots, \xi_m]$ it is clear that, if the extension exists, then it is unique. As before, we consider the isomorphism $s: a \to a + (aD)t$ of $\mathfrak{A}$ in $\mathrm{P} \otimes \mathfrak{T}$ satisfying $a^{s\pi} = a$. Then D can be extended to a derivation of $\mathfrak{A}[\xi_1, \cdots, \xi_m]$ into P so that $\xi_i \to \eta_i$ if and only if s can be extended to an isomorphism of $\mathfrak{A}[\xi_1, \cdots, \xi_m]$ into $\mathrm{P} \otimes \mathfrak{T}$ so that $\xi_i \to \xi_i + \eta_i t$. Clearly the condition is necessary and, if it holds, we have $a^{s\pi} = a$ and $\xi_i^{s\pi} = \xi_i$ for the extension s. Hence s will give rise to a derivation of $\mathfrak{A}[\xi_1, \xi_2, \cdots, \xi_m]$, as before. The conditions for the extension of s have been given in IV′ of the Introduction. We recall that the set $\mathfrak{R}$ of polynomials $f(x_1, x_2, \cdots, x_m) \varepsilon \mathfrak{A}[x_1, x_2, \cdots, x_m]$, x_i indeterminates, such that $f(\xi_1, \xi_2, \cdots, \xi_m) = 0$ is an ideal in $\mathfrak{A}[x_1, x_2, \cdots, x_m]$. The condition IV′ for an extension s of s such that $\xi_i^s = \zeta_i$, $1 \leq i \leq m$, is that $g^s(\zeta_1, \zeta_2, \cdots, \zeta_m) = 0$ for every $g \varepsilon \mathfrak{X}$, a set of generators of $\mathfrak{R}$. Hence we see that D can be extended to a derivation of $\mathfrak{A}[\xi_1, \cdots, \xi_m]$ into P such that $\xi_i \to \eta_i$, $i = 1, 2, \cdots, m$, if and only if

$$g^s(\xi_1 + \eta_1 t, \quad \xi_2 + \eta_2 t, \cdots, \xi_m + \eta_m t) = 0$$

for every g in a set $\mathfrak{X}$ of generators of the ideal $\mathfrak{R}$ of polynomials $f(x_1, \cdots, x_m) \varepsilon \mathfrak{A}[x_1, \cdots, x_m]$ such that $f(\xi_1, \cdots, \xi_m) = 0$.

We proceed to work out these conditions in detail. Let $a \varepsilon \mathfrak{A}$ and consider the monomial $M(x_1, \cdots, x_m) = ax_1^{k_1}x_2^{k_2} \cdots x_m^{k_m}$. Then

$$\begin{aligned} M^s(\xi_1 + \eta_1 t, \quad & \xi_2 + \eta_2 t, \cdots, \xi_m + \eta_m t) \\ &= a^s(\xi_1 + \eta_1 t)^{k_1}(\xi_2 + \eta_2 t)^{k_2} \cdots (\xi_m + \eta_m t)^{k_m} \\ &= (a + (aD)t)(\xi_1 + \eta_1 t)^{k_1}(\xi_2 + \eta_2 t)^{k_2} \cdots (\xi_m + \eta_m t)^{k_m} \\ &= a\xi_1^{k_1}\xi_2^{k_2} \cdots \xi_m^{k_m} + (aD)\xi_1^{k_1}\xi_2^{k_2} \cdots \xi_m^{k_m}t \\ &\qquad + (k_1 a\xi_1^{k_1-1}\xi_2^{k_2} \cdots \xi_m^{k_m}\eta_1 \\ &\qquad + k_2 a\xi_1^{k_1}\xi_2^{k_2-1}\xi_3^{k_3} \cdots \xi_m^{k_m}\eta_2 \\ &\qquad + \cdots + k_m a\xi_1^{k_1} \cdots \xi_{m-1}^{k_{m-1}}\xi_m^{k_m-1}\eta_m)t. \end{aligned}$$

If we define the formal partial derivative of $f = \Sigma a_{k_1 \cdots k_m} x_1^{k_1} \cdots x_m^{k_m}$ relative to x_i as

$$\frac{\partial f}{\partial x_i} = \Sigma k_i a_{k_1 \cdots k_m} x_1^{k_1} \cdots x_i^{k_i-1} \cdots x_m^{k_m},$$

and denote its value at $(\xi_1, \xi_2, \cdots, \xi_m)$ by $\left(\frac{\partial f}{\partial x_i}\right)_{x_j=\xi_j}$, then the above calculation shows that

$$M^s(\xi_1 + \eta_1 t, \cdots, \xi_m + \eta_m t)$$
$$= M(\xi_1, \cdots, \xi_m) + \left[M^D(\xi_1, \cdots, \xi_m) + \sum_1^m \left(\frac{\partial M}{\partial x_i}\right)_{x_j=\xi_j} \eta_i\right] t,$$

where, in general $f^D(x_1, \cdots, x_m)$ is the polynomial obtained from f by replacing the coefficients by their images under D. Hence if $f \in \mathfrak{A}[x_1, \cdots, x_m]$, then we have

$$(12) \quad f^s(\xi_1 + \eta_1 t, \xi_2 + \eta_2 t, \cdots, \xi_m + \eta_m t)$$
$$= f(\xi_1, \cdots, \xi_m) + f^D(\xi_1, \cdots, \xi_m)t + \sum_1^m \left(\frac{\partial f}{\partial x_i}\right)_{x_j=\xi_j} \eta_i t.$$

It is now clear that $f^s(\xi_1 + \eta_1 t, \cdots, \xi_m + \eta_m t) = 0$ if and only if $f(\xi_1, \cdots, \xi_m) = 0$ and

$$(13) \qquad f^D(\xi_1, \cdots, \xi_m) + \sum_{i=1}^m \left(\frac{\partial f}{\partial x_i}\right)_{x_j=\xi_j} \eta_i = 0.$$

The criterion which we gave can now be stated in the following manner.

Theorem 14. *Let $\mathfrak{A}$ be a subalgebra over Φ of the field P/Φ and let $\xi_1, \xi_2, \cdots, \xi_m, \eta_1, \eta_2, \cdots, \eta_m$ be elements of* P, *D a derivation of $\mathfrak{A}$ into* P. *Let $\mathfrak{K}$ be the ideal of polynomials $f(x_1, \cdots, x_m) \in \mathfrak{A}[x_1, \cdots, x_m]$ such that $f(\xi_1, \cdots, \xi_m) = 0$ and let $\mathfrak{X}$ be any set of generators for $\mathfrak{K}$. Then D can be extended to a derivation D of $\mathfrak{A}[\xi_1, \cdots, \xi_m]$ into* P *such that $\xi_i D = \eta_i$, $i = 1, 2, \cdots, m$, if and only if*

$$(14) \qquad g^D(\xi_1, \cdots, \xi_m) + \sum_{i=1}^m \left(\frac{\partial g}{\partial x_i}\right)_{x_j=\xi_j} \eta_i = 0$$

for every $g \in \mathfrak{X}$. If the extension exists, then it is unique.

A special case of the result is the following: If the ξ_i are algebraically independent over $\mathfrak{A}$, then there exists a derivation D extending D on $\mathfrak{A}$ and mapping $\xi_i \to \eta_i$ where $\eta_1, \cdots, \eta_m$ are arbitrary in P. This is clear since in this case the ideal $\mathfrak{K} = 0$, so the condition for extension is trivially satisfied.

We consider next an arbitrary algebra $\mathfrak{B}$, a subalgebra $\mathfrak{A}$, and the set $\mathfrak{D}_\Phi(\mathfrak{A}, \mathfrak{B})$ of derivations of $\mathfrak{A}/\Phi$ into $\mathfrak{B}/\Phi$. If $D_1, D_2 \in \mathfrak{D}_\Phi(\mathfrak{A}, \mathfrak{B})$ and $\alpha \in \Phi$, then αD_1 and $D_1 + D_2$ are linear mappings of $\mathfrak{A}$ into $\mathfrak{B}$. Moreover, if $a, b \in \mathfrak{A}$,

$$\begin{aligned}(ab)(\alpha D_1) &= \alpha((ab)D_1) = \alpha((aD_1)b + a(bD_1))\\ &= (a(\alpha D_1))b + a(b(\alpha D_1))\\ (ab)(D_1 + D_2) &= (ab)D_1 + (ab)D_2\\ &= (aD_1)b + a(bD_1) + (aD_2)b + a(bD_2)\\ &= (a(D_1 + D_2))b + a(b(D_1 + D_2)).\end{aligned}$$

This shows that αD_1 and $D_1 + D_2$ are derivations. Hence $\mathfrak{D}_\Phi(\mathfrak{A}, \mathfrak{B})$ is a subspace of the space $\mathfrak{L}_\Phi(\mathfrak{A}, \mathfrak{B})$ of linear mappings of $\mathfrak{A}/\Phi$ into $\mathfrak{B}/\Phi$. Next let c be an element of the center of $\mathfrak{B}$ and, as usual, let c_R denote the mapping $x \to xc = cx$ in $\mathfrak{B}$. We assert that, if D is a derivation of $\mathfrak{A}$ into $\mathfrak{B}$, then Dc_R is also a derivation of $\mathfrak{A}$ into $\mathfrak{B}$. For, it is clear that Dc_R is linear and we have

$$\begin{aligned}(ab)Dc_R &= ((aD)b + a(bD))c_R\\ &= (aDc_R)b + a((bD)c_R).\end{aligned}$$

Hence $Dc_R \in \mathfrak{D}_\Phi(\mathfrak{A}, \mathfrak{B})$.

Next let $\mathfrak{B} = \mathfrak{A}$ and let $\mathfrak{D}_\Phi(\mathfrak{A}) = \mathfrak{D}_\Phi(\mathfrak{A}, \mathfrak{A})$ the set of derivations in $\mathfrak{A}$. Let $D_1, D_2 \in \mathfrak{D}_\Phi(\mathfrak{A})$. Then D_1D_2 is a linear transformation of the space $\mathfrak{A}$. However,

$$\begin{aligned}(ab)D_1D_2 &= (a(bD_1) + (aD_1)b)D_2\\ &= a(bD_1D_2) + (aD_2)(bD_1) + (aD_1)(bD_2) + (aD_1D_2)b.\end{aligned}$$

Since $(aD_2)(bD_1) + (aD_1)(bD_2)$ may be $\neq 0$, it is clear that D_1D_2 need not be a derivation. The "obstruction" $(aD_2)(bD_1) + (aD_1)(bD_2)$ is symmetric in D_1 and D_2 so we obtain the same obstruction for D_2D_1. These cancel off if we form $[D_1D_2] \equiv D_1D_2 - D_2D_1$. Hence it is clear that $[D_1D_2] \in \mathfrak{D}_\Phi(\mathfrak{A})$. The expression $[D_1D_2]$ is called the *Lie* or *additive commutator* of D_1 and D_2. Our result is that $\mathfrak{D}_\Phi(\mathfrak{A})$ is a subspace of the space of linear transformations of $\mathfrak{A}$ closed under Lie commutators, that is, if $D_1, D_2 \in \mathfrak{D}_\Phi(\mathfrak{A})$, then $[D_1D_2] \in \mathfrak{D}_\Phi(\mathfrak{A})$. A subspace of $\mathfrak{L}_\Phi(\mathfrak{A})$ having this property is called a *Lie algebra* of linear transformations. The Lie product $[D_1D_2]$ is bilinear but it is not associative. The basic properties which it has are

$$(15)\quad [DD] = 0,\quad [[D_1D_2]D_3] + [[D_2D_3]D_1] + [[D_3D_1]D_2] = 0.$$

The first of these is clear and the second follows from a straightforward calculation which we leave to the reader. We note next the following Leibniz formula for the k-th power of a derivation:

$$(16)\qquad (ab)D^k = \sum_{i=0}^{k} \binom{k}{i} (aD^i)(bD^{k-i}),\quad k = 1, 2, \cdots.$$

This is readily proved by induction on k. Now suppose the base field is of characteristic $p \neq 0$. Then $\binom{p}{i} a = 0$ for $i = 1, 2, \cdots, p - 1$ and any $a \in \mathfrak{A}$, so in this case, (16) for $k = p$ reduces to

$$(17)\qquad (ab)D^p = (aD^p)b + a(bD^p),$$

which shows that $\mathfrak{D}_\Phi(\mathfrak{A})$ is also closed under p-th powers, that is, if $D \in \mathfrak{D}_\Phi(\mathfrak{A})$, then $D^p \in \mathfrak{D}_\Phi(\mathfrak{A})$. A Lie algebra of linear transformations in a vector space over a field Φ of characteristic $p \neq 0$ having this extra closure property is called a *restricted Lie algebra of characteristic p.*

EXERCISES

1. Let $\mathfrak{A}$ be an algebra over Φ and let $d \in \mathfrak{A}$. Verify that the mapping $a \rightarrow [ad] = ad - da$ is a derivation in $\mathfrak{A}$. Such a derivation is called an *inner derivation* of $\mathfrak{A}$. Prove that, if I_d denotes the inner derivation determined by d, then $I_{\alpha_1 d_1 + \alpha_2 d_2} = \alpha_1 I_{d_1} + \alpha_2 I_{d_2}$, α_i in Φ and $I_{[d_1 d_2]} = [I_{d_1} I_{d_2}]$. Show also that, if Φ is of characteristic $p \neq 0$, then $I_{d^p} = (I_d)^p$.

2. Let $\mathfrak{A}$ be a subalgebra of an algebra $\mathfrak{B}$. Verify that a mapping D of $\mathfrak{A}$ into $\mathfrak{B}$ is a derivation if and only if the mapping

$$a \rightarrow \begin{bmatrix} a & aD \\ 0 & a \end{bmatrix}$$

if $\mathfrak{A}$ into the matrix algebra $\mathfrak{B}_2$ of 2×2 matrices over $\mathfrak{B}$ is an isomorphism.

7. Derivations, separability and p-independence. We shall now take up the study of derivations in a field P/Φ. We note first that, if $\mathfrak{A}$ is a subalgebra of P/Φ, then any derivation D of $\mathfrak{A}/\Phi$ into P/Φ has a unique extension to a derivation D of the subfield E of P generated by $\mathfrak{A}$. This is a special case of Theorem 13 since $E = \mathfrak{A}_M$ for M the set of non-zero elements of $\mathfrak{A}$. Suppose next that E is a subfield of P/Φ and D is a derivation of E/Φ into P/Φ. Let $\xi \in P$. Then if ξ is transcendental over P, D can be extended to $E[\xi]$ so that $\xi D = \eta$ is any element of P. This is a consequence of Theorem 14. Moreover, D can be extended to the field $E(\xi)$ so that $\xi D = \eta$. Next assume ξ is algebraic over E, so $E[\xi] = E(\xi)$ and let $f(x)$ be the minimum polynomial of ξ over E. Then the ideal $\mathfrak{R}$ in $E[x]$ of polynomials $h(x)$ such that $h(\xi) = 0$ is the principal ideal $(f(x))$. Hence Theorem 14 shows that D can be extended to a derivation of $E(\xi)$ such that $\xi \to \eta$ if and only if

$$f^D(\xi) + f'(\xi)\eta = 0, \tag{18}$$

$f'(x)$ the usual derivative of $f(x)$ (cf. V of Introd.). If ξ is separable, then $f'(\xi) \neq 0$ and (18) gives $\eta = -f^D(\xi)f'(\xi)^{-1}$. Hence there is only one choice possible for η to give an extension of D. Thus we see that, if $E(\xi)$ is separable algebraic over E, then a derivation of E/Φ into P/Φ can be extended in one and only one way to a derivation of $E(\xi)$ over Φ. In particular, if $D = 0$ on E, then the only extension of D to a derivation in $E(\xi)$ is $D = 0$ on $E(\xi)$. If ξ is inseparable, then $f'(\xi) = 0$. Hence D can be extended to a derivation in $E(\xi)$ if and only if $f^D(\xi) = 0$ and, when this condition is fulfilled, then η is arbitrary so D can be extended to $E(\xi)$ in such a way that $\xi D = \eta$ is any chosen element of P. If $f(x) = x^n + \alpha_1 x^{n-1} + \cdots$, then $f^D(x) = (\alpha_1 D)x^{n-1} + (\alpha_2 D)x^{n-2} + \cdots$ and since $f(x)$ is the minimum polynomial, the condition $f^D(\xi) = 0$ holds if and only if every $\alpha_i D = 0$. Thus, a necessary and sufficient condition for the extendability of D to $E(\xi)$, ξ inseparable algebraic over E is that the coefficients of the minimum polynomial of ξ over E are D-constants. We shall need this criterion particularly in the case $f(x) = x^p - \alpha$. Then the condition is simply that $\alpha D = 0$.

Now let $P = \Phi(\xi_1, \xi_2, \cdots, \xi_m)$ a finitely generated extension field of Φ (that is, a field of algebraic functions). Let $\mathfrak{R}$ be the ideal in $\Phi[x_1, x_2, \cdots, x_m]$ of polynomials $f(x_1, x_2, \cdots, x_m)$ such that $f(\xi_1, \xi_2, \cdots, \xi_m) = 0$ and let $\mathfrak{X}$ be a basis for $\mathfrak{R}$. If D is a derivation of the algebra $\Phi[\xi_1, \xi_2, \cdots, \xi_m]/\Phi$ into P/Φ, D has a unique extension to P/Φ. Theorem 14, applied to the derivation

$D = 0$ on Φ, shows that there exists a derivation D of $\Phi[\xi_1, \xi_2, \cdots, \xi_m]/\Phi$ into P/Φ—hence of P/Φ into itself—such that $\xi_i D = \eta_i$, $i = 1, 2, \cdots, m$, if and only if

$$\sum_i \left(\frac{\partial g}{\partial x_i}\right)_{x_j=\xi_j} \eta_i = 0 \qquad (19)$$

for every $g \in \mathfrak{X}$.

We have considered the system $\mathfrak{D}_\Phi(P)$ of derivations of P/Φ in the last section and we have seen that this is a Lie algebra of linear transformations which is restricted in the characteristic $p \neq 0$ case, and $\mathfrak{D}_\Phi(P)$ is closed under right multiplication by elements ρ_R, $\rho \in P$. Hence we see that $\mathfrak{D}_\Phi(P)$ is a subspace of the right vector space $\mathfrak{L}_\Phi(P)$ over P (see § 1.1). We shall now investigate $\mathfrak{D}_\Phi(P)$ as right vector space over P for $P = \Phi(\xi_1, \xi_2, \cdots, \xi_m)$.

For this purpose we introduce the right vector space $P^{(m)}$ of m-tuples $(\rho_1, \rho_2, \cdots, \rho_m)$, $\rho_j \in P$, with the usual addition and multiplication by elements of P. If $D \in \mathfrak{D} \equiv \mathfrak{D}_\Phi(P)$, then we map D into the element $(\xi_1 D, \xi_2 D, \cdots, \xi_m D) \in P^{(m)}$. This mapping is P-linear and so its image $\mathfrak{D}'$ is a subspace of $P^{(m)}/P$. If $\xi_i D = 0$, $1 \leq i \leq m$, then $D = 0$ since the ξ_i are generators of $P = \Phi(\xi_1, \xi_2, \cdots, \xi_m)$. This shows that the kernel of the mapping $D \to (\xi_i D)$ of $\mathfrak{D}$ onto $\mathfrak{D}'$ is 0 and so the mapping is a P-linear isomorphism of $\mathfrak{D}$ onto $\mathfrak{D}'$.

Next we shall give a description of the subspace $\mathfrak{D}'$ of $P^{(m)}$ in terms of the ideal $\mathfrak{K}$ defined before. We note first that, if $f \in \Phi[x_1, x_2, \cdots, x_m]$, then the mapping

$$d_f: (\eta_1, \eta_2, \cdots, \eta_m) \to \sum \left(\frac{\partial f}{\partial x_i}\right)_{x_j=\xi_j} \eta_i \qquad (20)$$

is a linear function on $P^{(m)}$, that is, an element of the conjugate space $P^{(m)*}$ of $P^{(m)}$. Let $d\mathfrak{X}$ denote the subspace of $P^{(m)*}$ spanned by the elements d_g, $g \in \mathfrak{X}$, a set of generators for $\mathfrak{K}$. The condition (19) on $(\eta_1, \eta_2, \cdots, \eta_m)$ is that $(\eta_i)d_g = 0$ for all $g \in \mathfrak{X}$. Hence we see that there exists an element $D \in \mathfrak{D}_\Phi(P)$ such that $\xi_i D = \eta_i$, $1 \leq i \leq m$, if and only if $(\eta_i)d_g = 0$ for all $g \in \mathfrak{X}$. This clearly implies that $\mathfrak{D}'$ is the subspace of $P^{(m)}$ of vectors incident with the subspace $d\mathfrak{X}$ of $P^{(m)*}$ (Vol. II, p. 55). We recall that the sum of the dimensionalities of $\mathfrak{D}'$ and $d\mathfrak{X}$ is m. If we replace $\mathfrak{X}$ by the complete ideal $\mathfrak{K}$, then we have $d\mathfrak{X} \subseteq d\mathfrak{K}$; but, since both spaces have the same dimensionality $m - [\mathfrak{D}' : \mathrm{P}]_R$, it is clear that $d\mathfrak{X} = d\mathfrak{K}$. This shows that $d\mathfrak{X}$ is the same for any two sets of generators, a fact which is easy to see directly also. The result that we have obtained is the following

Theorem 15. *Let* $\mathrm{P} = \Phi(\xi_1, \xi_2, \cdots, \xi_m)$ *a field of algebraic functions over* Φ. *Let* $\mathfrak{X}$ *be a set of generators for the ideal* $\mathfrak{K}$ *of polynomials* $f(x_1, x_2, \cdots, x_m)$ *such that* $f(\xi_1, \xi_2, \cdots, \xi_m) = 0$ *and let* $\mathfrak{D}_\Phi(\mathrm{P})$ *be the right* P*-vector space of derivations in* P/Φ. *Then*

$$[\mathfrak{D}_\Phi(\mathrm{P}) : \mathrm{P}] = m - [d\mathfrak{X} : \mathrm{P}]_R \tag{21}$$

where $d\mathfrak{X}$ *is the set of linear functions* d_g, $g \in \mathfrak{X}$, *defined by* (20).

If $\mathfrak{X} = \{g_1, g_2, \cdots, g_r\}$, then it is clear from the definition of d_f and from the relation between dimensionality and determinantal rank (Vol. II, p. 22) that $[d\mathfrak{X} : \mathrm{P}]_R$ is the rank of the matrix

$$\begin{bmatrix} \left(\dfrac{\partial g_1}{\partial x_1}\right)_{x_j=\xi_j} & \left(\dfrac{\partial g_1}{\partial x_2}\right)_{x_j=\xi_j} & \cdots & \left(\dfrac{\partial g_1}{\partial x_m}\right)_{x_j=\xi_j} \\ \vdots & \vdots & & \vdots \\ \left(\dfrac{\partial g_r}{\partial x_1}\right)_{x_j=\xi_j} & \left(\dfrac{\partial g_r}{\partial x_2}\right)_{x_j=\xi_j} & \cdots & \left(\dfrac{\partial g_r}{\partial x_m}\right)_{x_j=\xi_j} \end{bmatrix}. \tag{22}$$

Hence the rank of this "Jacobian" matrix and (21) give the dimensionality of $\mathfrak{D}_\Phi(\mathrm{P})$ over P.

We shall now look at these questions in a different way from the point of view of the structure of P/Φ and we prove first the following

Lemma. *Let* $\mathrm{P} = \Phi(\xi_1, \xi_2, \cdots, \xi_m)$. *Then* 0 *is the only derivation of* P/Φ *into itself if and only if* P *is separable algebraic over* Φ.

Proof. If ρ is a separable algebraic element of P and D is a derivation of P/Φ, then we have seen that $\rho D = 0$. Hence it is clear that, if P/Φ is separable algebraic, then $D = 0$ is the only derivation in P/Φ. Next suppose P is not separable algebraic over

Φ. We may suppose $\{\xi_1, \xi_2, \cdots, \xi_r\}$ is a transcendency basis ($r = 0$ if P is algebraic). If P is not separable over $\Phi(\xi_1, \xi_2, \cdots, \xi_r)$, then the characteristic is $p \neq 0$ and, if Σ is the subfield of elements of P which are separable over $\Phi(\xi_1, \xi_2, \cdots, \xi_r)$, then $P \supset \Sigma$ and P is purely inseparable over Σ. We assert that there exists a subfield $E \supseteq \Sigma$ such that $P = E(\rho)$, where the minimum polynomial of ρ over E is $x^p - \beta$, $\beta \varepsilon E$. We have $[P:\Sigma] < \infty$ and we can take E to be a maximal proper subfield of P containing Σ. If $\sigma \varepsilon P, \notin E$, $P = E(\sigma)$ by the maximality of E. Since P is purely inseparable over Σ, hence over E, the minimum polynomial of σ over E has the form $x^{p^k} - \beta$, $k > 0$. Then $\rho = \sigma^{p^{k-1}} \notin E$ so $E(\rho) \supset E$. By the maximality of E we have $E(\rho) = P$. Moreover, $\rho^p = \sigma^{p^k} = \beta$, so $x^p - \beta$ is the minimum polynomial of ρ over E. Now we have seen that there exists a derivation D of P/E such that ρD is any chosen element of P. If we take $\rho D \neq 0$, D is a non-zero derivation of P/Φ. Next assume P is separable algebraic over $\Phi(\xi_1, \xi_2, \cdots, \xi_r)$. Then since P is not separable algebraic over Φ, $r > 0$, and there exists a non-zero derivation of $\Phi[\xi_1, \cdots, \xi_r]$ over Φ into P over Φ. This can be extended to P; hence in this case also we obtain a non-zero derivation in P/Φ.

We can now prove the following result on the dimensionality of $\mathfrak{D}_\Phi(P)$ over P.

Theorem 16. *If* $P = \Phi(\xi_1, \xi_2, \cdots, \xi_m)$, *then* $[\mathfrak{D}_\Phi(P):P]_R$ *is the smallest integer* s *such that there exists a subset* $\{\xi_{i_1}, \xi_{i_2}, \cdots, \xi_{i_s}\}$ *of* $\{\xi_1, \xi_2, \cdots, \xi_m\}$ *such that* P *is separable algebraic over* $\Phi(\xi_{i_1}, \xi_{i_2}, \cdots, \xi_{i_s})$.

Proof. As before, we consider the mapping $D \rightarrow (\xi_1 D, \xi_2 D, \cdots, \xi_m D)$ of $\mathfrak{D} = \mathfrak{D}_\Phi(P)$ into $P^{(m)}$. We know that this is a P isomorphism into $P^{(m)}/P$. Let $(D_1, D_2, \cdots, D_s)$ be a right basis of $\mathfrak{D}$ over P. Then $s \leq m$ and the image of $\mathfrak{D}$ in $P^{(m)}$ has the basis $(\xi_1 D_j, \xi_2 D_j, \cdots, \xi_m D_j)$, $1 \leq j \leq s$. The rank of the $s \times m$ matrix $(\xi_i D_j)$ is s, so we can choose the order of the ξ's so that $\det(\xi_i D_j) \neq 0$, $1 \leq i, j \leq s$. Set $E = \Phi(\xi_1, \xi_2, \cdots, \xi_s)$ and let D be a derivation of P/E into itself. Then $D \varepsilon \mathfrak{D}$ so $D = \sum_{j=1}^{s} D_j \mu_j$, $\mu_j \varepsilon P$. Also $\xi_i D = \sum_{j=1}^{s} (\xi_i D_j)\mu_j = 0$ for $i = 1, 2, \cdots, s$. Since $\det(\xi_i D_j)$

$\neq 0$, this implies that every $\mu_j = 0$ so $D = 0$. We therefore see that the only derivation of P/E is $D = 0$. Hence, by the lemma, P is separable algebraic over $E = \Phi(\xi_1, \xi_2, \cdots, \xi_s)$. Next suppose $\{\xi_{i_1}, \xi_{i_2}, \cdots, \xi_{i_t}\}$ is a subset of the ξ's such that P is separable algebraic over $\Phi(\xi_{i_1}, \cdots, \xi_{i_t})$. If we re-order the ξ's we may assume the given set is $\{\xi_1, \xi_2, \cdots, \xi_t\}$. We now use these ξ's to map $\mathfrak{D}$ into $P^{(t)}$ by means of the mapping $D \to (\xi_j D)$, $1 \leq j \leq t$. Again this is P-linear. If $(\xi_j D) = 0$, then D maps $E = \Phi(\xi_1, \xi_2, \cdots, \xi_t)$ into 0 and so D is a derivation of P/E into itself. Since P is separable algebraic over E, the lemma shows that $D = 0$. Hence we see that the mapping $D \to (\xi_j D)$ is an isomorphism and consequently $s = [\mathfrak{D}:P]_R \leq t$. This completes the proof.

Corollary. *If* $P = \Phi(\xi_1, \xi_2, \cdots, \xi_m)$, *then* $[\mathfrak{D}_\Phi(P):P]_R \geq r =$ tr. d. P/Φ *and equality holds if and only if* P *is separably generated over* Φ.

Proof. The theorem shows that, if $s = [\mathfrak{D}:P]_R$, then we may assume that P is separable algebraic over $E = \Phi(\xi_1, \xi_2, \cdots, \xi_s)$. Since P is algebraic over E, it follows that $\{\xi_1, \xi_2, \cdots, \xi_s\}$ contains a transcendency basis; hence $s \geq r$. If $s = r$, then since P is separable over E, the set $\{\xi_1, \xi_2, \cdots, \xi_r\}$ is a separating transcendency basis. Conversely, suppose P is separably generated. Then we know that we may select a separating transcendency basis from the set of ξ's. We may assume this is $\{\xi_1, \xi_2, \cdots, \xi_r\}$. Then P is separable algebraic over $\Phi(\xi_1, \xi_2, \cdots, \xi_r)$ and the theorem shows that $r \geq [\mathfrak{D}:P]_R$. Since we have shown that $[\mathfrak{D}:P]_R \geq r$ always, we see that $[\mathfrak{D}:P]_R = r$.

In the remainder of this section we shall assume the characteristic of the field P is $p \neq 0$. We shall see that the theory of derivations in this case is closely connected with the study of purely inseparable extensions of a simple type. We assume P is purely inseparable (algebraic) over Φ. If $\rho \in P$, then the minimum polynomial of ρ over Φ has the form $x^{p^e} - \beta$ (Lemma 2, § 1.9). We call e the *exponent* of the purely inseparable element ρ. Evidently the exponent is 0 if and only if $\rho \in \Phi$. If there exists a maximum k for the exponents of the elements of P, then we say that P is of *exponent* k over Φ; otherwise, the exponent of P/Φ is infinite.

We shall be interested particularly in purely inseparable extensions of exponent ≤ 1. P has this property relative to Φ if and only if $P^p \subseteq \Phi$ where P^p is the subfield of p-th powers of elements of P. Hence it is clear that, if P is any extension of Φ of characteristic p, then P is purely inseparable of exponent ≤ 1 over $\Phi' = \Phi(P^p)$. We shall now say that an element $\rho \in P$ is *p-dependent in* P *over* Φ *on the subset* S *of* P if $\rho \in \Phi'(S)$ where $\Phi' = \Phi(P^p)$. We indicate this relation by $\rho <_p S$ (assuming P and Φ are fixed in our discussion). We proceed to show that this is a dependence relation in the sense of § 3. First, it is clear that, if $\rho \in S$, then $\rho \in \Phi'(S)$; hence $\rho <_p S$. If $\rho <_p S$ we have $\rho \in \Phi'(S)$ and, since $\Phi'(S)$ is the union of its subfields $\Phi'(F)$ where F is a finite subset of S, then $\rho <_p F$ for some finite subset F of S. If $\rho \in \Phi'(S)$ and every $\sigma \in S$ is contained in $\Phi'(T)$, then $\rho \in \Phi'(T)$. Hence if $\rho <_p S$ and every $\sigma \in S$ satisfies $\sigma <_p T$, then $\rho <_p T$. It remains to check the exchange axiom. This states that, if $\rho \in \Phi'(S)$ and $\rho \notin \Phi'(S - \{\sigma\})$ for some σ in S, then $\sigma \in \Phi'((S - \{\sigma\}) \cup \{\rho\})$. Set $T = S - \{\sigma\}$ and consider the subfields $\Phi'(T, \rho, \sigma)$, $\Phi'(T, \rho)$, $\Phi'(T, \sigma)$, $\Phi'(T)$ for which we have the diagram:

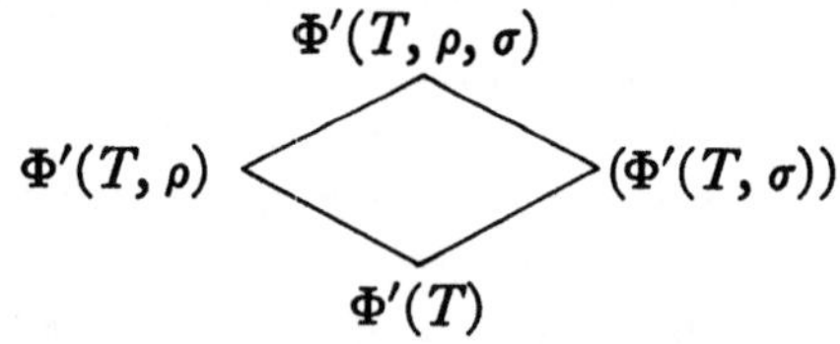

We have $\Phi'(T, \rho) \supset \Phi'(T)$ and $\Phi'(T, \sigma) \supset \Phi'(T)$. Also $\rho^p \in \Phi'(T)$ and $\sigma^p \in \Phi'(T)$. It follows that $[\Phi'(T, \sigma):\Phi'(T)] = p = [\Phi'(T, \rho):\Phi'(T)]$. Since $\rho \in \Phi'(T, \sigma)$, $\Phi'(T, \rho, \sigma) = \Phi'(T, \sigma)$ so $[\Phi'(T, \rho, \sigma):\Phi'(T)] = p$. It follows that $\Phi'(T, \rho, \sigma) = \Phi'(T, \rho) = \Phi'(T, \sigma)$ so $\sigma <_p T \cup \{\rho\} = (S - \{\sigma\}) \cup \{\rho\}$. This completes the verification of the axioms for a dependence relation.

We can now apply the general theory of dependence relations. Accordingly, we call a subset S of P *p-independent* if $\sigma \not<_p S - \{\sigma\}$ for every $\sigma \in S$. The general basis theorem implies that there exists a p-independent subset B of P such that every element is p-dependent on B. The latter condition is equivalent to $P = \Phi'(B)$. The set B is called a *p-basis* for P over Φ. Any two p-bases have the same cardinal number.

If $F = \{\rho_1, \rho_2, \cdots, \rho_m\}$ is a p-independent set, then $\rho_i^p = \beta_i \varepsilon \Phi'$ and $\rho_i \notin \Phi'(\rho_1, \rho_2, \cdots, \rho_{i-1})$. Hence $[\Phi'(\rho_1, \cdots, \rho_i):\Phi'(\rho_1, \cdots, \rho_{i-1})] = p$ and $[\Phi'(\rho_1, \cdots, \rho_m):\Phi'] = p^m$. It follows that the p^m elements

$$(23) \qquad \rho_1^{k_1}\rho_2^{k_2} \cdots \rho_m^{k_m}, \quad 0 \leq k_i < p,$$

form a basis for $\Phi'(\rho_1, \rho_2, \cdots, \rho_m)$ over Φ'. Conversely, if this condition holds, then it is immediate that F is a p-independent set. We shall find it useful to apply this criterion in the following equivalent form: F is p-independent if and only if the only relation of the form

$$(24) \qquad \Sigma\alpha_{k_1\cdots k_m}\rho_1^{k_1} \cdots \rho_m^{k_m} = 0, \quad 0 \leq k_i < p$$

with the α's in Φ' is the trivial one in which every $\alpha_{k_1\cdots k_m} = 0$.

We note also that any p-independent subset A can be imbedded in a maximal p-independent set B and such a set is necessarily a basis.

We return to the consideration of derivations in any field P/Φ of characteristic p. If E is a subfield of P/Φ and D is a derivation of E/Φ into P/Φ, then $\epsilon^p D = p\epsilon^{p-1}(\epsilon D) = 0$ for any ϵ in E. The set of D-constants is a subfield Γ of E over Φ and the remark just made shows that $\Gamma \supseteq \Phi(\mathrm{E}^p)$. If $\gamma \varepsilon \Gamma$ and $\epsilon \varepsilon \mathrm{E}$, then $(\gamma\epsilon)D = \gamma(\epsilon D)$. This shows that D is a derivation of E/Γ into P/Γ. Since $\Phi(\mathrm{E}^p) \subseteq \Gamma$, every derivation of E/Φ into P/Φ is a derivation of $\mathrm{E}/\Phi(\mathrm{E}^p)$ into $\mathrm{P}/\Phi(\mathrm{E}^p)$ and the converse is clear. Hence, in considering the derivations of E over Φ into P over Φ, we may as well replace Φ by $\Phi(\mathrm{E}^p)$ and so we may assume $\mathrm{E}^p \subseteq \Phi$. In other words, *we may assume* E *is purely inseparable of exponent* ≤ 1 over Φ. It is now an easy matter to determine the derivations of E over Φ into P over Φ. This is given in the following

Theorem 17. *Let* P *be an arbitrary field of characteristic* $\neq 0$, Φ *a subfield and* E *an intermediate field. Let* B *be a p-basis of* E *over* Φ. *Let* δ *be an arbitrary mapping of* B *into* P. *Then there exists one and only one derivation* D *of* E *over* Φ *into* P *over* Φ *such that* $\epsilon D = \delta(\epsilon)$ *for every* $\epsilon \varepsilon B$.

Proof. As we indicated, there is no loss in generality in assuming E is purely inseparable of exponent ≤ 1 over Φ. Also, we may suppose $E \supset \Phi$ which means that B is non-vacuous and the exponent of E/Φ is exactly one. Let $\epsilon \in B$ and set $B_\epsilon = B - \{\epsilon\}$. Then $\epsilon \notin \Phi(B_\epsilon)$ so the minimum polynomial of ϵ over $\Phi(B_\epsilon)$ is of the form $x^p - \beta$. Hence there exists a derivation D_ϵ of $E = \Phi(B_\epsilon, \epsilon)$ over $\Phi(B_\epsilon)$ into P/Φ sending ϵ into the element $\delta(\epsilon)$. If $F = \{\epsilon_1, \epsilon_2, \cdots, \epsilon_r\}$ is a finite subset F of B, then $D_F = D_{\epsilon_1} + D_{\epsilon_2} + \cdots + D_{\epsilon_r}$ is a derivation of E/Φ into P/Φ such that $\epsilon_i D_F = \delta(\epsilon_i)$, $i = 1, 2, \cdots, r$. If G is a finite subset of B containing F, then the restriction of D_G to $\Phi(F)$ coincides with the restriction of D_F to $\Phi(F)$. Now if ξ is an arbitrary element of E, we can choose a finite subset F such that $\xi \in \Phi(F)$ and we can map $\xi \to \xi D_F$. Then it is clear that ξD_F is the same for any finite subset such that $\xi \in \Phi(F)$. Hence the mapping $D: \xi \to \xi D_F$ is single-valued. It is immediate that D is a derivation of E/Φ into P/Φ such that $\epsilon D = \delta(\epsilon)$ for every $\epsilon \in B$. Since $E = \Phi(B)$, D is unique.

Let $\mathfrak{D}_\Phi(E, P)$ denote the set of derivations of E/Φ into P/Φ. We consider $\mathfrak{D}_\Phi(E, P)$ as right vector space over P as we did before for $\mathfrak{D}_\Phi(P)$ (cf. § 1.1 and p. 176). Then we have

Corollary 1. $[\mathfrak{D}_\Phi(E, P):P]_R < \infty$ *if and only if* E/Φ *has a finite p-basis. Then* $[\mathfrak{D}_\Phi(E, P):P]_R = |B|$.

Proof. Let B be a p-basis for E over Φ. Let $\Delta(B, P)$ be the set of mappings of B into P which we consider as a right vector space over P in the obvious way: $(\delta_1 + \delta_2)(\beta) = \delta_1(\beta) + \delta_2(\beta)$, $\delta_i \in \Delta$, $\beta \in B$ and $(\delta\rho)(\beta) = \delta(\beta)\rho$, $\delta \in \Delta$, $\beta \in B$, $\rho \in P$. We now map $\mathfrak{D}_\Phi(E, P)$ into $\Delta(B, P)$ by sending $D \in \mathfrak{D}_\Phi(E, P)$ into its restriction δ to B. This mapping is linear and, since $E = \Phi(B)$, $D \to \delta$ is an isomorphism. Moreover, the theorem shows that the mapping is surjective. Now it is clear that $[\Delta(B, P):P]_R$ is infinite (even uncountable) if B is infinite. Moreover, if B is finite, say, $B = \{\beta_1, \beta_2, \cdots, \beta_r\}$, then the r-mappings δ_i such that $\delta_i(\beta_j) = \delta_{ij}$ (the Kronecker δ_{ij}) form a basis for $[\Delta(B, P):P]_R$. Hence $r = [\Delta(B, P):P]_R = [\mathfrak{D}_\Phi(E, P):P]_R$.

In the special case $P = E = \Phi(\xi_1, \xi_2, \cdots, \xi_m)$ this corollary gives, in addition to Theorem 15 and 16, still a third way of evaluating $[\mathfrak{D}_\Phi(P):P]_R$ in the characteristic $p \neq 0$ case, namely, this dimensionality is the number of elements in a p-basis for

P/Φ. A second consequence of Theorem 17 is the following

Corollary 2. *Every derivation of* E/Φ *into* P/Φ *can be extended to a derivation of* P/Φ *if and only if the elements of any p-basis B of* E/Φ *are p-independent in* P/Φ.

Proof. If the condition holds, then B can be imbedded in a p-basis C of P over Φ. If D is a derivation of E/Φ into P/Φ, then the restriction δ_B of D to B can be extended to a mapping δ_C of C to P. The corresponding derivation D' of P/Φ into itself is an extension of D. On the other hand, suppose B is not p-independent in P over Φ and let β be an element of B which is p-dependent in P on $B_\beta = B - \{\beta\}$. If D' is any derivation in P such that $\beta' D' = 0$ for all $\beta' \varepsilon B_\beta$, then since $\beta \varepsilon \Phi(\mathrm{P}^p, B_\beta)$, $\beta D' = 0$. The theorem shows that there exist derivations D of E over Φ into P over Φ such that $\beta' D = 0$, $\beta' \varepsilon B_\beta$ but $\beta D \neq 0$. Clearly, such a derivation cannot be extended to P.

Our next two corollaries will deal with the special case E = P. The proofs are quite similar to those we have just given so we leave these as exercises.

Corollary 3. *Let* P *be any field of characteristic $p \neq 0$ over* Φ. *Then an element $\rho \varepsilon$* P *is in* $\Phi' = \Phi(\mathrm{P}^p)$ *if and only if $\rho D = 0$ for every derivation D of* P *over* Φ.

Corollary 4. *A subset S of* P *is p-independent if and only if for every $\rho \varepsilon S$ there exists a derivation D of* P *over* Φ *such that $\rho D \neq 0$ and $\sigma D = 0$ for every $\sigma \neq \rho$ in S.*

We shall now specialize our results by taking Φ to be the prime field Φ_0 ($\cong I_p$). A derivation of E/Φ_0 into P/Φ_0 will simply be called a *derivation of* E *into* P. We remark that, if D is a mapping of E into P such that $(\epsilon_1 + \epsilon_2)D = \epsilon_1 D + \epsilon_2 D$ and $(\epsilon_1\epsilon_2)D = (\epsilon_1 D)\epsilon_2 + \epsilon_1(\epsilon_2 D)$, then D is a derivation of E into P in the present sense, since $(\alpha\epsilon)D = \alpha(\epsilon D)$ for $\alpha \varepsilon \Phi_0$ is a consequence of the first property. We note also that $\Phi_0(\mathrm{E}^p) = \mathrm{E}^p$. Hence Corollary 3 gives a criterion for an element to be a p-th power. We shall now investigate the criterion given in Corollary 2 that the derivations of E into P be extendable to derivations in P. We proceed to show that the condition given is equivalent to separability, in the general sense, of P over E. First assume the condition: every p-basis of E (over Φ_0) is p-independent in P. Let $\rho_1, \rho_2, \cdots, \rho_n \in P$ and suppose we have $\Sigma\epsilon_i\rho_i^p = 0$ for $\epsilon_i \neq 0$ in E. If B is a p-basis for E, then we can write $\epsilon_i = \Sigma\gamma_{ik_1k_2\cdots k_r}\beta_1^{k_1}\beta_2^{k_2}\cdots\beta_r^{k_r}$ where the $\beta_j \in B$, $0 \leq k_j < p$, $\gamma_{ik_1\cdots k_r} \in E^p$. We have $\Sigma\delta_{k_1\cdots k_r}\beta_1^{k_1}\cdots\beta_r^{k_r} = 0$ where

$$\delta_{k_1\cdots k_r} = \sum_i \gamma_{ik_1\cdots k_r}\rho_i^p \in P^p.$$

Since the β's are p-independent in P we have $\delta_{k_1\cdots k_r} = 0$. We can write $\gamma_{ik_1\cdots k_r} = \eta_{ik_1\cdots k_r}{}^p$, $\eta_{ik_1\cdots k_r}$ in E. Then $0 = \delta_{k_1\cdots k_r} = \sum_i \eta_{ik_1\cdots k_r}{}^p\rho_i^p$ gives $\sum_i \eta_{ik_1\cdots k_r}\rho_i = 0$ for all k_j. Since the $\epsilon_i \neq 0$ one of these relations is non-trivial, so we have shown that any non-trivial relation of the form $\Sigma\epsilon_i\rho_i^p = 0$, $\epsilon_i \in E$, $\rho_i \in P$ implies one of the form $\Sigma\eta_i\rho_i = 0$, $\eta_i \in E$. This is equivalent to separability of P/E. Conversely, assume P separable over E and let $\beta_1, \beta_2, \cdots, \beta_r$ be elements of E for which we have a relation $\Sigma\gamma_{k_1\cdots k_r}\beta_1^{k_1}\beta_2^{k_2}\cdots\beta_r^{k_r} = 0$, $\gamma_{k_1\cdots k_r} = \eta_{k_1\cdots k_r}{}^p \in P^p$, $0 \leq k_i < p$. Let $\{\rho_\alpha\}$ be a basis for P/E and write $\eta_{k_1\cdots k_r} = \sum_{i=1}^n \lambda_{k_1\cdots k_ri}\rho_i$, λ's in E. Then we have

$$0 = \Sigma\gamma_{k_1\cdots k_r}\beta_1^{k_1}\cdots\beta_r^{k_r} = \sum_i \mu_i\rho_i^p$$

where $\mu_i = \sum_k \lambda_{k_1\cdots k_ri}{}^p\beta_1^{k_1}\beta_2^{k_2}\cdots\beta_r^{k_r}$. Since P is separable over E, $\Sigma\mu_i\rho_i^p = 0$ implies every $\mu_i = 0$ so

$$\Sigma\lambda_{k_1\cdots k_ri}{}^p\beta_1^{k_1}\beta_2^{k_2}\cdots\beta_r^{k_r} = 0, \quad i = 1, 2, \cdots, n.$$

If some $\gamma_{k_1\cdots k_r} \neq 0$, one of the $\lambda_{k_1\cdots k_ri} \neq 0$ and so we obtain a non-trivial relation with coefficients in E^p involving the powers $\beta_1^{k_1}\cdots\beta_r^{k_r}$. This implies that, if $\{\beta_1, \beta_2, \cdots, \beta_r\}$ is p-dependent in P, then it is p-dependent in E. It is clear that this implies the condition of Corollary 2. This corollary therefore gives the following

Theorem 18. *The following two conditions on a field* P/E *of characteristic p are equivalent:* (1) P/E *is separable,* (2) *every derivation of* E *into* P *can be extended to a derivation in* P.

EXERCISES

1. Let $P = \Phi(\rho)$ where $\rho^{p^2} \in \Phi$ but $\rho^p \notin \Phi$. Show that $\{\rho^p\}$ is a p-independent subset of $E = \Phi(\rho^p)/\Phi$ but $\{\rho^p\}$ is not a p-independent subset of P/Φ.

2. Let B be a p-basis for P over Φ of characteristic $p \neq 0$. Show that for every positive integer k, $P = \Phi(P^{p^k}, B)$.

In ex. 3, 4, P is purely inseparable of exponent one over Φ and $[P:\Phi] = p^m < \infty$.

3. (Baer) Show that there exists a derivation D of P/Φ such that the only D-constants are the elements of Φ. (Hint: Let E be a proper subfield of P and suppose we already have a derivation D in E/Φ satisfying the condition. Let $\rho \in P$, $\notin E$. We can choose a $\beta \in E$ which is not of the form ϵD, ϵ in E and extend D to $E(\rho)$ by specifying that $\rho D = \beta$. Then the only D-constants of $E(\rho)$ are the elements of Φ.)

4. Show that the D in ex. 3 can be chosen so that D is nilpotent.

5. (Faith) Let P be a field of algebraic functions over Φ (any characteristic) and let E over Φ be a subfield. Show that $[\mathfrak{D}_\Phi(P):P]_R \geq [\mathfrak{D}_\Phi(E):E]_R$.

6. Let $P = \Phi(\xi_1, \xi_2, \cdots, \xi_m)$, Φ of characteristic $p \neq 0$. Show that tr. d. P/Φ does not exceed the number of elements in a p-basis for P/Φ.

7. Let P and Φ be as in ex. 6. Show that, if $(D_1, D_2, \cdots, D_r)$ is a right P-basis for $\mathfrak{D}_\Phi(P)$, then the elements $\rho_1, \rho_2, \cdots, \rho_r$ form a p-basis for P over Φ if and only if the matrix $(\rho_i D_j)$ is non-singular. Show also that, if $\rho_1, \rho_2, \cdots, \rho_r$ is a p-basis, then the elements $D_1, D_2, \cdots, D_r$ form a right P-basis of $\mathfrak{D}_\Phi(P)$ if and only if $(\rho_i D_j)$ is non-singular.

8. Let $P = \Phi(\xi_1, \xi_2, \cdots, \xi_m)$. Show that P/Φ is separable algebraic if and only if there exist m polynomials $g_1(x_1, x_2, \cdots, x_m), \cdots, g_m(x_1, x_2, \cdots, x_m)$ in $\Phi[x_1, x_2, \cdots, x_m]$ such that $g_i(\xi_1, \xi_2, \cdots, \xi_m) = 0$ and the Jacobian

$$\det\left(\left(\frac{\partial g_i}{\partial x_j}\right)_{x_k = \xi_k}\right) \neq 0.$$

9. Let D be a derivation in P/Φ, Γ the subfield of D-constants. Prove that the elements $\rho_1, \rho_2, \cdots, \rho_m$ are Γ-dependent if and only if the *Wronskian* determinant

$$\begin{vmatrix} \rho_1 & \rho_2 & \cdots & \rho_m \\ \rho_1 D & \rho_2 D & \cdots & \rho_m D \\ \vdots & \vdots & \ddots & \vdots \\ \rho_1 D^{m-1} & \rho_2 D^{m-1} & \cdots & \rho_m D^{m-1} \end{vmatrix} = 0.$$

8. Galois theory for purely inseparable extensions of exponent one. In this section we shall develop a Galois theory for purely inseparable extensions of exponent one in which the role of the Galois group of the classical theory is taken by the Lie algebra of derivations.

First, let P be purely inseparable of exponent ≤ 1 over Φ and suppose P has a finite p-basis $B = \{\rho_1, \rho_2, \cdots, \rho_m\}$ over Φ. Then

$[P:\Phi] = p^m$ and the elements $\rho_1^{k_1}\rho_2^{k_2} \cdots \rho_m^{k_m}$, $0 \leq k_i < p$, form a basis for P/Φ. We have $\rho_i^p = \beta_i \varepsilon \Phi$. As before, we let $\mathfrak{D}_\Phi(P)$ denote the set of derivations of P/Φ and we recall that $\mathfrak{D}_\Phi(P)$ is a restricted Lie algebra of linear transformations in P over Φ. This means that $\mathfrak{D}_\Phi(P)$ is a subspace of the space $\mathfrak{L}_\Phi(P)$ of linear transformations of the vector space P over Φ such that, if $D_1, D_2 \varepsilon \mathfrak{D}_\Phi(P)$, then $[D_1D_2] = D_1D_2 - D_2D_1 \varepsilon \mathfrak{D}_\Phi(P)$ and $D_1^p \varepsilon \mathfrak{D}_\Phi(P)$. We have seen also that $\mathfrak{D}_\Phi(P)$ is a right vector space over P relative to $D\rho = D\rho_R$ for ρ in P. Also we know that $[\mathfrak{D}_\Phi(P):P]_R = m$ (Cor. 1 to Th. 17) and, if ρ is an element of P such that $\rho D = 0$ for every $D \varepsilon \mathfrak{D}_\Phi(P)$, then $\rho \varepsilon \Phi$ (Cor. 3 to Th. 17). This last result gives one half of the Galois correspondence which we shall establish.

To obtain the second half of this correspondence we now suppose that P is any field of characteristic $p \neq 0$ and we do not specify any subfield as base field. As at the end of the last section, we consider derivations in P, which can be defined either as derivations of P over its prime field or as endomorphisms D of $(P, +)$ such that $(\rho\sigma)D = (\rho D)\sigma + \rho(\sigma D)$, $\rho, \sigma \varepsilon P$. We suppose now that we are given a set $\mathfrak{D}$ of derivations in P with the following closure properties: (1) $\mathfrak{D}$ is closed under addition. (2) $\mathfrak{D}$ is closed under Lie commutation $[D_1D_2]$. (3) $\mathfrak{D}$ is closed under p-th powers. (4) $\mathfrak{D}$ is closed under right multiplication by elements ρ_R, $\rho \varepsilon P$. The conditions (1) and (4) amount to saying that $\mathfrak{D}$ is a subspace of the right vector space of endomorphisms of the additive group $(P, +)$ considered as a space over P relative to $A\rho = A\rho_R$. Any set of endomorphisms of $(P, +)$ which satisfy (1) to (4) will be called a *restricted* P-*Lie algebra* of endomorphisms of $(P, +)$.* We can now state the following theorem.

Theorem 19 (Jacobson). *Let* P *be a field of characteristic* $p \neq 0$ *and let* $\mathfrak{D}$ *be a restricted* P-*Lie algebra of derivations in* P *such that* $[\mathfrak{D}:P]_R = m < \infty$. *Then:* (1) *if* Φ *is the subfield of* $\mathfrak{D}$-*constants, then* P *is purely inseparable of exponent* ≤ 1 *over* Φ *and* $[P:\Phi] = p^m$; (2) *if* D *is any derivation of* P *over* Φ, *then* $D \varepsilon \mathfrak{D}$; (3) *if* $(D_1, D_2, \cdots, D_m)$ *is any right basis for* $\mathfrak{D}$ *over* P, *then the set of monomials*

* It should not be inferred from this terminology that $\mathfrak{D}$ is an algebra over P as base field. One of the conditions for an algebra is that $[D_1D_2]\rho = [D_1\rho, D_2] = [D_1, D_2\rho]$ and this does not hold for every ρ (see equation (26) given subsequently).

(25) $D_1^{k_1}D_2^{k_2}\cdots D_m^{k_m},\quad 0 \le k_i < p,\quad (D_i^0 = 1)$

is a right basis for the ring $\mathfrak{L}_\Phi(\mathrm{P})$ *of linear transformations of* P *over* Φ *considered as a right vector space over* P.

Proof. The idea of the proof we shall give is basically the same as that we used for the Galois theory of automorphisms: we shall use the given set $\mathfrak{D}$ to define a set of endomorphisms $\mathfrak{A}$ satisfying the hypotheses of the Jacobson-Bourbaki theorem (Th. 1.2). In the present case we let $\mathfrak{A}$ be the set of right P-linear combinations of the endomorphisms given in (25). Then it is clear that $\mathfrak{A}$ is a right vector space over P and that $[\mathfrak{A}:\mathrm{P}]_R < \infty$. It remains to show that $\mathfrak{A}$ is a subring of the ring of endomorphisms of (P, +) and for this it is enough to show that $1 \in \mathfrak{A}$ and that $\mathfrak{A}$ is closed under multiplication. The first of these is clear since $\mathfrak{A}$ contains $D_1^0D_2^0 \cdots D_m^0 = 1$. To show closure under multiplication it is enough to prove that every product $(D_1^{k_1}D_2^{k_2} \cdots D_m^{k_m}\rho)D_j \in \mathfrak{A}$ for $\rho \in \mathrm{P}$; for, if this holds, then one sees easily that every product $(D_1^{k_1} \cdots D_m^{k_m}\rho)(D_1^{l_1} \cdots D_m^{l_m}\sigma)$, $\sigma \in \mathrm{P}$, is contained in $\mathfrak{A}$. If D is a derivation in P, the condition $(\xi\rho)D = (\xi D)\rho + \xi(\rho D)$ can be written in operator form as:

(26) $$\rho_R D = D\rho_R + (\rho D)_R.$$

This implies that $(D_1^{k_1} \cdots D_m^{k_m}\rho)D_j = D_1^{k_1} \cdots D_m^{k_m}D_j\rho + D_1^{k_1} \cdots D_m^{k_m}(\rho D_j)$. Hence to show that $\mathfrak{A}$ is closed under multiplication it is enough to show that $D_1^{k_1} \cdots D_m^{k_m}D_j \in \mathfrak{A}$ for every $j = 1, \cdots, m$ and $0 \le k_i \le p - 1$. We shall now assign an (apparent) *degree* $N = k_1 + k_2 + \cdots + k_m$ to the monomial $D_1^{k_1}D_2^{k_2} \cdots D_m^{k_m}$ and we shall show that $D_1^{k_1} \cdots D_m^{k_m}D_j$ is a right P-linear combination of monomials (25) of degree $\le N + 1$. This is clear if $N = 0$ so we assume it holds for every $D_1^{l_1} \cdots D_m^{l_m}$ of degree $\Sigma l_i < N$. Suppose first that $j = m$. Then if $k_m < p - 1$, $D_1^{k_1} \cdots D_m^{k_m}D_m$ is one of the monomials (25) of degree $N + 1$ so the result holds in this case. If $k_m = p - 1$, then $(D_1^{k_1} \cdots D_m^{k_m})D_m = D_1^{k_1} \cdots D_{m-1}^{k_{m-1}}D_m^p$ and $D_m^p = \Sigma D_i\mu_{im}$ since D is closed under p-th powers. Hence

$$D_1^{k_1} \cdots D_m^{k_m}D_m = \Sigma D_1^{k_1} \cdots D_{m-1}^{k_{m-1}}D_i\mu_{im}$$

so the induction hypothesis applies to show that $D_1^{k_1} \cdots D_m^{k_m}D_m$

is a right P-linear combination of monomials (25) of degree $\leq N + 1$. Thus we have the result if $j = m$ and so we can now make an additional induction hypothesis, namely, that the result asserted holds for $D_1^{k_1} \cdots D_m^{k_m} D_l$ for all $l > j$. Since $N = \Sigma k_i > 0$, some $k_i \neq 0$ so we may assume $k_r \neq 0$ and $k_s = 0$ if $s > r$. Then we have $(D_1^{k_1} \cdots D_m^{k_m})D_j = (D_1^{k_1} \cdots D_r^{k_r})D_j$. If $j > r$, the product is a monomial (25) so the result holds in this case. If $j = r$, the argument given before for $j = m$ is applicable to prove the assertion. Hence it remains to consider the case: $j < r$. Since $\mathfrak{D}$ is closed under commutation, $D_r D_j = D_j D_r + \Sigma D_h \nu_{hrj}$, $\nu_{hrj} \in \mathrm{P}$. Then

$$D_1^{k_1} \cdots D_r^{k_r} D_j = D_1^{k_1} \cdots D_r^{k_r - 1} D_j D_r + \sum_h D_1^{k_1} \cdots D_r^{k_r - 1} D_h \nu_{hjr}$$

and every $D_1^{k_1} \cdots D_r^{k_r - 1} D_h$ is a P-linear combination of monomials (25) of total degree $< N$. Also this holds for $D_1^{k_1} \cdots D_r^{k_r - 1} D_j$ and, since $r > j$, multiplication on the right by D_r gives a P-linear combination of terms (25) of total degree $\leq N + 1$. This completes the proof of our assertion and shows that $\mathfrak{A}$ is a subring of the ring of endomorphisms of (P, +). It is clear from the definition of $\mathfrak{A}$ that $[\mathfrak{A}:\mathrm{P}]_R \leq p^m$ and equality holds only if the monomials (25) are right P-independent and thus form a basis. We can now apply the Jacobson-Bourbaki theorem (Th. 1.2) to $\mathfrak{A}$ and we obtain the following conclusions: If Φ is the subfield of elements α of P such that $\alpha_R A = A\alpha_R$ for all $A \in \mathfrak{A}$, then $[\mathrm{P}:\Phi] = [\mathfrak{A}:\mathrm{P}]_R$ and $\mathfrak{A} = \mathfrak{L}_\Phi(\mathrm{P})$. Now it is clear that $\alpha_R A = A\alpha_R$ holds for all $A \in \mathfrak{A}$ if and only if $\alpha_R D = D\alpha_R$ for all D in $\mathfrak{D}$, and since $\alpha_R D = D\alpha_R + (\alpha D)_R$ the condition for this is $\alpha D = 0$ for all $D \in \mathfrak{D}$. Hence we see that Φ is the subfield of $\mathfrak{D}$-constants. If ρ is any element of P, then ρ^p is a $\mathfrak{D}$-constant. Hence P is purely inseparable of exponent ≤ 1 over Φ so we have $[\mathrm{P}:\Phi] = p^{m'}$ where m' is the number of elements in a p-basis of P/Φ, and $m' \leq m$ since $[\mathrm{P}:\Phi] = [\mathfrak{A}:\mathrm{P}]_R \leq p^m$. Also we know that, if $\mathfrak{D}_\Phi(\mathrm{P})$ is the set of derivations of P/Φ, then $[\mathfrak{D}_\Phi(\mathrm{P}):\mathrm{P}]_R = m'$. If $\alpha \in \Phi$ and $D \in \mathfrak{D}$, then $(\alpha\rho)D = \alpha(\rho D) + (\alpha D)\rho = \alpha(\rho D)$ so $D \in \mathfrak{D}_\Phi(\mathrm{P})$. Thus $\mathfrak{D} \subseteq \mathfrak{D}_\Phi(\mathrm{P})$ and, since $[\mathfrak{D}:\mathrm{P}]_R = m$, we must have $\mathfrak{D} = \mathfrak{D}_\Phi(\mathrm{P})$ and $m = m'$. Then $\mathfrak{D}$ contains every derivation of P/Φ and $[\mathrm{P}:\Phi] = p^m$. This completes the proof.

We can now establish a Galois type correspondence between the following two collections determined by an arbitrary field P of characteristic p: Let $\mathscr{E}$ be the collection of subfields Φ of P such that P is purely inseparable of exponent ≤ 1 over Φ and $[P:\Phi] < \infty$ and let $\mathscr{D}$ denote the collection of derivation algebras $\mathfrak{D}$ in P which are restricted P-Lie algebras of finite dimensionality over P. If $\mathfrak{D} \in \mathscr{D}$, let $C(\mathfrak{D})$ be the subfield of $\mathfrak{D}$-constants, and, if $\Phi \in \mathscr{E}$, let $\mathfrak{D}_\Phi(P)$ be the set of derivations of P/Φ. Then $C(\mathfrak{D}_\Phi(P)) = \Phi$ and $\mathfrak{D}_{C(\mathfrak{D})}(C(\mathfrak{D})) = \mathfrak{D}$. In particular, we obtain a 1–1 correspondence between the collection of intermediate fields of P/Φ, P purely inseparable of exponent ≤ 1 over Φ, $[P:\Phi] < \infty$ and the restricted P-Lie subalgebras of the Lie algebra $\mathfrak{D}_\Phi(P)$ of derivations in P over Φ.

EXERCISES

1. Let $\Phi[x, y]$ be the polynomial ring in indeterminates x, y over a field of characteristic p, $\mathfrak{A}$ any algebra over Φ. Use the identities $(x - y)^p = x^p - y^p$ and $(x - y)^{p-1} = \sum_{i+j=p-1} x^i y^j$ in $\Phi[x, y]$ to prove the following identities in $\mathfrak{A}$:

$$[\cdots \overbrace{[ba]a] \cdots a}^{p}] = [ba^p] \tag{27}$$

$$[\cdots \overbrace{[ba]a] \cdots a}^{p-1}] = \sum_{i+j=p-1} a^i b a^j. \tag{28}$$

(Hint: Note that $[ba] = b(a_R - a_L)$ where a_R and a_L are the right and left multiplications determined by a in $\mathfrak{A}$. Specialize the indicated identities by taking $x = a_R$, $y = a_L$ in the commutative algebra of linear transformations generated by a_R and a_L.)

2. Let $\mathfrak{A}$ be as in ex. 1, $\mathfrak{A}[x]$ the polynomial algebra over $\mathfrak{A}$ in an indeterminate x. Let $a, b \in \mathfrak{A}$ and write

$$(a + bx)^p = a^p + \sum_1^{p-1} s_i(a,b)x^i + b^p x^p. \tag{29}$$

Use the fact that $\Sigma a_i x^i \to \Sigma i a_i x^{i-1}$ is a derivation in $\mathfrak{A}[x]$ and (29) to obtain

$$\sum_{i+j=p-1} (a + bx)^i b(a + bx)^j = \sum_{i=1}^{p-1} i s_i(a, b)x^{i-1}. \tag{30}$$

Use this relation and (28) to prove the following identity

$$(a + b)^p = a^p + b^p + \sum_1^{p-1} s_i(a, b) \tag{31}$$

where $i s_i(a, b)$ is the coefficient of x^{i-1} in

$$[\cdots \overbrace{[[b, a + bx]a + bx] \cdots a + bx}^{p-1}].$$

3. Let $P = \Phi(\rho_1, \cdots, \rho_m)$, Φ of characteristic $p \neq 0$, $\rho_i^p = \beta_i \in \Phi$, $[P:\Phi] = p^m$. Let D be a derivation in P/Φ such that Φ is the subfield of D-constants (see ex. 3 in § 7). Show that the minimum polynomial of D as linear transformation in P over Φ is a p-polynomial of the form

$$x^{p^m} + \beta_1 x^{p^{m-1}} + \beta_2 x^{p^{m-2}} + \cdots + \beta_m x, \quad \beta_i \in \Phi. \tag{32}$$

Show that there exists an element $\rho \in P$ such that $(\rho, \rho D, \cdots, \rho D^{p^m-1})$ is a basis for P over Φ. (This is an analogue of the normal basis theorem for separable normal extensions.) Show that every element in the algebra $\mathfrak{L}_\Phi(P)$ of linear transformations in P over Φ can be written in one and only one way in the form

$$1\sigma_0 + D\sigma_1 + D^2\sigma_2 + \cdots + D^{p^m-1}\sigma_{p^m-1}, \quad \sigma_i \in P. \tag{33}$$

4. Let P, Φ be as in ex. 3, $\mathfrak{D}_\Phi(P)$ the set of derivations of P/Φ. Let $\mathfrak{F}$ be a subspace of the right vector space $\mathfrak{D}_\Phi(P)$ over P which is closed under p-th powers. Prove that $\mathfrak{F}$ is also closed under commutation so $\mathfrak{F}$ satisfies all the conditions of Theorem 19.

5. Show that if D is a derivation in P and $\eta \in P$, then $\eta_R D^i = \sum_{j=0}^{i} \binom{i}{j} D^j (\eta D^{i-j})_R$.

6. Let D be a derivation in an algebra $\mathfrak{A}$ and let $\mathfrak{A}[t, D]$ be the set of formal polynomials $\sum_0^m t^i a_i$, $a_i \in \mathfrak{A}$. Equality, addition, and multiplication by elements of Φ are defined as for ordinary polynomials. Multiplication is defined by

$$\left(\sum_i t^i a_i\right)\left(\sum_j t^j b_j\right) = \sum_{i,j,k} \binom{j}{k} t^{i+k}(a_i D^{j-k}) b_j. \tag{34}$$

Verify the associative law and hence show that $\mathfrak{A}[t, D]$ is an algebra.

7. Let D be a derivation in a field P of characteristic $p \neq 0$, Φ the subfield of D-constants and assume $[P:\Phi] = p^m < \infty$. Then ex. 3 implies that there exists a p-polynomial (32) such that $D^{p^m} + \beta_1 D^{p^{m-1}} + \cdots + \beta_m D = 0$, $\beta_i \in \Phi$. Let $P[t, D]$ be the algebra of *differential polynomials* defined as in 6. Verify that if γ is any element of Φ, then $\pi(\gamma) \equiv t^{p^m} + t^{p^{m-1}}\beta_1 + \cdots + t\beta_m - \gamma$ is in the center of $P[t, D]$. Let $(\pi(\gamma))$ denote the ideal generated by $\pi(\gamma)$. Show that, if $\mathfrak{A}_\gamma = P[t, D]/(\pi(\gamma))$, then $[\mathfrak{A}_\gamma:\Phi] = p^{2m}$. Show that $\mathfrak{A}_0 \cong \mathfrak{L}_\Phi(P)$.

8. Same notations as 7. Let ρ be any element of P. Show that there exists an automorphism of $P[t, D]$ such that $t \rightarrow t + \rho$ and $\eta \rightarrow \eta$ for every $\eta \in P$. Note that $[\rho, t] = \rho t - t\rho = \rho D$ by (34) and deduce from this and (31) that $(t + \rho)^p = t^p + (\rho^p + \rho D^{p-1})$. More generally prove that

$$(t + \rho)^{p^j} = t^{p^j} + \rho^{[p^j]} \tag{35}$$

where

$$\rho^{[p^j]} = \rho^{p^j} + (\rho D^{p-1})^{p^{j-1}} + (\rho D^{p^2-1})^{p^{j-2}} + \cdots + \rho D^{p^j-1}. \tag{36}$$

9. Continuation of ex. 7 and 8. Show that the automorphism of $P[t, D]$ such that $t \rightarrow t + \rho$, $\eta \rightarrow \eta$, $\eta \in P$, sends the ideal generated by $\pi(\gamma)$ into itself if and only if ρ satisfies

$$\rho^{[p^m]} + \beta_1 \rho^{[p^{m-1}]} + \beta_2 \rho^{[p^{m-2}]} + \cdots + \beta_m \rho = 0. \tag{37}$$

10. Continuation of ex. 7 through 9. Prove that there exists an automorphism of $\mathfrak{L}_\Phi(P)$ sending every $\eta \in P$ into itself and sending $D \to D + 1\rho$, $\rho \in P$, if and only if ρ satisfies (37). Use this to prove the following analogue of Hilbert's Satz 90: An element ρ satisfies (37) if and only if it is a "logarithmic derivative" $(\sigma D)\sigma^{-1}$ of some σ in P.

11. Prove the following analogue of the result that the first cohomology group $H^1(G, P^*) = 0$ in the Galois case (cf. § 1.15). Let P be a purely inseparable extension of exponent one over Φ, $[P:\Phi] = p^m < \infty$ and let $\mathfrak{D}$ be the restricted P-Lie algebra of derivations of P over Φ. Let $D \to \mu(D)$ be a P-linear mapping of $\mathfrak{D}$ into P (thus an element of the conjugate space $\mathfrak{D}^*$ of $\mathfrak{D}$) such that

$$\mu(D^p) = \mu(D)^p + \mu(D)D^{p-1}. \tag{38}$$

Then there exists a σ in P such that $\mu(D) = (\sigma D)\sigma^{-1}$ for all D.

12. Show that, if $\mathfrak{A}_\gamma$ is as in ex. 7, then $\mathfrak{A}_\gamma \cong \mathfrak{A}_\delta$ if

$$\delta - \gamma = \rho^{[p^m]} + \beta_1\rho^{[p^{m-1}]} + \beta_2\rho^{[p^{m-2}]} + \cdots + \beta_m\rho \tag{39}$$

(as in (37)) for some $\rho \in P$. Hence use ex. 7 to show that $\mathfrak{A} \cong \Phi_{p^m}$ if there exists a $\rho \in P$ such that $\gamma = \rho^{[p^m]} + \beta_1\rho^{[p^m-1]} + \cdots + \beta_m\rho$. (The conditions given here are also necessary.)

13. Apply ex. 1 to prove the following result on polynomials with integer coefficients: Let $g(x)$ be any such polynomial and define $g_k(x) = g_{k-1}(x)g'(x)$, $g_1(x) = g(x)$ where $'$ is the standard derivative. Show that for any prime p, $g_{p-1}'(x) \equiv \eta(x^p) \pmod p$ where $\eta(x)$ is a polynomial with integer coefficients.

14. Let γ and δ be elements of Φ which are not p-th powers in Φ of characteristic $p \neq 0$. Use ex. 12 (both necessity and sufficiency) to prove that

$$(x_0^p + x_{p-1}) + x_1^p\gamma + x_2^p\gamma^2 + \cdots + x_{p-1}^p\gamma^{p-1} = \delta \tag{40}$$

has a solution for $x_i \in \Phi$ if and only if

$$(y_0^p + y_{p-1}) + y_1^p\delta + y_2^p\delta^2 + \cdots + y_{p-1}^p\delta^{p-1} = \gamma \tag{41}$$

has a solution for $y_i \in \Phi$.

15. Let D be a non-zero derivation in a field P of characteristic p. Show that the operators $1, D, \cdots, D^{p-1}$ are right linearly independent over P and that, if $\rho_i \in P$, then $\rho_0 + D\rho_1 + \cdots + D^{p-1}\rho_{p-1}$ is a derivation only if every $\rho_i = 0$, $i \neq 1$. Show that, if $\rho \in P$, then $(D\rho)^k = D^k\rho^k + D(\rho E)^{k-1} + \sum_{2}^{k-1} D^i\rho_i$ where $\rho_i \in P$ and $E = D\rho$ $(\equiv D\rho_R)$. Use these results to prove the following formula which is due to Hochschild:

$$E^p = (D\rho)^p = D^p\rho^p + D(\rho E^{p-1}).$$

16. Investigate the possibility of a Krull type Galois theory for purely inseparable extensions of exponent one of infinite dimensionality.

9. Higher derivations. The notion of a derivation can be generalized in the following way.

Definition 5. *Let $\mathfrak{A}$ be a subalgebra of an algebra $\mathfrak{B}$ over Φ. Then a sequence of mappings $D^{(m)} = \{D_0 = 1, D_1, \cdots, D_m\}$ of $\mathfrak{A}$ into*

$\mathfrak{B}$ *is called a* higher derivation of rank m *of* $\mathfrak{A}$ *into* $\mathfrak{B}$ *if every* D_i *is* Φ*-linear and*

$$(42) \qquad (ab)D_j = \sum_{i=0}^{j} (aD_i)(bD_{j-i}), \quad j = 0, 1, \cdots, m$$

holds for every $a, b \in \mathfrak{A}$. *A* higher derivation of infinite rank *is an infinite sequence* $\{D_0 = 1, D_1, \cdots\}$ *of linear mappings of* $\mathfrak{A}$ *into* $\mathfrak{B}$ *such that* (42) *holds for all* $j = 0, 1, 2, \cdots$.

Clearly, if $\{D_0, D_1, D_2, \cdots\}$ is a higher derivation of infinite rank, then the section $\{D_0, D_1, \cdots, D_m\}$ is a higher derivation of rank m and any section $\{D_0, D_1, \cdots, D_q\}$, $q \leq m$, of the higher derivation $\{D_0, \cdots, D_m\}$ is a higher derivation. The mapping D_1 is a derivation of $\mathfrak{A}$ into $\mathfrak{B}$.

Let $\mathfrak{A} = \mathfrak{B} = \Phi[x]$ where x is transcendental and let D_i be the linear mapping in $\mathfrak{A}$ whose effect on the basis $(1, x, x^2, \cdots)$ is given by

$$(43) \qquad x^m D_i = \binom{m}{i} x^{m-i}$$

where we agree that $\binom{m}{i} = 0$ if $i > m$. Then

$$x^{m+n} D_j = \binom{m+n}{j} x^{m+n-j}$$

and

$$(x^m D_i)(x^n D_{j-i}) = \binom{m}{i}\binom{n}{j-i} x^{m+n-j}.$$

Since $\sum_{i=0}^{j} \binom{m}{i}\binom{n}{j-i} = \binom{m+n}{j}$, we have $\sum_{i=0}^{j} (x^m D_i)(x^n D_{j-i}) = x^{m+n} D_j$. This shows that $(1, D_1, D_2, \cdots)$ is a higher derivation of infinite rank in $\Phi[x]$.

If Φ is of characteristic 0, then (43) shows that $i!D_i = D_1^i$ where D_1 is the usual standard derivation in $\Phi[x]$. Thus $D_i = \frac{1}{i!} D_1^i$. More generally, if D_1 is a derivation in any algebra of characteristic 0 and we define $D_i = \frac{1}{i!} D_1^i$, then $\{1, D_1, D_2, \cdots\}$ is a higher derivation of infinite rank in $\mathfrak{A}$. This follows immediately from Leibniz' formula: $(ab)D^j = \sum_{i=0}^{j} \binom{j}{i} (aD^i)(bD^{j-i})$ which gives $(ab)(D^j/j!) = \Sigma(aD^i/i!)(bD^{j-i}/(j-i)!)$. This is (42) for $D_i = \frac{1}{i!} D_1{}^i$.

The device we used for reducing the study of derivations to homomorphisms can be generalized so as to apply to higher derivations. Let $\mathfrak{T}^{(m)}$ be the algebra over Φ with basis $(1, t, \cdots, t^m)$ such that $t^{m+1} = 0$. Hence $\mathfrak{T}^{(m)} \cong \Phi[x]/(x^{m+1})$. Let $\mathfrak{B}^{(m)} = \mathfrak{B} \otimes_\Phi \mathfrak{T}^{(m)}$. If $D^{(m)} = \{1, D_1, \cdots, D_m\}$ is a higher derivation of rank m of $\mathfrak{A}$ into $\mathfrak{B}$, then we introduce the mapping $s(D^{(m)})$ of $\mathfrak{A}$ into $\mathfrak{B}^{(m)}$ as

$$(44) \qquad a \to a + (aD_1)t + (aD_2)t^2 + \cdots + (aD_m)t^m.$$

Evidently $s = s(D^{(m)})$ is linear. Also

$$\begin{aligned} a^s b^s &= \sum_0^m (aD_i)t^i \sum_0^m (bD_k)t^k \\ &= \sum_0^m \sum_{i=0}^{j} (aD_i)(bD_{j-i})t^j \\ &= \sum_0^m (ab)D_j t^j \\ &= (ab)^s. \end{aligned}$$

This shows that s is a homomorphism of $\mathfrak{A}$ into $\mathfrak{B}^{(m)}$. We have the homomorphism $\pi: a_0 + a_1t + a_2t^2 + \cdots + a_mt^m \to a_0$, $a_i \in \mathfrak{B}$ and $a^{s\pi} = a$ for every $a \in \mathfrak{A}$. As in the special case of derivations, this property is characteristic of the homomorphisms s obtained from higher derivations of rank m.

Similar considerations apply to higher derivations of infinite rank. The place of the algebra $\mathfrak{B}^{(m)}$ is now taken by the algebra $\mathfrak{B}[[t]]$ of power series

$$(45) \qquad a_0 + a_1t + a_2t^2 + \cdots$$

where the $a_i \in \mathfrak{B}$ (cf. Vol. I, p. 95). As before, if $\{1, D_1, \cdots\}$ is a higher derivation of infinite rank, then the mapping $s: a \to a + (aD_1)t + (aD_2)t^2 + \cdots$ is a homomorphism of $\mathfrak{A}$ into $\mathfrak{B}[[t]]$ such that $a^{s\pi} = a$ for all $a \varepsilon \mathfrak{A}$ where π is the homomorphism $\Sigma a_i t^i \to a_0$. Conversely, if $a \to a^s$ is a homomorphism of $\mathfrak{A}$ into $\mathfrak{B}[[t]]$ such that $a^{s\pi} = a$, $a \varepsilon \mathfrak{A}$, then we write $a^s = a + (aD_1)t + (aD_2)t^2 + \cdots$ and $\{D_0 = 1, D_1, D_2, \cdots\}$ is a higher derivation of $\mathfrak{A}$ into $\mathfrak{B}$.

If $\{D_i\}$ is a higher derivation of rank m (infinite rank) of $\mathfrak{A}$ into $\mathfrak{B}$, an element $a \varepsilon \mathfrak{A}$ is a *constant* relative to the higher derivation if $aD_i = 0$ for all $i > 0$. This simply means that $a^s = a$ for the homomorphism associated with the higher derivation. Hence it is clear that the set of constants is a subalgebra of the algebra $\mathfrak{A}$.

Our purpose in this section is to give just an introduction to higher derivations and to examine briefly higher derivations of purely inseparable fields. We suppose now that P/Φ is a field of characteristic $p \neq 0$. Let E be a subfield of P/Φ and let $D^{(m)} = \{1, D_1, \cdots, D_m\}$ be a higher derivation of rank m of E/Φ into P/Φ. In general, if $D_1 = D_2 = \cdots = D_{q-1} = 0$ but $D_q \neq 0$, then we shall say that the higher derivation is of *order* q and $D^{(m)}$ is called *proper* if $D_1 \neq 0$. If the order is q, the associated homomorphism $s = s(D^{(m)})$ of E into $\mathrm{P}^{(m)}$ has the form

$$\epsilon \to \epsilon + (\epsilon D_q)t^q + (\epsilon D_{q+1})t^{q+1} + \cdots + (\epsilon D_m)t^m, \tag{46}$$

where $\epsilon D_q \neq 0$ for some ϵ in E. We shall use this to prove the following

Theorem 20. *Let* P/Φ *be a field of characteristic* $p \neq 0$, E *a subfield of* P/Φ, $D^{(m)}$ *a higher derivation of rank* m *and order* q *of* E/Φ *into* P/Φ. *Let* Γ *be the subfield of* $D^{(m)}$*-constants of* E *and let* p^e *be the smallest power of* $p > \frac{m}{q}$. *Then* E *is purely inseparable of exponent* e *over* Γ.

Proof. We have to show that $\epsilon^{p^e} \varepsilon \Gamma$ for every $\epsilon \varepsilon \mathrm{E}$ and that there exists an $\epsilon \varepsilon \mathrm{E}$ such that $\epsilon^{p^{e-1}} \notin \Gamma$. The first is clear from (46) since

$$(\epsilon^{p^e})^s = (\epsilon^s)^{p^e} = (\epsilon + (\epsilon D_q)t^q + \cdots)^{p^e}$$
$$= \epsilon^{p^e} + (\epsilon D_q)^{p^e} t^{p^e q} + \cdots = \epsilon^{p^e}.$$

Hence $\epsilon^{p^e} \varepsilon \Gamma$. Now choose ϵ so that $\epsilon D_q \neq 0$. Then $(\epsilon^{p^{e-1}})^s = \epsilon^{p^{e-1}} + (\epsilon D_q)^{p^{e-1}} t^{p^{e-1}q} + \cdots$. Since $p^{e-1}q \leq m$ it is clear that $(\epsilon^{p^{e-1}})^s \neq \epsilon^{p^{e-1}}$. Hence $\epsilon^{p^{e-1}} \notin \Gamma$.

We consider next a purely inseparable simple extension field $P = \Phi(\xi)$ where $x^{p^e} - \alpha$ is the minimum polynomial of ξ over Φ. Let $\{D_i\}$ be the higher derivation in the polynomial algebra $\Phi[x]$ defined by (43) and let $D^{(p^e-1)} = \{1, D_1, \cdots, D_{p^e-1}\}$ be the higher derivation of rank $p^e - 1$ which is a section of this higher derivation. We have $(x^{p^e} - \alpha)D_j = 0$ for $1 \leq j \leq p^e - 1$ which together with the defining relations (42) imply that the principal ideal $\mathfrak{J} = (x^{p^e} - \alpha)$ is mapped into itself by every D_j. Hence every D_j induces a linear mapping, which we denote again by D_j, in $P = \Phi(\xi) \cong \Phi[x]/\mathfrak{J}$. The conditions in $\Phi[x]$ for D_j go over to the same conditions (42) for the D_j in $\Phi(\xi)$. Hence we obtain a higher derivation $D^{(p^e-1)}$ in $\Phi(\xi)$ such that

$$(47) \qquad \xi^m D_i = \binom{m}{i}\xi^{m-i}, \quad m = 0, 1, \cdots, p^e - 1.$$

We shall now show that the subfield Γ of $\{D_i\}$-constants for $D^{(p^e-1)}$ is Φ. Thus suppose $\Phi \subset \Gamma$. Then the minimum polynomial of ξ over Γ is $x^{p^f} - \beta$ with $f < e$ and β in Γ. Then $\xi^{p^f} \in \Gamma$. On the other hand, the definition (47) gives $\xi^{p^f}D_{p^f} = 1$. This proves our assertion.

We assume next that P is a purely inseparable extension of Φ which is a tensor product of simple extensions. $P_1, P_2, \cdots, P_r$, $P_i = \Phi(\xi_i)$. This means that $P = \Phi(\xi_1, \xi_2, \cdots, \xi_r)$ and the monomials $\xi_1^{k_1}\xi_2^{k_2} \cdots \xi_r^{k_r}$, $0 \leq k_i < p^{e_i}$, form a basis for P over Φ. If we set $\Phi_i = \Phi(\xi_1, \cdots, \xi_{i-1}, \xi_{i+1}, \cdots, \xi_r)$, then $P = \Phi_i(\xi_i)$ and $\Phi_1 \cap \Phi_2 \cap \cdots \cap \Phi_r = \Phi$. There exists a higher derivation in P whose constants are the elements of Φ_i. Hence it is clear that Φ is the subset of P of elements which are constants relative to all the higher derivations of finite rank in P over Φ.

EXERCISES

1. Let $\{D_i\}$ be the higher derivation in $\Phi[x]$, x transcendental, defined by (43). Show that

$$f(x + \alpha) = f(\alpha) + (fD_1)(\alpha)x + (fD_2)(\alpha)x^2 + \cdots.$$

2. Let $\Phi[x_1, x_2, \cdots, x_m]$ be the algebra of polynomials in indeterminates x_i over a field Φ. If $(k_1, k_2, \cdots, k_m)$ is a sequence of non-negative integers k_i we define a linear operator $D_{k_1k_2\cdots k_m}$ in $\Phi[x_1, x_2, \cdots, x_m]$ by its action in the basis

$(x_1^{n_1}x_2^{n_2}\cdots x_m^{n_m})$ as follows:

$$(x_1^{n_1}x_2^{n_2}\cdots x_m^{n_m})D_{k_1k_2\cdots k_m} = \begin{cases} 0 \text{ if any } k_i > n_i \\ \binom{n_1}{k_1}\binom{n_2}{k_2}\cdots\binom{n_m}{k_m} x_1^{n_1-k_1}x_2^{n_2-k_2}\cdots x_m^{n_m-k_m} \\ \qquad \text{if } k_i \leq n_i \end{cases}$$

Show that, if $f(x_1, x_2, \cdots, x_m) \in \Phi[x_1, x_2, \cdots, x_m]$ and $\alpha_1, \alpha_2, \cdots, \alpha_m \in \Phi$, then

$$f(x_1 + \alpha_1, x_2 + \alpha_2, \cdots, x_m + \alpha_m) = \sum_{k_i} (fD_{k_1k_2\cdots k_m})_{x_j=\alpha_j} x_1^{k_1}x_2^{k_2}\cdots x_m^{k_m}.$$

3. Let $f(x_1, \cdots, x_m)$ be a homogeneous polynomial of degree $n \leq m$ in $\Phi[x_1, x_2, \cdots, x_m]$. Suppose there exists an $\alpha = (\alpha_1, \alpha_2, \cdots, \alpha_m)$, $\alpha_i \in \Phi$, such that $(fD_{k_1k_2\cdots k_m})_{x_j=\alpha_j} = 0$ if $\Sigma k_i \leq n - 2$ and

$$\sum_{k_1+k_2+\cdots+k_m=n-1} (fD_{k_1k_2\cdots k_m})_{x_j=\alpha_j} x_1^{k_1}x_2^{k_2}\cdots x_m^{k_m} \neq 0.$$

Show that the equation $f(x_1, x_2, \cdots, x_m) = \beta$ has a solution in Φ for any $\beta \in \Phi$. Use this to prove that

$$x^3 + y^3 + z^3 - 3xyz = \beta$$

is solvable in any field of characteristic $\neq 3$.

4. A higher derivation in $\mathfrak{A}$ of infinite rank is called *iterative* if $D_iD_j = \binom{i+j}{i} D_{i+j}$ and a higher derivation $D^{(m)} = \{D_i\}$ is called *iterative* if $D_iD_j = \binom{i+j}{i} D_{i+j}$ for $i + j \leq m$ and $D_iD_j = 0$ if $i + j > m$. Verify that the higher derivations defined by (43) and (47) are iterative.

5. Let $P = \Phi(\xi)$ where Φ is of characteristic $p \neq 0$ and the minimum polynomial of ξ over Φ is $x^{p^e} - \alpha$. Show that the subfields of P/Φ are the fields $\Phi(\xi^{p^f})$ where $0 \leq f \leq e$, and that the indicated $e + 1$ subfields are distinct.

6. (Weisfeld). Let Φ_0 be a field of characteristic $p \neq 0$, $\Phi = \Phi_0(\alpha, \beta, \gamma)$ where $\alpha^p, \beta^p, \gamma^p \in \Phi_0$ and these elements are p-independent over $\Phi_0([\Phi:\Phi_0] = p^3)$. Let $P = \Phi(\xi, \eta)$ where $\xi^{p^2} = \alpha$, $\eta^p = \beta\xi^p + \gamma$. Show that $[P:\Phi] = p^3$. Show that $[\Phi(\xi):\Phi] = p^2$, $[\Phi(\eta):\Phi] = p^2$ and $\Phi(\xi) \cap \Phi(\eta) = \Phi$. Show that $P \neq \Phi(\xi, \zeta)$ and $P \neq \Phi(\eta, \zeta)$ where ζ is any element such that $\zeta^p \in \Phi$. Hence show that P/Φ is not a tensor product of simple extensions.

7. Show that $\{D_i\}$ is a higher derivation if and only if $a_R D_j = \sum_{i=0}^{j} D_i(aD_{j-i})_R$, $j = 0, 1, \cdots$. Show that, if $D_1 \neq 0$ in the higher derivation $\{1, D_1, D_2, \cdots, D_m\}$, then the endomorphisms $(1, D_1, \cdots, D_m)$ in $(P, +)$ are right P-independent.

8. Let $D^{(p^e-1)}$ be an iterative higher derivation of rank $p^e - 1$ in a field P of characteristic p. Assume $D^{(p^e-1)}$ is proper and that Φ is the subfield of constants. Show that every linear transformation in P over Φ has the form $\sum_0^{p^e-1} D_i\rho_i = \Sigma D_i\rho_{iR}$, $\rho_i \in P$, and that $P = \Phi(\xi)$ where the minimum polynomial of ξ over Φ is $x^{p^e} - \alpha$.

9. Continuation of 8. Show that a sequence of linear transformations $\{d_0, d_1, \cdots, d_{p^e-1}\}$ in P/Φ satisfies $\rho_R d_j = \sum_{i=0}^{j} d_i(\rho D_{j-i})$, $j = 0, 1, \cdots, p^e - 1$, if and

only if there exists a vector $(\sigma_0, \sigma_1, \cdots, \sigma_{p^e-1})$, $\sigma_0 = 1$, $\sigma_i \in P$ such that $d_i = D_i\sigma_0 + D_{i-1}\sigma_1 + \cdots + 1\sigma_i$. Use this to obtain necessary and sufficient conditions that a vector $(\sigma_0, \sigma_1, \cdots, \sigma_{p^e-1})$ be a "logarithmic derivative" in the sense that there exists a $\rho \in P$ such that $\sigma_i = \rho^{-1}(\rho D_i)$, $i = 0, 1, \cdots, p^e - 1$.

10. Tensor products of fields. In Chapter I we considered tensor products of two fields, one of which was finite dimensional over the base field. We saw that it was necessary to know the maximal ideals of $P \otimes_\Phi E$ in order to survey the composites of the field P/Φ and E/Φ where $[P:\Phi] < \infty$. In this section and the next we shall obtain the extension of these results to arbitrary fields. We shall first collect a number of results for the case in which one of the fields is algebraic. In our statements, separability will mean separability in the general sense defined on p. 166; pure inseparability will be used for an extension which is purely inseparable algebraic. Also we shall say that a subfield Φ is algebraically closed (separably algebraically closed) in P if every algebraic (separable algebraic) element of P/Φ is contained in Φ. We can now state the following

Theorem 21. *Let* P/Φ *and* E/Φ *be extension fields of* Φ.

(1) *If* P/Φ *is separable and* E/Φ *is purely inseparable, then* $P \otimes_\Phi E$ *is a field. On the other hand, if* P/Φ *is not separable, then there exists a purely inseparable extension* E/Φ *of exponent* 1 *such that* $P \otimes_\Phi E$ *contains a non-zero nilpotent element.*

(2) *If* P/Φ *is separable algebraic, then* $P \otimes_\Phi E$ *has no non-zero nilpotent elements for arbitrary* E/Φ, *and* $P \otimes_\Phi E$ *is a field if* Φ *is separably algebraically closed in* E.

(3) *The elements of* $P \otimes_\Phi E$ *are either units or nilpotents if either* P/Φ *is purely inseparable and* E/Φ *is arbitrary, or* P/Φ *is algebraic and* Φ *is separably algebraically closed in* E.

Proof. In (1) and the first part of (3) we may assume the characteristic is $p \neq 0$. In all cases we write $P \otimes E$ for $P \otimes_\Phi E$ and we identify P and E with subalgebras of $P \otimes E = PE$. These are linearly disjoint and consequently they satisfy the various linear independence properties which we have noted for this relation.

(1) Assume P/Φ is separable and E/Φ is purely inseparable. The separability implies that, if $\rho_1, \rho_2, \cdots, \rho_m$ are Φ-independent ele-

ments of P, then the elements $\rho_1^{p^e}, \rho_2^{p^e}, \cdots, \rho_m^{p^e}$ are Φ-independent for every $e = 0, 1, 2, \cdots$. Now let $z = \sum_1^m \rho_i\sigma_i \in P \otimes E$ where the $\rho_i \in P$ and $\sigma_i \in E$. We may assume the ρ_i are Φ-independent and, if $z \neq 0$, then we may assume also that every $\sigma_i \neq 0$. Since E/Φ is purely inseparable there exists a positive integer e such that $\sigma_i^{p^e} = \alpha_i \in \Phi$ for $1 \leq i \leq m$. Then $z^{p^e} = \Sigma\alpha_i\rho_i^{p^e} \in P$ and, if $z \neq 0$, then the $\alpha_i \neq 0$ and z^{p^e} is a non-zero element of P. Hence z^{p^e} and consequently z has an inverse. Thus $P \otimes E$ is a field. Next assume P/Φ is not separable. Then there exist elements $\rho_1, \rho_2, \cdots, \rho_m$ in P which are Φ-independent but for which there exist $\gamma_i \neq 0$ in Φ such that $\Sigma\gamma_i\rho_i^p = 0$. Not all the γ_i are p-th powers in Φ so an extension field of the form $E = \Phi(\sigma_1, \sigma_2, \cdots, \sigma_m)$, $\sigma_i^p = \gamma_i$, is of exponent 1 over Φ. The element $z = \Sigma\rho_i\sigma_i$ of $P \otimes E$ is not zero since the ρ_i are Φ-independent and the $\sigma_i \in E$. On the other hand, $z^p = \Sigma\rho_i^p\sigma_i^p = \Sigma\gamma_i\rho_i^p = 0$.

(2) Assume P/Φ is separable algebraic, E/Φ is arbitrary. We have to show that $P \otimes E$ has no non-zero nilpotents and that $P \otimes E$ is a field if Φ is separably algebraically closed in E. If $z \in P \otimes E$, $z = \sum_1^m \rho_i\sigma_i$ where the $\rho_i \in P$ and $\sigma_i \in E$. Since P/Φ is algebraic, the ρ_i generate a finite dimensional extension and we may clearly replace P by this extension in proving our result. Hence it suffices to assume that $[P:\Phi] < \infty$. Then the separability of P implies that $P = \Phi(\theta) \cong \Phi[x]/(f(x))$ where $f(x)$ is separable and irreducible in $\Phi[x]$. As we saw in Chapter I (p. 87) $P \otimes E \cong E[x]/(f(x))$. Hence our result will follow if we can show that $E[x]/(f(x))$ has no non-zero nilpotents and this is a field if Φ is separably algebraically closed in E. Now we have seen in Chapter I that $E[x]/(f(x))$ is a direct sum of fields, and it is easy to verify that an algebra having this structure contains no non-zero nilpotent elements. This proves the first statement. Next assume $E[x]/(f(x))$ is not a field. Then $f(x) = g(x)h(x)$ in $E[x]$ where $\deg g > 0$ and $\deg h > 0$. Let Ω be a splitting field over Φ of $f(x)$ and let $f(x) = \Pi(x - \omega_i)$ in $\Omega[x]$. Since the ω_i are roots of $f(x)$, they are separable algebraic over Φ. It follows that the coefficients of $g(x)$ and $h(x)$ are separable algebraic over Φ. These are elements of E and they are not all contained in Φ

since $f(x)$ is irreducible in $\Phi[x]$. Thus Φ is not separably algebraically closed in E.

(3) Assume first that P/Φ is purely inseparable and E/Φ is arbitrary. Let $z = \sum_1^m \rho_i\sigma_i \in \mathrm{P} \otimes \mathrm{E}$ where $\rho_i \in \mathrm{P}$, $\sigma_i \in \mathrm{E}$. Choose $e > 0$ so that $\rho_i^{p^e} = \alpha_i \in \Phi$. Then $z^{p^e} = \Sigma\alpha_i\sigma_i^{p^e} \in \mathrm{E}$. Either $z^{p^e} = 0$ or z^{p^e} has an inverse in E. In the latter case z is a unit in P $\otimes$ E. Next assume P/Φ is algebraic and Φ is separably algebraically closed in E. Let Σ/Φ be the maximum separable subfield of P/Φ. The subalgebra ΣE of PE = P $\otimes$ E over Φ is the tensor product of Σ/Φ and E/Φ. Since Φ is separably algebraically closed in E, we have, by (2), that ΣE = $\Sigma \otimes$ E is a field. Let $\{\rho_\alpha\}$ be a basis for P/Σ, $\{\sigma_\beta\}$ a basis for Σ/Φ. Then $\{\rho_\alpha\sigma_\beta\}$ is a basis for P/Φ and these elements are E-independent in P $\otimes$ E. It follows that the elements ρ_α are ΣE-independent. This implies that, if P and ΣE are regarded as algebras over Σ, then P(ΣE) = P $\otimes_\Sigma \Sigma$E. On the other hand, P(ΣE) is the same algebra over Φ as PE = P $\otimes_\Phi$ E; hence it suffices to show that every element of P $\otimes_\Sigma \Sigma$E is either nilpotent or a unit. Since P/Σ is purely inseparable, this follows from the first part of the present proof.*

Our next task is to obtain some information on tensor products of two fields, one of which is purely transcendental. The result we shall prove for these in the following

Theorem 22. *Let* P *be purely transcendental over* Φ, *say*, P = $\Phi(B)$ *where* B *is a transcendency basis and let* E/Φ *be arbitrary. Then* P $\otimes_\Phi$ E *has no zero-divisors, and if* Ω *is its field of fractions, then* Ω = E(B) *is purely transcendental over* E *with* B *as transcendency basis. Moreover, if* Φ *is algebraically closed (separably algebraically closed) in* E, *then* P = $\Phi(B)$ *is algebraically closed (separably algebraically closed) in* Ω = E(B).

Proof. As usual, we consider P and E as subalgebras of P $\otimes_\Phi$ E. Since B is an algebraically independent set, the set M of distinct monomials $\beta_1^{k_1}\beta_2^{k_2}\cdots\beta_r^{k_r}$, $k_i \geq 0$ in the $\beta \in B$ forms a basis for the subalgebra $\Phi[B]$ generated by B. Since $\Phi[B]$ and

* The identification of P$\otimes_\Phi$ E with P$\otimes_\Phi$ ($\Sigma\otimes_\Phi$ E) which was used in the proof can be established also by general formulas on tensor products. One has the associativity: P$\otimes_\Sigma$ ($\Sigma\otimes_\Phi$ E) $\cong$ (P$\otimes_\Sigma\Sigma$)$\otimes_\Phi$ E (cf. ex. 5, p. 15). Moreover, P$\otimes_\Sigma\Sigma \cong$ P. Hence P$\otimes_\Sigma$ ($\Sigma\otimes_\Phi$ E) $\cong$ P$\otimes_\Phi$ E.

E are linearly disjoint, the set M is E-independent. Hence B is algebraically independent in $E[B]$. Then we know that, if F is a finite subset of B, $E[F]$ has no zero divisors (Vol. I, p. 106). Hence $E[B]$ is an integral domain and so it has a quotient field Ω whose elements have the form PQ^{-1} where $P, Q \in E[B]$. Thus we see that $\Omega = E(B)$ and, since B is an algebraically independent set over E, clearly Ω is purely transcendental over E with B as a transcendency basis. We observe next that Ω contains the subalgebra Ω_1 of elements of the form Pq^{-1} where $P \in E[B]$ and $q \in \Phi[B]$. We proceed to show that this subalgebra can be identified with $P \otimes_\Phi E$. First, we have the identity isomorphism of $E[B] \subseteq \Omega$ into $E[B] \subseteq P \otimes_\Phi E$ and this can be extended, by I of the Introduction, to a unique isomorphism of $\Omega_1 = \{Pq^{-1} | P \in E[B], q \neq 0 \text{ in } \Phi[B]\}$ into $P \otimes_\Phi E$, since the element q^{-1} exists in $P = \Phi(B)$. Let z be any element of $P \otimes_\Phi E$ and write $z = \Sigma \rho_i \epsilon_i$, $\rho_i \in P = \Phi(B)$, $\epsilon_i \in E$. We can write $\rho_i = p_i q^{-1}$ where $p_i, q \in \Phi[B]$. Then $z = (\Sigma p_i \epsilon_i) q^{-1} = Pq^{-1}$ where $P \in E[B]$. It follows that z is in the image of the isomorphism of Ω_1, so Ω_1 is isomorphic to $P \otimes_\Phi E$. Hence if we identify $P \otimes_\Phi E$ with Ω_1 and observe that Ω is also the field of fractions of Ω_1 since $\Omega_1 \supseteq \Phi[B]$, we obtain the first statement. To prove the second we shall show that, if $\Omega = E(B)$ contains an element which is algebraic (separable algebraic) over $\Phi(B)$ which is not contained in $\Phi(B)$, then E contains an element which is algebraic (separable algebraic) over Φ not contained in Φ. Clearly, if an element of the type indicated exists in $\Omega = E(B)$, then it exists in $E(F)$ for a finite subset F of B. Hence we may take B finite and an induction argument shows that it is enough to prove the following result: Let E/Φ be arbitrary and let ξ be transcendental over E. If $E(\xi)$ contains an element which is algebraic (separable algebraic) over $\Phi(\xi)$ and not contained in $\Phi(\xi)$, then E contains an element which is algebraic (separable algebraic) over Φ and not contained in Φ. Thus let η be an element of $E(\xi)$ which is algebraic over $\Phi(\xi)$ and let $x^n + \beta_1 x^{n-1} + \cdots + \beta_n$ be its minimum polynomial over $\Phi(\xi)$. We can write $\beta_i = p_i q^{-i}$ where $p_i, q \in \Phi[\xi]$ (e.g., q can be taken to be the product of the denominators of the β_i). Then $H = q\eta$ is algebraic over $\Phi(\xi)$ with minimum polynomial $x^n + p_1 x^{n-1} + p_2 x^{n-2} + \cdots + p_n$. If $H = PQ^{-1}$ where P and $Q \in E[\xi]$ and are relatively prime polynomials, then the equation for H gives

$$P^n = -p_1P^{n-1}Q - p_2P^{n-2}Q^2 - \cdots - p_nQ^n.$$

If Q is of positive degree, then Q has an irreducible factor and the displayed relation shows that this is a factor of P^n, hence of P, contrary to the assumption on P and Q. It follows that Q is a unit and so $H \in \mathrm{E}[\xi]$. We now write $H = \epsilon_0 + \epsilon_1\xi + \epsilon_2\xi^2 + \cdots + \epsilon_m\xi^m$ where the $\epsilon_i \in \mathrm{E}$ and we shall show that the relation $0 = H(\xi)^n + p_1(\xi)H(\xi)^{n-1} + \cdots + p_n(\xi)$, $H = H(\xi)$, $p_i = p_i(\xi) \in \Phi[\xi]$, implies that the coefficients ϵ_j are algebraic over Φ. Thus let $\alpha \in \Phi$ and consider the homomorphism of $\mathrm{E}[\xi]$ over E into E sending $\xi \to \alpha$. Such a homomorphism exists since ξ is transcendental. As usual we denote the image of $Q(\xi)$ by $Q(\alpha)$. Then we have the relation $H(\alpha)^n + p_1(\alpha)H(\alpha)^{n-1} + \cdots + p_n(\alpha) = 0$. Since the $p_i(\alpha) \in \Phi$, this shows that the element $\beta = H(\alpha)$ is algebraic over Φ. Suppose first that Φ contains $m + 1$ distinct elements $\alpha_1, \alpha_2, \cdots, \alpha_{m+1}$. Then $H(\alpha_k) = \sum_{j=0}^{m} \epsilon_j\alpha_k{}^j = \beta_k$ is algebraic over Φ for $k = 1, 2, \cdots, m + 1$. Since the Vandermonde determinant $\det(\alpha_k{}^j) \neq 0$, these equations for the ϵ_j have a unique solution which is given by the usual determinant formulas. These show that the ϵ's are algebraic over Φ. If Φ does not have $m + 1$ elements, we have to modify this argument slightly in the following manner. If p is the characteristic, we choose r so that $p^r > m$ and we let $\bar{\mathrm{E}}$ be a splitting field over E of $x^{p^r} - 1$. We let $\bar{\Phi}$ be the subfield of $\bar{\mathrm{E}}$ of elements which are algebraic over Φ. Evidently this contains $m + 1$ distinct α_k. We now make the argument with these elements using $\bar{\mathrm{E}}$ in place of E, $\bar{\Phi}$ in place of Φ. Then we can conclude as before that the ϵ_j are algebraic over $\bar{\Phi}$, hence, over Φ. Now it is clear that, if the η we started with $\notin \Phi(\xi)$, then $H \notin \Phi(\xi)$ and consequently not every ϵ_i in $H = \Sigma\epsilon_i\xi^i$ is in Φ. Thus there exists an ϵ in E algebraic over Φ which is not contained in E. Next assume $\eta \notin \Phi(\xi)$ and η is separable over $\Phi(\xi)$. Then $H \notin \Phi(\xi)$ and is separable algebraic over $\Phi(\xi)$. Then the ϵ_i are algebraic and the field $\Phi(\epsilon_1, \epsilon_2, \cdots, \epsilon_m)$ contains a separable algebraic element not in Φ. Otherwise, the characteristic is p and we have $\epsilon_i{}^{p^e} \in \Phi$ for some $e = 1, 2, \cdots$. Then we have $H^{p^e} \in \Phi(\xi)$ contrary to the separability of H over $\Phi(\xi)$. This completes the proof.

We are now ready to handle the "mixed" cases in which the fields need not be either algebraic or purely transcendental. We prove first the following extension of a part of Theorem 21:

Theorem 23. *If* P/Φ *is separable and* E/Φ *is arbitrary, then* $\mathrm{P} \otimes_\Phi \mathrm{E}$ *has no non-zero nilpotent elements.*

Proof. It is clear that it suffices to prove this result under the additional assumption that P is finitely generated. Then P is separably generated, so that P has a transcendency basis B such that P is separable algebraic over $\Phi(B)$. We now consider the subalgebra $\Phi(B)\mathrm{E} = \Phi(B) \otimes_\Phi \mathrm{E}$ generated by $\Phi(B)$ and E and we regard this as well as P as an algebra over the field $\Phi(B)$. One sees easily that, if $\{\rho_\alpha\}$ is a basis for P over $\Phi(B)$, then the only relations of the form $\Sigma c_i\rho_i = 0$, $c_i \in \Phi(B)\mathrm{E}$ are the trivial ones for which every $c_i = 0$. This implies that $\mathrm{P} \otimes_\Phi \mathrm{E} = \mathrm{P} \otimes_{\Phi(B)} \Phi(B)\mathrm{E}$.* We now apply Theorem 22 to the factor $\Phi(B)\mathrm{E} = \Phi(B) \otimes_\Phi \mathrm{E}$. According to this result $\Phi(B)\mathrm{E}$ can be imbedded in a field $\Omega = \mathrm{E}(B)$. Then $\mathrm{P} \otimes_{\Phi(B)} \Phi(B)\mathrm{E}$ is a subalgebra of $\mathrm{P} \otimes_{\Phi(B)} \Omega$ where Ω is a field over $\Phi(B)$ and it suffices to prove that $\mathrm{P} \otimes_{\Phi(B)} \Omega$ has no non-zero nilpotent elements. Since P is separable algebraic over $\Phi(B)$, this follows from Theorem 21 (2).

We assume next that P is arbitrary and that Φ is separably algebraically closed in E. Let B be a transcendency basis for P over Φ. As in the foregoing proof we have $\mathrm{P} \otimes_\Phi \mathrm{E} = \mathrm{P} \otimes_{\Phi(B)} \Phi(B)\mathrm{E}$ and this is a subalgebra of $\mathrm{P} \otimes_{\Phi(B)} \Omega$ where Ω is a field $\mathrm{E}(B)$. By Theorem 22 we know that $\Phi(B)$ is separably algebraically closed in Ω. Since P is algebraic over $\Phi(B)$, Theorem 21(3) shows that every element of $\mathrm{P} \otimes_{\Phi(B)} \Omega$ is either nilpotent or a unit. Now let z be any element of $\mathrm{P} \otimes_\Phi \mathrm{E} \subseteq \mathrm{P} \otimes_{\Phi(B)} \Omega$. Either z is nilpotent or it is a unit in $\mathrm{P} \otimes_{\Phi(B)} \Omega$. In the latter case z is not a zero divisor in $\mathrm{P} \otimes_\Phi \mathrm{E}$. We can therefore state the following

Theorem 24. *If* P *is an arbitrary extension field of a field* Φ *and* Φ *is separably algebraically closed in* E *then every zero divisor of* $\mathrm{P} \otimes_\Phi \mathrm{E}$ *is nilpotent.*

* A more sophisticated argument can be used to establish this. See the footnote on p. 199.

Clearly the last two theorems have the following immediate consequence.

Corollary 1. *Let* P *and* E *be extension fields of* Φ *such that* (1) *either* P/Φ *or* E/Φ *is separable,* (2) Φ *is separably algebraically closed in either* P *or* E. *Then* P $\otimes_\Phi$ E *is an integral domain.*

In particular, we see that, if P/Φ is separable and Φ is algebraically closed in P, then P $\otimes_\Phi$ E is an integral domain for any E/Φ. An extension P/Φ satisfying these two conditions is called *regular.* If Φ is algebraically closed, then it is perfect so any extension P/Φ is separable. Moreover, it is clear that Φ is algebraically closed in P. Hence every extension of an algebraically closed field is regular and consequently we have

Corollary 2. *If* Φ *is algebraically closed, then* P $\otimes_\Phi$ E *is an integral domain for arbitrary extension fields* P *and* E *of* Φ.

11. Free composites of fields. We recall that a composite of two fields E and P over Φ is a triple (Γ, s, t) where Γ is a field over Φ and s and t are isomorphisms of E over Φ and P over Φ respectively into Γ such that Γ is generated by the images E^s and P^t (§ 1.16). The composites (Γ, s, t) and (Γ', s', t') of E and P are equivalent if there exists an isomorphism u of Γ onto Γ' such that $s' = us$, $t' = ut$. In § 1.16 we studied composites of a finite dimensional extension P and another extension. In algebraic geometry one is interested in composites of fields which need not be algebraic but one restricts the notion in the following way.

Definition 6. *A field composite* (Γ, s, t) *of* E/Φ *and* P/Φ *is called* free *if for any algebraically independent subsets* C *and* D *of* E *and* P *respectively, the sets* C^s, D^t *are non-overlapping and* $C^s \cup D^t$ *is algebraically independent in* Γ/Φ.

Since any algebraically independent set can be imbedded in a transcendency basis, it is clear that the condition that (Γ, s, t) is free is equivalent to the following: for every pair of transcendency bases B and B' of E/Φ and P/Φ, respectively, B^s and B'^t are non-overlapping and $B^s \cup B'^t$ is algebraically independent. We now observe that the word "every" can be replaced by "some" in this criterion. Thus suppose there exists a transcendency basis B for E/Φ and a transcendency basis B' for P/Φ such that B^s and B'^t are non-overlapping and $B^s \cup B'^t$ is algebraically independent. We assert that this implies that the composite Γ is free. It clearly suffices to establish the condition of the definition for finite sets, C, D. Now we can find a finite subset F of B such that C is algebraically dependent over Φ on F. Since F is a subset of B, F^s is algebraically independent in Γ over P^t. Hence F^s is algebraically independent in $\Phi(F^s, D^t)$ over $\Phi(D^t)$. This implies that the transcendency degree of $\Phi(F^s, C^s, D^t)$ over Φ is $f + d$ where f is the cardinal number $|F|$ and $d = |D|$ (cf. ex. 3, § 3). Since the transcendency degree of $\Phi(F^s, C^s)$ over Φ is f and C^s is algebraically independent, the transcendency degree of $\Phi(F^s, C^s)$ over $\Phi(C^s)$ is $f - c$ where $c = |C|$. It follows that the transcendency degree of $\Phi(F^s, C^s, D^t)$ over $\Phi(C^s, D^t)$ does not exceed $f - c$. This and the formula for the transcendency degree of $\Phi(F^s, C^s, D^t)$ over Φ imply that the transcendency degree of $\Phi(C^s, D^t)$ over Φ is at least $(f + d) - (f - c) = d + c$. It follows that C^s, D^t are not overlapping and $C^s \cup D^t$ is algebraically independent. We state this result as the following

Lemma 1. *Let* (Γ, s, t) *be a field composite of the fields* E *over* Φ *and* P *over* Φ. *Suppose there exists a transcendency basis* B *for* E *over* Φ *and a transcendency basis* B' *for* P *over* Φ *such that* B^s, B'^t *are non-overlapping and* $B^s \cup B'^t$ *is algebraically independent. Then* (Γ, s, t) *is a free composite of* E/Φ *and* P/Φ.

We remark also that if the condition of the lemma holds for B and B', then $B^s \cup B'^t$ is a transcendency basis for Γ. For, it is clear that the elements of E^s and of P^t are algebraic over $\Phi(B^s \cup B'^t)$. Since Γ is generated by E^s and P^t, it follows that Γ is algebraic over $\Phi(B^s \cup B'^t)$. Hence $B^s \cup B'^t$ is a transcendency basis.

We can use the criterion of the lemma to prove the existence of a free composite for any two fields E and P over Φ. Let B and B' be transcendency bases for E and P over Φ respectively. If B and B' are finite, say $B = \{\xi_1, \cdots, \xi_m\}$, $B' = \{\eta_1, \cdots, \eta_n\}$, then we construct the polynomial algebra $\Phi[x_1, x_2, \cdots, x_{m+n}]$ in $m + n$ indeterminates $x_1, x_2, \cdots, x_{m+n}$, and we form the field of fractions $\Phi(x_1, x_2, \cdots, x_{m+n})$. We have an isomorphism s of $\Phi(B)$ into $\Phi(x_1, x_2, \cdots, x_{m+n})$ such that $\xi_i \to x_i$, $i = 1, 2, \cdots, m$ and an isomorphism t of $\Phi(B')$ into $\Phi(x_1, x_2, \cdots, x_{m+n})$ such that $\eta_j \to x_{m+j}$, $j = 1, 2, \cdots, n$. Now let Ω be an algebraic closure of $\Phi(x_1, x_2, \cdots, x_{m+n})$. Then we know that the isomorphisms s and t can be extended to isomorphisms s and t of the algebraic extensions E and P of $\Phi(B)$ and $\Phi(B')$ into Ω (cf. ex. 1, p. 147). From the lemma, then, if Γ is the subfield of Ω generated by E^s and P^t, (Γ, s, t) is a free composite of E and P. If either B or B' is infinite a similar procedure can be employed, or we can modify it slightly by defining 1–1 mappings of B and of B' into the one of these, say B, which has the larger cardinal number, in such a way that the images are disjoint. These mappings can be extended to isomorphisms s and t of $\Phi[B]$ and $\Phi[B']$ into $\Phi(B)$. Then they can be extended to isomorphisms s and t of $\Phi(B)$ and $\Phi(B')$ into $\Phi(B)$, which can then be extended to isomorphisms s and t of E and P into an algebraic closure Ω of $\Phi(B)$. Then (Γ, s, t), where Γ is the subfield generated by E^s and P^t, is a free composite of P and E.

We shall now extend the considerations of § 1.16 to obtain a survey of all the composites and all free composites (in the sense of equivalence) of two given fields E and P over Φ. As before, we form the tensor product $E \otimes_\Phi P$ and we identify E and P with their images in $E \otimes_\Phi P$. Let $\mathfrak{P}$ be a prime ideal in $E \otimes_\Phi P$ (Vol. I, p. 173); hence $(E \otimes_\Phi P)/\mathfrak{P}$ is an integral domain as well as an algebra over Φ. We can imbed this in its field of fractions Γ. Let s denote the canonical homomorphism $\epsilon \to \epsilon + \mathfrak{P}$ of E ($\subseteq E \otimes P$) into $(E \otimes_\Phi P)/\mathfrak{P}$. Since E is a field and $1^s = 1$, this is an isomorphism. Also since $(E \otimes_\Phi P)/\mathfrak{P} \subseteq \Gamma$, we can consider s as an isomorphism of E/Φ into Γ/Φ. Similarly, we have the isomorphism $t{:}\rho \to \rho + \mathfrak{P}$ of P into Γ. Now E and P generate $E \otimes P$. Consequently E^s and P^t generate the algebra $(E \otimes P)/\mathfrak{P}$. Since Γ is the field of fractions of $(E \otimes P)/\mathfrak{P}$ we see that the field Γ is generated by its subfields E^s and P^t. Hence (Γ, s, t) is a composite of E/Φ and P/Φ.

Next let $\mathfrak{P}'$ be a second prime ideal in $E \otimes_\Phi P$ and let (Γ', s', t') be the corresponding composite constructed in the manner just given. Suppose (Γ', s', t') is equivalent to (Γ, s, t). Then we have an isomorphism u of Γ onto Γ' so that $s' = su$, $t' = tu$.

Thus u maps $\epsilon^s = \epsilon + \mathfrak{P} \to \epsilon^{s'} = \epsilon + \mathfrak{P}', \rho^t = \rho + \mathfrak{P} \to \rho + \mathfrak{P}'$. Consequently the restriction of u to the subalgebra $E^s P^t/\mathfrak{P}$ sends $\Sigma\epsilon_i\rho_i + \mathfrak{P} \to \Sigma\epsilon_i\rho_i + \mathfrak{P}'$ for $\epsilon_i \in E$, $\rho_i \in P$. It follows as in § 1.16 that $\Sigma\epsilon_i\rho_i \in \mathfrak{P}$ implies $\Sigma\epsilon_i\rho_i \in \mathfrak{P}'$. Hence $\mathfrak{P} \subseteq \mathfrak{P}'$, and if we repeat the argument with u^{-1} we see that $\mathfrak{P}' \subseteq \mathfrak{P}$. Thus we see that *distinct prime ideals in* $E \otimes_\Phi P$ *give rise to inequivalent composites of* E/Φ *and* P/Φ.

Now let (Γ', s', t') be any composite of E/Φ and P/Φ. Then we can combine the isomorphisms s', t' of E/Φ and P/Φ into Γ' to obtain the homomorphism $\Sigma\epsilon_i\rho_i \to \Sigma\epsilon_i^{s'}\rho_i^{t'}$ of $E \otimes_\Phi P$ into Γ'. The image under this homomorphism is the subalgebra $E^{s'}P^{t'}$ generated by $E^{s'}/\Phi$ and $P^{t'}/\Phi$. This is an integral domain. Hence if $\mathfrak{P}$ is the kernel of the homomorphism, then $(E \otimes P)/\mathfrak{P} \cong E^{s'}P^{t'}$ and $(E \otimes P)/\mathfrak{P}$ is an integral domain. Hence $\mathfrak{P}$ is a prime ideal in $E \otimes P$ so this can be used to construct the composite (Γ, s, t) as before. Now the homomorphism of $E \otimes P$ onto $E^{s'}P^{t'}$ gives rise to the isomorphism u of $(E \otimes P)/\mathfrak{P}$ onto $E^{s'}P^{t'}$ such that $\Sigma\epsilon_i\rho_i + \mathfrak{P} \to \Sigma\epsilon_i^{s'}\rho_i^{t'}$. This has a unique extension to an isomorphism u of the field of fractions Γ of $(E \otimes P)/\mathfrak{P}$ onto Γ'. We have $\epsilon^{su} = (\epsilon + \mathfrak{P})^u = \epsilon^{s'}$, $\epsilon \in E$ and $\rho^{tu} = (\rho + \mathfrak{P})^u = \rho^{t'}$, $\rho \in P$. Hence u is an equivalence of (Γ, s, t) and (Γ', s', t'). Our considerations therefore establish a 1–1 surjective mapping from the set of prime ideals $\mathfrak{P}$ in $E \otimes_\Phi P$ to the set of equivalence classes of composites in E/Φ and P/Φ.

In § 1.16 we established a 1–1 surjective correspondence between the set of maximal ideals in $E \otimes_\Phi P$ for $[P:\Phi] < \infty$ and the equivalence classes of composites of E/Φ and P/Φ. We can now see that this is a special case of the present more general considerations. We recall that an integral domain which is a finite dimensional algebra is a field (Introd., p. 8). This implies that any prime ideal in a finite dimensional algebra is maximal. If P/Φ is finite dimensional, then $E \otimes_\Phi P$ can be considered as a finite dimensional algebra over E. Hence the prime ideals in this algebra are maximal and the present correspondence reduces to the earlier one for $[P:\Phi] < \infty$.

It remains to sort out the prime ideals $\mathfrak{P}$ in $E \otimes P$ for which the corresponding composites (Γ, s, t) are free. Let B and B' be transcendency bases for E and P respectively. We know that the set M of monomials in the $\beta \varepsilon B$ are Φ-independent. A similar statement holds for the set M' of monomials in the $\beta' \varepsilon B'$. Moreover, if $M = \{m_i\}$ and $M' = \{n_j\}$, then the set of products $\{m_i n_j\}$ is Φ-independent. This implies that the sets B and B' are not overlapping and $B \cup B'$ is an algebraically independent set. The same statement can be made about the images $B^s = \{\beta + \mathfrak{P}\}$ and $B'^t = \{\beta' + \mathfrak{P}\}$ if and only if no non-zero element of the subalgebra $\Phi[B \cup B']$ is mapped into 0 in the canonical homomorphism of $\mathrm{E} \otimes \mathrm{P}$ into $(\mathrm{E} \otimes \mathrm{P})/\mathfrak{P}$. This is equivalent to the condition that $\Phi[B \cup B'] \cap \mathfrak{P} = 0$. Hence we obtain our first condition: The composite (Γ, s, t) determined by the prime ideal $\mathfrak{P}$ in $\mathrm{E} \otimes \mathrm{P}$ is free if and only if $\Phi[B \cup B'] \cap \mathfrak{P} = 0$. It is convenient to change this slightly by replacing $\Phi[B \cup B']$ by the subalgebra $\Phi(B)\Phi(B')$ generated by the subfields $\Phi(B)$ and $\Phi(B')$ of E and P respectively. It is easily seen that the elements of this subalgebra of $\mathrm{E} \otimes \mathrm{P}$ have the form $Pq^{-1}r^{-1}$ where $P \varepsilon \Phi[B \cup B']$, $q \varepsilon \Phi[B]$, $r \varepsilon \Phi[B']$. It is clear that $\Phi[B \cup B']$ is an integral domain and this and the form of the elements of $\Phi(B)\Phi(B')$ imply that $\Phi(B)\Phi(B')$ is an integral domain. If $Pq^{-1}r^{-1} \neq 0$ is in $\mathfrak{P} \cap \Phi(B)\Phi(B')$, then $P \neq 0$ and $P \varepsilon \Phi(B)\Phi(B') \cap \mathfrak{P}$. Hence $\mathfrak{P} \cap \Phi(B)\Phi(B') \neq 0$ implies $\mathfrak{P} \cap \Phi[B \cup B'] \neq 0$. Since the converse is clear the foregoing condition gives the following

Lemma 2. *The composite field* (Γ, s, t) *defined by a prime ideal* $\mathfrak{P}$ *in* $\mathrm{E} \otimes \mathrm{P}$ *is free if and only if* $\mathfrak{P} \cap \Phi(B)\Phi(B') = 0$ *where* B *and* B' *are transcendency bases for* E/Φ *and* P/Φ *respectively.*

We recall that if $\mathfrak{o}$ is a commutative ring and $\mathfrak{S}$ is a subring, then an element $a \varepsilon \mathfrak{o}$ is called integral over $\mathfrak{S}$ if there exists a polynomial $g(x) \varepsilon \mathfrak{S}[x]$ such that $g(x)$ has leading coefficient 1 and $g(a) = 0$ (Vol. I, p. 181). We have proved in Vol. I, p. 182, that, if $\mathfrak{S}$ is Noetherian, then the set of $\mathfrak{S}$-integral elements of $\mathfrak{o}$ form a subring containing $\mathfrak{S}$. We shall see later (§ 5.13) that this result is valid also for any commutative integral domain $\mathfrak{o}$. However, the Noetherian case is adequate to prove the following result which we require.

Lemma 3. *Let* B *and* B' *be transcendency bases for* E/Φ *and* P/Φ *respectively. Then every element of* $\mathrm{E} \otimes_\Phi \mathrm{P}$ *is integral over* $\Phi(B)\Phi(B')$.

Proof. Since E and P are algebraic over $\Phi(B)$ and $\Phi(B')$ respectively, it is clear that the elements of E and of P are integral over $\Phi(B)\Phi(B')$. Since E $\otimes$ P is generated by E and P, the result will follow if we can show that the set of $\Phi(B)\Phi(B')$-integral elements is a subring. Hence we have to show that, if α, β are $\Phi(B)\Phi(B')$-integral, then so are $\alpha - \beta$ and $\alpha\beta$. Since any pair α, β are both integral over a subalgebra $\Phi(F)\Phi(F')$ where F and F' are finite subsets of B and B', it suffices to prove this for B and B' finite. In this case we can apply Hilbert's basis theorem for polynomial rings (Vol. I, p. 172) to conclude that $\Phi(B)[B']$ is Noetherian. We shall show next that $\Phi(B)\Phi(B')$ is Noetherian. Thus let $\mathfrak{J}$ be an ideal in $\Phi(B)\Phi(B')$. Then $\mathfrak{J}' = \mathfrak{J} \cap \Phi(B)[B']$ is an ideal in $\Phi(B)[B']$, so it has a finite set of generators $P_1, P_2, \cdots, P_m$. Any element of $\Phi(B)\Phi(B')$ has the form Pq^{-1} where $P \in \Phi(B)[B']$ and $q \in \Phi[B']$. If this element is in $\mathfrak{J}$, then $P = (Pq^{-1})q \in \mathfrak{J}'$ so $P = \Sigma A_iP_i$ where $A_i \in \Phi(B)[B']$. Hence $Pq^{-1} = \Sigma(A_iq^{-1})P_i$. This shows that $P_1, P_2, \cdots, P_m$ is a set of generators for $\mathfrak{J}$. Hence $\Phi(B)\Phi(B')$ is Noetherian. It follows that $\alpha - \beta$ and $\alpha\beta$ are $\Phi(B)\Phi(B')$-integral and this completes the proof.

We can now prove the following

Theorem 25. *The composite* (Γ, s, t) *of* E *and* P *over* Φ *determined by the prime ideal* $\mathfrak{P}$ *in* $E \otimes_\Phi P$ *is free if and only if all the elements of* $\mathfrak{P}$ *are zero divisors in* $E \otimes_\Phi P$.

Proof. In view of Lemma 2 one has to show that

$$\mathfrak{P} \cap \Phi(B)\Phi(B') = 0$$

for B and B' transcendency bases for E/Φ and P/Φ if and only if every element of $\mathfrak{P}$ is a zero divisor. Suppose first that $\mathfrak{P}$ contains only zero divisors and let $P \in \mathfrak{P} \cap \Phi(B)\Phi(B')$. Then P is an element of $\Phi(B)\Phi(B')$ which is a zero divisor in E $\otimes$ P. We shall show that P is a zero divisor in $\Phi(B)\Phi(B')$. For this purpose we choose a basis $\{u_\alpha\}$ for E over $\Phi(B)$ and a basis $\{v_\beta\}$ for P over $\Phi(B')$. Then it is easily seen that every element of E $\otimes$ P can be written as a sum $\Sigma Q_{\alpha\beta}u_\alpha v_\beta$, $Q_{\alpha\beta} \in \Phi(B)\Phi(B')$ and $\Sigma Q_{\alpha\beta}u_\alpha v_\beta = 0$ only if every $Q_{\alpha\beta} = 0$. (We leave this as an exercise.) Since P is a zero divisor in E $\otimes$ P we have an element $\Sigma Q_{\alpha\beta}u_\alpha v_\beta \neq 0$ such that $P(\Sigma Q_{\alpha\beta}u_\alpha v_\beta) = 0$. Then $\Sigma PQ_{\alpha\beta}u_\alpha v_\beta = 0$ and since

$PQ_{\alpha\beta} \in \Phi(B)\Phi(B')$ we have $PQ_{\alpha\beta} = 0$ and some $Q_{\alpha\beta} \neq 0$. Thus P is a zero divisor in $\Phi(B)\Phi(B')$. Since $\Phi(B)\Phi(B')$ is an integral domain, this implies $P = 0$. Hence $\mathfrak{P} \cap \Phi(B)\Phi(B') = 0$. Conversely, suppose $\mathfrak{P} \cap \Phi(B)\Phi(B') = 0$. Let P be any element of $\mathfrak{P}$. Then Lemma 3 implies that there exists a relation of the form $P^n + c_1P^{n-1} + c_2P^{n-2} + \cdots + c_n = 0$ where the $c_i \in \Phi(B)\Phi(B')$. We may assume n minimal. This relation shows that $c_n = -P^n - c_1P^{n-1} - \cdots - c_{n-1}P \in \mathfrak{P} \cap \Phi(B)\Phi(B')$. Hence $c_n = 0$. Then we have $P(P^{n-1} + c_1P^{n-2} + \cdots + c_{n-1}) = 0$ and since n was minimal $P^{n-1} + c_1P^{n-2} + \cdots + c_{n-1} \neq 0$. Hence P is a zero divisor and we have shown that any $P \in \mathfrak{P}$ is a zero divisor. This completes the proof.

The set of nilpotent elements of a commutative ring $\mathfrak{o}$ forms an ideal called the (nil)radical $\mathfrak{R}$ of $\mathfrak{o}$ (Vol. I, p. 173). If $\mathfrak{P}$ is a prime ideal in $\mathfrak{o}$ and $z \in \mathfrak{R}$, then $z^m \in \mathfrak{P}$ for some integer m. This implies that $z \in \mathfrak{P}$. Hence $\mathfrak{R}$ is contained in every prime ideal $\mathfrak{P}$ of $\mathfrak{o}$.* We have shown in the last section that, if E is any field over Φ and Φ is separably algebraically closed in P, then the zero divisors of $\mathrm{E} \otimes_{\Phi} \mathrm{P}$ are nilpotent. This and the result just noted implies that the radical $\mathfrak{R}$ of $\mathrm{E} \otimes \mathrm{P}$ is the only prime ideal in $\mathrm{E} \otimes_{\Phi} \mathrm{P}$ all of whose elements are zero divisors. Hence we can conclude from Th. 25 and the fact that every composite of E and P over Φ is equivalent to one determined by a prime ideal in $\mathrm{E} \otimes \mathrm{P}$ the following

Theorem 26. *If* E *is an arbitrary extension field of* Φ *and* Φ *is separably algebraically closed in* P, *then in the sense of equivalence there is only one free composite of* E/Φ *and* P/Φ.

* We shall see in Chapter V that $\mathfrak{R}$ is the intersection of all the prime ideals of $\mathfrak{o}$.

Chapter V

VALUATION THEORY

The notion of a valuation of a field arises when one attempts to assign magnitudes to the elements of a field. The classical case is that of the absolute value $|\alpha|$ in the field of real numbers or in the field of rational numbers. Of basic importance for the study of arithmetic properties of the rational and more generally of number fields (finite algebraic extensions of the rationals) are the p-adic valuations of the field of rational numbers. For a given prime p the valuation $\varphi_p(\alpha)$ of the rational number α indicates the power of p which divides the rational number α. Valuations play a fundamental role also in the study of algebraic function fields. For these it is necessary to generalize the notion somewhat so that it becomes equivalent to the notion of a place, which was first introduced by Dedekind and Weber in giving a purely algebraic definition of Riemann surfaces for algebraic functions. Valuation theory forms a solid link between algebra and analysis. On the one hand, it permits a precise study of algebraic functions and, on the other hand, it leads to the introduction of analytic notions (convergence, integration) in the study of arithmetic questions.

We shall begin our discussion with real valued valuations. One can distinguish two types of these: archimedean and non-archimedean. The latter lead to the extension in which the values are taken from an ordered commutative group rather than the field of real numbers. We shall determine the valuations of the simplest types of fields and consider in some detail the problem of extension of valuations. Applications to the Hilbert Nullstellensatz and to the study of the integral closure of a commutative integral domain will be given.

1. Real valuations. We shall consider first valuations which are real valued and we shall call these real valuations. It is possible to give a development of the theory which gives at the same time a development of the real number system from the point of view of convergence. This adds a small complication, so we shall avoid it and assume familiarity on the part of the reader with the basic notions on real numbers which will be needed.

Definition 1. *A* real valuation φ *of a field* Φ *is a mapping* $\alpha \to \varphi(\alpha)$ *of* Φ *into the field of real numbers such that*

(i) $\varphi(\alpha) \geq 0$, $\varphi(\alpha) = 0$ *if and only if* $\alpha = 0$
(ii) $\varphi(\alpha\beta) = \varphi(\alpha)\varphi(\beta)$
(iii) $\varphi(\alpha + \beta) \leq \varphi(\alpha) + \varphi(\beta)$.

Examples.

(1) Φ the field of complex numbers, $\varphi(\alpha)$ the usual absolute value $\sqrt{a^2 + b^2}$ of the complex number $\alpha = a + b\sqrt{-1}$, a, b real. This gives a valuation on any subfield, in particular, on the field of real numbers and on the rational field.

(2) Φ the field of rational numbers. Let p be a prime integer. If $\alpha \neq 0$ in Φ, we write $\alpha = \alpha' p^k$ where $k = 0, \pm 1, \pm 2, \cdots$ and α' is a rational number prime to p (notation: $(\alpha', p) = 1$) in the sense that its numerator and denominator in some representation are prime to p. The integer k is uniquely determined by α and we write $\nu_p(\alpha) = k$, $\varphi_p(\alpha) = p^{-\nu_p(\alpha)}$. Also, we set $\nu_p(0) = \infty$, $\varphi_p(0) = 0$. Then (i) is evident and (ii) and (iii) are valid. This is obvious if either $\alpha = 0$ or $\beta = 0$. Suppose $\alpha \neq 0$, $\beta \neq 0$, and let $\alpha = \alpha' p^k$, $\beta = \beta' p^l$ where $(\alpha', p) = 1 = (\beta', p)$. Then $\alpha\beta = \alpha'\beta' p^{k+l}$ and $(\alpha'\beta', p) = 1$. Hence $\nu_p(\alpha\beta) = k + l = \nu_p(\alpha) + \nu_p(\beta)$, so $\varphi_p(\alpha\beta) = \varphi_p(\alpha)\varphi_p(\beta)$. If $k \leq l$, then $\alpha + \beta = p^k(\alpha' + \beta' p^{l-k})$ and $\nu_p(\alpha + \beta) \geq \min\ (\nu_p(\alpha), \nu_p(\beta))$. Hence $\varphi_p(\alpha + \beta) \leq \max\ (\varphi_p(\alpha), \varphi_p(\beta))$ which is a stronger relation than (iii). Hence $\varphi_p(\alpha)$ is a valuation. This is called the *p-adic valuation* of the rational field.

(3) $\mathrm{P} = \Phi(x)$ the extension field of Φ by a transcendental element x. Let $\pi(x)$ be an irreducible polynomial in $\Phi[x]$. If α is a non-zero rational expression, we write $\alpha = \pi(x)^k \alpha'$ where k is an integer and α' is a rational expression which is prime to π ($(\alpha', \pi) = 1$) in the sense that it has a representation with numerator and denominator prime to π. We set $\nu_\pi(\alpha) = k$ and $\varphi_\pi = c^k$ where c is a real number, $0 < c < 1$. Also we set $\nu_\pi\ (0) = \infty, \varphi_\pi(0) = 0$. One checks as in example 2 that φ_π is a valuation. A classical case of this type of valuation is that in which Φ is the field of complex numbers and $\Phi(x)$ is identified with the field of rational functions on Φ. Here $\pi(x)$ has the form $x - r$ and $\nu_\pi\ (\alpha(x))$ describes the behavior of the rational function $\alpha(x)$ in the neighborhood of the point $x = r$. One sees that, if $\nu_\pi(\alpha) = k > 0$, then α has a zero of order k at r and, if $\nu_\pi(\alpha) = -k$, $k > 0$, then $\alpha(x)$ has a pole of order k at $x = r$. If $\nu_\pi(\alpha) = 0$, then α has neither a zero nor pole at $x = r$. It is of interest also to consider the behavior of $\alpha(x)$ at infinity. This can be done by introducing another valuation in $\Phi(x)$. If $\alpha(x) \neq 0$, we write $\alpha(x) = (\beta_0 + \beta_1 x + \cdots + \beta_m x^m)(\gamma_0 + \gamma_1 x + \cdots + \gamma_n x^n)^{-1}$

where $\beta_m \neq 0$, $\gamma_n \neq 0$. Then $\alpha(x) = \left(\frac{1}{x}\right)^{n-m}(\beta_0\left(\frac{1}{x}\right)^m + \cdots + \beta_m)(\gamma_0\left(\frac{1}{x}\right)^n + \gamma_1\left(\frac{1}{x}\right)^{n-1} + \cdots + \gamma_n)^{-1}$ and $\alpha(x)$ has a zero of order $n - m$ at infinity if $n - m > 0$, a pole of order $m - n$ if $m - n > 0$, and has neither a zero nor pole at infinity if $n = m$. We define $\nu_\infty(\alpha(x)) = n - m$, $\varphi_\infty(\alpha(x)) = c^{n-m}$, $0 < c < 1$, $\nu_\infty(0) = \infty$, $\varphi_\infty(0) = 0$. This gives a valuation. This procedure is applicable to any $\Phi(x)$, x transcendental.

(4) Any field Φ with $\varphi(\alpha) = 1$ if $\alpha \neq 0$ and $\varphi(0) = 0$. Such a valuation is called *trivial*. We remark that the valuations φ_π and φ_∞ of example 3 are all trivial on Φ.

We now list some immediate consequences of the definition of a real valuation. We note first that (ii) implies that $\varphi(1) = 1$, $\varphi(-1) = 1$, and $\varphi(-\alpha) = \varphi(\alpha)$. Also $\varphi(\alpha^{-1}) = \varphi(\alpha)^{-1}$ if $\alpha \neq 0$, and $\varphi(\zeta) = 1$ if ζ is a root of unity. This implies that the only valuation in a finite field is the trivial one. Also we note that

$$|\varphi(\alpha) - \varphi(\beta)| \leq \varphi(\alpha - \beta) \tag{1}$$

where $|\quad|$ is the ordinary absolute value. All these assertions are readily established and we leave their verification to the reader.

Definition 2. *The real valuations φ_1 and φ_2 are called* equivalent *if $\varphi_1(\alpha) > \varphi_1(\beta)$ holds for $\alpha, \beta \in \Phi$ if and only if $\varphi_2(\alpha) > \varphi_2(\beta)$.*

It is natural from the point of view of convergence which we shall consider in § 4 to identify valuations that are related as in the foregoing definition. This relation leads to the following somewhat surprising consequence.

Theorem 1. *If φ_1 is equivalent to φ_2, then there exists a positive real number s such that $\varphi_2(\alpha) = \varphi_1(\alpha)^s$ for all $\alpha \in \Phi$.*

Proof. We may assume that one of the valuations is non-trivial and, since the conclusion is symmetric in φ_1 and $\varphi_2 (\varphi_1 = \varphi_2^{s^{-1}})$, we may suppose that φ_1 is non-trivial. Then there exists an α_0 in Φ such that $0 < \varphi_1(\alpha_0) < 1 = \varphi_1(1)$. Then also $0 < \varphi_2(\alpha_0) < 1$, so φ_2 is non-trivial. Moreover, we can write $\varphi_2(\alpha_0) = \varphi_1(\alpha_0)^s$ where $s > 0$. In fact, this relation is equivalent to $s = \log \varphi_2(\alpha_0)/\log \varphi_1(\alpha_0)$ which is positive since $\log \varphi_1(\alpha_0) < 0$ and $\log \varphi_2(\alpha_0) < 0$. We wish to show that

$$\frac{\log \varphi_2(\alpha)}{\log \varphi_2(\alpha_0)} = \frac{\log \varphi_1(\alpha)}{\log \varphi_1(\alpha_0)} \tag{2}$$

if α is any element in Φ such that $0 < \varphi_1(\alpha) < 1$ and so $0 < \varphi_2(\alpha) < 1$. The two ratios in (2) are positive. Let m and n be positive integers such that $m/n > \log \varphi_1(\alpha)/\log \varphi_1(\alpha_0)$. Then $m \log \varphi_1(\alpha_0) < n \log \varphi_1(\alpha)$, $\log \varphi_1(\alpha_0{}^m) < \log \varphi_1(\alpha^n)$ and $\varphi_1(\alpha_0{}^m) < \varphi_1(\alpha^n)$. Hence $\varphi_2(\alpha_0{}^m) < \varphi_2(\alpha^n)$ so, if we re-trace the steps, then we see that $m/n > \log \varphi_2(\alpha)/\log \varphi_2(\alpha_0)$. By symmetry ($\varphi_2$ is non-trivial), if $m/n > \log \varphi_2(\alpha)/\log \varphi_2(\alpha_0)$, then $m/n > \log \varphi_1(\alpha)/\log \varphi_1(\alpha_0)$. Since these relations hold for all positive rationals $r = m/n$, we have the equality (2). Hence

$$\frac{\log \varphi_2(\alpha)}{\log \varphi_1(\alpha)} = \frac{\log \varphi_2(\alpha_0)}{\log \varphi_1(\alpha_0)} = s$$

and $\varphi_2(\alpha) = \varphi_1(\alpha)^s$ holds for all α with $\varphi_1(\alpha) < 1$. By taking α^{-1} we see that this holds also if $\varphi_1(\alpha) > 1$. Moreover, it is clear that, if $\varphi_1(\alpha) = 1 = \varphi_1(1)$, then $\varphi_2(\alpha) = 1$. Hence $\varphi_2(\alpha) = \varphi_1(\alpha)^s$ for all α.

Definition 3. *A real valuation is called* archimedean *if* $\varphi(n) > 1$ *for some integer* $n(= n1 = 1 + 1 \cdots +1$, *n times) in the prime field. Otherwise the valuation is* non-archimedean.

If Φ has characteristic $p \neq 0$, then any $n \neq 0$ in the prime field is a root of unity; hence $\varphi(n) = 1$. Consequently, every valuation of a field of characteristic p is non-archimedean. We note also that any valuation which satisfies $\varphi(\alpha + \beta) \leq \max(\varphi(\alpha), \varphi(\beta))$ is non-archimedean. For, this can be extended by induction to give $\varphi(\alpha_1 + \alpha_2 + \cdots + \alpha_n) \leq \max(\varphi(\alpha_1), \cdots, \varphi(\alpha_n))$ and this implies that $\varphi(n) \leq \varphi(1) = 1$. The converse of this result is valid also since we have

Theorem 2. *If* φ *is a non-archimedean real valuation, then* $\varphi(\alpha + \beta) \leq \max(\varphi(\alpha), \varphi(\beta))$ *for every* α, β *in* Φ.

Proof. We have

$$\begin{aligned}\varphi(\alpha + \beta)^n &= \varphi\left(\alpha^n + \binom{n}{1}\alpha^{n-1}\beta + \cdots + \beta^n\right)\\ &\leq \varphi(\alpha)^n + \varphi(\alpha)^{n-1}\varphi(\beta) + \cdots + \varphi(\beta)^n\\ &\leq (n + 1)\max(\varphi(\alpha)^n, \varphi(\beta)^n).\end{aligned}$$

Hence we have $\varphi(\alpha + \beta) \leq (n + 1)^{1/n} \max (\varphi(\alpha), \varphi(\beta))$. Since $\lim_{n\to\infty} (n + 1)^{1/n} = 1$, this implies that

$$(3) \qquad \varphi(\alpha + \beta) \leq \max (\varphi(\alpha), \varphi(\beta)).$$

EXERCISES

In these exercises "valuation" will mean "real valuation."

1. Show that, if φ is a valuation and s is a real number such that $0 < s < 1$, then $\alpha \to \varphi(\alpha)^s$ is a valuation. Show also that, if φ is non-archimedean, then $\alpha \to \varphi(\alpha)^s$ is a valuation for any $s > 0$.

2. Establish the following properties of non-archimedean valuations:

$$(4) \qquad \varphi(\alpha + \beta) = \varphi(\alpha) \text{ if } \varphi(\alpha) > \varphi(\beta).$$

$$(5) \qquad \text{If } \alpha_1 + \alpha_2 + \cdots + \alpha_n = 0, \text{ then } \varphi(\alpha_i) = \varphi(\alpha_j) \text{ for some } i \neq j.$$

3. Let φ be a valuation in P such that φ is trivial on a subfield Φ of P such that P is algebraic over Φ. Show that φ is trivial on P.

4. Let φ be a non-trivial valuation of Φ and let β be a non-zero element of Φ such that $\varphi(\beta) < 1$. Show that $\varphi(\alpha) \leq 1$ if and only if $\varphi(\beta\alpha^n) < 1, n = 1, 2, \cdots$. Use this to prove that, if ψ is a valuation such that $\varphi(\gamma) < 1$ implies $\psi(\gamma) < 1$, then also $\varphi(\gamma) > 1$ implies $\psi(\gamma) > 1$ and $\varphi(\gamma) = 1$ implies $\psi(\gamma) = 1$. Hence show that φ and ψ are equivalent.

5. Show that, if $\varphi_1, \varphi_2, \cdots, \varphi_n$ are inequivalent non-trivial valuations of a field Φ, then there exists an α in Φ such that $\varphi_1(\alpha) > 1$ and $\varphi_i(\alpha) < 1$ for $i = 2, 3, \cdots, n$. (Hint: The case $n = 2$ is an easy consequence of ex. 4. Using this and induction one obtains β such that $\varphi_1(\beta) > 1$, $\varphi_j(\beta) < 1, j = 2, \cdots, n - 1$, and γ such that $\varphi_1(\gamma) > 1$, $\varphi_n(\gamma) < 1$. If $\varphi_n(\beta) \leq 1$, one can take $\alpha = \beta^k\gamma$ for a sufficiently large integer k. If $\varphi_n(\beta) > 1$, one can take $\alpha = \gamma\beta^k(1 + \beta^k)^{-1}$ for k sufficiently large.)

2. Real valuations of the field of rational numbers. We begin by determining the archimedean valuations of the rationals. The result is the following

Theorem 3. *Any archimedean real valuation of the rationals is equivalent to the absolute value valuation.*

Proof (Artin). Let n and n' be integers > 1 and write $n' = a_0 + a_1 n + \cdots + a_k n^k$, $0 \leq a_i < n$, $a_k \neq 0$. Then,

$$\varphi(n') \leq \varphi(a_0) + \varphi(a_1)\varphi(n) + \cdots + \varphi(a_k)\varphi(n)^k.$$

Since $0 \leq \varphi(a_i) \leq a_i < n$, this gives

$$\varphi(n') < n(1 + \varphi(n) + \cdots + \varphi(n)^k) < n(k + 1) \max (1, \varphi(n)^k).$$

We have $n' \geq n^k$ so $k \leq \log n'/\log n$ and

$$(6)\qquad \varphi(n') < n\left(\frac{\log n'}{\log n} + 1\right)\max\left(1, \varphi(n)^{\frac{\log n'}{\log n}}\right).$$

If we replace n' by $(n')^r$, r a positive integer, we obtain from (6):

$$\varphi(n')^r < n\left(\frac{r\log n'}{\log n} + 1\right)\max\left(1, \varphi(n)^{\frac{r\log n'}{\log n}}\right).$$

Taking r-th roots we obtain

$$(7)\qquad \varphi(n') < \left[n\left(\frac{r\log n'}{\log n} + 1\right)\right]^{1/r}\max\left(1, \varphi(n)^{\frac{\log n'}{\log n}}\right).$$

Since $\lim_{r\to\infty} (ra + b)^{1/r} = 1$ if $a \neq 0$, (7) implies

$$(8)\qquad \varphi(n') \leq \max\left(1, \varphi(n)^{\frac{\log n'}{\log n}}\right).$$

Since φ is archimedean, n' can be chosen so that $\varphi(n') > 1$; hence by (8),

$$(9)\qquad 1 < \varphi(n') \leq \varphi(n)^{\frac{\log n'}{\log n}}.$$

Hence $\varphi(n) > 1$, so we can interchange the roles of n and n' to obtain

$$(10)\qquad \varphi(n)^{\frac{1}{\log n}} = \varphi(n')^{\frac{1}{\log n'}}$$

for any two positive integers n, n'. Then $\log \varphi(n)/\log n$ is a positive real number s independent of n and $\varphi(n) = n^s$. It follows that $\varphi(\alpha) = |\alpha|^s$ for every rational number α. Evidently $\varphi(\alpha)$ is equivalent to the absolute value valuation.

Theorem 4. *Any non-trivial non-archimedean real valuation of the rationals is equivalent to a p-adic valuation for some prime p.*

Proof. We have $\varphi(n) \leq 1$ for every integer n. If $\varphi(n) = 1$ for every integer, then φ is trivial. Hence there exist non-zero integers b such that $\varphi(b) < 1$. Let $\mathfrak{P}$ be the collection of integers b satisfying this condition. This set is an ideal in the ring of integers I since $\varphi(b_1 - b_2) \leq \max(\varphi(b_1), \varphi(b_2)) < 1$ if $b_i \in \mathfrak{P}$, and $\varphi(nb) = \varphi(n)\varphi(b) < 1$ if $n \in I$, $b \in \mathfrak{P}$. Also $\mathfrak{P}$ is prime since $\varphi(n) = 1 = \varphi(n')$ implies $\varphi(nn') = 1$. Hence $\mathfrak{P} = (p)$ where p is a prime.

We can write $\varphi(p) = p^{-s}$ where $s > 0$ since $0 < \varphi(p) < 1$. Let n be any integer and write $n = n'p^k$ where $k \geq 0$, $(n', p) = 1$. Then $n' \notin \mathfrak{P}$ so $\varphi(n') = 1$; hence $\varphi(n) = p^{-ks}$. It follows that φ is the s-th power of the p-adic valuation determined by p.

3. Real valuations of $\Phi(x)$ which are trivial on Φ. We suppose x is transcendental in $\mathrm{P} = \Phi(x)$ and we shall determine the real valuations φ which are trivial on Φ. Since the prime field is contained in Φ, $\varphi(n) = 1$ for every integer $\neq 0$ in the prime field. Hence φ is non-archimedean. We distinguish two cases:

I. $\varphi(x) \leq 1$. In this case $\varphi(f(x)) \leq 1$ for every $f(x) = \alpha_0 + \alpha_1 x + \cdots + \alpha_n x^n \in \Phi[x]$. This is clear from the non-archimedean property of φ. From now on we assume φ is non-trivial and this implies that there exists a polynomial $f(x)$ such that $\varphi(f) < 1$. Let $\mathfrak{P}$ be the subset of $\Phi[x]$ of polynomials f such that $\varphi(f) < 1$. As in the proof of Theorem 4 one sees that $\mathfrak{P}$ is a prime ideal, $\mathfrak{P} = (\pi(x))$, in $\Phi[x]$. We have $\varphi(\pi(x)) = c$, $0 < c < 1$. If $f(x) = \pi(x)^k g(x)$ where $(\pi(x), g(x)) = 1$, then $\varphi(f) = c^k$. Hence φ is the valuation φ_π discussed in example 3 of § 1.

II. $\varphi(x) > 1$. Let $f(x) = \alpha_0 + \alpha_1 x + \cdots + \alpha_m x^m$ where $\alpha_m \neq 0$. Then $\varphi(\alpha_m x^m) = \varphi(x)^m > \varphi(\alpha_i x^i)$ for $i < m$. Hence $\varphi(f) = \varphi(x)^m$ (cf. ex. 2 in § 1). If we set $\varphi(x) = c^{-1}$, $0 < c < 1$, then $\varphi(f) = c^{-m}$. It is easy to check that φ is a valuation φ_∞ as defined in example 3 of § 1.

4. Completion of a field. One of the most important aspects of a real valuation is that it leads to the introduction of metric space notions for a field. The most convenient form for these is based on sequences and convergence. The basic definitions are patterned after those of ordinary analysis.

Definition 4. *Let Φ be a field with a real valuation φ. A sequence $\{\alpha_k\}$, $k = 1, 2, \cdots$ is said to* converge *in Φ (relative to φ) if there exists an α in Φ such that for any real $\epsilon > 0$ there exists an integer $N = N(\epsilon)$ such that*

$$\varphi(\alpha - \alpha_n) < \epsilon \tag{11}$$

for all $n \geq N$. Then α is unique and is called the limit *of $\{\alpha_k\}$. If $\alpha = 0$, $\{\alpha_k\}$ is a* null sequence. *A sequence $\{\alpha_k\}$ is called a* Cauchy sequence *if for any $\epsilon > 0$ there exists an integer $N = N(\epsilon)$ such that*

(12) $$\varphi(\alpha_m - \alpha_n) < \epsilon$$

for all $m, n \geq N(\epsilon)$.

Convergence of series $\sum_1^\infty a_k$ is defined as usual by the convergence of the sequence $\{s_k\}$ of partial sums $s_k = \sum_1^k a_j$. For example, in the rational field with the p-adic valuation, the series $\sum_1^\infty p^{k-1}$ converges to $1/(1 - p)$ since

$$\varphi_p\left(\frac{1}{1-p} - s_n\right) = \varphi_p(p^n)/\varphi_p(1 - p) = p^{-n} < \epsilon$$

if n is sufficiently large.

It is easy to see, as in the real case, that any convergent sequence is a Cauchy sequence, but the converse need not hold. This leads to the following

Definition 5. *A field Φ is said to be* complete *with respect to a real valuation φ if every Cauchy sequence of elements of Φ is convergent in Φ.*

We shall now carry out for any field Φ with a real valuation φ a construction of a *completion* $\bar{\Phi}$ of Φ. This is a field $\bar{\Phi}$ with the following properties:

1. $\bar{\Phi}$ is an extension field of Φ and has a real valuation $\bar{\varphi}$ which is an extension of the valuation φ of Φ.
2. $\bar{\Phi}$ is $\bar{\varphi}$-complete.
3. The subfield Φ is dense in $\bar{\Phi}$ in the sense that every element of $\bar{\Phi}$ is a limit of a convergent sequence of elements of Φ.

We consider first the set C of Cauchy sequences $\{\alpha_k\}$ of elements $\alpha_k \in \Phi$. We shall show that C is a ring relative to the compositions $\{\alpha_k\} + \{\beta_k\} = \{\alpha_k + \beta_k\}$, $\{\alpha_k\}\{\beta_k\} = \{\alpha_k\beta_k\}$. For this and a later application we require the following

Lemma 1. *If $\{\alpha_k\}, \{\beta_k\} \in C$, then $\{\alpha_k + \beta_k\}$ and $\{\alpha_k\beta_k\} \in C$. If $\{\alpha_k\} \in C$ and is not a null sequence, then there exists $\eta > 0$ and an integer N such that $\varphi(\alpha_n) > \eta$ for all $n \geq N$.*

Proof. Given $\epsilon > 0$, determine N_1 so that $\varphi(\alpha_m - \alpha_n) < \epsilon/2$ if $m, n \geq N_1$, and N_2 such that $\varphi(\beta_p - \beta_q) < \epsilon/2$ if $p, q \geq N_2$.

Let $N = \max(N_1, N_2)$. Then $\varphi(\alpha_m + \beta_m - \alpha_n - \beta_n) \leq \varphi(\alpha_m - \alpha_n) + \varphi(\beta_m - \beta_n) < \epsilon/2 + \epsilon/2 = \epsilon$ if $m, n \geq N$. Hence $\{\alpha_k + \beta_k\} \varepsilon C$. We note next that there exist positive real numbers s, t such that $\varphi(\alpha_k) < s$ for all k and $\varphi(\beta_k) < t$ for all k; for, we have $\varphi(\alpha_m - \alpha_N) < 1$ if $m \geq N$ and N is sufficiently large. Hence $\varphi(\alpha_m) - \varphi(\alpha_N) \leq \varphi(\alpha_m - \alpha_N) < 1$ so $\varphi(\alpha_m) < \varphi(\alpha_N) + 1$ for all $m \geq N$. Then if $s = \max(\varphi(\alpha_i) + 1)$, $i = 1, 2, \cdots, N$, then $\varphi(\alpha_k) < s$ for all k. Similarly we can find $t > 0$ such that $\varphi(\beta_k) < t$ for all k. Then

$$\begin{aligned}(13)\quad \varphi(\alpha_m\beta_m - \alpha_n\beta_n) & \\ &= \varphi(\alpha_m\beta_m - \alpha_m\beta_n + \alpha_m\beta_n - \alpha_n\beta_n) \\ &\leq \varphi(\alpha_m)\varphi(\beta_m - \beta_n) + \varphi(\beta_n)\varphi(\alpha_m - \alpha_n) \\ &< s\varphi(\beta_m - \beta_n) + t\varphi(\alpha_m - \alpha_n).\end{aligned}$$

If we take N_1 so that $\varphi(\beta_m - \beta_n) < \epsilon/2s$ for $m, n \geq N_1$, and N_2 so that $\varphi(\alpha_m - \alpha_n) < \epsilon/2t$ for $m, n \geq N_2$, then (13) shows that $\varphi(\alpha_m\beta_m - \alpha_n\beta_n) < \epsilon$ if $m, n \geq N = \max(N_1, N_2)$. Hence $\{\alpha_k\beta_k\} \varepsilon C$. Now suppose $\{\alpha_k\} \varepsilon C$ and this is not a null sequence. Then there exists $\epsilon > 0$ such that $\varphi(\alpha_k) > \epsilon$ for an infinite number of k. Also there is an N such that $\varphi(\alpha_m - \alpha_n) < \epsilon/2$ for all m, $n \geq N$. There is a $p \geq N$ such that $\varphi(\alpha_p) > \epsilon$. Then if $n \geq p$, $\varphi(\alpha_n) = \varphi(\alpha_p - (\alpha_p - \alpha_n)) \geq \varphi(\alpha_p) - \varphi(\alpha_p - \alpha_n) > \epsilon/2 = \eta$. This completes the proof.

To see that C is a ring under the indicated compositions we recall that the set of unrestricted sequences of elements of Φ is a ring under component addition and multiplication. This is just the complete direct sum of a countable number of copies of Φ. The 0 element of the ring is $\{0\}$ and the identity is $\{1\}$ where here we write $\{\alpha\}$ for the sequence $\{\alpha_k\}$ with $\alpha_k = \alpha$ for $k = 1, 2, \cdots$. We call this the *constant sequence* $\{\alpha\}$. It is clear that the set of constant sequences is a subring of the ring of sequences and this subring is isomorphic to Φ under the mapping $\alpha \rightarrow \{\alpha\}$. Lemma 1 implies that the set C of Cauchy sequences is a subring of the ring of sequences and evidently C contains the ring of constant sequences. Thus we see that C is a commutative ring with an element $1 = \{1\}$ and C contains the subring of constant sequences which is isomorphic to Φ.

We consider next the subset Z of C consisting of the null sequences. We have the following

Lemma 2. *Z is a maximal ideal in C.*

Proof. It is easy to see that the difference $\{\alpha_k\} - \{\beta_k\} = \{\alpha_k - \beta_k\}$ of two null sequences is a null sequence. Now let $\{\alpha_k\}$ be a null sequence and let $\{\gamma_k\}$ be a Cauchy sequence. The proof of Lemma 1 shows that there exists a positive real s such that $\varphi(\gamma_k) < s$ for all k. If $\epsilon > 0$ we choose N so that $\varphi(\alpha_n) < \epsilon/s$ for all $n \geq N$. Then $\varphi(\alpha_n\gamma_n) < \epsilon$ for all $n \geq N$, so $\{\alpha_k\gamma_k\}$ is a null sequence. Hence Z is an ideal in C. To show that Z is maximal we have to show two things: $Z \neq C$ and, if B is any ideal in C containing Z and an element $\{\alpha_k\} \notin Z$, then $B = C$. The first of these is clear since no constant sequence $\neq \{0\}$ is contained in Z. Next let B be an ideal in C containing Z and containing the element $\{\alpha_k\} \notin Z$. Lemma 1 shows that there exists a positive η and an integer p such that $\varphi(\alpha_n) > \eta$ for all $n \geq p$. Let $\beta_k = 1$ if $k < p$ and $\beta_k = \alpha_k$ if $k \geq p$. Then $\{\alpha_k\} - \{\beta_k\} \in Z$. Consider the sequence $\{\beta_k^{-1}\}$. We have $\varphi(\beta_m^{-1} - \beta_n^{-1}) = \dfrac{1}{\varphi(\beta_m\beta_n)}\varphi(\beta_m - \beta_n) < \dfrac{1}{\eta^2}\varphi(\alpha_m - \alpha_n)$ if $m, n \geq p$. This implies that $\{\beta_k^{-1}\} \in C$. Since $\{\alpha_k\} - \{\beta_k\} \in Z \subseteq B$ and $\{\alpha_k\} \in B$, $\{\beta_k\} \in B$. Hence $1 = \{\beta_k^{-1}\}\{\beta_k\} \in B$ and so $B = C$.

Lemma 2 implies that the difference ring $\bar{\Phi} = C/Z$ is a field. We proceed to show that $\bar{\Phi}$ is a field with a valuation which has the properties of a completion of Φ.

Theorem 5. *Let Φ be a field with a real valuation φ and let $\bar{\Phi} = C/Z$ the difference ring of the ring of Cauchy sequences with respect to the ideal Z of null sequences. Then $\bar{\Phi}$ is a field which contains a subfield isomorphic to Φ such that, if Φ is identified with this subfield, then $\bar{\Phi}$ is a completion of Φ.*

Proof. We have seen that the mapping $\alpha \to \{\alpha\}$ is an isomorphism of Φ with a subring of C. The canonical homomorphism $\{\alpha\} \to \{\alpha\} + Z$ is an isomorphism since the only constant sequence contained in Z is $\{0\}$. Hence we have the isomorphism $\alpha \to \{\alpha\} + Z$ of Φ into $\bar{\Phi} = C/Z$. From now on we shall identify α with $\{\alpha\} + Z$, Φ with its image in $\bar{\Phi}$. We show next that $\bar{\Phi}$ has a valuation $\bar{\varphi}$ which is an extension of the valuation φ in Φ. Now let $\{\alpha_k\} \varepsilon C$. Then $\{\varphi(\alpha_k)\}$ is a Cauchy sequence of real numbers since $|\varphi(\alpha_m) - \varphi(\alpha_n)| \leq \varphi(\alpha_m - \alpha_n)$ and $\{\alpha_k\}$ is a Cauchy sequence. Hence, by the completeness of the field of real numbers with respect to the absolute value valuation, $\lim \varphi(\alpha_k)$ exists. Next let $\{\alpha_k'\}$ be another Cauchy sequence such that $\{\alpha_k\} + Z = \{\alpha_k'\} + Z$. This means that $\varphi(\alpha_n - \alpha_n') < \epsilon$ for a given $\epsilon > 0$ provided $n \geq N(\epsilon)$. Then $|\varphi(\alpha_n) - \varphi(\alpha_n')| < \varphi(\alpha_n - \alpha_n') < \epsilon$ if $n \geq N(\epsilon)$. Hence $\lim \varphi(\alpha_k) = \lim \varphi(\alpha_k')$, so this real number is independent of the choice of the element $\{\alpha_k\}$ in the coset $A = \{\alpha_k\} + Z$. We now set $\bar{\varphi}(A) = \lim \varphi(\alpha_k)$ and we proceed to show that $\bar{\varphi}$ is a valuation in $\bar{\Phi}$. First, it is clear that $\bar{\varphi}(A) \geq 0$. If $A = \{\alpha_k\} + Z$ and $\bar{\varphi}(A) = 0$, then $\lim \varphi(\alpha_k) = 0$; hence $\{\alpha_k\}$ is a null sequence, so $\{\alpha_k\} \varepsilon Z$ and $A = 0$. If $B = \{\beta_k\} + Z$, then $AB = \{\alpha_k\beta_k\} + Z$ and $\bar{\varphi}(AB) = \lim \varphi(\alpha_k\beta_k) = \lim \varphi(\alpha_k)\varphi(\beta_k) = \bar{\varphi}(A)\bar{\varphi}(B)$. Also $A + B = \{\alpha_k + \beta_k\} + Z$ and $\bar{\varphi}(A + B) = \lim \varphi(\alpha_k + \beta_k) \leq \lim (\varphi(\alpha_k) + \varphi(\beta_k)) = \bar{\varphi}(A) + \bar{\varphi}(B)$. Hence $\bar{\varphi}$ is a valuation. If $A = \alpha \varepsilon \Phi$, so $A = \{\alpha\} + Z$, then $\bar{\varphi}(A) = \lim \varphi(\alpha) = \varphi(\alpha)$; hence $\bar{\varphi}$ is an extension of the valuation φ on Φ. We shall show next that Φ is dense in $\bar{\Phi}$. Let $A = \{\alpha_k\} + Z$ be an element of $\bar{\Phi}$. Let α_k' be the constant sequence all of whose terms are α_k. Then we have identified $A_k = \alpha_k' + Z$ with α_k. We assert that $\lim A_k = A$. For, if $\epsilon > 0$ is given, we can find N such that $\varphi(\alpha_m - \alpha_n) < \epsilon$ if $m, n \geq N$. Then $\lim_{n\to\infty} \varphi(\alpha_m - \alpha_n)$ exists and this is $\leq \epsilon$. On the other hand, $\bar{\varphi}(A - A_m) = \lim_{n\to\infty} \varphi(\alpha_m - \alpha_n)$, so $\bar{\varphi}(A - A_m) \leq \epsilon$ if $m \geq N$. Hence $\lim A_k = A$ and Φ is dense in $\bar{\Phi}$. It remains to show that $\bar{\Phi}$ is complete. Let $\{A_k\}$ be a Cauchy sequence of elements of $\bar{\Phi}$. For each k we can choose $\alpha_k \varepsilon \Phi \subseteq \bar{\Phi}$ such that $\bar{\varphi}(A_k - \alpha_k) < \frac{1}{2^k}\cdot$ Then $\varphi(\alpha_m - \alpha_n) = \bar{\varphi}(\alpha_m - A_m + A_m - A_n + A_n - \alpha_n) \leq \bar{\varphi}(\alpha_m - A_m) + \bar{\varphi}(A_m - A_n) + \bar{\varphi}(A_n - \alpha_n) \leq \bar{\varphi}(A_m - A_n) + \frac{1}{2^m} + \frac{1}{2^n}\cdot$ Since $\{A_k\}$ is a Cauchy sequence, this shows that $\{\alpha_k\}$ is a Cauchy sequence of elements of Φ. If we now go back to the original Φ and take the element $A = \{\alpha_k\} + Z$, α_k in the original Φ, then we see easily that $\lim A_k = A$.

From now on we shall write φ for the valuation $\bar{\varphi}$ in $\bar{\Phi}$.

We now take up the question of uniqueness of the field $\bar{\Phi}$. More generally, let $\bar{\Phi}_i$, $i = 1, 2$, be a complete field with a valuation φ_i and let Φ_i be a dense subfield. Suppose s is an isomorphism of Φ_1 onto Φ_2 which is *isometric* in the sense that $\varphi_2(\alpha^s) = \varphi_1(\alpha)$, $\alpha \varepsilon \Phi_1$. Let $A \varepsilon \bar{\Phi}_1$ and let $\{\alpha_k\}$ be a sequence of elements of Φ_1 such that $\lim \alpha_k = A$. Then $\{\alpha_k{}^s\}$ is a Cauchy sequence in Φ_2, so it has a limit B. If $\{\alpha_k'\}$ is a second sequence such that $\lim \alpha_k' = A$, then $\lim (\alpha_k - \alpha_k') = 0$, $\lim \varphi_1(\alpha_k - \alpha_k') = 0$; hence $\lim \varphi_2(\alpha_k{}^s - \alpha_k'^s) = 0$ and $\lim (\alpha_k{}^s - \alpha_k'^s) = 0$. This implies that $\lim \alpha_k'^s = B$. Hence the mapping $\bar{s}: A \to B$ of $\bar{\Phi}_1$ into $\bar{\Phi}_2$ is single-valued. It is easy to check that this is a homomorphism. Clearly $\bar{s} = s$ on Φ_1. Similarly, we can extend s^{-1} to a homomorphism $\overline{s^{-1}}$ of $\bar{\Phi}_2$ into $\bar{\Phi}_1$ which is defined in the same way as $\bar{s}$. Then one sees that $A^{\bar{s}\,\overline{s^{-1}}} = A$ for all $A \varepsilon \bar{\Phi}_1$ and $B^{\overline{s^{-1}}\,\bar{s}} = B$ for all $B \varepsilon \bar{\Phi}_2$. This implies that $\bar{s}$ is surjective and an isomorphism. We remark finally that, if $\bar{s}_1$ and $\bar{s}_2$ are isometric isomorphisms of $\bar{\Phi}_1$ onto $\bar{\Phi}_2$ which coincide on Φ_1, then $\bar{s}_1 = \bar{s}_2$. The proof is clear. We have therefore established the following

Theorem 6. *Let $\bar{\Phi}_i$, $i = 1, 2$, be a complete field with a valuation φ_i and Φ_i a dense subfield of $\bar{\Phi}_i$. Let s be an isometric isomorphism of Φ_1 onto Φ_2. Then s has a unique extension to an isometric isomorphism of $\bar{\Phi}_1$ onto $\bar{\Phi}_2$.*

This result implies, in particular, that, if $\bar{\Phi}_1$ and $\bar{\Phi}_2$ are completions of the same field Φ, then there exists an isometric isomorphism of $\bar{\Phi}_1/\Phi$ onto $\bar{\Phi}_2/\Phi$. We just have to apply the theorem to the identity mapping in Φ. In this sense the completion is unique and we have the right to use the term: *the* completion of the field Φ relative to the real valuation φ.

EXERCISES

1. Let Φ be a field with a real valuation φ. Show that the sum, product, and difference are continuous functions on Φ in the usual sense. Show also that the mapping $\alpha \to \alpha^{-1}$ is continuous on Φ^*, the set of non-zero elements of Φ.

2. Let $\bar{\Phi}$ be the completion of Φ. Show that the identity mapping is the only continuous automorphism of $\bar{\Phi}$ over Φ.

5. Some properties of the field of *p*-adic numbers. We look first at some properties of any field relative to a non-trivial non-archimedean real valuation φ. If Φ is such a field, then the subset $\mathfrak{o}$ of elements $\alpha \varepsilon \Phi$ such that $\varphi(\alpha) \leq 1$ is a subring of Φ; for, if α, $\beta \varepsilon \mathfrak{o}$, then $\varphi(\alpha\beta) = \varphi(\alpha)\varphi(\beta) \leq 1$ and $\varphi(\alpha - \beta) \leq \max(\varphi(\alpha), \varphi(\beta)) \leq 1$. The ring $\mathfrak{o}$ is called the *valuation ring* of φ. The subset $\mathfrak{p}$ of $\mathfrak{o}$ of elements β such that $\varphi(\beta) < 1$ is an ideal in $\mathfrak{o}$, since $\varphi(\beta_1) < 1$, $\varphi(\beta_2) < 1$, $\varphi(\alpha) \leq 1$ imply $\varphi(\beta_1 - \beta_2) < 1$ and $\varphi(\alpha\beta_1) < 1$. The elements α in $\mathfrak{o}$ which are not in $\mathfrak{p}$ satisfy $\varphi(\alpha) = 1$; hence $\varphi(\alpha^{-1}) = 1$ and $\alpha^{-1} \varepsilon \mathfrak{o}$. Conversely, if α is a unit in $\mathfrak{o}$, then $\varphi(\alpha) \leq 1$, $\varphi(\alpha^{-1}) \leq 1$ and $\varphi(\alpha)\varphi(\alpha^{-1}) = 1$ imply that $\varphi(\alpha) = 1$, so $\alpha \not\varepsilon \mathfrak{p}$. Thus we see that $\mathfrak{p}$ is the set of non-units of $\mathfrak{o}$. This implies that any ideal of $\mathfrak{o}$ properly containing $\mathfrak{p}$ contains a unit and so coincides with $\mathfrak{o}$. Hence $\mathfrak{p}$ is a maximal ideal in $\mathfrak{o}$. If $\mathfrak{q}$ is any ideal properly contained in $\mathfrak{o}$, then $\mathfrak{q}$ contains no units of $\mathfrak{o}$; consequently $\mathfrak{q} \subseteq \mathfrak{p}$. Hence $\mathfrak{p}$ is the only maximal ideal of $\mathfrak{o}$. The difference ring $\mathfrak{o}/\mathfrak{p}$ is a field which is called the *residue field* of Φ relative to φ.

The set $\Gamma = \{\varphi(\alpha), \alpha \neq 0 \text{ in } \Phi\}$ is clearly a subgroup of the multiplicative group of positive real numbers. Γ is called the *value group* of φ. The valuation is called *discrete* if Γ is a cyclic group. It is easy to see that a subgroup $\Gamma \neq 1$ of the positive reals is cyclic if and only if the subset Γ' of elements < 1 has a maximal element. This element is a generator of Γ. Let φ be discrete and let π be an element of Φ such that $\varphi(\pi)$ is the largest element < 1 in Γ. Then $\pi \varepsilon \mathfrak{p}$ the maximal ideal of the valuation ring $\mathfrak{o}$ and, if β is any element of $\mathfrak{p}$, then $\varphi(\beta) \leq \varphi(\pi)$, $\varphi(\beta\pi^{-1}) \leq 1$, so $\beta\pi^{-1} = \alpha \varepsilon \mathfrak{o}$ and $\beta = \alpha\pi$. Then $\mathfrak{p}$ is the principal ideal (π). Conversely, if $\mathfrak{p}$ is a principal ideal: $\mathfrak{p} = (\pi)$, then any $\beta \varepsilon \mathfrak{p}$ has the form $\alpha\pi$, $\alpha \varepsilon \mathfrak{o}$, so $\varphi(\beta) = \varphi(\alpha)\varphi(\pi) \leq \varphi(\pi)$. Hence $\varphi(\pi)$ is the largest element < 1 in Γ and φ is discrete. Since $\varphi(\pi)$ is a generator of Γ, we have for any non-zero α in Φ, $\varphi(\alpha) = \varphi(\pi)^k$ for some integer k. Then if $\epsilon = \alpha\pi^{-k}$, $\varphi(\epsilon) = \varphi(\alpha)\varphi(\pi)^{-k} = 1$, so ϵ is a unit in $\mathfrak{o}$. Consequently, any non-zero element of Φ has the form $\epsilon\pi^k$, $k = 0, \pm 1, \pm 2, \cdots$ where ϵ is a unit in $\mathfrak{o}$.

Let Φ be any field with a non-archimedean real valuation φ and let $\bar{\Phi}$ be the completion of Φ relative to φ. We shall now show that the value group of Φ and $\bar{\Phi}$ are the same and in a certain sense the same statement can be made for the residue fields. Let $\bar{\alpha} \varepsilon \bar{\Phi}$. Then the density of Φ in $\bar{\Phi}$ implies that there exists an α in Φ such that $\varphi(\bar{\alpha} - \alpha) < \varphi(\bar{\alpha})$. Since the valuation is non-archimedean we have $\varphi(\alpha) = \varphi(\bar{\alpha} + (\alpha - \bar{\alpha})) = \max(\varphi(\bar{\alpha}), \varphi(\alpha - \bar{\alpha})) = \varphi(\bar{\alpha})$. Thus we have an $\alpha \varepsilon \Phi$ such that $\varphi(\alpha) = \varphi(\bar{\alpha})$ and clearly this means that Φ and $\bar{\Phi}$ have the same value group Γ. Next let $\bar{\mathfrak{o}}$ be the valuation ring of $\bar{\Phi}$, $\bar{\mathfrak{p}}$ its maximal ideal of non-units. If $\mathfrak{o}$ and $\mathfrak{p}$ are the corresponding subsets of Φ, then $\mathfrak{o} = \bar{\mathfrak{o}} \cap \Phi$, $\mathfrak{p} = \bar{\mathfrak{p}} \cap \Phi$. If $\bar{\alpha} \varepsilon \bar{\mathfrak{o}}$ we choose $\alpha \varepsilon \Phi$ so that $\varphi(\bar{\alpha} - \alpha) < 1$. Then $\bar{\alpha} - \alpha \varepsilon \bar{\mathfrak{p}}$ and so $\alpha \varepsilon \mathfrak{o}$. Hence $\bar{\alpha} \equiv \alpha \pmod{\bar{\mathfrak{p}}}$ which shows that $\mathfrak{o} + \bar{\mathfrak{p}} = \bar{\mathfrak{o}}$. We have the standard isomorphism:

$$\bar{\mathfrak{o}}/\bar{\mathfrak{p}} = (\mathfrak{o} + \bar{\mathfrak{p}})/\bar{\mathfrak{p}} \cong \mathfrak{o}/(\mathfrak{o} \cap \bar{\mathfrak{p}}) = \mathfrak{o}/\mathfrak{p}.$$

By means of this isomorphism we can identify the residue field of $\bar{\Phi}$ with that of Φ.

The theory of convergence of series in a complete field with a non-archimedean valuation is strikingly simple. The completeness implies that $\sum_1^\infty \alpha_k$ converges if and only if for any $\epsilon > 0$ there exists an integer N such that $\varphi(\alpha_{m+1} + \cdots + \alpha_{m+k}) < \epsilon$ if $m \geq N$ and $k = 1, 2, \cdots$. Since the valuation is non-archimedean, $\varphi(\alpha_{m+1} + \cdots + \alpha_{m+k}) \leq \max \varphi(\alpha_{m+i})$. Hence the condition is equivalent to $\varphi(\alpha_{m+i}) < \epsilon$ for $m \geq N$, $i = 1, 2, \cdots$. This is equivalent to $\lim \alpha_n = 0$. This shows that a series converges if and only if its n-th term converges to 0. Since $\sum_1^\infty \alpha_k = \sum_1^m \alpha_i + \sum_{m+1}^\infty \alpha_j$ we have $\varphi\left(\sum_1^\infty \alpha_k\right) \leq \max\left(\varphi\left(\sum_1^m \alpha_i\right), \varphi\left(\sum_{m+1}^\infty \alpha_j\right)\right)$ and, since $\varphi\left(\sum_{m+1}^\infty \alpha_j\right)$ can be made arbitrarily small by taking m large enough, we have $\varphi\left(\sum_1^\infty \alpha_k\right) = \varphi\left(\sum_1^m \alpha_i\right)$ if m is large enough. If, in addition, we have $\varphi(\alpha_1) > \varphi(\alpha_2) > \varphi(\alpha_3) > \cdots$, then $\varphi\left(\sum_1^m \alpha_i\right) = \varphi(\alpha_1)$. Hence $\varphi\left(\sum_1^\infty \alpha_k\right) = \varphi(\alpha_1)$ in this case.

We now consider the special case of the field $R^{(p)}$ which is the completion of the rational field R_0 with respect to the p-adic valuation $\varphi_p(\alpha) = p^{-k}$ for $\alpha = p^k\alpha'$, $(\alpha', p) = 1$. The field $\bar{R}^{(p)}$ is called the *field of p-adic numbers*. Evidently the value group of R_0 relative to φ_p is the cyclic group generated by p^{-1}; hence the same result holds for $\bar{R}^{(p)}$ and the valuations of R_0 and $\bar{R}^{(p)}$ are discrete. Let $\bar{\mathfrak{o}}$ be the valuation ring of $\bar{R}^{(p)}$. The elements of $\bar{\mathfrak{o}}$ are the p-adic numbers $\bar{\alpha}$ such that $\varphi_p(\bar{\alpha}) \leq 1$ and these are called *p-adic integers*. It is clear from the definition of φ_p that the rational numbers α which are p-adic integers are those which can be written in the form m/n where $(n, p) = 1$. If $\mathfrak{o}$ is the valuation ring of R_0, $\mathfrak{p}$ its maximal ideal, then $\mathfrak{o} = \bar{\mathfrak{o}} \cap R_0$, $\mathfrak{p} = \bar{\mathfrak{p}} \cap R_0$. We have seen that $\bar{\mathfrak{o}} = \mathfrak{o} + \bar{\mathfrak{p}}$. Hence, if $\bar{\alpha}$ is any p-adic integer, then there exists a rational number m/n with $(n, p) = 1$ such that $\bar{\alpha} - m/n \in \bar{\mathfrak{p}}$. There exist integers a, b such that $na + pb = 1$. Then $m/n = ma + p(bm/n)$ so $m/n \equiv ma \pmod{\bar{\mathfrak{p}}}$. Hence $\bar{\alpha} \equiv ma \pmod{\bar{\mathfrak{p}}}$, which shows that we have $\bar{\mathfrak{o}} = I + \bar{\mathfrak{p}}$, where I is the ring of integers. It is clear that $\bar{\mathfrak{p}} \cap I = (p)$. Hence the residue field $\bar{\mathfrak{o}}/\bar{\mathfrak{p}} \cong I/(p)$ is just the field of p elements.

Let $\bar{\alpha}$ be a p-adic integer. Then our argument shows that there exists an element a of I (that is, an ordinary integer) such that $\bar{\alpha} - a \in \bar{\mathfrak{p}}$. If $a \equiv b \pmod{p}$, then $a - b \in \bar{\mathfrak{p}}$ and so $\bar{\alpha} - b \in \bar{\mathfrak{p}}$. This shows that for every p-adic integer $\bar{\alpha}$ we can choose a_0 in $\{0, 1, 2, \cdots, p - 1\}$ such that $\bar{\alpha} - a_0 \in \bar{\mathfrak{p}}$. We assert that $\bar{\alpha}_1 = \frac{1}{p}(\bar{\alpha} - a_0)$ is a p-adic integer. We note first that p is an element of $\mathfrak{p}$ such that $\varphi(p)$ is maximal. Since the value groups of R_0 and $\bar{R}^{(p)}$ are identical, it follows that p is an element of $\bar{\mathfrak{p}}$ with maximal $\varphi(p)$. Consequently, the ideal $\bar{\mathfrak{p}}$ is principal with p as generator, so if $\bar{\beta}$ satisfies $\varphi(\bar{\beta}) < 1$, then $\bar{\beta} = \bar{\gamma}p$ where $\bar{\gamma}$ is a p-adic integer. In particular, $\bar{\alpha} - a_0 = p\bar{\alpha}_1$ where $\bar{\alpha}_1 \in \bar{\mathfrak{o}}$. Hence $\bar{\alpha}_1 = \frac{1}{p}(\bar{\alpha} - a_0) \in \bar{\mathfrak{o}}$. We can repeat the argument with $\bar{\alpha}_1$. Thus we can find $a_1 = 0, 1, 2, \cdots, p - 1$ such that $\bar{\alpha}_1 - a_1 \in \bar{\mathfrak{p}}$ and $\bar{\alpha}_2 = \frac{1}{p}(\bar{\alpha}_1 - a_1) \in \bar{\mathfrak{o}}$. Then $\bar{\alpha} = a_0 + p\bar{\alpha}_1 = a_0 + a_1p + \bar{\alpha}_2p^2$. Continuing this process we obtain

$$\bar{\alpha} = a_0 + a_1p + a_2p^2 + \cdots + a_kp^k + \bar{\alpha}_{k+1}p^{k+1}$$

where $0 \leq a_i \leq p - 1$ and $\bar{\alpha}_{k+1} \varepsilon \bar{\mathfrak{o}}$. Then $\bar{\alpha}_{k+1}p^{k+1} \to 0$ and so we have

$$(14) \qquad \bar{\alpha} = a_0 + a_1 p + a_2 p^2 + \cdots, \qquad 0 \leq a_i \leq p - 1.$$

Conversely, consider any series of this form. We may suppose this is $a_m p^m + a_{m+1} p^{m+1} + \cdots$ where $m \geq 0$, $a_m \neq 0$. Then this series converges. If $\bar{\alpha}$ is its limit, then $\varphi_p(\bar{\alpha}) = \varphi_p(p^m) = p^{-m}$. Hence $\bar{\alpha} \varepsilon \bar{\mathfrak{o}}$. We see also that $\bar{\alpha}$ is a unit in $\bar{\mathfrak{o}}$ if and only if $m = 0$. Hence the units of $\bar{\mathfrak{o}}$ are the elements (14) with $a_0 \neq 0$. We have seen that the ideal $\bar{\mathfrak{p}}$ is the principal ideal generated by p. It follows that every element of $\bar{R}^{(p)}$ has the form $p^k \bar{\epsilon}$ where $\bar{\epsilon}$ is a unit and $k = 1, \pm 1, \pm 2, \cdots$. Hence every element has the form $p^k(a_0 + a_1 p + \cdots)$ where $0 \leq a_i \leq p - 1$.

Let U be the multiplicative group of units in $\bar{\mathfrak{o}}$. We wish to analyze the structure of U, and first we shall show that U contains a subgroup isomorphic to the multiplicative group of non-zero elements of $I/(p)$, that is, a cyclic group of order $p - 1$. Let a be one of the numbers $1, 2, \cdots, p - 1$. We know that $a^p = a + xp$ where $x \varepsilon I$. It follows by induction that $a^{p^k} \equiv a^{p^{k-1}} \pmod{p^k}$ so $\dfrac{a^{p^k} - a^{p^{k-1}}}{p^k} \varepsilon I \subseteq \bar{\mathfrak{o}}$. It follows that

$$(15) \qquad \zeta_a = a + \left(\frac{a^p - a}{p}\right)p + \left(\frac{a^{p^2} - a^p}{p^2}\right)p^2 + \cdots$$

is a well-defined element of $\bar{R}^{(p)}$. This is a limit of the sequence $\{\zeta_a^{(k)}\}$ where

$$\zeta_a^{(k)} = a + \left(\frac{a^p - a}{p}\right)p + \cdots + \left(\frac{a^{p^k} - a^{p^{k-1}}}{p^k}\right)p^k = a^{p^k}.$$

Now in any field with a valuation one can prove as for the reals that $\lim a_k = \bar{\alpha}$, $\lim b_k = \bar{\beta}$ imply $\lim (a_k \pm b_k) = \bar{\alpha} \pm \bar{\beta}$ and $\lim a_k b_k = \bar{\alpha}\bar{\beta}$ (cf. ex. 1, § 4). Hence $\lim \zeta_a^{(k)} = \zeta_a$ implies $\lim (\zeta_a^{(k)})^p = \zeta_a^p$. Since $\zeta_a^{(k)} = a^{p^k}$ we have $\lim a^{p^k} = \zeta_a$ and $\lim (a^{p^k})^p = \lim a^{p^{k+1}} = \zeta_a^p$. But evidently $\lim a^{p^{k+1}} = \zeta_a$. Hence we have $\zeta_a^p = \zeta_a$. Also $\zeta_a \equiv a \pmod{\bar{\mathfrak{p}}}$ is clear from (15), and since $a \not\equiv 0 \pmod{\bar{\mathfrak{p}}}$, $\zeta_a \neq 0$. Hence $\zeta_a^{p-1} = 1$. The same argument shows that, if $a \neq b$ in the set $\{1, 2, \cdots, p - 1\}$, then $\zeta_a \neq \zeta_b$. Hence we have constructed $p - 1$ distinct $(p - 1)$-st roots of 1. This is all we can have in a field. We know also that $\{\zeta_a\}$ is a cyclic group (Lemma 1, § 1.13).

Let $\bar{\epsilon} \in U$, so that $\bar{\epsilon} = a + a_1 p + a_2 p^2 + \cdots$ where $0 < a < p$ and $0 \leq a_i < p$. Then $\zeta_a \equiv \bar{\epsilon} \pmod{\mathfrak{p}}$ and $\bar{\epsilon}_1 = \zeta_a^{-1}\bar{\epsilon} \equiv 1 \pmod{\mathfrak{p}}$. We shall call a p-adic integer a 1-*unit* (*Einseinheit*) if it is congruent to 1 modulo $\mathfrak{p}$. We have shown that every unit in $\bar{\mathfrak{o}}$ has the form $\bar{\epsilon} = \bar{\epsilon}_1 \zeta_a$ where $\bar{\epsilon}_1$ is a 1-unit. Let U_1 be the set of 1-units. Then U_1 is a subgroup of U. To see this, let $\bar{\eta}_1, \bar{\eta}_2 \in U_1$ so $\bar{\eta}_i = 1 + \bar{\beta}_i, \bar{\beta}_i \in \mathfrak{p}$. Then we have $\bar{\eta}_1\bar{\eta}_2 = 1 + \bar{\beta}_1 + \bar{\beta}_2 + \bar{\beta}_1\bar{\beta}_2 \equiv 1 \pmod{\mathfrak{p}}$, since $\bar{\beta}_1 + \bar{\beta}_2 + \bar{\beta}_1\bar{\beta}_2 \in \mathfrak{p}$. Also, it is easy to see that $1 - \bar{\beta}_1 + \bar{\beta}_1^2 - \cdots = (1 + \bar{\beta}_1)^{-1}$. Clearly, $1 - \bar{\beta}_1 + \bar{\beta}_1^2 - \cdots \equiv 1 \pmod{\mathfrak{p}}$. Hence $\bar{\eta}_1^{-1} = (1 + \bar{\beta}_1)^{-1} \in U_1$.

In order to study the subgroup U_1 of U more closely we find it convenient to introduce the exponential function in the field of p-adic numbers. We define this by means of the series:

$$\exp x = 1 + \frac{x}{1!} + \frac{x^2}{2!} + \cdots \tag{16}$$

which we shall show converges for all $x \in \mathfrak{p}$ if $p \neq 2$. As in § 1, we write $\varphi_p(x) = p^{-\nu_p(x)}$ and we know that $\nu_p(x)$ is an integer. The condition $\nu_p(x) = l > 0$ is equivalent to: $x \in \mathfrak{p}^l$. For a rational number, $\nu_p(x)$ is the power of p which divides x in the sense that $x = p^{\nu_p(x)}y$ where $(y, p) = 1$. In order to prove convergence of (16) we evidently need a formula for $\nu_p(k!)$. For this we note that $\left[\frac{k}{p}\right]$ of the numbers $1, 2, \cdots, k$ are divisible by p where, as usual $[z]$ denotes the integral part of the real number z. (These are $p, 2p, 3p, \cdots, \left[\frac{k}{p}\right] p$.) Similarly, $\left[\frac{k}{p^2}\right]$ of the numbers $1, 2, \cdots, k$ are divisible by p^2, $\left[\frac{k}{p^3}\right]$ are divisible by p^3, etc. This implies that

$$\nu_p(k!) = \left[\frac{k}{p}\right] + \left[\frac{k}{p^2}\right] + \left[\frac{k}{p^3}\right] + \cdots.$$

Hence $\nu_p(k!) < k\left(\frac{1}{p} + \frac{1}{p^2} + \cdots\right) = k\left(\frac{1}{p-1}\right)$. We now assume $p \neq 2$. Then if $\nu_p(x) = l \geq 1$ and $k > 0$, $\nu_p(x^k/k!) > k\left(l - \frac{1}{p-1}\right) > 0$. We can include $x = 0$ in this by taking $\nu_p(0) = \infty$. The inequalities show that $x^k/k! \in \mathfrak{p}$ if $k > 0$ and $\lim x^k/k! = 0$. Hence (16) is convergent and $\exp x$ is defined for all $x \in \mathfrak{p}$. Moreover, since $x^k/k! \in \mathfrak{p}$ for $k > 0$, $\exp x$ is an element of $\bar{\mathfrak{o}}$ and $\exp x \equiv 1 \pmod{\mathfrak{p}}$. Hence $\exp x \in U_1$. If $x \neq 0$ and $\nu_p(x) = l \geq 1$, then $\nu_p(x^2/2!) = 2l > l$ and, if $k > 2$, then $\nu_p(x^k/k!) > k\left(l - \frac{1}{p-1}\right) > l$. It follows that, if $x \in \mathfrak{p}^l$, $l \geq 1$, then

$$\exp x - 1 - x \in \mathfrak{p}^{l+1}. \tag{17}$$

We shall now show that, if $x, y \in \mathfrak{p}$, then

$$\exp(x + y) = (\exp x)(\exp y). \tag{18}$$

Let $\quad X_n = \sum_0^n \frac{x^k}{k!}, \quad Y_n = \sum_0^n \frac{y^k}{k!}, \quad Z_n = \sum_0^n \frac{(x+y)^k}{k!}.$

Since
$$\frac{1}{k!}(x+y)^k = \sum_{l=0}^{k} \frac{x^k}{l!}\frac{y^{k-l}}{(k-l)!},$$

$$Z_{2n} - X_n Y_n = \sum_{\substack{l+k \leq 2n \\ l>n \text{ or} \\ k>n}} \frac{x^l}{l!}\frac{y^k}{k!}.$$

The inequalities noted before imply $\lim (Z_{2n} - X_n Y_n) = 0$. Since $\lim X_n = \exp x$, $\lim Y_n = \exp y$, $\lim Z_{2n} = \exp(x + y)$, this gives (18). This equation and the fact that $\exp x \in U_1$ establish a homomorphism of the additive group $(\mathfrak{p}, +)$ into U_1. We shall show that the mapping $x \to \exp x$ is in fact an isomorphism of $(\mathfrak{p}, +)$ onto U_1. To see that the mapping is an isomorphism, it suffices to show that $x \neq 0$ implies $\exp x \neq 1$. Thus suppose $\nu_p(x) = l \neq \infty$, so $x \in \mathfrak{p}^l$, $\notin \mathfrak{p}^{l+1}$. Then it is clear from (17) that $\exp x \neq 1$.

Next consider any element of U_1. This has the form $1 + y$, $y \in \mathfrak{p}$. Set $x_1 = y$ and consider $(1 + y)\exp(-x_1)$. By (17),

$\exp(-x_1) = 1 - x_1 + z_1$ where $z_1 \varepsilon \bar{\mathfrak{p}}^2$; hence

$$(1 + y)\exp(-x_1) = (1 + x_1)(1 - x_1 + z_1)$$
$$= 1 + (z_1 - x_1^2 + x_1 z_1)$$
$$= 1 + x_2$$

where $x_2 = z_1 - x_1^2 + x_1 z_1 \varepsilon \bar{\mathfrak{p}}^2$. Suppose we have already determined elements $x_1, x_2, \cdots, x_k$ such that $x_i \varepsilon \bar{\mathfrak{p}}^i$ and

$$(1 + y)\exp(-x_1 - x_2 - \cdots - x_k) = 1 + x_{k+1}$$

where $x_{k+1} \varepsilon \bar{\mathfrak{p}}^{k+1}$. Then

$$(1 + y)\exp(-x_1 - x_2 - \cdots - x_{k+1})$$
$$= (1 + y)\exp(-x_1 - x_2 - \cdots - x_k)\exp(-x_{k+1})$$
$$= (1 + x_{k+1})\exp(-x_{k+1})$$
$$= (1 + x_{k+1})(1 - x_{k+1} + z_{k+1}),$$

where $z_{k+1} \varepsilon \bar{\mathfrak{p}}^{k+2}$ and $(1 + x_{k+1})(1 - x_{k+1} + z_{k+1}) = 1 + x_{k+2}$ where $x_{k+2} = z_{k+1} - x_{k+1}^2 + x_{k+1}z_{k+1} \varepsilon \bar{\mathfrak{p}}^{k+2}$. This shows that for any integer $n \geq 1$ we have $x_1, x_2, \cdots, x_n$, $x_i \varepsilon \bar{\mathfrak{p}}^i$, such that $(1 + y)\exp(-\sum_1^n x_i) \equiv 1 \pmod{\bar{\mathfrak{p}}^{n+1}}$. Then $x = \sum_1^\infty x_k$ is an element of $\bar{\mathfrak{p}}$ and we assert that $\exp x = 1 + y$. Let $X_n = \sum_1^n x_i$. Then $\exp(-x)\exp X_n = \exp(X_n - x) \equiv 1 \pmod{\bar{\mathfrak{p}}^{n+1}}$ since $X_n - x \varepsilon \bar{\mathfrak{p}}^{n+1}$. Now one verifies as for $\bar{\mathfrak{p}}$ that, if $z_1 \equiv 1 \pmod{\bar{\mathfrak{p}}^{n+1}}$ and $z_2 \equiv 1 \pmod{\bar{\mathfrak{p}}^{n+1}}$, then $z_1 z_2 \equiv 1 \pmod{\bar{\mathfrak{p}}^{n+1}}$. Hence we can conclude from $(1 + y)\exp(-X_n) \equiv 1 \pmod{\bar{\mathfrak{p}}^{n+1}}$ and $\exp(-x)\exp X_n \equiv 1 \pmod{\bar{\mathfrak{p}}^{n+1}}$ that

$$(1 + y)\exp(-x) \equiv 1 \pmod{\bar{\mathfrak{p}}^{n+1}}.$$

Since n is arbitrary, this gives $(1 + y)\exp(-x) = 1$ and $1 + y = \exp x$ as required. This shows that $x \to \exp x$, $x \varepsilon \bar{\mathfrak{p}}$ is surjective on U_1. Hence we have proved the following

Theorem 7. *Let $\bar{\mathfrak{p}}$ be the maximal ideal in the ring $\bar{\mathfrak{o}}$ of p-adic integers, $p \neq 2$, and let U_1 be the group of elements $\equiv 1 \pmod{\bar{\mathfrak{p}}}$. Then the exponential mapping $x \to \exp x$ is an isomorphism of the additive group $(\bar{\mathfrak{p}}, +)$ onto the multiplicative group U_1 of 1-units of $\bar{\mathfrak{o}}$.*

Remark. It is natural to establish the fact that $x \to \exp x$ is surjective by giving the inverse function $\log(1 + y) = y - \frac{y^2}{2} + \frac{y^3}{3} - \cdots$, which is defined for all $y \in \bar{\mathfrak{p}}$ (ex. 4 below). Then one has to show that $\exp(\log(1 + y)) = 1 + y$. The details of this are somewhat lengthy. For this reason we have preferred the above proof that $x \to \exp x$ is surjective since it does not require the explicit definition of the inverse. The reader may refer to Hasse's *Zahlentheorie*, Berlin 1949, pp. 188–199, for a complete treatment of these questions.

It is clear from Theorem 7 that the group U_1 has no elements of finite order. Hence if Z denotes the group of $(p - 1)$-st roots of 1 which we constructed before, then $U_1 \cap Z = 1$. We have seen that every element of the group of units U of $\bar{\mathfrak{o}}$ is a product of an element of Z and an element of U_1. Hence we have $U = U_1 \times Z$ (direct product).

As an application of these results we consider the question of solvability of equations of the form $x^2 = m$ in p-adic fields where m is an ordinary integer prime to p and $p \neq 2$. Then $m \in U$ and we can write $m = \bar{\eta}\zeta_a$ where $\bar{\eta} \in U_1$ and $m \equiv a \pmod{p}$, $0 < a < p$. It is clear that, if $\bar{\alpha}^2 = m$ for $\bar{\alpha} \in \bar{R}^{(p)}$, then $\varphi_p(\bar{\alpha}) = 1$, so if a solution of $x^2 = m$ exists in $\bar{R}^{(p)}$, then this solution must belong to U. Hence it has the form $\bar{\lambda}\zeta_b$ where ζ_b is one of the $(p - 1)$-st roots of 1 and $\bar{\lambda} \in U_1$. It follows from $U = U_1 \times Z$ that $\bar{\lambda}^2 = \bar{\eta}$, $\zeta_b^2 = \zeta_a$. We now note that the equation $x^2 = \bar{\eta}$ has a solution for any $\bar{\eta} \in U_1$. Using the isomorphism of U_1 with $(\bar{\mathfrak{p}}, +)$, it suffices to see that the mapping $x \to 2x$ is an automorphism of the latter group. This is clear since $2^{-1} \in U$ and $x \to 2^{-1}x$ maps $\bar{\mathfrak{p}}$ into itself and is the inverse of the mapping $x \to 2x$. Hence we see that the equation $x^2 = m$ is solvable in $\bar{R}^{(p)}$ if and only if $\zeta_b^2 = \zeta_a$ is solvable. It is easy to see that the condition for this is that $x^2 \equiv m$ or $x^2 \equiv a \pmod{p}$ is solvable, that is, m is a quadratic residue modulo p. Hence $x^2 = m$ is solvable in $\bar{R}^{(p)}$, $p \neq 2$ $(m, p) = 1$ if and only if $x^2 \equiv m \pmod{p}$ is solvable in integers, that is, if and only if $\left(\frac{m}{p}\right) = 1$ where $\left(\frac{m}{p}\right)$ is the Legendre symbol.

For example, if $p = 5$ and $m = -1$, then $2^2 \equiv -1 \pmod{5}$ so

$\left(\frac{-1}{5}\right) = 1$. Hence $\sqrt{-1}$ exists in the 5-adic field. On the other hand, $\left(\frac{3}{5}\right) = -1$, so $\sqrt{3}$ does not exist in this field.

EXERCISES

1. Obtain the 5-adic expansion of the form (14) for the 5-adic integer $\frac{2}{3}$.
2. Show that the field of p-adic numbers is uncountable for any $p = 2, 3, 5, \cdots$. Use this to prove the existence of p-adic numbers which are transcendental over the rational subfield.
3. Use the binomial expansion of $(1 - 2x)^{1/2}$ to obtain a convergent series for $\sqrt{-1} = \frac{1}{3}(1 - 10)^{1/2}$ in the 5-adic field.
4. Define $\log(1 + y) = y - \frac{y^2}{2} + \frac{y^3}{3} - \cdots$. Show that this series converges for all $y \in \bar{\mathfrak{p}}$ if $p \neq 2$. Show that $\log(1 + y_1)(1 + y_2) = \log(1 + y_1) + \log(1 + y_2)$, $y_i \in \bar{\mathfrak{p}}$.
5. Show that the equation $x^3 = 4$ is solvable in the field of 5-adic numbers.
6. Show that in the field of 2-adic numbers the exponential mapping is an isomorphism of $\bar{\mathfrak{p}}^2$ onto the group of elements of $\mathfrak{o}$ which are $\equiv 1 \pmod{\bar{\mathfrak{p}}^2}$.

6. Hensel's lemma. There is another, more powerful, method for handling equations in p-adic fields and more generally in complete fields with a discrete non-archimedean real valuation. This is based on a fundamental reducibility criterion for polynomials which is known as

Hensel's lemma. *Let Φ be a complete field relative to a non-archimedean discrete real valuation φ. Let $\mathfrak{o}$ be the valuation ring of Φ, $\mathfrak{p}$ its maximal prime ideal, $\Delta = \mathfrak{o}/\mathfrak{p}$ the residue field and let $\alpha \to \alpha^* = \alpha + \mathfrak{p}$ be the canonical homomorphism of $\mathfrak{o}$ onto Δ. Suppose $f(x) \in \mathfrak{o}[x]$ has the property that its image $f^*(x) = \gamma(x)\eta(x)$ in $\Delta[x]$ where $(\gamma(x), \eta(x)) = 1$ and the leading coefficient of $\gamma(x)$ is* 1. *Then $f(x) = g(x)h(x)$ in $\mathfrak{o}[x]$ where $g^*(x) = \gamma(x)$, $h^*(x) = \eta(x)$*, $\deg g(x) = \deg \gamma(x)$ *and $g(x)$ has leading coefficient* 1.

Proof. Let $\deg f(x) = n$, $\deg \gamma(x) = r \leq n$. We can choose $g_1(x), h_1(x) \in \mathfrak{o}[x]$ so that $g_1^*(x) = \gamma(x)$, $h_1^*(x) = \eta(x)$, $\deg g_1(x) = r$, $\deg h_1(x) \leq n - r$, leading coefficient of $g_1(x)$ is 1. Then we have $f(x) \equiv g_1(x)h_1(x) \pmod{\mathfrak{p}}$ in the sense that the coefficients are congruent $(\text{mod } \mathfrak{p})$. We proceed to determine two sequences of polynomials $\{g_k(x)\}$, $\{h_k(x)\}$, $k = 1, 2, \cdots$, in $\mathfrak{o}[x]$ such that: (i) $g_k(x) \equiv g_{k+1}(x) \pmod{\mathfrak{p}^k}$, $h_k(x) \equiv h_{k+1}(x) \pmod{\mathfrak{p}^k}$, (ii) $f(x) \equiv$

$g_k(x)h_k(x) \pmod{\mathfrak{p}^k}$, (iii) $\deg g_k(x) = r$, $\deg h_k(x) \leq n - r$, leading coefficient of $g_k(x) = 1$. We can begin these sequences with the $g_1(x)$ and $h_1(x)$ which we have chosen. Hence we may suppose that the sequences have already been constructed for $k \leq s$. We set $g_{s+1}(x) = g_s(x) + u(x)\pi^s$, $h_{s+1}(x) = h_s(x) + v(x)\pi^s$, where $\mathfrak{p} = (\pi)$ (as in § 5). Then (i) will hold for any choice of $u(x)$ and $v(x)$ in $\mathfrak{o}[x]$. We seek to satisfy (ii). This requires that

$$\begin{aligned} f(x) &\equiv [g_s(x) + u(x)\pi^s][h_s(x) + v(x)\pi^s] \\ &\equiv g_s(x)h_s(x) + [g_s(x)v(x) + h_s(x)u(x)]\pi^s \pmod{\mathfrak{p}^{s+1}} \end{aligned}$$

or

$$f(x) - g_s(x)h_s(x) \equiv [g_s(x)v(x) + h_s(x)u(x)]\pi^s \pmod{\mathfrak{p}^{s+1}}. \tag{19}$$

Since $f(x) \equiv g_s(x)h_s(x) \pmod{\mathfrak{p}^s}$ we can write $f(x) - g_s(x)h_s(x) = \pi^s\omega(x)$ where $\omega(x) \in \mathfrak{o}[x]$. Since $\deg f(x) = n$ and $\deg g_s(x)h_s(x) \leq n$ we may suppose $\deg \omega(x) \leq n$. It is clear that (19) will hold if

$$g_s(x)v(x) + h_s(x)u(x) \equiv \omega(x) \pmod{\mathfrak{p}}. \tag{20}$$

Now it is clear from (i) that $g_s^*(x) = g_1^*(x) = \gamma(x)$ and $h_s^*(x) = \eta(x)$, so we consider the equation

$$\gamma(x)v^*(x) + \eta(x)u^*(x) = \omega^*(x) \tag{21}$$

in $\Delta[x]$. Since $(\gamma(x), \eta(x)) = 1$, there exist polynomials $\alpha(x)$, $\beta(x)$ in $\Delta[x]$ such that $\alpha(x)\gamma(x) + \beta(x)\eta(x) = 1$. Multiplication by $\omega^*(x)$ gives polynomials $\kappa(x)$, $\lambda(x)$ such that $\kappa(x)\gamma(x) + \lambda(x)\eta(x) = \omega^*(x)$. We can write $\lambda(x) = \gamma(x)\mu(x) + \rho(x)$ where $\deg \rho(x) < r$ and then we obtain

$$\begin{aligned} \omega^*(x) &= \kappa(x)\gamma(x) + (\gamma(x)\mu(x) + \rho(x))\eta(x) \\ &= (\kappa(x) + \mu(x)\eta(x))\gamma(x) + \rho(x)\eta(x). \end{aligned}$$

Then $\deg \rho(x)\eta(x) < n$ while $\deg \omega^*(x) \leq n$. Since $\deg \gamma(x) = r$, the foregoing relation shows that the degree of $(\kappa(x) + \mu(x)\eta(x))$ does not exceed $n - r$. If we call this polynomial $\sigma(x)$, we have

$$\sigma(x)\gamma(x) + \rho(x)\eta(x) = \omega^*(x),$$

where $\deg \rho(x) < r$ and $\deg \sigma(x) \leq n - r$. Then we can choose $u(x)$ and $v(x) \in \mathfrak{o}[x]$ so that $u^*(x) = \rho(x)$, $v^*(x) = \sigma(x)$, $\deg u(x)$

$= \deg\ \rho(x) < r$, $\deg\ v(x) \leq n - r$. Then (21) holds and $g_{s+1}(x) = g_s(x) + u(x)\pi^s$, $h_{s+1}(x) + h_s(x) + v(x)\pi^s$ satisfy also (iii). This completes the proof of the existence of the sequence $\{g_k(x)\}$, $\{h_k(x)\}$, $k = 1, 2, \cdots$ satisfying (i), (ii), and (iii). The conditions (i) and (iii) and the completeness of Φ imply that the sequences $\{g_k(x)\}$, $\{h_k(x)\}$ converge to polynomials $g(x)$, $h(x)$ in the sense that the sequences of coefficients of like powers of x converge to those of $g(x)$, $h(x)$. Moreover, we have $\deg g(x) = r$, $\deg h(x) \leq n - r$, leading coefficient of $g(x) = 1$. It follows also from (ii) that $f(x) = g(x)h(x)$ and this completes the proof.

EXERCISES

1. Use Hensel's lemma to prove that in the field of p-adic numbers there exists a ζ_a such that $\zeta_a^{p-1} = 1, \zeta_a \equiv a \pmod{\bar{\mathfrak{p}}}$ where a is any integer prime to p. Also use this to obtain another proof of the existence of $\sqrt{-1}$ and $\sqrt[3]{4}$ in the 5-adic field.

2. Hypotheses on Φ as in Hensel's lemma. Let $f(x) = a_0x^n + a_1x^{n-1} + \cdots + a_n \in \mathfrak{o}[x]$ satisfy: $a_0, a_n \in \mathfrak{p}$ but there exists a_r, $1 \leq r \leq n - 1$, such that $a_r \notin \mathfrak{p}$. Then $f(x)$ is reducible in $\mathfrak{o}[x]$. Use this to show that, if $g(x) = x^n + \alpha_1x^{n-1} + \cdots + \alpha_n$ is an irreducible polynomial in $\Phi[x]$ and $\alpha_n \in \mathfrak{o}$, then all the $\alpha_i \in \mathfrak{o}$.

7. Construction of complete fields with given residue fields. Let Δ be a given field. We consider the problem of constructing complete fields with non-archimedean real valuations such that the residue field is the given field Δ. We shall give two constructions: the first, in which the complete field contains Δ and so has the same characteristic as Δ; the second, in which Δ is perfect of characteristic $p \neq 0$ and the complete field is of characteristic 0. A special case of the latter is $\Delta = I_p$ and the complete field is the field of p-adic numbers.

We consider first the field $\Phi = \Delta(\xi)$ where ξ is transcendental over Δ. We introduce the order function ν by $\nu(\alpha(\xi)) = k$ if $\alpha(\xi) = \xi^k\beta(\xi)\gamma(\xi)^{-1}$ where $\beta(\xi)$ and $\gamma(\xi)$ are polynomials not divisible by ξ. We define a valuation φ by $\varphi(\alpha(\xi)) = c^{\nu(\alpha(\xi))}$, c a fixed real number $0 < c < 1$ (cf. example 3, § 1). Let $\bar{\Phi}$ be the completion of Φ relative to φ. Since φ is trivial on Δ, it is clear that φ is non-archimedean. Hence its extension to $\bar{\Phi}$, which we shall denote by φ also, is non-archimedean. The value group Γ of Φ and of $\bar{\Phi}$ consists of the powers of c, so the valuation is discrete. Let $\bar{\mathfrak{o}}$ be the valuation ring of $\bar{\Phi}$, $\bar{\mathfrak{p}}$ its maximal ideal, and let $\mathfrak{o} = \bar{\mathfrak{o}} \cap \Phi$,

$\mathfrak{p} = \bar{\mathfrak{p}} \cap \Phi$. It is clear that ξ is an element of $\bar{\mathfrak{p}}$ for which $\varphi(\xi) = c$ is maximal. Hence, as in the p-adic case, every element $\bar{\alpha}$ of $\bar{\Phi}$ has the form $\xi^k \bar{\epsilon}$ where $\bar{\epsilon} \varepsilon \bar{\mathfrak{o}}, \notin \bar{\mathfrak{p}}$ and k is an integer. We can therefore define $\nu(\xi^k \bar{\epsilon}) = k$ and it is clear that this coincides on Φ with the order ν which we defined originally in Φ.

We have seen in § 5 that $\bar{\mathfrak{o}} = \mathfrak{o} + \bar{\mathfrak{p}}$ and this permits us to identify the residue fields $\bar{\mathfrak{o}}/\bar{\mathfrak{p}}$ and $\mathfrak{o}/\mathfrak{p}$. The ring $\mathfrak{o}$ is the set of rational expressions in ξ with coefficients in Δ which are "finite at 0" in the sense that $\alpha(\xi) = \beta(\xi)\gamma(\xi)^{-1}$ where β and γ are polynomials and $\gamma(0) \neq 0$. The argument in the p-adic case showing that $\bar{\mathfrak{o}} = I + \bar{\mathfrak{p}}$ (p. 224) can be used in the present situation to prove that $\bar{\mathfrak{o}} = \Delta[\xi] + \bar{\mathfrak{p}}$. Since $\xi \varepsilon \bar{\mathfrak{p}}$, this gives $\bar{\mathfrak{o}} = \Delta + \bar{\mathfrak{p}}$ and, since $\bar{\mathfrak{p}} \cap \Delta = 0$, we have the isomorphism $\delta \rightarrow \delta + \bar{\mathfrak{p}}$ in $\bar{\mathfrak{o}}/\bar{\mathfrak{p}}$ of Δ with the residue field $\bar{\mathfrak{o}}/\bar{\mathfrak{p}}$. In this sense we can say that Δ is the residue field of $\bar{\Phi}$.

Now let $\bar{\alpha}$ be any element of $\bar{\mathfrak{o}}$. Then $\bar{\mathfrak{o}} = \Delta + \bar{\mathfrak{p}}$ shows that we can find $\delta_0 \varepsilon \Delta$ such that $\bar{\alpha} - \delta_0 \varepsilon \bar{\mathfrak{p}}$. Then $\bar{\alpha}_1 = (\bar{\alpha} - \delta_0)\xi^{-1} \varepsilon \bar{\mathfrak{o}}$ and we can repeat the argument with this obtaining $\delta_1 \varepsilon \Delta$ such that $\bar{\alpha}_1 - \delta_1 \varepsilon \bar{\mathfrak{p}}$ and $\bar{\alpha}_2 = (\bar{\alpha}_1 - \delta_1)\xi^{-1} \varepsilon \bar{\mathfrak{o}}$. We have $\bar{\alpha} = \delta_0 + \delta_1\xi + \bar{\alpha}_2\xi^2$, $\bar{\alpha}_2 \varepsilon \bar{\mathfrak{o}}$. As in the p-adic case, we can continue this process and obtain

$$\bar{\alpha} = \delta_0 + \delta_1\xi + \delta_2\xi^2 + \cdots + \delta_k\xi^k + \bar{\alpha}_{k+1}\xi^{k+1} \tag{22}$$

where the $\delta_i \varepsilon \Delta$ and $\bar{\alpha}_{k+1} \varepsilon \bar{\mathfrak{o}}$. Since $\nu(\bar{\alpha}_{k+1}\xi^{k+1}) \geq k + 1$, it is clear that the sequence $\{\bar{\alpha}_k\xi^k\}$ is a null sequence. Hence we have

$$\bar{\alpha} = \delta_0 + \delta_1\xi + \delta_2\xi^2 + \cdots, \quad \delta_i \varepsilon \Delta, \tag{23}$$

for any $\bar{\alpha} \varepsilon \bar{\mathfrak{o}}$. If $\bar{\beta}$ is any element of $\bar{\Phi}$ we can write $\bar{\beta} = \bar{\alpha}\xi^{-k}$ where k is a non-negative integer and $\bar{\alpha} \varepsilon \bar{\mathfrak{o}}$. Then we have

$$\bar{\beta} = \xi^{-k}(\delta_0 + \delta_1\xi + \delta_2\xi^2 + \cdots). \tag{24}$$

This shows that $\bar{\Phi}$ is the set of power series of the form (24) in ξ with coefficients in the field Δ. It is easy to see that the expression (24) for $\bar{\beta}$ is unique, that is, k and the $\delta_i \varepsilon \Delta$ are uniquely determined by $\bar{\beta}$. Moreover, the addition and multiplication of elements of $\bar{\Phi}$ are the usual ones for formal power series based on the compositions in Δ. For example, we have $(\delta_0 + \delta_1\xi + \cdots) + (\epsilon_0 + \epsilon_1\xi + \cdots) = (\delta_0 + \epsilon_0) + (\delta_1 + \epsilon_1)\xi + \cdots$ for δ_i, ϵ_i in Δ and

$\left(\sum_0^\infty \delta_i\xi^i\right)\left(\sum_0^\infty \epsilon_j\xi^j\right) = \sum_0^\infty \eta_k\xi^k$ where $\eta_k = \sum_{i=0}^k \delta_i\epsilon_{k-i}$. It is clear that we have a good hold on the field Φ as the field of formal power series (24).

We consider next the case in which Δ is a perfect field of characteristic $p \neq 0$ and we shall use this to construct a field Φ which is a generalization of the field of p-adic numbers. The construction we shall give is based on Witt vectors which we considered in § 3.4. We begin with the definition of the ring $\mathfrak{W}(\mathfrak{A})$ of *Witt vectors* (*of infinite length*) based on a commutative algebra $\mathfrak{A}$ over I_p. The elements of $\mathfrak{W}(\mathfrak{A})$ are the infinite sequences

$$(a_0, a_1, a_2, \cdots), \quad a_i \in \mathfrak{A}, \tag{25}$$

where equality is defined component-wise. We can define addition and multiplication by the formulas (22) of Chap. III which were used to define these compositions in $\mathfrak{W}_m(\mathfrak{A})$ the ring of Witt vectors of length m defined by $\mathfrak{A}$. Then one can verify that $\mathfrak{W}(\mathfrak{A})$ is a ring. It is more convenient, however, to adopt an equivalent but slightly different approach which is a special case of the definition of an inverse limit of rings. In the present case we are dealing with such a limit for the rings $\mathfrak{A} = \mathfrak{W}_1(\mathfrak{A}), \mathfrak{W}_2(\mathfrak{A}), \cdots$ with the restriction homomorphism R of $\mathfrak{W}_m(\mathfrak{A})$ into $\mathfrak{W}_{m-1}(\mathfrak{A})$. We associate with the element $A = (a_0, a_1, \cdots)$ of $\mathfrak{W}(\mathfrak{A})$ its projection $A^{\pi_m} = (a_0, a_1, \cdots, a_{m-1})$ in $\mathfrak{W}_m(\mathfrak{A})$. Then $A^{\pi_m R} = (a_0, \cdots, a_{m-2}) = A^{\pi_{m-1}}$. On the other hand, let $\{A_m | m = 0, 1, 2, \cdots\}$ be any sequence of elements A_m where $A_m \in \mathfrak{W}_m(\mathfrak{A})$ and $A_m{}^R = A_{m-1}$, $m = 1, 2, \cdots$. Then it is clear that $\{A_m\} = A^{\pi_m}$ for a unique $A \in \mathfrak{W}(\mathfrak{A})$. Hence we can identify the elements of $\mathfrak{W}(\mathfrak{A})$ with the sequences $\{A_m\}$, $A_m \in \mathfrak{W}_m(\mathfrak{A})$ such that $A_m{}^R = A_{m-1}$. If $A = \{A_m\}$ and $B = \{B_m\}$ are two such sequences, we define $A + B = \{A_m + B_m\}$, $AB = \{A_mB_m\}$. Since R is a ring homomorphism, $(A_m + B_m)^R = A_m{}^R + B_m{}^R = A_{m-1} + B_{m-1}$ and $(A_mB_m)^R = A_m{}^R B_m{}^R = A_{m-1}B_{m-1}$. Hence $A + B$ and $AB \in \mathfrak{W}(\mathfrak{A})$. It is trivial to check that $\mathfrak{W}(\mathfrak{A})$ is a commutative ring relative to these compositions and that $0 = (0, 0, \cdots)$, $1 = (1, 0, 0, \cdots)$. For a fixed m, the mapping π_m: $A \to A^{\pi_m}$ is a homomorphism of $\mathfrak{W}(\mathfrak{A})$ onto $\mathfrak{W}_m(\mathfrak{A})$. Since 1^{π_m} has order p^m, it is clear that the identity 1 of $\mathfrak{W}(\mathfrak{A})$ has infinite order in the additive group $(\mathfrak{W}(\mathfrak{A}), +)$.

Let $\mathfrak{N}_m$ denote the kernel of π_m. Then $\mathfrak{N}_m$ is the set of elements of $\mathfrak{W}(\mathfrak{A})$ of the form $(0, \cdots, 0, a_m, a_{m+1}, \cdots)$. Hence

$$(26) \qquad \mathfrak{N}_1 \supset \mathfrak{N}_2 \supset \mathfrak{N}_3 \supset \cdots, \quad \bigcap_{m=1}^{\infty} \mathfrak{N}_m = 0.$$

We can use the set $\{\mathfrak{N}_m\}$ to define convergence in $\mathfrak{W}(\mathfrak{A})$. If $\{A_k | k = 1, 2, \cdots\}$ is a sequence of elements of $\mathfrak{W}(\mathfrak{A})$, then we say that $\{A_k\}$ *converges* to the element A of $\mathfrak{W}(\mathfrak{A})$ ($A_k \to A$) if for any positive integer m there exists a positive integer $N(m)$ such that $A - A_k \in \mathfrak{N}_{N(m)}$ for all $k \geq N(m)$. It is easy to see that the limit A is unique and that $A_k \to A$, $B_k \to B$ imply $A_k \pm B_k \to A \pm B$ and $A_k B_k \to AB$. Suppose $\{C_k | k = 0, 1, 2, \cdots\}$ is a sequence such that $C_k \in \mathfrak{N}_k$, $k = 1, 2, \cdots$. Set $A_k = C_0 + C_1 + \cdots + C_k$. Then $A_m^{\pi_{m+1}R} = A_m^{\pi_m} = A_{m-1}^{\pi_m}$; hence the sequence of elements $\{A_m^{\pi_{m+1}}, m = 0, 1, \cdots\}$ where $A_m^{\pi_{m+1}} \in \mathfrak{W}_{m+1}(\mathfrak{A})$ can be identified with an element $A \in \mathfrak{W}(\mathfrak{A})$. One checks that $A_k \to A$. Since $A_k = C_0 + C_1 + \cdots + C_k$, we shall indicate the convergence $A_k \to A$ by writing $\sum_{k=0}^{\infty} C_k = A$.

We recall that, if $\mathfrak{N}$ is the ideal of elements $(0, a_1, \cdots, a_{m-1})$ in $\mathfrak{W}_m(\mathfrak{A})$, then $\mathfrak{N}$ is nilpotent (Th. 3.12). In fact, the proof of this result shows that $\mathfrak{N}^k$ is contained in the set of vectors of the form $(0, \cdots, 0, a_{k+1}, \cdots, a_{m-1})$. This implies that $\mathfrak{N}_1{}^k \subseteq \mathfrak{N}_k$ in $\mathfrak{W}(\mathfrak{A})$. Hence, if $Z \in \mathfrak{N}_1$, then $\sum_{k=0}^{\infty} Z^k$ is defined. Since $\left(\sum_0^m Z^k\right)(1 - Z) = 1 - Z^{m+1}$, it follows that $1 - Z$ is a unit in $\mathfrak{W}(\mathfrak{A})$ with $\sum_0^{\infty} Z^k$ as inverse. Since $(a_0, \cdots)(a_0{}^{-1}, \cdots) = (1, \cdots)$ this implies that, if a_0 is a unit in $\mathfrak{A}$, then $(a_0, a_1, \cdots)$ is a unit in $\mathfrak{W}(\mathfrak{A})$.

Now let $\mathfrak{A} = \Delta$ a perfect field of characteristic p. The formula $p(a_0, a_1, \cdots, a_{m-1}) = (0, a_0{}^p, a_1{}^p, \cdots, a_{m-2}{}^p)$ in $\mathfrak{W}_m(\mathfrak{A})$ which we established in § 3.4 (Equation (27)) implies that $p(a_0, a_1, \cdots) = (0, a_0{}^p, a_1{}^p, \cdots)$ holds in $\mathfrak{W}(\mathfrak{A})$. Iteration of this formula gives

$$(27) \qquad p^k(a_0, a_1, \cdots) = (\overbrace{0, \cdots, 0}^{k}, a_0{}^{p^k}, a_1{}^{p^k}, \cdots).$$

Since $\mathfrak{A} = \Delta$ is perfect, the elements $a_i{}^{p^k}$ can be taken to be arbitrary elements of Δ. Hence we have $p^k\mathfrak{W}(\Delta) = \mathfrak{N}_k$ where $\mathfrak{N}_k$ is the ideal we defined before. Also (27) shows that, if $A \neq 0$, then $p^kA \neq 0$ for $k = 1, 2, \cdots$. Now let A and B be any non-zero elements of $\mathfrak{W}(\Delta)$. Then we can write $A = p^kC$, $B = p^lD$ where $C, D \not\in \mathfrak{N}_1$. Then $C = (c_0, \cdots)$ and $D = (d_0, \cdots)$ where $c_0 \neq 0$, $d_0 \neq 0$. Hence $CD = (c_0d_0, \cdots) \neq 0$ and $AB = p^{k+l}CD \neq 0$. This shows that $\mathfrak{W}(\Delta)$ is an integral domain. Let Φ be the field of fractions of $\mathfrak{W}(\Delta)$ and consider the subset Φ' of Φ of elements of the form p^kC where $C \in \mathfrak{W}(\Delta)$ and $k = 0, \pm 1, \pm 2, \cdots$. Since any $C \in \mathfrak{W}(\Delta)$, $\not\in p\mathfrak{W}(A)$ is a unit in $\mathfrak{W}(\Delta)$, it is clear that the non-zero elements of Φ' form a group under multiplication. Since Φ' is a subring of Φ which contains $\mathfrak{W}(\Delta)$, it follows that $\Phi' = \Phi$.

If $A = p^kC$, $C \in \mathfrak{W}(\Delta)$, $\not\in \mathfrak{N}_1$, then we define the order $\nu(A) = k$ and we define $\varphi(A) = p^{-k}$, $\varphi(0) = 0$. Then φ is a real non-archimedean valuation of Φ. The subring $\mathfrak{W}(\Delta)$ is the set of elements satisfying $\varphi(A) \leq 1$ and $\mathfrak{N}_1$ is the ideal of elements B of $\mathfrak{W}(\Delta)$ such that $\varphi(B) < 1$. The residue ring is $\mathfrak{W}(\Delta)/\mathfrak{N}_1$ which is isomorphic to Δ. The result we noted before on convergence of sequences in $\mathfrak{W}(\mathfrak{A})$ implies that Φ is complete relative to the valuation φ. We leave it to the reader to check this. Since 1 is of infinite order, Φ is of characteristic 0. Thus Φ has all the properties we required: completeness relative to a non-archimedean real valuation, characteristic 0, residue field the given perfect field Δ of characteristic p. If we start with $\Delta = I_p$, then the field Φ we obtain in this way is the field of p-adic numbers.

8. Ordered groups and valuations. A non-archimedean real valuation satisfies $\varphi(\alpha + \beta) \leq \max(\varphi(\alpha), \varphi(\beta))$, $\varphi(\alpha\beta) = \varphi(\alpha)\,\varphi(\beta)$. Hence it is clear that in considering such a valuation the addition of the reals plays no role. Only the multiplication and order of the non-negative reals are involved in the defining properties. As we shall see, this leads to a generalization of the concept of a non-archimedean real valuation to (non-archimedean) valuations with values in any ordered commutative group. Besides the increased generality which results from this extension, the generalization is essentially simpler and more natural than the original concept. We consider first the notion of an ordered commutative group.

Definition 6. *An* ordered (commutative) group G *is a commutative group* G *together with a subset* H *satisfying the three conditions:* 1) $1 \notin H$, 2) *if* $a \in G$ *either* $a \in H$, $a = 1$ *or* $a^{-1} \in H$, 3) H *is closed under the multiplication in* G.

If (G, H) is an ordered group, then we let $H^{-1} = \{b^{-1} | b \in H\}$. Then condition 2 states that $G = H \cup \{1\} \cup H^{-1}$. Moreover, these sets are non-overlapping. This is assumed for H and $\{1\}$ in condition 1 and it follows for H^{-1} and $\{1\}$ on observing that, if $1 \in H^{-1}$, then $1 \in H$ contrary to condition 1. Finally, if $a \in H \cap H^{-1}$, then $a^{-1} \in H$ and $1 = aa^{-1} \in H$ by condition 3. This again contradicts condition 1.

The positive reals form an ordered group if we take H to be the set of elements < 1. We can take H equally well to be the set of elements > 1. In fact, if G is any ordered group, then H^{-1} is closed under multiplication and satisfies conditions 1 and 2 of Definition 6, so we can obtain another ordered group on replacing H by H^{-1}. In any ordered group G we define $a < b$ to mean that $ab^{-1} \in H$. This defines a linear ordering in G, that is, we have the following properties: 1. $a < b$, $b < c$ implies $a < c$. 2. For any pair (a, b), $a, b \in G$, one and only one of the following holds: $a < b$, $a = b$, $b < a$ (as usual we write $b > a$ for $a < b$). The order in G is invariant under multiplication, that is, we have: 3. If $a < b$, then $ac < bc$. Conversely, if a relation $a < b$ is defined in a group G so that properties 1, 2, and 3 hold, then G is ordered by the subset $H = \{a | a < 1\}$. Clearly condition 1 of Definition 6 holds for H. To prove conditions 2 and 3 we note first that, if $a < b$ and $c < d$, then $ac < bc < bd$ so $ac < bd$; hence, $a < b$ if and only if $a^{-1} > b^{-1}$. In particular, $a < 1$ if and only if $a^{-1} > 1$. Since any a satisfies one of the conditions: $a < 1$, $a = 1$, $a > 1$, it is clear that condition 2 of Definition 6 holds. Finally, $a < 1$, $b < 1$ imply $ab < 1$, so H is closed under the multiplication in G. We remark also that the ordering defined by H in the manner indicated: $a < b$ if $ab^{-1} \in H$ is the same as the original ordering since $ab^{-1} \in H$ means $ab^{-1} < 1$ and this holds if and only if $a < b$.

If G_1 is a subgroup of an ordered group G ordered by the set $H = \{a | a \in G, a > 1\}$, then G_1 has an induced ordering defined by $H_1 = G_1 \cap H$. This can be verified directly, or it can be seen by noting that the relation $>$ defined in G gives a relation in G_1 which satisfies the conditions stated before. If G is ordered by H and G' is a second ordered group, ordered by H', then an isomorphism η of G into G' is called an *order-isomorphism* if $H\eta \subseteq H'$. Also G and G' are *order-isomorphic* if there exists an order isomorphism η of G onto G'. In this case one necessarily has $H\eta = H'$. For example, the group of positive reals under multiplication with H defined as before is order isomorphic to the additive group of all the real numbers ordered by the set H' of negative reals. The mapping $a \to \log a$ (natural logarithm) is an order isomorphism of the first group onto the second one.

If G is an ordered group, G contains no elements $\neq 1$ of finite order; for, if $a < 1$ ($a > 1$), then $a^n < 1$ ($a^n > 1$), so $a^n \neq 1$ for every positive integer n. A consequence of this property of G is that for any fixed integer n the mapping $x \to x^n$ of G is an isomorphism of G onto a subgroup of G, which is order preserving if $n \geq 1$.

To define general valuations we shall need to consider *ordered groups* V *with* 0. We define such a system to be an ordered group G to which a 0 element has been adjoined: $V = G \cup \{0\}$. The ordering in G is extended to V by defining $0 < a$ for every $a \in G$ and we define $a0 = 0$ for all a. We can now give the following

Definition 7. *Let* Φ *be a field and let* V *be an ordered (commutative) group with* 0. *A mapping* $\varphi: \alpha \to \varphi(\alpha)$ *of* Φ *into* V *is called a* valuation *if*

(i) $\varphi(\alpha) = 0$ *if and only if* $\alpha = 0$.
(ii) $\varphi(\alpha\beta) = \varphi(\alpha)\varphi(\beta)$.
(iii) $\varphi(\alpha + \beta) \leq \max(\varphi(\alpha), \varphi(\beta))$.

The exact sweep of this definition will become apparent soon. At this point it is clear that real non-archimedean valuations are a special case in which V is the set of non-negative real numbers. On the other hand, it should be noted that the real archimedean valuations are not valuations in the present sense. This inconsistency in terminology will cause no real difficulty. We shall now give an example of a valuation for which V is not the non-negative reals.

Example. In this example we shall find it convenient to use the additive notation in the group G. The modifications in Definition 7 which are necessitated by this change are obvious, so we shall not write these down. The group G we shall consider is the additive group of integer pairs (k, l). We introduce the lexicographic order in G, that is, we define $(k, l) < (k', l')$ if either $k < k'$ or $k = k'$ and $l < l'$. One checks that this is a linear ordering preserved under addition; hence G is an ordered (additive) group. We let $V = G \cup \{\infty\}$ where the ordering is extended to V by setting $\infty > (k, l)$ for every $(k, l) \in G$. Also we define $(k, l) + \infty = \infty$. Now let $P = \Phi(\xi, \eta)$, a purely transcendental extension of a field Φ where $\{\xi, \eta\}$ is a transcendency basis for P over Φ. If $a \in P$ and $a \neq 0$, we can write $a = \xi^m \eta^n p(\xi, \eta) q(\xi, \eta)^{-1}$ where $p(\xi, \eta)$ and $q(\xi, \eta)$ are polynomials in ξ, η with non-zero constant terms, and m and n are integers. Then we define $\varphi(a) = (m, n)$. Also we set $\varphi(0) = \infty$. Then (i) holds. It is easy to check that $\varphi(ab) = \varphi(a) + \varphi(b)$ and $\varphi(a + b) \geq \min\,(\varphi(a), \varphi(b))$. The first of these is (ii) in the additive notation and the second can be changed to (iii) by reversing the ordering (writing $>$ for $<$). Hence our function is essentially a valuation.

EXERCISES

1. Let G be the additive ordered group of integer pairs (k, l) given in the foregoing example. Let c and e be real numbers such that $0 < c < 1$ and e is positive and irrational. Show that the mapping $(k, l) \rightarrow c^{k+el}$ is an isomorphism of G into the ordered multiplicative group of positive real numbers P. Show that G is not order isomorphic to a subgroup of P.

2. Let $P = \Phi(\xi, \eta)$ and $a = \xi^m \eta^n p(\xi, \eta) q(\xi, \eta)^{-1}$ where p and q are polynomials in ξ, η with non-zero constant terms, as in the example above. Define $\psi(a) = c^{m+en}$ where c and e are real numbers, $0 < c < 1$, e positive irrational. Show that ψ is a non-archimedean real valuation which is not discrete.

3. Define a valuation φ of an integral domain $\mathfrak{o}$ by replacing the field Φ in Definition 7 by the integral domain $\mathfrak{o}$. Show that any valuation ψ of $\mathfrak{o}$ into V has a unique extension to a valuation of the field of fractions Φ of $\mathfrak{o}$.

4. Let G be an arbitrary (commutative) ordered group and let $\mathfrak{o} = \Phi_0(G)$ be the group ring over a field Φ_0 of G (Vol. I, ex. 2, p. 95). Show that $\mathfrak{o}$ is an integral domain. If $a = \sum_1^r \alpha_i g_i$, $\alpha_i \neq 0$ in Φ_0, $g_i \in G$, define $\varphi(a) = \min g_i$ (in the ordering $<$ defined in G). Define $\varphi(0) = 0$. Show that φ is a valuation of $\mathfrak{o}$. Use exs. 3 and 4 to show that if V is any ordered group with 0, then there exists a field Φ with a valuation φ of Φ into V such that $\varphi(\Phi) = V$.

9. Valuations, valuation rings, and places. In this section we shall establish an equivalence between the concepts of a valuation in the sense of Definition 7 and two other concepts: valuation ring and place. The first of these, valuation ring, is an intrinsic notion in the sense that its definition does not require any system external to the given field Φ. Moreover, the valuation rings give the link between valuations and places. We have already encountered these for real non-archimedean valuations.

Now let Φ be any field and let φ be a valuation with values in the ordered group V with 0. We note first that $\varphi(1)^2 = \varphi(1^2) = \varphi(1)$ and, since G contains no elements of finite order $\neq 1$, $\varphi(1) = 1$. Also $\varphi(-1)^2 = \varphi(1) = 1$, so $\varphi(-1) = 1$ and $\varphi(-\alpha) = \varphi(-1)\varphi(\alpha) = \varphi(\alpha)$. From $\alpha\alpha^{-1} = 1$ we obtain $\varphi(\alpha^{-1}) = \varphi(\alpha)^{-1}$ and $\varphi(\alpha\beta^{-1}) = \varphi(\alpha)\varphi(\beta)^{-1}$. Now let $\mathfrak{o}$ be the subset of Φ of elements α such that $\varphi(\alpha) \leq 1$. Then, if $\alpha, \beta \in \mathfrak{o}$, $\varphi(\alpha - \beta) \leq \max(\varphi(\alpha), \varphi(\beta)) \leq 1$ and $\varphi(\alpha\beta) = \varphi(\alpha)\varphi(\beta) \leq 1$. Hence $\mathfrak{o}$ is a subring. Now suppose $\alpha \notin \mathfrak{o}$, then $\varphi(\alpha) > 1$ and $\varphi(\alpha^{-1}) = \varphi(\alpha)^{-1} < 1$. Hence $\alpha^{-1} \in \mathfrak{o}$. We therefore see that $\mathfrak{o}$ is a valuation ring (in Φ) in the sense of the following

Definition 8. *If Φ is a field, a* valuation ring $\mathfrak{o}$ *in Φ is a subring of Φ (containing* 1) *such that every element of Φ is either in $\mathfrak{o}$ or is the inverse of an element of $\mathfrak{o}$.*

If $\mathfrak{o}$ is the subring of elements α satisfying $\varphi(\alpha) \leq 1$ for the valuation φ, then $\mathfrak{o}$ is called *the valuation ring* of φ. This is a direct generalization of the definition we gave before for non-archimedean real valuations. We shall now show that any valuation ring gives rise to a valuation φ' for which the given ring is the valuation ring. Suppose $\mathfrak{o}$ is a valuation ring in Φ. Let U be the set of units of $\mathfrak{o}$, $\mathfrak{p}$ the set of non-units, $\mathfrak{p}^*$ the set of non-units $\neq 0$, Φ^* the multiplicative group of non-zero elements of Φ. Then U is a subgroup of the commutative group Φ^* and we shall take $G' = \Phi^*/U$ for our group. We introduce an ordering in G' by letting H' be the set of cosets βU, $\beta \in \mathfrak{p}^*$. It is clear that the product of a non-unit of $\mathfrak{o}$ with any element of $\mathfrak{o}$ is a non-unit. Hence if $\beta_1, \beta_2 \in \mathfrak{p}^*$, then $\beta_1\beta_2 \in \mathfrak{p}^*$; so if $\beta_1 U, \beta_2 U, \in H'$, then $(\beta_1 U)(\beta_2 U) = \beta_1\beta_2 U \in H'$. If βU is any element of $G' = \Phi^*/U$, then $\beta \neq 0$, and if $\beta \notin \mathfrak{p}^*$, then either $\beta \in U$ or $\beta \notin U$ and $\beta \notin \mathfrak{p}^*$. In the first case $\beta U = U$, and in the second $\beta \notin \mathfrak{o}$, so $\beta^{-1} \in \mathfrak{o}$ and, since $\beta^{-1} \in U$ implies $\beta \in U$, we have $\beta^{-1} \in \mathfrak{p}^*$. Hence $(\beta U)^{-1} = \beta^{-1} U \in H'$. Thus we see that $G' = H' \cup \{1\} \cup (H')^{-1}$ holds. Also $1 = U \notin H'$. Hence H' makes G' an ordered group as in Definition 6. Next we adjoin a 0 to G', obtaining $V' = G' \cup \{0\}$, and we define a mapping φ' of Φ into V' by

$$\varphi'(0) = 0, \quad \varphi'(\alpha) = \alpha U \in G' \quad \text{if} \quad \alpha \neq 0. \tag{28}$$

The conditions (i) and (ii) for a valuation are clearly satisfied. Also (iii) is clear if either $\alpha = 0$ or $\beta = 0$. If $\alpha \neq 0$, $\beta \neq 0$, either $\alpha\beta^{-1} \varepsilon \mathfrak{o}$ or $\beta\alpha^{-1} \varepsilon \mathfrak{o}$ and we may as well assume the former. Then we have $\alpha = \beta\gamma$ where $\gamma \varepsilon \mathfrak{o}$ and $\varphi'(\alpha) = \varphi'(\beta)\varphi'(\gamma) \leq \varphi'(\beta)$ since $\varphi'(\gamma) = \gamma U \leq 1 = U$. Also $\alpha\beta^{-1} + 1 \varepsilon \mathfrak{o}$, so $\varphi'(\alpha\beta^{-1} + 1) \leq 1$ and $\varphi'(\alpha + \beta) = \varphi'(\alpha\beta^{-1} + 1)\varphi'(\beta) \leq \varphi'(\beta) = \max(\varphi'(\alpha), \varphi'(\beta))$. Hence (iii) holds. It is clear from (28) and the definition of G' and H' that $\varphi'(\alpha) \leq 1$ is equivalent to $\alpha \varepsilon \mathfrak{o}$. Hence $\mathfrak{o}$ is the valuation ring of the valuation φ'. We shall call the valuation φ' the *canonical valuation* of the valuation ring $\mathfrak{o}$.

Now consider again an arbitrary valuation φ of Φ into $V = (G, 0)$ where G is a commutative group ordered by H. Let $\mathfrak{o}$ be the valuation ring of φ and φ' the canonical valuation of Φ into $V' = (G', 0)$ where $G' = \Phi^*/U$ is ordered by $H' = \{\beta U | \beta \varepsilon \mathfrak{p}^*\}$. The defin̄ition (28) gives $\varphi'(0) = 0$, $\varphi'(\alpha) = \alpha U$ if $\alpha \neq 0$. We have the homomorphism $\alpha \rightarrow \varphi(\alpha)$ of the multiplicative group Φ^* into G whose kernel is the subgroup U. Hence we have the induced isomorphism $\eta: \varphi'(\alpha) = \alpha U \rightarrow \varphi(\alpha)$ of $G' = \Phi^*/U$ into G. This is an order isomorphism since, if $\beta U \varepsilon H'$, then $\beta \varepsilon \mathfrak{p}^*$, so $\varphi(\beta) < 1$. We now see that the given valuation can be factored as $\varphi = \varphi'\eta$ where η is an order isomorphism of G' into G (more precisely, V' into V).

These considerations make it natural to lump together the valuations of Φ which have the same valuation ring $\mathfrak{o}$. Accordingly, we shall say that such valuations are *equivalent*.

There is a third concept, that of a place which is also equivalent to the concepts of valuation and valuation ring. We define this as follows:

Definition 9. *If Φ is a field, a* place $\mathscr{P}$ *is a homomorphism of a subring $\mathfrak{o}$ of Φ into a field Δ such that, if $\alpha \not\varepsilon \mathfrak{o}$ then $\alpha^{-1} \varepsilon \mathfrak{o}$ and $\mathscr{P}(\alpha^{-1}) = 0$. (We recall that $1 \varepsilon \mathfrak{o}$ and $\mathscr{P}(1) = 1$ by our conventions on subrings and homomorphisms.)*

It is clear from the definition that, if $\mathscr{P}$ is a place, then the subring $\mathfrak{o}$ given by $\mathscr{P}$ is a valuation ring. On the other hand, suppose $\mathfrak{o}$ is any valuation ring and let $\mathfrak{p}$ be the set of non-units of $\mathfrak{o}$. Then it is clear that, if $\beta \varepsilon \mathfrak{p}$ and $\alpha \varepsilon \mathfrak{o}$, $\alpha\beta \varepsilon \mathfrak{p}$. In particular, $-\beta = (-1)\beta \varepsilon \mathfrak{p}$. If β_1 and $\beta_2 \varepsilon \mathfrak{p}$, we may assume that $\beta_1\beta_2^{-1} \varepsilon \mathfrak{o}$.

Then $\beta_1\beta_2^{-1} + 1 \in \mathfrak{o}$, so $\beta_1 + \beta_2 = (\beta_1\beta_2^{-1} + 1)\beta_2 \in \mathfrak{p}$. Hence $\mathfrak{p}$ is an ideal in $\mathfrak{o}$. Since $\mathfrak{p}$ is the set of non-units of $\mathfrak{o}$, it is clear that $\mathfrak{p}$ is maximal and $\Delta' = \mathfrak{o}/\mathfrak{p}$ is a field. Let $\mathscr{P}'$ be the canonical homomorphism of $\mathfrak{o}$ onto $\Delta' = \mathfrak{o}/\mathfrak{p}$. Then it is clear that $\mathscr{P}'$ and $\mathfrak{o}$ satisfy the defining conditions for a place. We shall call this place the *canonical place* of the valuation ring $\mathfrak{o}$. The image of $\mathfrak{o}$ under $\mathscr{P}'$ is $\Delta' = \mathfrak{o}/\mathfrak{p}$ where $\mathfrak{p}$ is the ideal of non-units of $\mathfrak{o}$. As in the special case of real non-archimedean valuation, we shall call Δ' the *residue field* of the valuation ring $\mathfrak{o}$.

Now consider again an arbitrary place $\mathscr{P}$ of Φ into the field Δ and let $\mathfrak{o}$ be the valuation ring on which $\mathscr{P}$ is defined. Let $\mathfrak{p}$ be the ideal of non-units of $\mathfrak{o}$. If $\alpha \in \mathfrak{p}$ and $\alpha \neq 0$, then $\alpha^{-1} \notin \mathfrak{o}$, so the hypothesis on $\mathscr{P}$ gives $\mathscr{P}(\alpha) = 0$. This holds also if $\alpha = 0$. Hence we see that $\mathfrak{p}$ is contained in the kernel of $\mathscr{P}$. Since $\mathfrak{p}$ is a maximal ideal, this shows that $\mathfrak{p}$ is the kernel of $\mathscr{P}$. The homomorphism $\alpha \to \mathscr{P}(\alpha)$, $\alpha \in \mathfrak{o}$, therefore gives an isomorphism $\mathscr{P}'(\alpha) = \alpha + \mathfrak{p} \to \mathscr{P}(\alpha)$, and so the place $\mathscr{P}$ is the resultant of the canonical place $\mathscr{P}'$ and an isomorphism of Δ' into Δ. As for valuations, it is natural to consider as *equivalent* places that have the same valuation ring.

We have now established the procedures for passing from one of the concepts: valuation, valuation ring, place, to any other. Clearly, a result on one of these can be translated to the other two. In the sequel we shall apply this idea to obtain extensions of valuations via extensions of places. The latter amounts to extensions of homomorphisms, for which we have available the basic extension theorems of the Introduction.

EXERCISES

1. Let $\mathscr{P}$ be a place on Φ with values in Δ. Adjoin a new element ∞ to Δ and define $\infty + \delta = \infty = \delta + \infty$, $\delta \in \Delta$, $\infty\infty = \infty$, $\infty\delta = \infty = \delta\infty$ if $\delta \neq 0$ in Δ. Extend $\mathscr{P}$ to the whole of Φ by defining $\mathscr{P}(\alpha) = \infty$ if $\alpha \notin \mathfrak{o}$. Verify that

$$\mathscr{P}(\alpha + \beta) = \mathscr{P}(\alpha) + \mathscr{P}(\beta)$$
$$\mathscr{P}(\alpha\beta) = \mathscr{P}(\alpha)\mathscr{P}(\beta) \tag{29}$$

whenever the right-hand sides are defined. Conversely, assume that $\mathscr{P}$ is a function defined on Φ with values in (Δ, ∞), Δ a field where $\Delta \cap \{\infty\} = \varnothing$ and ∞ obeys the rules indicated. Assume (29) hold whenever the right-hand sides are defined. Let $\mathfrak{o}$ be the inverse image $\mathscr{P}^{-1}(\Delta)$. Show that the restriction of $\mathscr{P}$ to $\mathfrak{o}$ is a place. This gives an alternative definition of a place.

2. Let $\mathscr{P}$ be a place with valuation ring $\mathfrak{o}$. Assume $\mathscr{P}$ is an isomorphism. Show that $\mathfrak{o} = \Phi$ and that the canonical valuation of $\mathfrak{o}$ is trivial in the sense that $\varphi'(0) = 0$, $\varphi'(\alpha) = 1$ if $\alpha \neq 0$.

10. Characterization of real non-archimedean valuations. In order to apply the general theory of valuations to the case of non-archimedean real valuations it is necessary to characterize these among all possible valuations of a field. In view of the foregoing discussion this is equivalent to the problem of characterizing the ordered groups which are order isomorphic to subgroups of the multiplicative group of positive reals or, equivalently, to the additive group of all the real numbers with the usual order in this group. Hence we seek a characterization of the ordered groups which are order isomorphic to subgroups of the additive group of real numbers. It will be convenient to use the additive notation in all the groups which we shall consider in this section.

Let G be an ordered group: If $a \in G$, we define $|a| = a$ if $a \geq 0$ and $|a| = -a$ if $a < 0$. We define an *isolated subgroup* K of G as a subgroup such that, if $a \in K$ and $|b| \leq |a|$, then $b \in K$. Let K_1 and K_2 be isolated subgroups. Then we assert that either $K_1 \subseteq K_2$ or $K_2 \subseteq K_1$. For, if neither of these inclusions holds, then there exists a $b_1 \in K_1$, $\notin K_2$, and a $b_2 \in K_2$, $\notin K_1$, and we may suppose that $b_i > 0$. If $b_2 \geq b_1$, then $b_1 \in K_2$ contrary to assumption. Hence $b_2 \ngeq b_1$ and similarly $b_1 \ngeq b_2$ which contradicts the fact that G is an ordered group. Thus we have either $K_1 \supseteq K_2$ or $K_2 \supseteq K_1$ so the set of isolated subgroups is linearly ordered by the inclusion relation. The order type of the set of isolated subgroups is called the *rank* of G.* The simplest situation is that of a group of *rank one* in which $G \neq 0$ and G has no isolated subgroup $\neq 0, G$. These groups can be characterized by the archimedean property which is familiar for real numbers:

Lemma. *An ordered group G ($\neq 0$) is of rank one if and only if given any $a, b \in G$ with $a > 0$ there exists a positive integer n such that $na > b$.*

Proof. Suppose first that G contains two elements a, b such that $a > 0$ and $na \leq b$ for all positive integers n. Let K_+ denote the subset of G of elements u such that $0 < u < ma$ for some positive integer m. K_+ is not vacuous since $a < 2a$ and clearly

* Cf., for example, F. Hausdorff, *Mengenlehre*, 3rd Ed., Chap. 3, de Gruyter & Co., 1937.

K_+ is closed under addition. Moreover, K_+ contains every v such that $0 < v < u$ for some u in K_+. Hence if u_1 and $u_2 \in K_+$ and $u_1 < u_2$, then $0 < u_2 - u_1 < u_2$ so $u_2 - u_1 \in K_+$. It follows that the union of K_+, 0, and $-K_+$, the set of negatives of the elements of K_+, is a subgroup K of G. Now K is isolated, since, if $u \in K$ and $u > 0$, then every v such that $0 < v \leq u$ is in K. Also $K \neq G$ since $b \notin K$. Hence G is not of rank one. Conversely, assume G not of rank one and let K be an isolated subgroup $\neq 0$, G. Since $K \neq G$ there exists a positive element b such that $b > a$ for every $a \in K$. Choose $a > 0$ in K, then $na < b$ for all $n = 1, 2, 3, \cdots$. Hence the archimedean property fails in G.

It is clear from this criterion that if G is of rank one, then any non-zero subgroup of G is of rank one. In particular, any non-zero subgroup of the additive group of real numbers is of rank one. Moreover, these are essentially all the ordered groups of rank one, since we have the following

Theorem 8. *Any ordered group G of rank one is order isomorphic to a subgroup of the additive group of real numbers.*

Proof. We shall define an order isomorphism η of G into the additive group R of real numbers. For this purpose we choose a $u > 0$ in G. If $v > 0$, then there exist pairs (m, n) of positive integers m, n such that $nv \geq mu$. Thus we may take $m = 1$ and, by the archimedean property, determine n so that $nv > u = 1u$. If $q \in P$, the collection of positive integers, then $qnv \geq qmu$ if and only if $nv \geq mu$. Hence if $r = m/n = m'/n'$, $m, n, m', n' \in P$, then $nv \geq mu$ if and only if $n'v \geq m'u$. The rational numbers $r = m/n$ satisfying this condition form a set which we denote as R_v. If $r = m/n$ and $s = m'/n' < r$, $m', n' \in P$, then $m'n < mn'$. If $r \in R_v$, then $nv \geq mu$ and $nn'v \geq mn'u > m'nu$. Hence $n'v > m'u$ so that $s \in R_v$. We note next that the set of positive rationals R_v is bounded above. Otherwise, the result just proved implies that R_v is the complete set of positive rationals. Hence every positive integer k is in R_v which means that $v \geq ku$, $k \in P$. This contradicts the archimedean property of G. We now define v^η to be the positive real number sup R_v. Since R_v contains every $s \leq r$ for every $r \in R_v$, it is clear that R_v and its complementary set R_v' in the set of positive rationals defines a Dedekind cut. Hence

$\sup R_v = \inf R_v'$. Now let v_1, v_2 be positive elements of G and let $m_1/n_1 \in R_{v_1}$, $m_2/n_2 \in R_{v_2}$ where $m_i, n_i \in P$. Then $n_1v_1 \geq m_1u$, $n_2v_2 \geq m_2u$ and $n_1n_2v_1 \geq m_1n_2u$, $n_1n_2v_2 \geq n_1m_2u$. Hence $n_1n_2(v_1 + v_2) \geq (m_1n_2 + m_2n_1)u$ and so $m_1/n_1 + m_2/n_2 \in R_{v_1+v_2}$. This implies that $(v_1 + v_2)^\eta \geq v_1^\eta + v_2^\eta$. On the other hand, a repetition of the argument just given shows that, if $m_1/n_1 \in R_{v_1}'$ (that is, $n_1v_1 < m_1u$) and $m_2/n_2 \in R_{v_2}'$, then $m_1/n_1 + m_2/n_2 \in R_{v_1+v_2}'$. Since $v^\eta = \inf R_v'$, this implies that $(v_1 + v_2)^\eta \leq v_1^\eta + v_2^\eta$. Hence

$$(30) \qquad (v_1 + v_2)^\eta = v_1^\eta + v_2^\eta$$

holds for v_1, v_2 positive in G. We extend the mapping η to all of G by defining $0^\eta = 0$ and $(-v)^\eta = -v^\eta$ if v is positive. It is immediate that (30) holds if either $v_1 \geq 0$, $v_2 \geq 0$, or $v_1 \leq 0$, $v_2 \leq 0$. Suppose $v_1 > 0$ and $v_2 < 0$. If $v_1 + v_2 \geq 0$, we write $v_1 = (v_1 + v_2) + (-v_2)$ and obtain $v_1^\eta = (v_1 + v_2)^\eta + (-v_2)^\eta = (v_1 + v_2)^\eta - v_2^\eta$. Then $(v_1 + v_2)^\eta = v_1^\eta + v_2^\eta$. If $v_1 + v_2 < 0$, then we write $-v_2 = -(v_1 + v_2) + v_1$ and obtain $(-v_2)^\eta = (-(v_1 + v_2))^\eta + v_1^\eta$. Thus $-v_2^\eta = -(v_1 + v_2)^\eta + v_1^\eta$ and again (30) holds. Similarly, (30) holds if $v_1 < 0$ and $v_2 > 0$. Thus η is a group homomorphism of G into R. If $v > 0$, then $v^\eta > 0$; hence no positive element is in the kernel of η. It follows that the kernel is 0 and η is an isomorphism. Since positive elements are mapped into positive elements by η, η is an order isomorphism of G into R.

There are several observations which should be made on the foregoing proof. In the first place it is clear from the definition of the isomorphism η that $u^\eta = 1$. We note next that η is determined by this property, that is, if ζ is any order isomorphism of G into R such that $u^\zeta = 1$, then $\zeta = \eta$. Thus let $b > 0$ and let m/n, m, n positive integers, satisfy $m/n \geq v^\eta$. Then $m1 \geq nv^\eta$ and $mu^\eta \geq nv^\eta$, $(mu)^\eta \geq (nv)^\eta$. Hence $mu \geq nv$ and re-tracing the steps we obtain $m/n \geq v^\zeta$. Similarly, $m/n \geq v^\zeta$ implies $m/n \geq v^\eta$. Since this holds for arbitrary rationals it follows that $v^\eta = v^\zeta$; hence $\eta = \zeta$. If β is any positive real number, then the mapping $x \to \beta x$ is an order preserving automorphism of R mapping $1 \to \beta$. It follows from this that there exists an order isomorphism of G mapping the given positive element u into any positive β in R. Moreover, such an isomorphism is unique.

A group of rank one is called *discrete* if it is order isomorphic to the ordered group of integers (positivity as usual). We have noted before (§ 5) that a subgroup of the multiplicative group of positive reals is discrete if and only if it contains a largest element < 1. This and Theorem 8 imply that an ordered group of rank one is discrete if and only if it contains a least positive element.

EXERCISES

1. Let $R^{(n)}$ denote the additive group of n-tuples $x = (\zeta_1, \cdots, \zeta_n)$ of real numbers ζ_i. Define the set of positive elements of $R^{(n)}$ by the condition that $x > 0$ if the first non-zero ζ_i is > 0. Show that this gives an ordered group. Determine the isolated subgroups.

2. Call an ordered group G of *rank* n, n a positive integer, if n is the cardinal number of the set of non-zero isolated subgroups. Show that any ordered group of rank n is order isomorphic to a subgroup of the group $R^{(n)}$ of ex. 1.

3. Call an ordered group G of rank n *discrete* if the factor groups of successive isolated subgroups are all infinite cyclic groups. Show that any such group is isomorphic to the subgroup of $R^{(n)}$ of n-tuples $a = (\alpha_1, \alpha_2, \cdots, \alpha_n)$ such that the α_i are integers.

11. Extension of homomorphisms and valuations. In this section we shall prove a fundamental theorem on extension of a homomorphism defined on a subring of a field. This result leads to a general theorem on extension of valuations from a subfield to a field. We prove first the following key lemma.

Lemma 1. *Let* $\mathfrak{o}$ *be a subring of a field* Φ *and let* $\mathfrak{m}$ *be a proper ideal in* $\mathfrak{o}$. *If* α *is a non-zero element of* Φ *and* $\mathfrak{o}[\alpha]$ *is the subring of* Φ *generated by* $\mathfrak{o}$ *and* α, *then either* $\mathfrak{m}\mathfrak{o}[\alpha]$, *the ideal generated by* $\mathfrak{m}$ *in* $\mathfrak{o}[\alpha]$, *is proper in* $\mathfrak{o}[\alpha]$ *or* $\mathfrak{m}\mathfrak{o}[\alpha^{-1}]$ *is proper in* $\mathfrak{o}[\alpha^{-1}]$.

Proof. Suppose the contrary: $\mathfrak{m}\mathfrak{o}[\alpha] = \mathfrak{o}[\alpha]$, $\mathfrak{m}\mathfrak{o}[\alpha^{-1}] = \mathfrak{o}[\alpha^{-1}]$. Then $1 \in \mathfrak{m}\mathfrak{o}[\alpha]$ and $1 \in \mathfrak{m}\mathfrak{o}[\alpha^{-1}]$, so we have relations of the form:

$$1 = \mu_0\alpha^m + \mu_1\alpha^{m-1} + \cdots + \mu_m, \quad \mu_i \in \mathfrak{m}, \tag{31}$$

$$1 = \nu_0\alpha^{-n} + \nu_1\alpha^{-(n-1)} + \cdots + \nu_n, \quad \nu_j \in \mathfrak{m}. \tag{32}$$

Since $\mathfrak{m} \neq \mathfrak{o}$, we have $m > 0$ and $n > 0$ and we may assume m, n are minimal for the relations (31) and (32). Also we may assume $m \geq n$. Then (32) implies that $\alpha^m = \nu_0\alpha^{m-n} + \nu_1\alpha^{m-n+1} + \cdots + \nu_n\alpha^m$; hence

$$\alpha^m(1 - \nu_n) = \nu_0\alpha^{m-n} + \cdots + \nu_{n-1}\alpha^{m-1}. \tag{33}$$

Multiplication of (31) by $1 - \nu_n$ gives

$$(34)\quad 1 - \nu_n = \mu_0(1 - \nu_n)\alpha^m + \mu_1(1 - \nu_n)\alpha^{m-1} + \cdots + \mu_m(1 - \nu_n).$$

Hence, by (33),

$$1 - \nu_n = \mu_0(\nu_0\alpha^{m-n} + \cdots + \nu_{n-1}\alpha^{m-1}) + \mu_1(1 - \nu_n)\alpha^{m-1} + \cdots + \mu_m(1 - \nu_n).$$

Since the $\mu_i, \nu_j \in \mathfrak{m}$, this gives another relation like (31) with m replaced by $m - 1$ contrary to the minimality of m. Hence the proof is complete.

If $\mathscr{P}$ is a place which is a homomorphism of a subring $\mathfrak{o}$ of the field Φ into the field Δ, then we shall say that $\mathscr{P}$ is Δ-*valued*. Our main result is an extension theorem for homomorphisms to places, as follows.

Theorem 9. *Let $\mathfrak{o}_0$ be a subring of a field Φ and let $\mathscr{P}_0$ be a homomorphism of $\mathfrak{o}_0$ into an algebraically closed field Ω. Then $\mathscr{P}_0$ can be extended to an Ω-valued place $\mathscr{P}$ on Φ.*

Proof. We consider the collection of extensions $\mathscr{P}'$ of the homomorphism $\mathscr{P}_0$ where $\mathscr{P}'$ is a homomorphism into Ω of a subring $\mathfrak{o}'$ of Φ containing $\mathfrak{o}_0$. These can be partially ordered in the usual manner: $\mathscr{P}' < \mathscr{P}''$ if $\mathscr{P}''$ is an extension of $\mathscr{P}'$. Then, as usual, we can apply Zorn's lemma to obtain a maximal extension $\mathscr{P}$ which is defined on a subring $\mathfrak{o}$ of Φ. The proof will be completed by showing that $\mathfrak{o}$ is a valuation ring. Then $\mathscr{P}$ will be an Ω-valued place for Φ. Let $\mathfrak{m}$ be the kernel of $\mathscr{P}$. Since $1 \rightarrow 1$, $\mathfrak{m} \neq \mathfrak{o}$. Since Ω has no zero-divisors $\neq 0$, $\mathfrak{m}$ is a prime ideal on $\mathfrak{o}$. Consequently, the complementary set M of $\mathfrak{m}$ in $\mathfrak{o}$ is multiplicatively closed and $0 \notin M$. Let $\mathfrak{o}'$ be the subset of Φ of elements of the form $\alpha\beta^{-1}$ where $\alpha, \beta \in \mathfrak{o}$ and $\beta \in M$. Then $\mathfrak{o}'$ is a subring of Φ containing $\mathfrak{o}$ and $\mathscr{P}$ can be extended to a homomorphism $\mathscr{P}'$ of $\mathfrak{o}'$ into Ω by defining $\mathscr{P}'(\alpha\beta^{-1}) = \mathscr{P}(\alpha)\mathscr{P}(\beta)^{-1}$ (I of Introd.). Since $\mathscr{P}$ is maximal we have $\mathfrak{o}' = \mathfrak{o}$. This implies that the image of $\mathfrak{o}$ under $\mathscr{P}$ is a subfield E of Ω; for, if $0 \neq \gamma = \mathscr{P}(\beta)$, $\beta \in \mathfrak{o}$, then $\beta \in M$, so $\beta^{-1} \in \mathfrak{o}' = \mathfrak{o}$ and $\gamma^{-1} = \mathscr{P}(\beta^{-1})$ is in the image of $\mathfrak{o}$. Now let α be any element $\neq 0$ of Φ. We shall show that either α or $\alpha^{-1} \in \mathfrak{o}$, which is what is needed to prove that $\mathfrak{o}$ is a valuation ring and $\mathscr{P}$ is a place.

Now Lemma 1 shows that $\mathfrak{m}\mathfrak{o}[\alpha] \subset \mathfrak{o}[\alpha]$ or $\mathfrak{m}\mathfrak{o}[\alpha^{-1}] \subset \mathfrak{o}[\alpha^{-1}]$ and we may as well assume the former. Then we shall show that $\mathscr{P}$ can be extended to a homomorphism of $\mathfrak{o}[\alpha]$ into Ω. This and the maximality of $\mathscr{P}$ will imply that $\alpha \in \mathfrak{o}$. We consider the polynomial rings $\mathfrak{o}[x]$ and $E[x]$, x an indeterminate, and we extend $\mathscr{P}$ to a homomorphism of $\mathfrak{o}[x]$ onto $E[x]$ sending $x \to x$. Let $\mathfrak{A}$ be the ideal of polynomials $g(x) \in \mathfrak{o}[x]$ such that $g(\alpha) = 0$ and let $\mathfrak{A}'$ be its image in $E[x]$ under the extension of $\mathscr{P}$. Since the homomorphism of $\mathfrak{o}[x]$ is surjective, $\mathfrak{A}'$ is an ideal in $E[x]$. Also $\mathfrak{A}' \subset E[x]$. Otherwise, there exists a polynomial $\sum_0^r \beta_i x^i \in \mathfrak{o}[x]$ such that $\Sigma\beta_i\alpha^i = 0$ and $\sum_0^r \mathscr{P}(\beta_i)x^i = 1$. Then $\mathscr{P}(\beta_0) = 1$ and $\mathscr{P}(\beta_i) = 0$ if $i > 0$, so $1 - \beta_0 \in \mathfrak{m}$ and $\beta_i \in \mathfrak{m}$ for $i > 0$. Then the relation $\Sigma\beta_i\alpha^i = 0$ gives $1 = 1 - \Sigma\beta_i\alpha^i = (1 - \beta_0) + \sum_{i>0} (-\beta_i)\alpha^i$. Since $1 - \beta_0$, $\beta_i \in \mathfrak{m}$, this implies that $1 \in \mathfrak{m}\mathfrak{o}[\alpha]$ contrary to hypothesis. Hence we see that $\mathfrak{A}'$ is a proper ideal in $E[x]$ and, since $E[x]$ is a principal ideal domain, $\mathfrak{A}' = (f(x))$ where $f(x)$ is either 0 or a polynomial of positive degree. In the first case, we choose any element γ in Ω and in the second case we choose $\gamma \in \Omega$ so that $f(\gamma) = 0$. This can be done since Ω is algebraically closed. Now our choice of γ amounts to this: If $g(x)$ is any polynomial in $\mathfrak{o}[x]$ such that $g(\alpha) = 0$, then $g^{\mathscr{P}}(\gamma) = 0$ for the image $g^{\mathscr{P}}[x]$ in $E[x]$. Hence the extension theorem IV′ of the Introduction shows that $\mathscr{P}$ can be extended to a homomorphism of $\mathfrak{o}[\alpha]$ into Δ sending α into γ. This completes the proof.

Suppose now that φ_0 is a valuation of a subfield Φ_0 of the field Φ. Let $\mathfrak{o}_0$ be the valuation ring of φ_0, $\mathfrak{p}_0$ the ideal of non-units, U_0 the multiplicative group of units of $\mathfrak{o}_0$. We have seen that φ_0 is equivalent to the canonical valuation φ_0' into the group Φ_0^*/U_0 where the positive elements of this group are the cosets $\beta_0 U_0$, $\beta_0 \neq 0$ in $\mathfrak{p}_0$. We also have the canonical place $\mathscr{P}_0'$ of Φ_0 determined by $\mathfrak{o}_0$. This is the homomorphism $\alpha_0 \to \alpha_0 + \mathfrak{p}_0$ of $\mathfrak{o}_0$ into the residue field $\mathfrak{o}_0/\mathfrak{p}_0$. We can imbed $\mathfrak{o}_0/\mathfrak{p}_0$ in an algebraically closed field Ω. Then $\mathscr{P}_0'$ can be considered as an Ω-valued place $\mathscr{P}_0$ on Φ_0. Since Ω is algebraically closed, the extension theorem states that $\mathscr{P}_0$ can be extended to an Ω-valued place $\mathscr{P}$ on

Φ. Let $\mathfrak{o}$ be the valuation ring in Φ on which $\mathscr{P}$ is defined and let $\mathfrak{p}$ be the ideal of non-units of $\mathfrak{o}$. Since $\mathscr{P}$ is an extension of $\mathscr{P}_0, \mathfrak{o} \supseteq \mathfrak{o}_0$ and since $\mathfrak{p}$ and $\mathfrak{p}_0$ are respectively the kernels of $\mathscr{P}$ and $\mathscr{P}_0$, $\mathfrak{p} \supseteq \mathfrak{p}_0$. Hence we have $\mathfrak{o} \cap \Phi_0 \supseteq \mathfrak{o}_0$ and $\mathfrak{p} \cap \Phi_0 \supseteq \mathfrak{p}_0$. If $\beta \in \mathfrak{o} \cap \Phi_0$ and $\beta \notin \mathfrak{o}_0$, then $\beta^{-1} \in \mathfrak{p}_0 \subseteq \mathfrak{p}$, but this implies that $\beta \notin \mathfrak{o}$. Hence $\mathfrak{o} \cap \Phi_0 = \mathfrak{o}_0$. Since $\mathfrak{p}$ and $\mathfrak{p}_0$ are the ideals of non-units of $\mathfrak{o}$ and $\mathfrak{o}_0$ respectively, the relation $\mathfrak{o} \cap \Phi_0 = \mathfrak{o}_0$ implies $\mathfrak{p} \cap \Phi_0 \subseteq \mathfrak{p}_0$. Hence $\mathfrak{p} \cap \Phi_0 = \mathfrak{p}_0$ and $U \cap \Phi_0 = U_0$ where U is the set of units of $\mathfrak{o}$. These relations imply that $\beta_0 U_0 \rightarrow \beta_0 U$, $\beta_0 \in \Phi_0{}^*$, is an order isomorphism of the ordered group $\Phi_0{}^*/U_0$ into Φ^*/U ordered by the set of elements βU, $\beta \in \mathfrak{p}$. If we apply this isomorphism to the canonical valuation $\varphi_0{}'$, we obtain an equivalent valuation $\varphi_0{}''$ of Φ_0 into the group Φ^*/U. We also have the canonical valuation φ' of Φ into Φ^*/U and the definitions show that φ' is an extension of the valuation $\varphi_0{}''$. In this sense we have obtained an "extension" of the given valuation of Φ_0 to a valuation on Φ.

We shall be interested particularly in the case in which Φ is finite dimensional over Φ_0 and the given valuation φ_0 is of rank 1. In the general case, if φ is a valuation of a field Φ into $V = (G, 0)$, then the subgroup of G of values $\varphi(\alpha)$, $\alpha \neq 0$ in Φ, is called the *value group* of φ. We shall need the following

Lemma 2. *Let φ be a valuation of a field Φ, Φ_0 a subfield of finite co-dimension in Φ. Then the value group of Φ is order isomorphic to a subgroup of the value group of Φ_0 (relative to the restriction of φ).*

Proof. Let $\xi \in \Phi$ and let $\alpha_1\xi^{n_1} + \alpha_2\xi^{n_2} + \cdots + \alpha_k\xi^{n_k} = 0$ where the $\alpha_i \neq 0$ in Φ_0 and $n_1 > n_2 > \cdots > n_k$. As in the case of non-archimedean real valuations, if $\varphi(\beta_1) > \varphi(\beta_j)$, $j \neq 1$, then $\varphi(\Sigma\beta_i) = \varphi(\beta_1)$. Hence our relation implies that there exist $i < j$ such that $\varphi(\alpha_i\xi^{n_i}) = \varphi(\alpha_j\xi^{n_j})$. Then $\varphi(\xi^{n_i - n_j}) = \varphi(\alpha_j\alpha_i{}^{-1})$. If $[\Phi:\Phi_0] = n$, then we may assume that $n_i - n_j \leq n$; hence $\varphi(\xi)^{n!}$ is in the value group of Φ_0. This shows that for any a in the value group G of Φ, $a^{n!}$ is in the value group G_0 of Φ_0. On the other hand, we have seen that $a \rightarrow a^{n!}$ is an order preserving isomorphism of G onto a subgroup. Hence G is order isomorphic to a subgroup of G_0.

This result and Theorem 8 imply that the value group of Φ is of rank 1 (discrete of rank 1) if and only if the same is true for the value group of Φ_0.

We shall now see how all of this applies to real valuations. Let φ_0 be a non-trivial non-archimedean valuation of a field Φ_0 into the non-negative reals and let Φ be a finite dimensional extension of Φ_0. Then we know that $\varphi_0 = \varphi_0'\eta$ where φ_0' is the canonical valuation of Φ_0 associated with the valuation ring $\mathfrak{o}_0$ of φ_0 and η is an order isomorphism of the value group G_0' of φ_0' into the positive reals P. Also we have just seen that we have a valuation ring $\mathfrak{o}$ of Φ and an order isomorphism ζ of G_0' into the value group G' of the canonical valuation φ' determined by $\mathfrak{o}$ such that $\varphi'(\alpha_0) = (\varphi'\zeta)(\alpha_0)$ for all $\alpha_0 \varepsilon \Phi_0$. Since G_0' is of rank 1 the same is true of G' and consequently we have an order isomorphism λ of G' into P. Thus we have the following diagram of mappings:

$$\begin{array}{ccccc} \Phi_0 & \xrightarrow{\varphi_0'} & G_0' & \xrightarrow{\eta} & P \\ \downarrow{\scriptstyle i} & & \downarrow{\scriptstyle \zeta} & & \downarrow{\scriptstyle 1} \\ \Phi & \xrightarrow[\varphi']{} & G' & \xrightarrow[\lambda]{} & P \end{array}$$

where i is the inclusion mapping and the first rectangle is commutative: $i\varphi' = \varphi_0'\zeta$. Assume $G_0' \neq 1$ and let δ be some element $\neq 1$ in G_0'. Then we can choose λ so that $\delta^{\zeta\lambda} = \delta^{\eta}$ and then we shall have $\gamma_0'^{\zeta\lambda} = \gamma_0'^{\eta}$ for all $\gamma_0' \varepsilon G_0'$ (§ 10). This means that the second rectangle in our diagram is also commutative. Then $\varphi = \varphi'\lambda$ is a real non-archimedean valuation which extends the given valuation φ_0 on Φ_0; for if $\alpha_0 \varepsilon \Phi_0$, then $\varphi_0(\alpha_0) = (\varphi_0'\eta)(\alpha_0) = (\varphi_0'\zeta\lambda)(\alpha_0) = (\varphi'\lambda)(\alpha_0) = \varphi(\alpha_0)$. If $G_0' = 1$, then Lemma 2 shows that necessarily $G' = 1$. Then η and λ are unique and commutativity holds. This case is, of course, trivial at the outset, since it is the one in which φ_0 is a trivial valuation. We have therefore proved the following

Theorem 10. *Let φ_0 be a non-archimedean real valuation on a field Φ_0 and let Φ be a finite dimensional extension field of Φ. Then there exists a real valuation on Φ which is an extension of φ.*

12. Application of the extension theorem: Hilbert Nullstellensatz. Before continuing our study of valuations we digress slightly to take up some important applications of the homomorphism extension theorem (Th. 9). The first of these, Hilbert's Nullstellensatz, plays an important role in algebraic geometry. We shall give it in its original ideal-theoretic form.

We consider a polynomial algebra $\Phi[x_1, x_2, \cdots, x_n]$ in indeterminates x_i over a field Φ. Let Ω be the algebraic closure of Φ. If $f(x_1, \cdots, x_n) \varepsilon \Phi[x_1, \cdots, x_n]$ and the ξ_i are elements of Ω such that $f(\xi_1, \cdots, \xi_n) = 0$, then we shall call $(\xi_1, \cdots, \xi_n)$ an *(algebraic) zero* of $f(x_1, \cdots, x_n)$. If S is a set of polynomials contained in $\Phi[x_1, \cdots, x_n]$, then we define a *zero of* S to be an n-tuple $(\xi_1, \cdots, \xi_n)$, $\xi_i \varepsilon \Omega$, which is a zero for every $f \varepsilon S$. Our main result concerns the zeros of a proper prime ideal $\mathfrak{P}$ in $\Phi[x_1, \cdots, x_n]$. This is the following

Theorem 11. *Let $\mathfrak{P}$ be a prime ideal in $\Phi[x_1, \cdots, x_n]$, Φ a field, and suppose $\mathfrak{P} \neq (1)$ $(= \Phi[x_1, \cdots, x_n])$. Let $g(x_1, \cdots, x_n)$ be a polynomial not contained in $\mathfrak{P}$. Then there exist ξ_i in the algebraic closure Ω of Φ such that $(\xi_1, \cdots, \xi_n)$ is a zero for $\mathfrak{P}$ and is not a zero for $g(x_1, \cdots, x_n)$.*

Proof. Since $\mathfrak{P} \neq (1)$, $\Phi[x_1, \cdots, x_n]/\mathfrak{P}$ is an algebra over Φ which is $\neq 0$ and this is generated over Φ by the cosets $\gamma_i = x_i + \mathfrak{P}$, $i = 1, 2, \cdots, n$. Also $\Phi[\gamma_1, \cdots, \gamma_n] = \Phi[x_1, \cdots, x_n]/\mathfrak{P}$ is an integral domain so this can be imbedded in its field of fractions $\mathrm{P} = \Phi(\gamma_1, \gamma_2, \cdots, \gamma_n)$. Suppose first that all the γ_i are algebraic. Then P is an algebraic extension of Φ so we have an isomorphism of P/Φ into the algebraic closure Ω/Φ. Suppose $\gamma_i \rightarrow \xi_i$ in this isomorphism. Then if $f(x_1, \cdots, x_n) \varepsilon \mathfrak{P}, f(\gamma_1, \cdots, \gamma_n) = 0$ and so $f(\xi_1, \cdots, \xi_n) = 0$. Hence $(\xi_1, \xi_2, \cdots, \xi_n)$ is a zero of $\mathfrak{P}$. On the other hand, $g(x_1, \cdots, x_n) \notin \mathfrak{P}$ so $g(\gamma_1, \cdots, \gamma_n) \neq 0$; hence $g(\xi_1, \cdots, \xi_n) \neq 0$. This proves the theorem in this case. Next assume that not all the γ_i are algebraic. We may suppose that the γ's are ordered so that $\{\gamma_1, \gamma_2, \cdots, \gamma_r\}$ $(r \geq 1)$ is a transcendency basis for P/Φ. Since $g(x_1, \cdots, x_n) \notin \mathfrak{P}, g(\gamma_1, \cdots, \gamma_n) \neq 0$ in P so $g(\gamma_1, \cdots, \gamma_n)^{-1}$ exists in P. This element and the elements $\gamma_{r+1}, \cdots, \gamma_n$ are algebraic over $\Phi(\gamma_1, \cdots, \gamma_r)$ and so they satisfy algebraic equations of the form

$$(35)\quad a_0(\gamma_1, \cdots, \gamma_r)x^m + a_1(\gamma_1, \cdots, \gamma_r)x^{m-1} + \cdots + a_m(\gamma_1, \cdots, \gamma_r) = 0$$

where the a_i are polynomials in the γ_j, $j = 1, \cdots, r$, and $a_0(\gamma_1, \cdots, \gamma_r) \neq 0$. For each $\gamma_{r+1}, \cdots, \gamma_n$ and $g(\gamma_1, \cdots, \gamma_n)^{-1}$ we choose such an equation and we let $a(\gamma_1, \cdots, \gamma_r)$ be the product of the leading coefficients of these equations. Since $a(x_1, \cdots, x_r) \neq 0$, we may choose $\xi_1, \cdots, \xi_r$ in the infinite field Ω so that $a(\xi_1, \cdots, \xi_r) \neq 0$ (Vol. I, p. 112). Since the γ_j, $1 \leq j \leq r$, are algebraically independent, we have an algebra homomorphism of $\Phi[\gamma_1, \cdots, \gamma_r]$ into Ω/Φ such that $\gamma_j \to \xi_j$. By the extension theorem (Theorem 9) this homomorphism can be extended to an Ω-valued place $\mathscr{P}$ on P. Since $\mathscr{P}$ is an extension of an algebra homomorphism, $\mathscr{P}$ is the identity on Φ and so $\mathscr{P}$ is an algebra homomorphism into Ω/Φ. We note next that the γ_k, $r + 1 \leq k \leq n$, are in the valuation ring $\mathfrak{o}$ of $\mathscr{P}$. Otherwise, $\mathscr{P}(\gamma_k^{-1}) = 0$. On the other hand, we have an equation of the form

$$a_0(\gamma_1, \cdots, \gamma_r) + a_1(\gamma_1, \cdots, \gamma_r)\gamma_k^{-1} + \cdots + a_m(\gamma_1, \cdots, \gamma_r)\gamma_k^{-m} = 0,$$

and applying $\mathscr{P}$, we obtain $a_0(\xi_1, \cdots, \xi_r) = \mathscr{P}(a_0(\gamma_1, \cdots, \gamma_r)) = 0$. This contradicts the facts that $a(\xi_1, \cdots, \xi_r) \neq 0$ and $a(\gamma_1, \cdots, \gamma_r)$ has $a_0(\gamma_1, \cdots, \gamma_r)$ as a factor. A similar argument shows that $\mathscr{P}(g(\gamma_1, \cdots, \gamma_n)) \neq 0$. Now let $\xi_k = \mathscr{P}(\gamma_k)$, $r + 1 \leq k \leq n$. Then we assert that $(\xi_1, \xi_2, \cdots, \xi_n)$ satisfies the conditions of the theorem. In the first place, if $f(x_1, \cdots, x_n) \in \mathfrak{P}$, then $f(\gamma_1, \cdots, \gamma_n) = 0$ and applying $\mathscr{P}$ we have $f(\xi_1, \cdots, \xi_n) = 0$. Next we see that $g(\xi_1, \cdots, \xi_n) = \mathscr{P}(g(\gamma_1, \cdots, \gamma_n)) \neq 0$.

The Hilbert Nullstellensatz is the extension of Theorem 11 from prime ideals to arbitrary ideals in $\Phi[x_1, \cdots, x_n]$. To obtain this we need a characterization of the (nil)radical of an ideal of a commutative ring (Vol. I, p. 173). The result we require is that, if $\mathfrak{A}$ is an ideal in a commutative ring $\mathfrak{o}$, then the radical $\mathfrak{R}(\mathfrak{A})$ is the intersection $\cap\, \mathfrak{P}$ of the prime ideals $\mathfrak{P}$ containing $\mathfrak{A}$. If $\mathfrak{o}$ is Noetherian, this result is an easy consequence of the decomposition theorem for ideals into primary ideals (Vol. I, p. 176, ex. 2, p.

181). Although this is all we need here it is of interest to establish this result in the general case. We prove first the following

Lemma 1. *Let* $\mathfrak{o}$ *be a commutative ring,* $\mathfrak{A}$ *an ideal in* $\mathfrak{o}$ *and* S *a non-vacuous multiplicatively closed subset of* $\mathfrak{o}$ *such that* $\mathfrak{A} \cap S = \emptyset$. *Then there exists a prime ideal* $\mathfrak{P}$ *in* $\mathfrak{o}$ *such that* $\mathfrak{P} \supseteq \mathfrak{A}$ *and* $\mathfrak{P} \cap S = \emptyset$.

Proof. Let U be the collection of ideals $\mathfrak{B}$ in $\mathfrak{o}$ such that: 1. $\mathfrak{B} \supseteq \mathfrak{A}$, 2. $\mathfrak{B} \cap S = \emptyset$. Then U is non-vacuous since $\mathfrak{A} \varepsilon U$. We order the elements of U by inclusion. Let V be a linearly ordered subset of U and let $\mathfrak{C} = \bigcup_{\mathfrak{B} \varepsilon V} \mathfrak{B}$. Then $\mathfrak{C} \cap S = \emptyset$ and $\mathfrak{C} \supseteq \mathfrak{A}$. Moreover, it is easy to check that $\mathfrak{C}$ is an ideal. Hence $\mathfrak{C} \varepsilon U$ and $\mathfrak{C}$ is an upper bound for the set V. Thus U is an inductive set and so we can apply Zorn's lemma to conclude that U contains a maximal element $\mathfrak{P}$. Let a_i, $i = 1, 2$, be elements of $\mathfrak{o}$ not contained in $\mathfrak{P}$. Then the ideal $\mathfrak{A}_i$ generated by a_i and $\mathfrak{P}$ properly contains $\mathfrak{P}$ and contains $\mathfrak{A}$. Since $\mathfrak{P}$ is maximal in U, it follows that $\mathfrak{A}_i \not\varepsilon U$ which means that $\mathfrak{A}_i \cap S \neq \emptyset$. Let $s_i \varepsilon \mathfrak{A}_i \cap S$. If we take into account the form of the elements of $\mathfrak{A}_i$ we see that $s_i = x_i a_i + p_i$ where $x_i \varepsilon \mathfrak{o}$ and $p_i \varepsilon \mathfrak{P}$. Then

$$s = s_1 s_2 = x_1 x_2 a_1 a_2 + p \tag{36}$$

where $p \varepsilon \mathfrak{P}$. Since S is multiplicatively closed, $s \varepsilon S$. If $a_1 a_2 \varepsilon \mathfrak{P}$, then (36) implies that $s \varepsilon \mathfrak{P}$ contrary to $\mathfrak{P} \cap S = \emptyset$. Hence we see that $a_1 a_2 \not\varepsilon \mathfrak{P}$ so we have shown that $a_1 \not\varepsilon \mathfrak{P}$, $a_2 \not\varepsilon \mathfrak{P}$ implies $a_1 a_2 \not\varepsilon \mathfrak{P}$. Hence $\mathfrak{P}$ is a prime ideal satisfying the required conditions.

We can now prove

Theorem 12. *Let* $\mathfrak{A}$ *be an ideal in the commutative ring* $\mathfrak{o}$. *Then the radical* $\mathfrak{R}(\mathfrak{A}) = \bigcap \mathfrak{P}$ *the intersection of the prime ideals* $\mathfrak{P}$ *containing* $\mathfrak{A}$.

Proof. Let $a \varepsilon \mathfrak{R}(\mathfrak{A})$ and let $\mathfrak{P}$ be a prime ideal containing $\mathfrak{A}$. A suitable power $a^n \varepsilon \mathfrak{A}$ so $a^n \varepsilon \mathfrak{P}$. Since $\mathfrak{P}$ is prime, this implies that $a \varepsilon \mathfrak{P}$. Hence $\mathfrak{R}(\mathfrak{A}) \subseteq \mathfrak{P}$ and $\mathfrak{R}(\mathfrak{A}) \subseteq \bigcap \mathfrak{P}$ for the prime ideals $\mathfrak{P}$ containing $\mathfrak{A}$. Next let $a \not\varepsilon \mathfrak{R}(\mathfrak{A})$ and let $S = \{a^n, n = 1, 2, \cdots\}$. Then $S \cap \mathfrak{A} = \emptyset$ and S is multiplicatively closed. Hence the lemma implies that there exists a prime ideal $\mathfrak{P}$ containing $\mathfrak{A}$

such that $a \notin \mathfrak{P}$. Hence a is not contained in the intersection of the prime ideals containing $\mathfrak{A}$. Thus we have proved that $\cap \mathfrak{P} \subseteq \mathfrak{R}(\mathfrak{A})$ and so by the earlier inclusion, $\mathfrak{R}(\mathfrak{A}) = \cap \mathfrak{P}$.

Theorems 11 and 12 imply the

Hilbert Nullstellensatz. *Let $\mathfrak{A}$ be an ideal in the polynomial algebra $\Phi[x_1, x_2, \cdots, x_n]$, Φ a field, x_i indeterminates, and let Ω be the algebraic closure of Φ. Then a polynomial $g(x_1, \cdots, x_n) \in \mathfrak{R}(\mathfrak{A})$ if and only if $g(\xi_1, \cdots, \xi_n) = 0$ for every zero $(\xi_1, \cdots, \xi_n)$, $\xi_i \in \Omega$, of the ideal $\mathfrak{A}$.*

Proof. Let V denote the set of zeros $(\xi_1, \cdots, \xi_n)$, $\xi_i \in \Omega$, of $\mathfrak{A}$. Suppose $g(x_1, \cdots, x_n) \in \mathfrak{R}(\mathfrak{A})$. Then $g^n \in \mathfrak{A}$ for some positive integer n. Hence $g(\xi_1, \cdots, \xi_n)^n = 0$ for every $(\xi_i) \in V$ and $g(\xi_1, \cdots, \xi_n) = 0$ for every $(\xi_i) \in V$. Conversely, let $g(x_1, \cdots, x_n)$ be a polynomial such that $g(\xi_1, \cdots, \xi_n) = 0$ for every $(\xi_i) \in V$. Let $\mathfrak{P}$ be a prime ideal containing $\mathfrak{A}$ and let W be the set of zeros of $\mathfrak{P}$. Since $\mathfrak{P} \supseteq \mathfrak{A}$, $W \subseteq V$ and consequently $g(\xi_1, \cdots, \xi_n) = 0$ for every $(\xi_i) \in W$. It follows from Theorem 11 that $g(x_1, \cdots, x_n) \in \mathfrak{P}$. Thus g is contained in every prime ideal containing $\mathfrak{A}$ and so, by Theorem 12, $g \in \mathfrak{R}(\mathfrak{A})$. This completes the proof.

We shall give next an application of the existence of an algebraic zero of a prime ideal to a theorem on finite generation of a field. We recall that we saw long ago (Lemma 2, § 1.5) that, if $\gamma_1, \gamma_2, \cdots, \gamma_n$ are algebraic over Φ, then the field $P = \Phi(\gamma_1, \gamma_2, \cdots, \gamma_n)$ coincides with the algebra $\Phi[\gamma_1, \gamma_2, \cdots, \gamma_n]$ generated by the γ_i. We can now prove the following converse of this result.

Theorem 13. *If the algebra $P = \Phi[\gamma_1, \gamma_2, \cdots, \gamma_n]$ over Φ generated by the γ_i is a field, then the γ_i are algebraic over Φ.*

Proof. Let $\Phi[x_1, x_2, \cdots, x_n]$ be the polynomial algebra over Φ in indeterminates x_i and consider the homomorphism of this algebra onto P/Φ mapping $x_i \to \gamma_i$, $1 \leq i \leq n$. Let $\mathfrak{P}$ be the kernel of the homomorphism. Since P is a field, $\mathfrak{P}$ is a maximal ideal. If Ω is the algebraic closure of Φ, then we have seen that we can find $(\xi_1, \xi_2, \cdots, \xi_n)$ in Ω such that $f(\xi_1, \cdots, \xi_n) = 0$ for every $f \in \mathfrak{P}$. By IV of the Introduction we have a homomorphism of $P = \Phi[\gamma_1, \gamma_2, \cdots, \gamma_n]$ over Φ onto $\Phi[\xi_1, \xi_2, \cdots, \xi_n]$ such that $\gamma_i \to \xi_i$, $1 \leq i \leq n$. Since P is a field, this homomorphism is an isomorphism. Since ξ_i is algebraic, it follows that γ_i is algebraic, $1 \leq i \leq n$.

EXERCISE

1. Let $P = \Phi[\gamma_1, \gamma_2, \cdots, \gamma_n]$ be a finitely generated commutative algebra over Φ and let $\mathfrak{N}$ be the ideal of nilpotent elements. Show that $\mathfrak{N}$ is the intersection of the maximal ideals of $\mathfrak{P}$.

13. Application of the extension theorem: integral closure. We shall apply the extension theorem next to obtain an important characterization of the integral closure of a subring of a field. Let $\mathfrak{g}$ be a subring of the field Φ. We recall that an element $\alpha \varepsilon \Phi$ is called integral over $\mathfrak{g}$ or $\mathfrak{g}$-integral if there exists a polynomial $f(x) \varepsilon \mathfrak{g}[x]$ with leading coefficient 1 such that $f(\alpha) = 0$. The set $\mathfrak{G}$ of elements of Φ which are $\mathfrak{g}$-integral is called the *integral closure* of $\mathfrak{g}$ in Φ. We shall characterize this set. In the proof we shall need the following

Lemma 1. *If* $\mathfrak{o}$ *is a commutative ring (with an identity* 1), *any proper ideal* $\mathfrak{A}$ *of* $\mathfrak{o}$ *can be imbedded in a maximal ideal.*

Proof. The proof is obtained as a special case of the argument in the proof of Lemma 1 of § 12. We let $S = \{1\}$, so S is multiplicatively closed and $S \cap \mathfrak{A} = \varnothing$. Let U be the set of ideals $\mathfrak{B}$ such that $\mathfrak{B} \supseteq \mathfrak{A}$ and $\mathfrak{B}$ is proper (so that $\mathfrak{B} \cap S = \varnothing$). Then U contains a maximal element $\mathfrak{P}$. It is immediate that $\mathfrak{P}$ is a maximal ideal containing $\mathfrak{A}$.

Theorem 14 (Krull). *Let* $\mathfrak{g}$ *be a subring containing* 1 *in a field* Φ. *Then the integral closure* $\mathfrak{G}$ *of* $\mathfrak{g}$ *in* Φ *is* $\bigcap \mathfrak{o}$, *the intersection of all the valuation rings of* Φ *which contain* $\mathfrak{g}$.

Proof. Let $\alpha \varepsilon \mathfrak{G}$ so that we have a relation $\alpha^n + \gamma_1\alpha^{n-1} + \cdots + \gamma_n = 0, n \geq 1, \gamma_i \varepsilon \mathfrak{g}$. Let φ be a valuation whose valuation ring $\mathfrak{o}$ contains $\mathfrak{g}$. If $\alpha \not\varepsilon \mathfrak{o}$, then $\varphi(\alpha^{-1}) < 1$. But $1 = -\gamma_1\alpha^{-1} - \cdots - \gamma_n\alpha^{-n}$ and $\varphi(\gamma_i) \leq 1$. Hence every $\varphi(\gamma_i\alpha^{-i}) < 1$ and this is impossible since the relation gives $1 = \varphi(1) \leq \max (\varphi(\gamma_i\alpha^{-i})) < 1$. Hence $\alpha \varepsilon \mathfrak{o}$ so we have proved that $\mathfrak{G}$ is contained in $\bigcap \mathfrak{o}$ for the valuation rings containing $\mathfrak{g}$. Next suppose $\alpha \not\varepsilon \mathfrak{G}$. Then α^{-1} is not a unit in the ring $\mathfrak{g}[\alpha^{-1}]$, since otherwise its inverse $\alpha = \gamma_0 1 + \gamma_1\alpha^{-1} + \cdots + \gamma_{n-1}\alpha^{-(n-1)}, \gamma_i \varepsilon \mathfrak{g}$, and hence $\alpha^n = \gamma_0\alpha^{n-1} + \gamma_1\alpha^{n-2} + \cdots + \gamma_{n-1}$ so $\alpha \varepsilon \mathfrak{G}$. Since α^{-1} is not a unit in $\mathfrak{g}[\alpha^{-1}]$, the principal ideal $\alpha^{-1}\mathfrak{g}[\alpha^{-1}]$ is properly contained in $\mathfrak{g}[\alpha^{-1}]$. By Lemma 1 there exists a maximal ideal $\mathfrak{m}$ in $\mathfrak{g}[\alpha^{-1}]$ containing $\alpha^{-1}\mathfrak{g}[\alpha^{-1}]$. Then $\mathfrak{g}[\alpha^{-1}]/\mathfrak{m}$ is a field which can be imbedded in an algebraically closed field Ω. The canonical homomorphism of $\mathfrak{g}[\alpha^{-1}]$ onto $\mathfrak{g}[\alpha^{-1}]/\mathfrak{m}$ can be considered as a homomorphism of $\mathfrak{g}[\alpha^{-1}]$ into Ω. The extension theorem gives an Ω-valued place $\mathscr{P}$ whose valuation ring $\mathfrak{o}$ contains $\mathfrak{g}[\alpha^{-1}]$. The ideal $\mathfrak{p}$ of non-units of $\mathfrak{o}$ contains $\mathfrak{m}$, hence, α^{-1}. It follows that $\alpha \notin \mathfrak{o}$. Thus $\alpha \notin \mathfrak{G}$ implies $\alpha \notin \bigcap \mathfrak{o}$ for the valuation rings $\mathfrak{o}$ containing $\mathfrak{g}$. We therefore have $\mathfrak{G} = \bigcap \mathfrak{o}$ and the proof is complete.

The subring $\mathfrak{g}$ is called integrally closed in Φ if $\mathfrak{G} = \mathfrak{g}$. Then we have the following

Corollary. *If $\mathfrak{g}$ is a subring of Φ, then the set $\mathfrak{G}$ of $\mathfrak{g}$-integral elements is a subring of Φ containing $\mathfrak{g}$ and $\mathfrak{G}$ is integrally closed in Φ.*

Proof. The first statement is clear since $\mathfrak{G}$ is an intersection of subrings of Φ and since $\mathfrak{G}$ certainly contains $\mathfrak{g}$. Also the set of $\mathfrak{G}$-integral elements is the intersection $\bigcap \mathfrak{o}$ for the valuation rings containing $\mathfrak{G}$ and hence containing $\mathfrak{g}$. On the other hand, if $\mathfrak{o}$ is a valuation ring containing $\mathfrak{g}$, then $\mathfrak{o} \supseteq \mathfrak{G}$. Hence the intersection of the valuation rings containing $\mathfrak{G}$ is the same as that of the valuation rings containing $\mathfrak{g}$, so this is $\mathfrak{G}$. Hence $\mathfrak{G}$ is integrally closed.

EXERCISES

1. (Artin). Let $\mathfrak{g}$ be a subring of a field and let $\alpha_1, \alpha_2, \cdots, \alpha_r$ be elements of Φ. Suppose that for each i there exists a positive integer n_i such that $\alpha_i^{n_i} = P_i(\alpha_1, \alpha_2, \cdots, \alpha_r)$ where P_i is a polynomial of total degree $< n_i$. Show that every α_i is $\mathfrak{g}$-integral.
2. (Artin). Let $\mathfrak{g}$ be as in ex. 1 and let $\mathfrak{M}$ be a subring of Φ which is a finitely generated $\mathfrak{g}$-module. Show that every element of $\mathfrak{M}$ is $\mathfrak{g}$-integral (cf. Vol. I, p. 182).
3. A commutative integral domain $\mathfrak{g}$ is called *integrally closed* if it is integrally closed in its field of fractions. Show that if $\mathfrak{g}$ is Gaussian (that is, unique factorization holds), then $\mathfrak{g}$ is integrally closed.
4. (Cohn). Show that a subalgebra $\mathfrak{A}$ of $\Phi[x]$, Φ a field, x an indeterminate, has a single generator if and only if $\mathfrak{A}$ is integrally closed. (Hint: Use Lüroth's theorem and Th. 14.)

14. Finite dimensional extensions of complete fields. In the remainder of this chapter we return to the consideration of real valuations (archimedean as well as non-archimedean). We shall begin by considering the problem of extending a valuation on a complete field Φ to a finite dimensional extension field. Our first objective is to prove uniqueness of the extension. For this we require

Lemma 1. *Let Φ be complete with respect to a non-trivial real valuation φ and let P be an extension field of Φ with a valuation φ which is an extension of that of Φ. Suppose $u_1, u_2, \cdots, u_r$ are elements of P which are Φ-independent. Then a sequence $\{a_n\}$, $a_n = \sum_{i=1}^{r} \alpha_{ni}u_i$, $\alpha_{ni} \in \Phi$, is a Cauchy sequence in P if and only if the r sequences $\{\alpha_{ni}\}$, $i = 1, 2, \cdots, r$, are Cauchy sequences in Φ.*

Proof. It is immediate that, if the $\{\alpha_{ni}\}$ are Cauchy sequences, then so is $\{a_n\}$. Conversely, suppose $\{a_n\}$ is Cauchy. If $r = 1$, then it is clear that $\{\alpha_{n1}\}$ is Cauchy. We shall now prove our assertion for arbitrary r by induction. If the sequence $\{\alpha_{nr}\}$ is a Cauchy sequence, then the sequence $\{b_n\}$, $b_n = a_n - \alpha_{nr}u_r$ is a Cauchy sequence. Since $b_n = \sum_{1}^{r-1} \alpha_{nj}u_j$ the required result follows by induction. The proof will now be completed by showing that the assumption that $\{\alpha_{nr}\}$ is not Cauchy leads to a contradiction. We make this assumption. Then there exists a real $\epsilon > 0$ such that for any positive N there exist $p, q > N$ such that $\varphi(\alpha_{pr} - \alpha_{qr}) > \epsilon$. Hence there exist pairs of positive integers (p_k, q_k), $p_1 < p_2 < \cdots$, $q_1 < q_2 < \cdots$ such that $\varphi(\alpha_{p_kr} - \alpha_{q_kr}) > \epsilon$. Then $(\alpha_{p_kr} - \alpha_{q_kr})^{-1}$ exists and we can form the sequence $\{b_k\}$ where

$$b_k = (\alpha_{p_kr} - \alpha_{q_kr})^{-1}(a_{p_k} - a_{q_k}). \tag{37}$$

We have $\varphi(\alpha_{p_kr} - \alpha_{q_kr})^{-1} < \dfrac{1}{\epsilon}$ and $\{a_{p_k} - a_{q_k}\}$ is a null sequence. Hence $\{b_k\}$ is a null sequence. On the other hand, $b_k = \sum_{j=1}^{r-1} \beta_{kj}u_j + u_r$ and this implies that, if $c_k = \Sigma\beta_{kj}u_j$, then $\{c_k\}$ is a Cauchy sequence. Then the $r - 1$ sequences $\{\beta_{kj}\}$, $j = 1, 2, \cdots, r - 1$, are Cauchy sequences. Since Φ is complete, $\lim \beta_{kj} = \beta_j$ exists. Since $\lim b_k = 0$ we get from $b_k = \sum_{1}^{r-1} \beta_{kj}u_j + u_r$, $0 = \sum_{1}^{r-1} \beta_ju_j$

$+ u_r$. This contradicts the linear independence of the u's and completes the proof.

We note two important consequences of this lemma: (1) If $\{a_n\}$ is a null sequence, then all the sequences $\{\alpha_{ni}\}$ are null sequences. (2) If $[\mathrm{P}:\Phi] < \infty$, then P is complete. The first of these is clear since the $\{\alpha_{ni}\}$ are Cauchy sequences. Hence $\lim \alpha_{ni} = \alpha_i$ exists and $\Sigma\alpha_i u_i = 0$. Hence every $\alpha_i = 0$ by the linear independence of the u_i. To prove the second statement we suppose that $(u_1, u_2, \cdots, u_r)$ is a basis. Then if $\{a_n\}$ is Cauchy, every $\{\alpha_{ni}\}$ is Cauchy and so $\lim \alpha_{ni} = \alpha_i$ exists and $\lim a_n = \Sigma\alpha_i u_i$.

We can now prove

Theorem 15. *Let* P *be a finite dimensional extension field of a field which is complete with respect to a non-trivial real valuation* φ. *Then if* φ *can be extended to a real valuation of* P, *this valuation is unique and is given by the formula*

$$\varphi(\rho) = \varphi(N_{\mathrm{P}|\Phi}(\rho))^{1/n}, \quad n = [\mathrm{P}:\Phi]. \tag{38}$$

Proof. Assume the extension φ exists and suppose there exists a $\rho \varepsilon \mathrm{P}$ such that (38) does not hold. Then $\varphi(\rho^n) \neq \varphi(N(\rho))$, so $\rho \neq 0$ and either $\varphi(\rho^n) < \varphi(N(\rho))$ or $\varphi(\rho^n) > \varphi(N(\rho))$. By replacing ρ by ρ^{-1}, if necessary, we may suppose $\varphi(\rho^n) < \varphi(N(\rho))$. Set $\sigma = \rho^n N(\rho)^{-1}$. Then $\varphi(\sigma) < 1$ and $N(\sigma) = N(\rho^n)N(\rho)^{-n} = 1$. Since $\varphi(\sigma) < 1$, we have $\lim \sigma^k = 0$. If $(u_1, u_2, \cdots, u_n)$ is a basis and $\sigma^k = \sum_{i=1}^{n} \alpha_{ki}u_i$, then $\lim \sigma^k = 0$ implies that $\lim \alpha_{ki} = 0$ for every i. Since the norm of an element $\epsilon = \Sigma\gamma_i u_i$, $\gamma_i \varepsilon \Phi$, is a homogeneous polynomial of the n-th degree in the γ_i with fixed coefficients, it is clear that $\lim \alpha_{ki} = 0$ for every i implies that $\lim N(\sigma^k) = 0$. This contradicts $N(\sigma^k) = N(\sigma)^k = 1$.

We have seen before that any non-archimedean real valuation on a subfield can be extended. Hence in the non-archimedean case the formula (38) provides a valuation for the finite dimensional extension P. It remains to consider the archimedean case. The extension theorem in this case will be obtained by a complete determination of the fields which are complete with respect to an archimedean real valuation. We shall show that the only such fields are the field of real numbers and the field of complex numbers.

Lemma 2. *Let* P *be a quadratic extension of a field* Φ *which is complete with respect to a real archimedean valuation* φ. *Then* φ *can be extended to a valuation of* P.

Proof. We recall that the existence of an archimedean valuation implies that the characteristic is 0. Hence P is Galois over Φ. Let $\alpha \to \bar{\alpha}$ be the automorphism of P/Φ which is not the identity. Then the trace and norm of $\alpha \in P$ are $T(\alpha) = \alpha + \bar{\alpha}$, $N(\alpha) = \alpha\bar{\alpha}$ and we have $\alpha^2 - T(\alpha)\alpha + N(\alpha) = 0$ for any $\alpha \in P$. We shall show that $\varphi(\alpha) \equiv \varphi(N(\alpha))^{1/2}$ defines a valuation of P. If $\alpha \in \Phi$, $N(\alpha) = \alpha^2$. This implies that the mapping φ defined on P is an extension of the φ which is given on Φ. We evidently have $\varphi(\alpha) = 0$ only if $\alpha = 0$ and the multiplicative property of the norm implies that $\varphi(\alpha\beta) = \varphi(\alpha)\varphi(\beta)$. Hence all one needs to show is: $\varphi(\alpha + \beta) \leq \varphi(\alpha) + \varphi(\beta)$. This will follow if we can show that $\varphi(\alpha + 1) \leq \varphi(\alpha) + 1$; for, $\varphi(\alpha + \beta) \leq \varphi(\alpha) + \varphi(\beta)$ is clear if $\beta = 0$, and if $\beta \neq 0$, then

$$\varphi(\alpha + \beta) = \varphi((\alpha\beta^{-1} + 1)\beta) = \varphi(\alpha\beta^{-1} + 1)\varphi(\beta).$$

Hence, if $\varphi(\alpha\beta^{-1} + 1) \leq \varphi(\alpha\beta^{-1}) + 1$, then

$$\begin{aligned}\varphi(\alpha + \beta) \leq (\varphi(\alpha\beta^{-1}) + 1)\varphi(\beta) &= (\varphi(\alpha)\varphi(\beta)^{-1} + 1)\varphi(\beta) \\ &= \varphi(\alpha) + \varphi(\beta).\end{aligned}$$

Now $\varphi(\alpha + 1) \leq \varphi(\alpha) + 1$ holds if $\alpha \in \Phi$ so we suppose that $\alpha \notin \Phi$. Then $P = \Phi(\alpha)$ and $x^2 - T(\alpha)x + N(\alpha)$ is the minimum polynomial of α and $N(\alpha + 1) = (\alpha + 1)(\bar{\alpha} + 1) = \alpha\bar{\alpha} + \alpha + \bar{\alpha} + 1 = N(\alpha) + T(\alpha) + 1$. Hence $\varphi(\alpha + 1) \leq \varphi(\alpha) + 1$ is equivalent to $\varphi(\alpha + 1)^2 \leq \varphi(\alpha)^2 + 2\varphi(\alpha) + 1$ and to

$$\varphi(1 + T(\alpha) + N(\alpha)) \leq 1 + 2\varphi(N(\alpha))^{1/2} + \varphi(N(\alpha)). \tag{39}$$

If we use the addition property of φ in Φ it is clear that (39) will hold if $\varphi(T(\alpha)) \leq 2\varphi(N(\alpha))^{1/2}$. Hence we suppose that $\varphi(T(\alpha)) > 2\varphi(N(\alpha))^{1/2}$, or $\varphi(T(\alpha))^2 > 4\varphi(N(\alpha))$. We write $a = T(\alpha)$, $b = N(\alpha)$, so we are assuming $\varphi(a)^2 > 4\varphi(b)$. We shall show that this implies that $\alpha \in \Phi$ which will contradict our assumption. Hence the proof will be completed by proving

Lemma 3. *Let* Φ *be a field which is complete relative to a real valuation* φ *and let* $x^2 - ax + b = 0$ *be an equation with coefficients* a, b *in* Φ *such that* $\varphi(a)^2 > 4\varphi(b)$. *Then the equation has roots in* Φ.

Proof. A non-zero root α of this equation will be a root of $\alpha = a - b\alpha^{-1}$. We shall obtain such a root as a limit of a sequence $\{a_n\}$ where a_n is defined recursively by $a_1 = \frac{1}{2}a$, $a_{n+1} = a - ba_n{}^{-1}$. We show first that no $a_n = 0$ so the definition works for all n. We have $\varphi(a_1) = \frac{1}{2}\varphi(a) > 0$ and we may suppose that $\varphi(a_n) \geq \frac{1}{2}\varphi(a)$. Then

$$\begin{aligned}\varphi(a_{n+1}) = \varphi(a - ba_n{}^{-1}) &\geq \varphi(a) - \varphi(b)\varphi(a_n)^{-1}\\ &\geq \varphi(a) - 2\varphi(b)\varphi(a)^{-1}\\ &\geq \varphi(a) - \tfrac{1}{2}\varphi(a)^2\varphi(a)^{-1} = \tfrac{1}{2}\varphi(a).\end{aligned}$$

Hence $\varphi(a_n) \geq \frac{1}{2}\varphi(a) > 0$ holds for all $n = 1, 2, 3, \cdots$ and every $a_n \neq 0$. Now we have $a_{n+2} - a_{n+1} = ba_{n+1}{}^{-1}a_n{}^{-1}(a_{n+1} - a_n)$; and $\varphi(a_{n+1})^{-1}\varphi(a_n)^{-1} \leq 4\varphi(a)^{-2}$; hence

$$\varphi(a_{n+2} - a_{n+1}) \leq \frac{4\varphi(b)}{\varphi(a)^2}\varphi(a_{n+1} - a_n). \tag{40}$$

If we set $r = 4\varphi(b)/\varphi(a)^2$ we have $0 \leq r < 1$ and we may iterate (40) to obtain $\varphi(a_{n+2} - a_{n+1}) \leq r^n c$ where $c = \varphi(a_2 - a_1)$. This inequality implies easily that $\{a_n\}$ is a Cauchy sequence. Hence $\alpha = \lim a_n$ exists and since $\varphi(a_n) \geq \frac{1}{2}\varphi(a) > 0$, $\alpha \neq 0$. Hence the recursion formula $a_{n+1} = a - ba_n{}^{-1}$ gives $\alpha = a - b\alpha^{-1}$ so $\alpha^2 - a\alpha + b = 0$.

We are now ready to prove

Theorem 16 (Ostrowski). *The only fields which are complete relative to a real archimedean valuation are the field of real numbers and the field of complex numbers.*

Proof. Let Φ be complete relative to the archimedean valuation φ. Then Φ is of characteristic 0 and so it contains the rationals. Since any real archimedean valuation of the rationals is equivalent to the absolute value valuation and Φ is complete, it is clear that Φ contains the field of real numbers. If Φ contains an element i such that $i^2 = -1$, then Φ contains the field C of complex numbers. Otherwise, we adjoin i to Φ and obtain $\Phi(i)$ which contains C. By Lemma 2, φ can be extended to a real valuation of $\Phi(i)$. Also we have seen that $\Phi(i)$ is complete. The theorem will therefore follow if we can show that, if Φ is complete with respect to an archimedean valuation and $\Phi \supseteq C$, then $\Phi = C$. Since the restriction of φ to the real subfield (the completion of the rationals) is equivalent to the absolute value valuation, Theorem 15 shows that φ is equivalent to the absolute value valuation on C.

Now suppose $\Phi \supset C$ and let $\alpha \varepsilon \Phi, \notin C$. Let $r = \inf \varphi(\alpha - c)$ for $c \varepsilon C$. Then we claim that there exists a $c_0 \varepsilon C$ such that $\varphi(\alpha - c_0) = r$. First, it is clear that $r = \inf \varphi(\alpha - c)$ for all c such that $\varphi(\alpha - c) \leq r + 1$ and, if c_1 and c_2 are two complex numbers satisfying $\varphi(\alpha - c_1) \leq r + 1$, $\varphi(\alpha - c_2) \leq r + 1$, then $\varphi(c_1 - c_2) \leq 2r + 2$. Hence the c satisfying $\varphi(\alpha - c) \leq r + 1$ form a closed and bounded set in C. Since $\varphi(\alpha - c)$ is a continuous function of c it is clear that there exists a c_0 such that $\varphi(\alpha - c_0) = r$. Since $\alpha \notin C$ we have $r > 0$. If we replace α by $\alpha - c_0$ we may assume that $c_0 = 0$. Then we have $\varphi(\alpha) = r > 0$ and $\varphi(\alpha - c) \geq r$ for every $c \varepsilon C$. We shall now show that we have $\varphi(\alpha - c) = r$ for every complex c with $\varphi(c) < r$. To see this we let n be any positive integer and we consider $\alpha^n - c^n = (\alpha - c)(\alpha - \epsilon c) \cdots (\alpha - \epsilon^{n-1}c)$ where ϵ is a primitive n-th root of 1 contained in C. Then

$$\varphi(\alpha - c)\varphi(\alpha - \epsilon c) \cdots \varphi(\alpha - \epsilon^{n-1}c)$$
$$= \varphi(\alpha^n - c^n) \leq \varphi(\alpha)^n + \varphi(c)^n.$$

Since $\varphi(\alpha - \epsilon^k c) \geq r$, we obtain

$$\varphi(\alpha - c)r^{n-1} \leq \varphi(\alpha)^n\left(1 + \frac{\varphi(c)^n}{\varphi(\alpha)^n}\right) = r^n\left(1 + \left(\frac{\varphi(c)}{r}\right)^n\right).$$

Hence

$$\varphi(\alpha - c) \leq r\left(1 + \left(\frac{\varphi(c)}{r}\right)^n\right),$$

so if $\varphi(c) < r$, then $\lim\left(1 + \left(\frac{\varphi(c)}{r}\right)^n\right) = 1$ gives the asserted relation $\varphi(\alpha - c) = r$. We can now replace α by $\alpha - c$ for any c such that $\varphi(c) < r$ and we obtain $\varphi(\alpha - 2c) = r$. If we repeat this process we obtain $\varphi(\alpha - nc) = r$ for all $n = 1, 2, \cdots$ and all c such that $\varphi(c) < r$. This amounts to saying that $\varphi(\alpha - c) = r$ if $\varphi(c) < nr$ and, since n is arbitrary, we have $\varphi(\alpha - c) = r$ for all

$c \in C$. Then if $c_1, c_2 \in C$, $\varphi(c_1 - c_2) \leq \varphi(\alpha - c_1) + \varphi(\alpha - c_2) = 2r$ which is absurd since φ is equivalent to the absolute value valuation on C. Thus we must have $C = \Phi$ and the theorem is proved.

The extension theorem for valuations for complete fields relative to an archimedean valuation becomes trivial in view of Ostrowski's theorem. If Φ is complete relative to an archimedean valuation, then Φ is either the reals or the complexes. In the first case, the only finite dimensional extensions are Φ and the field of complex numbers. In the second, the only possibility is Φ. In all cases the extension theorem is clear. If we combine this with the earlier results we obtain the following

Theorem 17. *If Φ is complete relative to a real valuation φ and* P *is a finite dimensional extension of Φ, then the valuation can be extended in one and only one way to* P. *The extension is given by the formula* (38). *Moreover,* P *is complete relative to its valuation.*

15. Extension of real valuations to finite dimensional extension fields. We now take up the problem of determining all the extensions of a real valuation defined in a field Φ to a finite dimensional extension field P/Φ. The case in which Φ is complete has been treated in the last section. We shall use the result obtained there to treat the general case. Let $\bar{\Phi}$ be the completion of Φ relative to φ and denote the valuation in $\bar{\Phi}$ which extends that in φ by $\bar{\varphi}$. Now suppose (E, s, t) is a field composite of P/Φ and $\bar{\Phi}/\Phi$: E is a field over Φ, s and t are isomorphisms of P/Φ and $\bar{\Phi}/\Phi$ respectively into E/Φ, and E is generated by P^s and $\bar{\Phi}^t$. Since $[P:\Phi] = n < \infty$ we have $[E:\bar{\Phi}^t] \leq n < \infty$. The valuation $\bar{\varphi}$ in $\bar{\Phi}$ can be transferred to $\bar{\Phi}^t$ by defining $\bar{\varphi}_t(\bar{\alpha}^t) = \bar{\varphi}(\bar{\alpha})$, $\bar{\alpha} \in \bar{\Phi}$. Clearly $\bar{\varphi}_t$ coincides with φ on Φ. Since $\bar{\Phi}$ is complete relative to $\bar{\varphi}$, it is clear that $\bar{\Phi}^t$ is complete relative to $\bar{\varphi}_t$. Since E is a finite dimensional extension of $\bar{\Phi}^t$, the real valuation $\bar{\varphi}_t$ has a unique extension to a real valuation $\bar{\psi}$ on E. Let ψ_s be the restriction of $\bar{\psi}$ to the subfield P^s and transfer ψ_s to P by $\psi(\rho) = \psi_s(\rho^s)$. Then it is clear that ψ is a real valuation on P which extends φ.

Thus we have a process for associating with every composite (E, s, t) of P and $\bar{\Phi}$ a real valuation ψ on P which extends φ. We shall show that this correspondence between the composites and the extensions of the valuations is 1–1 and surjective, if we identify equivalent composites. First, suppose the two composites (E_1, s_1, t_1) and (E_2, s_2, t_2) of P/Φ and $\bar{\Phi}/\Phi$ are equivalent. Then we have an isomorphism u of E_1/Φ onto E_2/Φ such that $\bar{\alpha}^{t_1 u} = \bar{\alpha}^{t_2}$, $\bar{\alpha} \varepsilon \bar{\Phi}$ and $\rho^{s_1 u} = \rho^{s_2}$, $\rho \varepsilon P$. For the valuations on $\bar{\Phi}^{t_1}$ and $\bar{\Phi}^{t_2}$ we have $\bar{\varphi}_{t_2}(\bar{\alpha}^{t_2}) = \bar{\varphi}(\bar{\alpha}) = \bar{\varphi}_{t_1}(\bar{\alpha}^{t_1})$. Hence $\bar{\varphi}_{t_2}(\bar{\alpha}^{t_1 u}) = \bar{\varphi}_{t_1}(\bar{\alpha}^{t_1})$. Let $\bar{\psi}_1$ and $\bar{\psi}_2$ be the valuations of E_1 and E_2 respectively, which extend $\bar{\varphi}_{t_1}$ and $\bar{\varphi}_{t_2}$. Now $\bar{\psi}_2'(\gamma_1{}^u) \equiv \bar{\psi}_1(\gamma_1)$, $\gamma_1 \varepsilon E_1$, defines a real valuation on E_2 such that for $\bar{\alpha}^{t_1 u}(\varepsilon \bar{\Phi}^{t_2})$ we have $\bar{\psi}_2'(\bar{\alpha}^{t_1 u}) = \bar{\psi}_1(\bar{\alpha}^{t_1}) = \bar{\varphi}_{t_1}(\bar{\alpha}^{t_1}) = \bar{\varphi}_{t_2}(\bar{\alpha}^{t_1 u})$. Thus $\bar{\psi}_2'$ is an extension of the valuation $\bar{\varphi}_{t_2}$ on $\bar{\Phi}^{t_2}$. Since $\bar{\Phi}^{t_2}$ is complete, this extension is unique and so it coincides with $\bar{\psi}_2$. Hence we have $\bar{\psi}_1(\gamma_1) = \bar{\psi}_2(\gamma_1{}^u)$ for every $\gamma_1 \varepsilon E_1$. This implies that the restrictions ψ_{s_1} and ψ_{s_2} to P^{s_1} and P^{s_2} satisfy $\psi_{s_1}(\rho^{s_1}) = \psi_{s_2}(\rho^{s_1 u}) = \psi_{s_2}(\rho^{s_2})$. Hence the corresponding valuations ψ_1 and ψ_2 on P satisfy $\psi_1(\rho) = \psi_{s_1}(\rho^{s_1}) = \psi_{s_2}(\rho^{s_2}) = \psi_2(\rho)$. Thus equivalent composites give the same valuation.

Conversely, assume $\psi_1(\rho) = \psi_2(\rho)$ for the valuations ψ_1, ψ_2 of P determined by the composites (E_1, s_1, t_1) and (E_2, s_2, t_2). Then we have $\psi_{s_1}(\rho^{s_1}) = \psi_{s_2}(\rho^{s_2})$, $\rho \varepsilon P$. Next we observe that E_i, $i = 1, 2$, is the closure of P^{s_i} in the topology defined by the valuation in E_i. Clearly, this closure contains $\bar{\Phi}^{t_i}$ and P^{s_i}, hence E_i, since this field is generated by $\bar{\Phi}^{t_i}$ and P^{s_i}. It is now clear that E_i is a completion of P^{s_i} relative to the valuation ψ_{s_i} in the sense of Definition 5. Consequently, by Theorem 6, the isomorphism $\rho^{s_1} \rightarrow \rho^{s_2}$ of P^{s_1} onto P^{s_2} has a unique extension to an isometric isomorphism u of E_1 onto E_2. We have $\psi_1(\gamma_1) = \psi_2(\gamma_1{}^u)$ for the valuations ψ_i of E_i and $\rho^{s_1 u} = \rho^{s_2}$. Since $\bar{\Phi}^{t_i}$ is the closure of Φ in E_i and since u is the identity on Φ, it is clear that u maps $\bar{\Phi}^{t_1}$ onto $\bar{\Phi}^{t_2}$. Hence the restriction of u to $\bar{\Phi}^{t_1}$ is an isometric isomorphism which is the identity on Φ. On the other hand, the mapping $\bar{\alpha}^{t_1} \rightarrow \bar{\alpha}^{t_2}$ has these same properties since $\bar{\varphi}_{t_1}(\bar{\alpha}^{t_1}) = \bar{\varphi}(\bar{\alpha}) = \bar{\varphi}_{t_2}(\bar{\alpha}^{t_2})$. Hence by Theorem 6, $\bar{\alpha}^{t_1} \rightarrow \bar{\alpha}^{t_2}$ coincides with the mapping u. Hence we have $\bar{\alpha}^{t_1 u} = \bar{\alpha}^{t_2}$ and so (E_1, s_1, t_1) and (E_2, s_2, t_2) are equivalent.

It remains to show that every valuation ψ on P which is an extension of φ can be obtained from a composite in the manner indicated. To see this we let E be the completion of P relative to ψ and let s denote the canonical imbedding (isomorphism) of P into E. Now we have an isomorphism t of the completion $\bar{\Phi}$ into the closure of Φ in E. The subfield of E generated by $\bar{\Phi}^t$ and P^s is a finite dimensional extension of $\bar{\Phi}^t$, so it is complete relative to the valuation obtained from E. It follows that this coincides with E. Hence we have a composite (E, s, t) and one checks that the valuation of P obtained from this composite is the given valuation ψ. We can now state the following

Theorem 18. *Let* P *be a finite dimensional extension field of a field* Φ *with a real valuation* φ *and let* $\bar{\Phi}$ *be the completion of* Φ. *Then the extensions of* φ *to valuations* ψ *in* P *are in* 1–1 *correspondence with the equivalence classes of composites* (E, s, t) *of* P/Φ *and* $\bar{\Phi}$.

In § 1.16 we have established a 1–1 correspondence between the equivalence classes of composites (E, s, t) and the maximal ideals of the algebra $\bar{\Phi} \otimes_\Phi P$. We have seen that, if $\mathfrak{J}$ is a maximal ideal in $\bar{\Phi} \otimes P$, then this determines a composite whose field is $E = (\bar{\Phi} \otimes P)/\mathfrak{J}$. Distinct $\mathfrak{J}$ give inequivalent composites and every composite is equivalent to one obtained from a maximal ideal $\mathfrak{J}$. We have seen also that the number of maximal ideals is finite and, if $\mathfrak{J}_1, \mathfrak{J}_2, \cdots, \mathfrak{J}_h$ are the distinct maximal ideals in $\bar{\Phi} \otimes P$ and $\mathfrak{R} = \bigcap \mathfrak{J}_j$, then $(\bar{\Phi} \otimes P)/\mathfrak{R} = E_1 \oplus E_2 \oplus \cdots \oplus E_h$ where $E_j \cong (\bar{\Phi} \otimes P)/\mathfrak{J}_j$. The field E_j is the completion of P relative to a valuation ψ_j. We shall call $[E_j:\bar{\Phi}] = n_j$ the *local dimensionality* of P determined by ψ_j. Then we have

$$
\begin{aligned}
\Sigma n_j &= [(\bar{\Phi} \otimes P):\bar{\Phi}] - [\mathfrak{R}:\bar{\Phi}] \\
&= [P:\Phi] - [\mathfrak{R}:\bar{\Phi}] \qquad (41)\\
&= n - [\mathfrak{R}:\bar{\Phi}] \leq n.
\end{aligned}
$$

Moreover, $\Sigma n_j = n$ if and only if $\mathfrak{R} = 0$. Since $\bar{\Phi} \otimes P$ can be considered a finite dimensional algebra over $\bar{\Phi}$, VII of the Introduction implies that $(\bar{\Phi} \otimes P)/\mathfrak{J}$ is a field if and only if it is an integral domain. Hence $\mathfrak{J}$ is maximal in $(\bar{\Phi} \otimes P)$ if and only if $\mathfrak{J}$ is prime. Hence, by Theorem 12, $\bigcap \mathfrak{J}_j = \mathfrak{R}$ is the radical of the algebra $\bar{\Phi} \otimes P$, that is, $\mathfrak{R}$ is the set of nilpotent elements of $\bar{\Phi} \otimes P$ and $\mathfrak{R} = 0$ if and only if $\bar{\Phi} \otimes P$ has no non-zero nilpotent elements. If P is separable over Φ we have $\bar{\Phi} \otimes_\Phi P = E_1 \oplus E_2 \oplus \cdots \oplus E_h$ where the $E_j/\bar{\Phi}$ are fields which can be determined explicitly from the minimum polynomial $f(x)$ of a primitive element θ of P over Φ (§ 1.16). Since a direct sum of fields contains no non-zero nilpotents, it is clear that $\bar{\Phi} \otimes$ P has zero radical $\mathfrak{R}$ if P is separable over Φ. Consequently, the formula (41) becomes

$$n = \Sigma n_j \tag{42}$$

in this case.

EXERCISE

1. Determine the number of extensions of the p-adic valuation of the rationals to the cyclotomic field of 5-th roots of 1 for $p = 3, 5, 11$.

16. Ramification index and residue degree. Let Φ be a field with a non-trivial non-archimedean real valuation φ and let γ be the value group, $\mathfrak{o}/\mathfrak{p}$ the residue field of Φ relative to φ (§ 5). Suppose P is a finite dimensional extension field, ψ an extension of the valuation φ to P, Γ the corresponding value group, $\mathfrak{O}/\mathfrak{P}$ the residue field of P. Since $\mathfrak{O}$ and $\mathfrak{P}$ are the sets of elements ρ satisfying $\psi(\rho) \leq 1$, $\psi(\rho) < 1$ respectively it is clear that $\mathfrak{o} \subseteq \mathfrak{O}$ and $\mathfrak{p} = \mathfrak{o} \cap \mathfrak{P}$. Hence we can identify the residue field $\mathfrak{o}/\mathfrak{p}$ with the subfield $(\mathfrak{o} + \mathfrak{P})/\mathfrak{P}$ of the residue field $\mathfrak{O}/\mathfrak{P}$. In this way we can consider the dimensionality $[\mathfrak{O}/\mathfrak{P}:\mathfrak{o}/\mathfrak{p}] = f$ which we shall call the *residue degree* of the valuation ψ of the extension P/Φ. It is clear also that the value group γ is a subgroup of Γ and we shall call the index e of γ in Γ the *ramification index* of ψ. If $\rho \in$ P then we can multiply ρ by a suitable non-zero element of $\mathfrak{p}$ to obtain an element of $\mathfrak{P}$. Hence we can choose elements of $\mathfrak{P}$ as representatives of the cosets of γ in Γ. Both the residue degree and the ramification index are finite and, in fact, we have

Lemma 1. $ef \leq n = [\mathrm{P}:\Phi]$.

Proof. Let $\rho_1, \rho_2, \cdots, \rho_{f_1}$ be elements of $\mathfrak{O}$ which are linearly independent over $(\mathfrak{o} + \mathfrak{P})/\mathfrak{P}$. Thus if α_i are elements of $\mathfrak{o}$ and $\Sigma\alpha_i\rho_i \in \mathfrak{P}$, then every $\alpha_i \in \mathfrak{p}$. Let $\pi_1, \pi_2, \cdots, \pi_{e_1}$ be elements of $\mathfrak{P}$ such that the cosets $\psi(\pi_1)\gamma, \cdots, \psi(\pi_{e_1})\gamma$ are distinct in Γ/γ. We assert that the $e_1 f_1$ elements $\rho_i\pi_j$ are Φ-independent. Thus suppose $\Sigma\alpha_{ij}\rho_i\pi_j = 0$ where the $\alpha_{ij} \in \Phi$. We shall show first that, if the $\alpha_i \in \Phi$ and $\Sigma\alpha_i\rho_i \neq 0$, then $\psi(\Sigma\alpha_i\rho_i) \in \gamma$. If $\Sigma\alpha_i\rho_i \neq 0$, then some $\alpha_i \neq 0$ and we may assume that $0 \neq \psi(\alpha_1) \geq \psi(\alpha_i)$. Then if $\beta_i = \alpha_i\alpha_1^{-1}$, $\psi(\beta_i) \leq 1$, so β_i is in the valuation ring $\mathfrak{o}$ of φ. We have $\Sigma\alpha_i\rho_i = \alpha_1(\Sigma\beta_i\rho_i)$. Also $\psi(\Sigma\beta_i\rho_i) \leq 1$ and since $\beta_1 = 1$ and the $\beta_i \in \mathfrak{o}$, it is clear that $\psi(\Sigma\beta_i\rho_i) < 1$ would contradict the linear independence of the ρ_i over $(\mathfrak{o} + \mathfrak{P})/\mathfrak{P}$. Hence we see that $\psi(\Sigma\beta_i\rho_i) = 1$ and so $\psi(\Sigma\alpha_i\rho_i) = \psi(\alpha_1)(\psi(\Sigma\beta_i\rho_i)) = \psi(\alpha_1) \in \gamma$. We now return to our relation $\Sigma\alpha_{ij}\rho_i\pi_j = 0$, $\alpha_{ij} \in \Phi$. Assume there exists a j so that $\psi(\Sigma\alpha_{ij}\rho_i) \neq 0$. Then we have distinct j, say $j = 1, 2$, so that $\psi(\Sigma\alpha_{i1}\rho_i\pi_1) = \psi(\Sigma\alpha_{i2}\rho_i\pi_2) \neq 0$ (ex. 2, § 1). Then $\psi(\Sigma\alpha_{i1}\rho_i)\psi(\pi_1) = \psi(\Sigma\alpha_{i2}\rho_i)\psi(\pi_2) \neq 0$ and the cosets $\gamma\psi(\pi_1) = \gamma\psi(\pi_2)$ by the result we have proved. This contradicts the choice of the π's. Hence we see that we must have $\psi(\Sigma\alpha_{ij}\rho_i) = 0$ or $\Sigma\alpha_{ij}\rho_i = 0$ for every j. The argument used before based on the linear independence over $(\mathfrak{o} + \mathfrak{P})/\mathfrak{P}$ of the ρ_i now implies that every $\alpha_{ij} = 0$. This proves our assertion that the e_1f_1 elements $\rho_i\pi_j$ are Φ-independent. Hence $e_1f_1 \leq n$. Evidently the definitions of e_1 and f_1 now imply that $ef \leq n$.

Lemma 2. *$ef = n$ if φ is discrete and Φ is complete relative to φ.*

Proof. Since φ is discrete the valuation ψ in P is discrete. Moreover, P is complete. The groups γ and Γ are cyclic and Γ/γ is cyclic of order e. Let π and β be elements of $\mathfrak{P}$ and $\mathfrak{p}$ respectively such that $\psi(\pi)$ and $\psi(\beta) = \varphi(\beta)$ are maximal. Any nonzero element of P has the form $\epsilon\pi^k$ where $\psi(\epsilon) = 1$ and $k = 0, \pm 1, \pm 2, \cdots$. Hence $\psi(\pi)$ is a generator of Γ. If $\beta = \eta\pi^{e'}$ where $\psi(\eta) = 1$ and $e' > 0$ since $\beta \in \mathfrak{p} \subseteq \mathfrak{P}$, then $\psi(\pi)^{e'} \in \gamma$ so e' is divisible by the order e of the coset $\psi(\pi)\gamma$. On the other hand, $\psi(\pi)^e = \psi(\pi^e) = \psi(\beta')$ for some $\beta' \in \mathfrak{p}$ and $\beta' = \zeta\beta^k$ where $\psi(\zeta) = 1$. Hence $\psi(\pi^e) = \psi(\beta^k) = \psi((\eta\pi^{e'})^k) = \psi(\pi^{e'k})$. Hence $e = e'k$. It follows that $k = 1$, $e' = e$, and so we have the relation $\beta = \eta\pi^e$, $\psi(\eta) = 1$, e the order of Γ/γ. Let $\rho_1, \rho_2, \cdots, \rho_f$ be elements of $\mathfrak{O}$ such that the cosets $\rho_i + \mathfrak{P}$ form a basis for the field $\mathfrak{O}/\mathfrak{P}$ over the subfield $(\mathfrak{o} + \mathfrak{P})/\mathfrak{P} \cong \mathfrak{o}/\mathfrak{p}$. We shall show that the elements $\rho_i\pi^j$, $1 \leq i \leq f$, $0 \leq j \leq e - 1$ form a basis for P over Φ. Since $\psi(\pi)\gamma$ is of order e, $\psi(1), \psi(\pi), \cdots, \psi(\pi^{e-1})$ are in distinct cosets relative to γ; hence the proof of Lemma 1 shows that the elements $\rho_i\pi^j$ are Φ-independent. It remains to show that every element of P is a Φ-linear combination of these elements and we shall show first that every element of $\mathfrak{O}$ is a linear combination with coefficients in $\mathfrak{o}$ of the elements $\rho_i\pi^j$. Let $\nu \in \mathfrak{O}$. Then $\psi(\nu) = \psi(\pi^k)$ for some $k \geq 0$. We can write $k = m_1e + j_1$ where $m_1 \geq 0$, $0 \leq j_1 \leq e - 1$. Then $\psi(\nu) = \psi(\beta^{m_1}\pi^{j_1})$ so $\mu = (\beta^{m_1}\pi^{j_1})^{-1}\nu$ satisfies $\psi(\mu) = 1$. The definition of the ρ_i shows that there exist elements $\alpha_{1i} \in \mathfrak{o}$ such that $\mu - \sum_1^f \alpha_{1i}\rho_i \in \mathfrak{P}$. Then $\psi(\Sigma\alpha_{1i}\rho_i) = \psi(\mu) = 1$ and, if $\nu_1 = \beta^{m_1}\pi^{j_1}(\mu - \Sigma\alpha_{1i}\rho_i)$, then $\psi(\nu_1) < \psi(\nu)$. We have

$$\nu = \beta^{m_1}\pi^{j_1}\mu = \beta^{m_1}\pi^{j_1}(\Sigma\alpha_{1i}\rho_i) + \nu_1. \tag{43}$$

We may repeat this argument with ν_1 and obtain a sequence ν_1, ν_2, $\cdots$ such that

$$\nu_{k-1} = \beta^{m_k}\pi^{j_k}(\Sigma\alpha_{ki}\rho_i) + \nu_k \tag{44}$$

where the $\alpha_{ki} \in \mathfrak{o}$, $m_k \geq 0$, $0 \leq j_k \leq e - 1$, $\psi(\Sigma\alpha_{ki}\rho_i) = 1$ and $\psi(\nu_k) < \psi(\nu_{k-1})$. Then (44) implies that $\psi(\nu_{k-1}) = \psi(\beta^{m_k}\pi^{j_k})$. It follows that $\nu_k \to 0$, $\beta^{m_k} \to 0$, and $(\Sigma\alpha_{ki}\rho_i)\beta^{m_k} \to 0$. The last implies that every infinite series whose terms form a subsequence of the sequence $(\Sigma\alpha_{ki}\rho_i)\beta^{m_k}$, $k = 1, 2, \cdots$, converges. By (43) and (44) we have

$$\begin{aligned}\nu = {} & \beta^{m_1}\pi^{j_1}(\Sigma\alpha_{1i}\rho_i) + \beta^{m_2}\pi^{j_2}(\Sigma\alpha_{2i}\rho_i) + \cdots \\ & + \beta^{m_k}\pi^{j_k}(\Sigma\alpha_{ki}\rho_i) + \nu_k.\end{aligned} \tag{45}$$

Since $\nu_k \to 0$ and the coefficients of the various powers π^j, $0 \leq j \leq e - 1$, in (45) converge, we obtain from (45) that $\nu = \Sigma\beta_{ij}\rho_i\pi^j$, $0 \leq j \leq e - 1$, where $\beta_{ij} \in \mathfrak{o}$. Now let ν be any element of P. Then we can find a power of β so that $\nu\beta^{-k} \in \mathfrak{O}$. Then we obtain $\nu = \beta^k(\Sigma\beta_{ij}\rho_i\pi^j)$ where $\beta_{ij} \in \mathfrak{o}$ so every element of P is a Φ-linear combination of the $\rho_i\pi^j$.

We can now prove

Theorem 19. *Let Φ be a field with a non-archimedean real valuation. Let P be a finite dimensional extension field of Φ, $\psi_1, \psi_2, \cdots, \psi_h$ the different valuations of P which extend φ and let e_i, f_i be the ramification index and residue degree of P/Φ relative to ψ_i. Then*

$$\sum_1^h e_if_i \leq n = [\mathrm{P}:\Phi] \tag{46}$$

and

$$\sum_{1}^{h} e_i f_i = n \tag{47}$$

holds if P *is separable over* Φ *and* φ *is discrete.*

Proof. Let E_i be the completion of P relative to ψ_i. Then for $\bar{\Phi}$ the completion of Φ and $n_i = [E_i:\bar{\Phi}]$ we have $\Sigma n_i \leq n$ and $\Sigma n_i = n$ for P separable over Φ. Also we have seen in § 5 that E_i and P have the same value group relative to ψ_i and $\bar{\Phi}$ and Φ have the same value group relative to φ. Hence the ramification index e_i of P over Φ relative to ψ_i is the same as that of E_i over $\bar{\Phi}$. Similarly, § 5 and the definitions show that the residue degree f_i of P/Φ relative to ψ_i is the same as that of E/$\bar{\Phi}$. By Lemmas 1 and 2 we have $e_i f_i \leq n_i$ and $e_i f_i = n_i$ if the valuation is discrete. Hence $\Sigma e_i f_i \leq \Sigma n_i \leq n$ in every case and $\Sigma e_i f_i = \Sigma n_i = n$ if P/Φ is separable and φ is discrete.

EXERCISE

1. Determine the residue degrees and ramification indices in the cases given in ex. 1 of § 15.

Chapter VI

ARTIN-SCHREIER THEORY

In this chapter we shall consider the theory of formally real fields which is due to Artin and Schreier. A basic algebraic property of the field of real numbers is that the only relations of the form $\Sigma\alpha_i^2 = 0$ which can hold in this field are the trivial ones: $0^2 + 0^2 + \cdots + 0^2 = 0$. This observation led Artin and Schreier to call any field having this property formally real. Any such field can be ordered and, on the other hand, any ordered field is formally real. Of central interest in the theory are the real closed fields, which are the formally real fields maximal under algebraic extension. A real closed field has a unique ordering which can be specified by the requirement that $\alpha > 0$ in such a field if and only if $\alpha = \beta^2 \neq 0$. Also, if P is real closed, then $P(\sqrt{-1})$ is algebraically closed. Any formally real field can be imbedded in a real closed field which is algebraic over the given field. Moreover, if the original field is ordered, then the imbedding can be made so that the (unique) ordering in the real closed algebraic extension is an extension of that of the given field. Such a real closed extension of an ordered field is essentially unique and is called the real closure of the ordered field.

The classical application of the Artin-Schreier theory is to the problem of determining which elements of a field are representable as sums of squares of elements of the field. For finite algebraic extensions of the rationals this has a simple answer which is due to Hilbert and to Landau (Th. 11). The theory of formally real fields led Artin to the solution of Hilbert's problem on the resolution of positive definite rational functions as sums of squares. We shall give a proof of Artin's theorem (Th. 12).

The most important development of the theory of formally real fields subsequent to the original work of Artin and Schreier is the metamathematical principle due to Tarski which asserts that any elementary statement of algebra which is valid for one real closed field is valid for every real closed field. This is based on an algorithm for deciding the solvability in a real closed field of a finite system of polynomial equations and inequalities with rational coefficients. Such a decision method was given originally by Tarski. We shall give an alternative one due to Seidenberg.

In the last section we shall establish the Artin-Schreier characterization of real closed fields as the fields which are not algebraically closed but are of finite co-dimension in algebraically closed fields.

1. Ordered fields and formally real fields. We have defined ordered groups in the last chapter (§ 5.7). In a similar manner one has the following

Definition 1. *An* ordered field Φ *is a field* Φ *together with a subset* P (*the set of* positive *elements*) *of* Φ *such that:* (1) $0 \notin P$, (2) *If* $\alpha \in \Phi$, *then either* $\alpha \in P$, $\alpha = 0$, *or* $-\alpha \in P$, (3) P *is closed under addition and multiplication.*

Since any field contains more than one element, it is clear that the subset P is not vacuous. If N denotes the set $\{-\alpha \mid \alpha \in P\}$, then (2) states that $\Phi = P \cup \{0\} \cup N$. Moreover, it is clear from (1) that $P \cap \{0\} = \varnothing$ and $N \cap \{0\} = \varnothing$. Also $P \cap N = \varnothing$ since, if $\alpha \in P \cap N$, then $-\alpha \in P \cap N$ and so $0 = \alpha + (-\alpha) \in P$ contrary to (1). Hence the decomposition $\Phi = P \cup \{0\} \cup N$ is one into non-overlapping sets. It is clear that N is closed under addition since $(-\alpha) + (-\beta) = -(\alpha + \beta) \in N$ if $\alpha, \beta \in P$. On the other hand, $(-\alpha)(-\beta) = \alpha\beta \in P$ if $-\alpha, -\beta \in N$.

We can introduce a partial ordering in the ordered field Φ (or more precisely Φ, P) by defining $\alpha > \beta$ if $\alpha - \beta \in P$. Then if α, β are any two elements of Φ, we have the trichotomy: one and only one of the relations $\alpha > \beta$, $\alpha = \beta$, $\beta > \alpha$ holds. Thus Φ is linearly ordered by the relation $\alpha > \beta$. If $\alpha > \beta$, then $\alpha + \gamma > \beta + \gamma$ and $\alpha\delta > \beta\delta$ if $\delta > 0$. Conversely, we can define an ordered field by means of a linear ordering $>$ such that $\alpha > \beta$ implies

$\alpha + \gamma > \beta + \gamma$ and $\alpha\delta > \beta\delta$ if $\delta > 0$. Let P denote the set of elements $\alpha > 0$. Then it is immediate that Φ, P is an ordered field in the original sense and that the relation $>$ defined by Φ, P is the given ordering relation.

As usual, it is convenient to write $\alpha < \beta$ for $\beta > \alpha$. The elementary properties of the ordering in the field of real numbers are readily established. We list some of these: $\alpha > 0$ implies $\alpha^{-1} > 0$ and $\alpha > \beta > 0$ implies $\beta^{-1} > \alpha^{-1} > 0$. If $\alpha > \beta$, then $-\alpha < -\beta$ and, if $\alpha > \beta$ and $\gamma > \delta$, then $\alpha + \gamma > \beta + \delta$. As usual, one defines $|\alpha| = \alpha$ if $\alpha \geq 0$ and $|\alpha| = -\alpha$ if $\alpha < 0$, and one proves that $|\alpha + \beta| \leq |\alpha| + |\beta|$ and $|\alpha\beta| = |\alpha|\ |\beta|$.

If Φ' is a subfield of an ordered field Φ, P, then Φ' is ordered relative to $P' = \Phi' \cap P$. We shall call this the *induced ordering* in Φ'. Evidently $\alpha' > \beta'$ in Φ', P' if and only if $\alpha' > \beta'$ in Φ, P. If Φ, P and Φ', P' are any two ordered fields, then an isomorphism s of Φ into Φ' is called an *order isomorphism* (or an isomorphism of the ordered fields) if $P^s \subseteq P'$. This implies that $N^s \subseteq N'$, the set of negatives of the elements of P' and, if s is surjective, then $P^s = P'$ and $N^s = N'$.

In any ordered field Φ, $\alpha \neq 0$ implies $\alpha^2 > 0$. Hence if $\alpha_1, \alpha_2, \cdots, \alpha_r$ are $\neq 0$, then $\Sigma\alpha_i^2 > 0$. This shows that any ordered field is formally real in the sense of the following

Definition 2. *A field Φ is called* formally real *if the only relations of the form $\sum_{i=1}^{r} \alpha_i^2 = 0$ in Φ are those for which every $\alpha_i = 0$.*

It is immediate that Φ is formally real if and only if -1 is not a sum of squares of elements of Φ. If the characteristic of Φ is $p \neq 0$, then $0 = 1^2 + 1^2 + \cdots + 1^2$ (p terms); hence it is clear that formally real fields are necessarily of characteristic 0.

In any field Φ let $\Sigma(\Phi)$ denote the subset of elements which are sums of squares. Evidently $\Sigma(\Phi)$ contains 0 and is closed under addition and multiplication. Moreover, we have seen that Φ is formally real if and only if $-1 \notin \Sigma(\Phi)$. If $\beta \neq 0$ is in $\Sigma(\Phi)$, then $\beta^{-1} \in \Sigma(\Phi)$; for, we have $\beta = \Sigma\beta_i^2$ and so $\beta^{-1} = \beta(\beta^{-1})^2 = \Sigma(\beta_i\beta^{-1})^2$. We note also that, if Φ is not formally real and not of characteristic two, then $\Sigma(\Phi) = \Phi$; for, $-1 \in \Sigma(\Phi)$ and, if α is any element of Φ, then

$$\alpha = \left(\frac{1+\alpha}{2}\right)^2 - \left(\frac{1-\alpha}{2}\right)^2 = \left(\frac{1+\alpha}{2}\right)^2 + (-1)\left(\frac{1-\alpha}{2}\right)^2$$

$\varepsilon\ \Sigma(\Phi)$ since $\Sigma(\Phi)$ is closed under addition and multiplication. It will be useful to state these results on $\Sigma(\Phi)$ in the following

Lemma. *Let Φ be a field and let $\Sigma(\Phi)$ be the subset of Φ of elements which are sums of squares. Then $\Sigma(\Phi)$ is closed under addition and multiplication and contains β^{-1} for every $\beta \neq 0$ in $\Sigma(\Phi)$. If Φ is not formally real and not of characteristic two, then $\Sigma(\Phi) = \Phi$.*

EXERCISES

1. Show that the field of rational numbers can be ordered in one and only one way.
2. Show that the field $R_0(\sqrt{2})$ where R_0 is the field of rational numbers has exactly two distinct orderings.
3. Let Φ be an ordered field, $f(x) = x^n + \alpha_1 x^{n-1} + \cdots + \alpha_n$ a polynomial with coefficients in Φ. Let $M = \max(1, |\alpha_1| + |\alpha_2| + \cdots + |\alpha_n|)$. Show that every root of $f(x)$ in Φ is contained in the interval $-M \leq x \leq M$.
4. Show that any purely transcendental extension of a formally real field is formally real.
5. Let R_0 be the rationals and let $\Phi = R_0(\xi)$ where ξ is transcendental. Show that Φ has a non-countable number of distinct orderings.
6. Let Φ be a formally real field and let $\mathfrak{H}(\Phi_n)$ denote the set of $n \times n$ symmetric matrices with entries in Φ. Show that $\mathfrak{H}(\Phi_n)$ is formally real in the sense that $\Sigma A_i^2 = 0$, $A_i\ \varepsilon\ \mathfrak{H}(\Phi_n)$ implies that every $A_i = 0$.
7. Let (x, y) be a symmetric bilinear form on an n dimensional vector space $\mathfrak{M}$ over Φ where Φ is an ordered field. Let $\{\beta_1, \beta_2, \cdots, \beta_n\}$ be a diagonal matrix for (x, y). Prove the following extension of Sylvester's theorem (Vol. II, p. 156): The number of positive β_i is an invariant of (x, y).
8. An ordered field is called *archimedean* if, given any $\alpha > 0$, $\beta > 0$, there exists an integer n such that $n\alpha > \beta$ (equivalently, given $\alpha > 0$, there exists an integer n such that $n > \alpha$). Let P be an ordered field, Φ a subfield with the induced ordering. Show that P is archimedean if: 1) Φ is archimedean and 2) $[\mathrm{P}:\Phi] < \infty$. (Hint: Use ex. 3.)
9. Prove that any archimedean ordered field is order isomorphic to a subfield of the field R of real numbers (cf. Th. 5.8).
10. (Cohn). Let Φ be ordered with P as the set of positive elements. Show that $\Phi(\xi)$, ξ transcendental over Φ can be ordered by choosing as set P_ξ of positive elements those elements which have the form $\beta\xi^r fg^{-1}$ where $\beta\ \varepsilon\ P$ and f and g are polynomials in ξ with constant term 1. Show that $\Phi(\xi)$ is not archimedean ordered.
11. (Cohn). Let Φ be ordered and let ξ, η be algebraically independent over Φ in $\Phi(\xi, \eta)$. Order $\Phi(\xi)$ as in ex. 10 and then repeat the process for $\Phi(\xi, \eta)$ con-

sidering this as the purely transcendental extension $\Phi(\xi)(\eta)$ of $\Phi(\xi)$. Show that every element of $\Phi(\xi, \eta)$ is majorized by an element of $\Phi(\eta)$ but that there exists no element of $\Phi(\eta)$ between ξ and ξ^2.

12. Let P be an ordered field, Φ a subfield. Let $\mathfrak{p}$ be the set of elements β of P such that $|\beta| < |\alpha|$ for every $\alpha \neq 0$ in Φ and let $\mathfrak{o} = \{\gamma \in P | \gamma\mathfrak{p} \subset \mathfrak{p}\}$. Show that $\mathfrak{o}$ is a valuation ring in P containing Φ and that $\mathfrak{p}$ is the ideal of non-units of $\mathfrak{o}$. Show that the residue field $\mathfrak{o}/\mathfrak{p}$ can be ordered by defining $\gamma + \mathfrak{p} > 0$ if $\gamma \notin \mathfrak{p}$ and $\gamma > 0$ in P. Show that $\mathfrak{o}/\mathfrak{p}$ is an extension of Φ (identified with $(\Phi + \mathfrak{p})/\mathfrak{p}$) which is an *archimedean extension* of Φ in the sense that every interval (a, b), $a, b \in \mathfrak{o}/\mathfrak{p}$, contains an element of Φ.

2. Real closed fields.

The deeper properties of formally real fields concern real closed fields which are defined as follows.

Definition 3. *A field Φ is called* real closed *if Φ is formally real and no proper algebraic extension of Φ is formally real.*

We shall show first that any real closed field can be ordered in one and only one way. This is an easy consequence of the following

Theorem 1. *If Φ is real closed, then any element of Φ is either a square or the negative of a square.*

Proof. Let α be an element of Φ which is not a square. Then we can construct the proper algebraic extension $\Omega = \Phi(\sqrt{\alpha})$. This field is not formally real, so there exist β_i, γ_i not all 0 in Φ such that $\Sigma(\beta_i + \gamma_i\sqrt{\alpha})^2 = 0$. This gives $\Sigma(\beta_i^2 + \gamma_i^2\alpha) + 2(\Sigma\beta_i\gamma_i)\sqrt{\alpha} = 0$. Since $\sqrt{\alpha} \notin \Phi$ we have $2\Sigma\beta_i\gamma_i = 0$ and $\Sigma\beta_i^2 + \alpha\Sigma\gamma_i^2 = 0$. Since Φ is formally real, $\Sigma\gamma_i^2 \neq 0$. Then $-\alpha = (\Sigma\beta_i^2)(\Sigma\gamma_i^2)^{-1}$. Using the properties of the set $\Sigma(\Phi)$ of sums of squares stated in the lemma of § 1, it follows that $-\alpha \in \Sigma(\Phi)$. Since $-1 \notin \Sigma(\Phi)$ by the formal reality of Φ this implies that $\alpha \notin \Sigma(\Phi)$. Thus we have shown that, if an element of Φ is not a square, then it is not a sum of squares. In other words, if $\alpha \in \Sigma(\Phi)$, then α is a square. Moreover, we have seen that, if α is not a square, then $-\alpha \in \Sigma(\Phi)$ and this now implies that $-\alpha$ is a square. This is what we wished to show.

We can now prove

Theorem 2. *Any real closed field can be ordered in one and only one way. Any automorphism of such a field is an order isomorphism.*

Proof. Let P be the subset of non-zero squares in the real closed field Φ. Then $0 \notin P$ and, if $\alpha \neq 0$ and $\alpha \notin P$, then $-\alpha \in P$

by Theorem 1. If $\alpha = \beta^2$ and $\gamma = \delta^2 \varepsilon P$, then $\alpha + \gamma \varepsilon P$. Otherwise, $\alpha + \gamma = -\epsilon^2$ where $\epsilon \varepsilon \Phi$. This gives $\beta^2 + \delta^2 + \epsilon^2 = 0$ contrary to the formal reality of Φ. Hence we see that the subset P satisfies the conditions 1, 2, 3 for an ordered field and Φ, P is such a field. Let P' be any subset of Φ which gives an ordering. If $\alpha \varepsilon P$, $\alpha = \beta^2 \neq 0$. Then $\alpha > 0$ in the ordering given by P'. Hence $P' \supseteq P$. This implies that $P' = P$ so the ordering in Φ is uniquely determined. If s is an automorphism of Φ, then it is clear that s maps the set P of non-zero squares into itself. Hence s is an order isomorphism of Φ.

The question of the existence of real closed fields is easily settled. In fact, we have the following

Theorem 3. *Let Φ be a formally real field and let Ω be an algebraic closure of Φ. Then Ω contains a real closed field Δ containing Φ.*

Proof. We consider the collection of formally real subfields of Ω containing Φ. This collection is not vacuous since it contains Φ. Moreover, it is clear that the collection is inductive, so, by Zorn's lemma, it contains a maximal element Δ. If Δ is not real closed, then it has a proper algebraic extension Δ' which is formally real. Since Ω is algebraically closed, we may suppose that $\Delta' \subseteq \Omega$ (ex. 1, p. 147). This contradicts the maximality of Δ in Ω. Hence Δ is real closed.

Evidently Theorems 2 and 3 and the existence of an algebraic closure for any field imply the following corollaries.

Corollary 1. *Any formally real field can be imbedded in a real closed field which is algebraic over the given field.*

Corollary 2. *Any formally real field can be ordered.*

If Φ is real closed, then -1 is not a square in Φ so $\Phi(\sqrt{-1}) \supset \Phi$. We shall show that $\Phi(\sqrt{-1})$ is algebraically closed and we shall see that this property is characteristic of real closed fields. For this purpose we prove first the following result.

Theorem 4. *If Φ is real closed, then every polynomial of odd degree with coefficients in Φ has a root belonging to Φ.*

Proof. The result is clear for polynomials of degree 1 and we use induction on the degree n of $f(x)$. If $f(x)$ is reducible, one of its factors is of odd degree so it has a root in Φ. Hence we may assume $f(x)$ is irreducible. Let $\Delta = \Phi(\theta)$ where $f(\theta) = 0$. Then $\Delta \supset \Phi$, so Δ is not formally real. Hence we have a relation $\Sigma\varphi_i(\theta)^2 = -1$ where $\varphi_i(x)$ is a polynomial in x of degree $\leq n - 1$. The relation indicated implies that $\Sigma\varphi_i(x)^2 = -1 + f(x)g(x)$. The leading coefficient of $\Sigma\varphi_i(x)^2$ is positive in the ordering in Φ and the degree of this polynomial is even and $\leq 2(n - 1)$. It follows that $\deg g(x)$ is odd and $\leq 2(n - 1) - n = n - 2$. Hence there exists a $\beta \in \Phi$ such that $g(\beta) = 0$. Substituting this β in the relation $\Sigma\varphi_i(x)^2 = -1 + f(x)g(x)$ gives $\Sigma\varphi_i(\beta)^2 = -1$ contrary to the formal reality of Φ.

We shall prove next the following generalization to real closed fields of the so-called fundamental theorem of algebra. The proof is patterned rather closely after one of Gauss' proofs of the classical result.

Theorem 5. *Let Φ be an ordered field such that:* (1) *positive elements in Φ have square roots in Φ,* (2) *any polynomial of odd degree with coefficients in Φ has a root in Φ. Then $\sqrt{-1} \notin \Phi$ and $\Phi(\sqrt{-1})$ is algebraically closed.*

Proof. Since Φ is real, it is clear that $\sqrt{-1} \notin \Phi$. Consider $\Phi(\sqrt{-1}) \supset \Phi$. Let $\rho \rightarrow \bar{\rho}$ be the automorphism of $\Phi(\sqrt{-1})$ over Φ such that $\bar{i} = -i$ for $i = \sqrt{-1}$. If $f(x) \in \Phi(\sqrt{-1})[x]$, then $f(x)\bar{f}(x) \in \Phi[x]$, and if this has a root in $\Phi(\sqrt{-1})$, then $f(x)$ has a root in $\Phi(\sqrt{-1})$. Hence the algebraic closure of $\Phi(\sqrt{-1})$ will follow if we can show that every non-constant polynomial with coefficients in Φ has a root in $\Phi(\sqrt{-1})$. This holds by (2) if the degree of the polynomial is odd. We show next that every element of $\Phi(\sqrt{-1})$ has a square root in this field. First, if $\alpha \in \Phi$ and $\alpha > 0$, then, by (1), $\alpha = \beta^2$, $\beta \in \Phi$. Next if $\alpha \in \Phi$ and $\alpha < 0$, then $-\alpha = \beta^2$ and $\alpha = (\sqrt{-1})^2\beta^2$. Now let $\rho = \alpha + \beta i$, $i = \sqrt{-1}$, α, β in Φ, $\beta \neq 0$. Consider the element $\xi + \eta i$, ξ, η in Φ. We have $(\xi + \eta i)^2 = \xi^2 - \eta^2 + 2\xi\eta i$ so $(\xi + \eta i)^2 = \alpha + \beta i$ is equivalent to

$$(1) \qquad \xi^2 - \eta^2 = \alpha, \quad 2\xi\eta = \beta.$$

Since $\beta \neq 0$ we may (by multiplying by a suitable element of Φ) assume that $\beta = 2$, so the second equation becomes $\xi\eta = 1$. This holds if $\eta = \xi^{-1}$. Then the first equation becomes $\xi^2 - \xi^{-2} = \alpha$ or $\lambda - \lambda^{-1} = \alpha$ for $\lambda = \xi^2$. Then we have $\lambda^2 - \alpha\lambda - 1 = 0$ which has the solution $(\alpha + \sqrt{\alpha^2 + 4})/2$ in Φ since $\alpha^2 + 4 > 0$. Also $\alpha + \sqrt{\alpha^2 + 4} > 0$ since $\alpha + \sqrt{\alpha^2 + 4} \leq 0$ leads to $4 \leq 0$. Hence there exists a $\xi \neq 0$ in Φ such that $\xi^2 = \frac{1}{2}(\alpha + \sqrt{\alpha^2 + 4})$. Then $\xi^4 - \alpha\xi^2 = 1$ and $\xi^2 - \xi^{-2} = \alpha$. Hence ξ and $\eta = \xi^{-1}$ satisfy (1) with $\beta = 2$. We have therefore proved that every element of $\Phi(\sqrt{-1})$ has a square root in this field. Consequently there exists no extension field Δ of $\Phi(\sqrt{-1})$ such that

$$[\Delta:\Phi(\sqrt{-1})] = 2.$$

We proceed to use this fact to prove that every polynomial of positive degree with coefficients in Φ has a root in $\Phi(\sqrt{-1})$. Let $f(x)$ be such a polynomial and let E be a splitting field over Φ of $(x^2 + 1)f(x)$. We may assume that $\mathrm{E} \supseteq \Phi(\sqrt{-1})$. Since the the characteristic is 0, E is Galois over Φ. Let G be its Galois group and let $(G:1) = 2^e m$ where m is odd. By Sylow's theorem G has a subgroup H of order 2^e. Let Δ be the subfield over Φ of H-invariants. Then $[\mathrm{E}:\Delta] = 2^e$ and $[\Delta:\Phi] = m$. Since Φ has no proper odd dimensional extension field we must have $\Delta = \Phi$ and $m = 1$. Hence $G = H$ has order 2^e. Such a group is solvable. If $e > 1$, it follows easily from Galois theory that E contains a subfield Γ over $\Phi(\sqrt{-1})$ such that $[\Gamma:\Phi(\sqrt{-1})] = 2$. This contradicts what we proved before. Hence $e = 1$, so $[\mathrm{E}:\Phi] = 2$ and $\mathrm{E} = \Phi(\sqrt{-1})$. This shows that $\Phi(\sqrt{-1})$ is a splitting field of $(x^2 + 1)f(x)$ and that $f(x)$ has a root of $\Phi(\sqrt{-1})$. Hence $\Phi(\sqrt{-1})$ is algebraically closed.

If Φ is a real closed field, then we have seen that Φ can be ordered in exactly one way. The proof of Theorem 2 shows that this ordering is obtained by specifying that $\alpha > 0$ if $\alpha = \beta^2$, $\beta \neq 0$. Hence we see that every real closed field is ordered and satisfies condition (1) of Theorem 5. Theorem 4 shows that every real closed field satisfies condition (2) of Theorem 5. Hence we have the following

Corollary. *If Φ is a real closed field, then $\sqrt{-1} \notin \Phi$ and $\Phi(\sqrt{-1})$ is algebraically closed.*

We shall prove next the converse of this, namely,

Theorem 6. *If Φ is a field such that $\sqrt{-1} \notin \Phi$ and $\Phi(\sqrt{-1})$ is algebraically closed, then Φ is real closed.*

Proof. Suppose Φ satisfies the conditions. We note first that the irreducible polynomials of positive degrees in $\Phi[x]$ have degree 1 or 2. Let $f(x)$ be such a polynomial and let θ be a root of $f(x)$ contained in $\Omega = \Phi(\sqrt{-1})$. Then $[\Phi(\theta):\Phi] = \deg f(x)$ and $[\Phi(\theta):\Phi] \leq [\Omega:\Phi] = 2$. Hence $\deg f(x) = 1$ or 2 as asserted. Now let $\alpha, \beta \neq 0 \in \Phi$ and consider the polynomial

$$
\begin{aligned}
(2) \qquad g(x) &= (x^2 - \alpha)^2 + \beta^2 = (x^2 - \alpha - \beta i)(x^2 - \alpha + \beta i) \\
&= (x - (\alpha + \beta i)^{1/2})(x + (\alpha + \beta i)^{1/2}) \cdot \\
&\qquad (x - (\alpha - \beta i)^{1/2})(x + (\alpha - \beta i)^{1/2}),
\end{aligned}
$$

where $i = \sqrt{-1}$. This polynomial belongs to $\Phi[x]$ and has no linear factors in $\Phi[x]$ since $\pm\alpha \pm \beta i \notin \Phi$. Hence $g(x)$ is a product of two irreducible quadratic polynomials. The one divisible by $x - (\alpha + \beta i)^{1/2}$ cannot be

$$(x - (\alpha + \beta i)^{1/2})(x + (\alpha + \beta i)^{1/2}) = x^2 - (\alpha + \beta i);$$

for, this would imply that $\alpha + \beta i \in \Phi$. Hence the polynomial in question is either

$$(x - (\alpha + \beta i)^{1/2})(x - (\alpha - \beta i)^{1/2})$$

or

$$(x - (\alpha + \beta i)^{1/2})(x + (\alpha - \beta i)^{1/2}).$$

Either possibility implies that $(\alpha^2 + \beta^2)^{1/2} \in \Phi$. Since α and β were arbitrary non-zero elements of Φ, we have proved that the sum of two squares of elements in Φ is a square. Induction shows that every sum of squares is a square in Φ. Since -1 is not a square, this implies that -1 is not a sum of squares in Φ and so Φ is formally real. If P is a proper algebraic extension of Φ, then P is isomorphic to $\Omega = \Phi(\sqrt{-1})$. Then P is not formally real and so Φ is real closed. This completes the proof of Theorem 6.

The corollary to Theorem 5 and Theorem 6 give the characterization of real closed fields by the properties that $\sqrt{-1} \notin \Phi$ and $\Phi(\sqrt{-1})$ is algebraically closed. We remark also that there is another characterization involved in our discussion, namely, an ordered field is real closed if and only if it satisfies conditions

(1) and (2) of Theorem 5, that is, positive elements of Φ have square roots in Φ and polynomials of odd degree with coefficients in Φ have roots in Φ. This is easily deduced from our results. We derive next the following useful consequence of one of our characterizations of real closed fields.

Corollary. *If* P *is a real closed extension field of a field* Φ, *then the subfield* A *of elements of* P *which are algebraic over* Φ *is real closed.*

Proof. Let $\Omega = P(\sqrt{-1})$. Then Ω is algebraically closed. Hence the subfield Γ of elements of Ω which are algebraic over Φ is algebraically closed. If $\alpha + \beta\sqrt{-1}$, $\alpha, \beta \in P$, is in Γ, then so is $\alpha - \beta\sqrt{-1}$. Hence $\alpha = \frac{1}{2}(\alpha + \beta\sqrt{-1} + \alpha - \beta\sqrt{-1}) \in \Gamma$. Then $\beta \in \Gamma$. Since $\alpha, \beta \in P$ we see that $\alpha, \beta \in A$. It follows that $\Gamma = A(\sqrt{-1})$. Since $\sqrt{-1} \notin A$, we see that A fulfills the conditions of Theorem 6. Hence A is a real closed field.

EXERCISE

1. Let Ω/Φ be algebraically closed, Φ formally real. Show that Ω/Φ contains a real closed subfield P/Φ such that $\Omega = P(\sqrt{-1})$. In particular, show that every algebraically closed field of characteristic 0 contains a real closed subfield P such that $\Omega = P(\sqrt{-1})$.

3. Sturm's theorem. In this section we shall derive a classical result, Sturm's theorem, which permits us to determine the exact number of roots in a real closed field of a polynomial equation $f(x) = 0$. This result is fundamental in the sequel. In deriving it we shall follow rather closely Weber's exposition in *Lehrbuch der Algebra* (1898), Vol. I, pp. 301–313. We shall need first the following basic result.

Lemma. *Let* Φ *be a real closed field and* $f(x)$ *a polynomial with coefficients in* Φ. *Suppose* α *and* β *are elements of* Φ *such that* $f(\alpha) < 0$ *while* $f(\beta) > 0$. *Then there exists a* γ *between* α *and* β *such that* $f(\gamma) = 0$.

Proof. We recall that the only irreducible polynomials in $\Phi[x]$ are the linear ones and the quadratic ones. Let $g(x) = x^2 + \mu x + \nu \in \Phi[x]$ be irreducible. We assert that necessarily $\mu^2 - 4\nu < 0$. This is clear from the formulas for the roots of a quadratic equation. We can now set $4\nu - \mu^2 = 4\delta^2$ where δ is a non-zero element of Φ and we have

$$(3) \qquad g(x) = x^2 + \mu x + \nu = \left(x + \frac{\mu}{2}\right)^2 + \delta^2.$$

Evidently this formula shows that $g(\eta) > 0$ for every η in Φ. Now let $f(x)$, α, β be as in the statement of the theorem. In $\Phi[x]$ we have the factorization

$$(4) \qquad f(x) = \rho(x - \rho_1)(x - \rho_2) \cdots (x - \rho_k)g_1(x) \cdots g_l(x)$$

where $g_i(x)$ is an irreducible quadratic with leading coefficient 1. Suppose none of the ρ_i is between α and β. Then for each i, $\alpha - \rho_i$ and $\beta - \rho_i$ have the same sign (both positive or both negative). Since $g_j(\alpha) > 0$ and $g_j(\beta) > 0$, $1 \leq j \leq l$, this implies that $f(\alpha)$ and $f(\beta)$ have the same sign, contrary to hypothesis. Hence there is a ρ_i between α and β. This completes the proof.

Let Φ be a real closed field and let $f(x)$ be a polynomial of positive degree with coefficients in Φ. Following Weber, we shall say that a sequence of polynomials

$$(5) \qquad f_0(x) = f(x), \quad f_1(x), \cdots, f_s(x)$$

is a *Sturm sequence* of polynomials for $f(x)$ for the interval $[\alpha, \beta]$ (that is, $\alpha \leq x \leq \beta$) if the $f_i(x) \in \Phi[x]$ and satisfy the following conditions:

(i) $f_s(x)$ has no roots in $[\alpha, \beta]$.
(ii) $f_0(\alpha) \neq 0$, $f_0(\beta) \neq 0$.
(iii) If $\gamma \in [\alpha, \beta]$ is a root of $f_j(x)$, $0 < j < s$, then

$$f_{j-1}(\gamma) f_{j+1}(\gamma) < 0$$

(iv) If $f(\gamma) = 0$, $\gamma \in [\alpha, \beta]$, then there exist intervals $\gamma_1 \leq x < \gamma$ and $\gamma < x \leq \gamma_2$ such that $f_0(x) f_1(x) < 0$ for x in the first of these and $f_0(x) f_1(x) > 0$ for x in the second. (This amounts to saying that $f_0(x) f_1(x)$ is an increasing function of x at $x = \gamma$.)

We shall establish the existence of such sequences for any polynomial with distinct roots, but first we shall see how such a sequence can be used to determine the number of roots of $f(x)$ in the open interval (α, β) (that is, $\alpha < x < \beta$). We consider the number of variations in sign of the sequence

$$(6) \qquad \begin{array}{c} f_0(\alpha), f_1(\alpha), \cdots, f_s(\alpha) \\ f_0(\beta), f_1(\beta), \cdots, f_s(\beta) \end{array}$$

of elements of Φ. If $\gamma = \{\gamma_1, \gamma_2, \cdots, \gamma_m\}$ is a finite sequence of non-zero elements of Φ, then we define the *number of variations in sign* of γ to be the number of i, $1 \leq i \leq m - 1$, such that $\gamma_i\gamma_{i+1} < 0$. If $\gamma = \{\gamma_1, \gamma_2, \cdots, \gamma_m\}$ is an arbitrary sequence of elements of Φ, then we define the number of variations in sign of γ to be the number of variations in sign of the abbreviated sequence γ' obtained by dropping the 0's in γ. For example,

$$\{1, 0, 0, 2, -1, 0, 3, 4, -2\}$$

has three variations in sign.

We can now state

Theorem 7. *Let $f(x)$ be a polynomial of positive degree with coefficients in a real closed field Φ and let $f_0(x) = f(x), f_1(x), \cdots, f_s(x)$ be a Sturm sequence for $f(x)$ for the interval $[\alpha, \beta]$. Then the number of distinct roots of $f(x)$ in (α, β) is $V_\alpha - V_\beta$ where, in general V_γ denotes the number of variations in sign of the sequence $\{f_0(\gamma), f_1(\gamma), \cdots, f_s(\gamma)\}$.*

Proof. The interval $[\alpha, \beta]$ is decomposed into subintervals by the roots of the polynomials $f_j(x)$ of the given Sturm sequence. Thus we have a sequence $\alpha = \alpha_0 < \alpha_1 < \cdots < \alpha_m = \beta$ such that none of the $f_j(x)$ has a root in (α_i, α_{i+1}). Choose $\alpha_i' \in (\alpha_{i-1}, \alpha_i)$, $1 \leq i \leq m$ (e.g., $\alpha_i' = \frac{1}{2}(\alpha_{i-1} + \alpha_i)$) and let $V_{\alpha_i'}$ be the number of variations in sign of the sequence $\{f_j(\alpha_i'), j = 0, 1, \cdots, s\}$. Evidently,

$$V_\alpha - V_\beta = V_\alpha - V_{\alpha_1'} + \sum_1^{m-1} (V_{\alpha_i'} - V_{\alpha_{i+1}'}) + V_{\alpha_m'} - V_\beta,$$

so we shall try to compute $V_\alpha - V_{\alpha_1'}$, $V_{\alpha_i'} - V_{\alpha_{i+1}'}$, $V_{\alpha_m'} - V_\beta$. We have $f_0(\alpha) \neq 0$, $f_0(\beta) \neq 0$, $f_s(\alpha_i) \neq 0$, $f_s(\alpha_i') \neq 0$. Suppose first that no $f_j(\alpha) = 0$, $0 < j < s$. Then $f_k(\alpha)f_k(\alpha_1') > 0$ for $k = 0, \cdots, s$, since, otherwise, by the lemma, one of the $f_k(x)$ has a root in (α, α_1') contrary to the property of the intervals (α_i, α_{i+1}). Hence we have $V_\alpha = V_{\alpha_1'}$ in the case under consideration.

Next let $f_j(\alpha) = 0$ for some j, $0 < j < s$. Then $f_{j-1}(\alpha)f_{j+1}(\alpha) < 0$, by (iii). Since $f_{j-1}(x)$ and $f_j(x)$ have no roots in (α, α_1), we have $f_{j-1}(\alpha)f_{j-1}(\alpha_1') > 0$ and $f_{j+1}(\alpha)f_{j+1}(\alpha_1') > 0$. Hence $f_{j-1}(\alpha_1')f_{j+1}(\alpha_1') < 0$. It follows that $f_{j-1}(\alpha)$, 0, $f_{j+1}(\alpha)$ and $f_{j-1}(\alpha_1')$, $f_j(\alpha_1')$, $f_{j+1}(\alpha_1')$ contribute the same number of variations of sign to V_α and $V_{\alpha_1'}$ respectively. Taking into account all the j we see that $V_\alpha - V_{\alpha_1'} = 0$. A similar argument shows that $V_{\alpha_m'} - V_\beta = 0$. Also the same argument shows that, if α_i, $1 \leq i \leq m - 1$, is not a root of $f(x) = f_0(x)$ then again $V_{\alpha_i'} - V_{\alpha_{i+1}'} = 0$. It remains to consider what happens if $f(\alpha_i) = 0$ for $1 \leq i \leq m - 1$. Then, by (iv) and the choice of the α_i', we have $f_0(\alpha_i')f_1(\alpha_i') < 0$ and $f_0(\alpha_{i+1}')f_1(\alpha_{i+1}') > 0$. Then the sequence $f_0(\alpha_i')$, $f_1(\alpha_i')$ has one variation in sign while the sequence $f_0(\alpha_{i+1}')$, $f_1(\alpha_{i+1}')$ has none. The argument used before shows that $f_{j-1}(\alpha_i')$, $f_j(\alpha_i')$, $f_{j+1}(\alpha_i')$ and $f_{j-1}(\alpha_{i+1}')$, $f_j(\alpha_{i+1}')$, $f_{j+1}(\alpha_{i+1}')$ have the same number of variations of sign if $j > 1$. Hence we see that $V_{\alpha_i'} - V_{\alpha_{i+1}'} = 1$ if $f(\alpha_i) = 0$. We have therefore shown that $V_{\alpha_i'} - V_{\alpha_{i+1}'} = 0$ or 1 according as $f(\alpha_i) \neq 0$ or $f(\alpha_i) = 0$. Hence

$$V_\alpha - V_\beta = V_\alpha - V_{\alpha_1'} + \sum_1^{m-1} (V_{\alpha_i'} - V_{\alpha_{i+1}'}) + V_{\alpha_m'} - V_\beta$$

is the number of α_i such that $f(\alpha_i) = 0$.

Now let $f(x)$ be an arbitrary polynomial. We define the *standard sequence* for $f(x)$ by

$$f_0(x) = f(x), \quad f_1(x) = f'(x) \quad (\text{formal derivative of } f(x)),$$

$$f_0(x) = q_1(x)f_1(x) - f_2(x), \quad \deg f_2 < \deg f_1$$

$$\vdots$$

$$f_{i-1}(x) = q_i(x)f_i(x) - f_{i+1}(x), \quad \deg f_{i+1} < \deg f_i \tag{7}$$

$$\vdots$$

$$f_{s-1}(x) = q_s(x)f_s(x).$$

Thus the $f_i(x)$ are obtained by modifying the Euclid algorithm for finding the highest common factor of $f(x)$ and $f'(x)$ in such a way that the last polynomial obtained at each stage is the negative of the remainder in the division process. Clearly, $f_s(x)$ is the highest common factor of $f(x)$ and $f'(x)$ and this is a divisor of all the $f_i(x)$. Now set $g_i(x) = f_i(x)f_s(x)^{-1}$ and consider the sequence

$$g_0(x), \quad g_1(x), \cdots, g_s(x). \tag{8}$$

We proceed to show that this is a Sturm sequence for $g_0(x)$ for any interval $[\alpha, \beta]$ such that $g_0(\alpha) \neq 0$, $g_0(\beta) \neq 0$. Clearly (ii) in the definition of Sturm sequences is satisfied. Also (i) holds since $g_s(x) = 1$. Dividing the polynomials in (7) by $f_s(x)$ gives the relation $g_{j-1}(x) = q_j(x)g_j(x) - g_{j+1}(x)$, $0 < j < s$. Suppose $g_j(\gamma) = 0$. Then $g_{j-1}(\gamma) \neq 0$ and $g_{j+1}(\gamma) \neq 0$, since otherwise the relations indicated would imply that all the $g_k(\gamma) = 0$ from a certain point on contrary to $g_s(x) = 1$. Thus $g_{j-1}(\gamma)g_{j+1}(\gamma) \neq 0$ and, since $g_{j-1}(\gamma) = g_j(\gamma)q_j(\gamma) - g_{j+1}(\gamma) = -g_{j+1}(\gamma)$, we have $g_{j-1}(\gamma)g_{j+1}(\gamma) < 0$ and (iii) holds. Now suppose that $g_0(\gamma) = 0$ for γ in $[\alpha, \beta]$. Then we have $f(x) = (x - \gamma)^e h(x)$, $e > 0$, $h(\gamma) \neq 0$ and $f'(x) = (x - \gamma)^e h'(x) + e(x - \gamma)^{e-1}h(x)$. Also $f_s(x) = (x - \gamma)^{e-1}k(x)$ where $k(\gamma) \neq 0$. Hence $h(x) = k(x)l(x)$ where $l(\gamma) \neq 0$ and $h'(x) = k(x)m(x)$. These relations give

$$\begin{aligned} g_0(x) &= (x - \gamma)l(x), \quad l(\gamma) \neq 0 \\ g_1(x) &= (x - \gamma)m(x) + el(x) \end{aligned} \tag{9}$$

so $g_1(\gamma) = el(\gamma) \neq 0$. Now choose an interval $[\gamma_1, \gamma_2]$ containing γ in its interior such that $l(x) \neq 0$ and $g_1(x) \neq 0$ in $[\gamma_1, \gamma_2]$. Then the lemma implies that $g_1(x)$ and $l(x)$ are either both positive or both negative in $[\gamma_1, \gamma_2]$ so $g_1(x)l(x) > 0$ in $[\gamma_1, \gamma_2]$. Hence $g_0(x)g_1(x) = (x - \gamma)g_1(x)l(x)$ has the same sign as $x - \gamma$ in $[\gamma_1, \gamma_2]$ so $g_0(x)g_1(x) < 0$ in $\gamma_1 \leq x < \gamma$ and $g_0(x)g_1(x) > 0$ in $\gamma < x \leq \gamma_2$. This shows that (iv) holds and so (8) is a Sturm sequence for $g_0(x)$.

If $f(x)$ has no multiple roots, then $f(x)$ and $f'(x)$ have 1 as highest common factor. Then the sequence $\{f_0(x), f_1(x), \cdots, f_s(x)\}$ differs from $\{g_0(x), g_1(x), \cdots, g_s(x)\}$ by a non-zero multiplier in Φ. Hence the sequence of $f_i(x)$ is a Sturm sequence for $f(x) = f_0(x)$. If $f(x)$ has multiple roots, then the standard sequence (7) will not be a Sturm sequence for an interval containing a multiple root. Nevertheless, we can still use the standard sequence to determine the number of distinct roots of $f(x)$ in (α, β). This is the content of

Sturm's theorem. *Let $f(x)$ be any polynomial of positive degree with coefficients in a real closed field Φ and let $\{f_0(x) = f(x), f_1(x) = f'(x), \cdots, f_s(x)\}$ be the standard sequence (7) for $f(x)$. Assume $[\alpha, \beta]$ is an interval such that $f(\alpha) \neq 0$, $f(\beta) \neq 0$. Then the number of distinct roots of $f(x)$ in (α, β) is $V_\alpha - V_\beta$ where V_γ denotes the number of variations in sign of $\{f_0(\gamma), f_1(\gamma), \cdots, f_s(\gamma)\}$.*

Proof. Let $g_i(x) = f_i(x)f_s(x)^{-1}$ as above. Then apart from multiplicities, the polynomials $f(x)$ and $g_0(x)$ have the same roots in in (α, β) (ex. 7, p. 40). Since the sequence $\{g_i(x)\}$ is a Sturm sequence for $g_0(x)$, the number of these roots is $V_\alpha(g) - V_\beta(g)$ where $V_\gamma(g)$ is the number of variations in sign in $\{g_i(\gamma)\}$. Since

$$f_i(\gamma) = g_i(\gamma)f_s(\gamma) \quad \text{and} \quad f_s(\alpha) \neq 0, \quad f_s(\beta) \neq 0$$

it is clear that $V_\alpha(g) = V_\alpha$ and $V_\beta(g) = V_\beta$. Hence $V_\alpha - V_\beta$ gives the number of distinct roots of $f(x)$ in (α, β).

We have seen that the roots of $f(x) = x^n + a_1x^{n-1} + \cdots + a_n$ in Φ are in the interval $[-M, M]$ where $M = \max(1, |a_1| + |a_2| + \cdots + |a_n|)$ (ex. 3, § 1). If we set $\mu = 1 + |a_1| + \cdots + |a_n|$, then the roots of $f(x)$ in Φ are in $(-\mu, \mu)$. If $f_0(x) = f(x), f_1(x), \cdots, f_s(x)$ is the standard sequence (7) for $f(x)$, then the number of roots of $f(x)$ in Φ is $V_{-\mu} - V_\mu$ where V_γ is the number of variations in sign in $\{f_0(\gamma), f_1(\gamma), \cdots, f_s(\gamma)\}$. This gives a constructive way of determining the number of roots of $f(x)$ in Φ. Sometimes it is preferable to use instead of μ a bound η which is a polynomial in the a_i. For this purpose we note that $1 + a_i^2 > |a_i|$, so we can take $\eta = 1 + \Sigma(1 + a_i^2) = (n + 1) + \Sigma a_i^2$. Then the roots in Φ lie in $(-\eta, \eta)$.

EXERCISES

In all of these exercises Φ is a real closed field.

1. Prove Rolle's theorem: If $f(x) \in \Phi[x]$ has roots α, β in Φ, $\alpha < \beta$, then there exists a γ in Φ, $\alpha < \gamma < \beta$ such that $f'(\gamma) = 0$.

2. Prove the mean value theorem for polynomials: If $\alpha < \beta$, then there exists a γ, $\alpha < \gamma < \beta$ such that $f(\beta) - f(\alpha) = (\beta - \alpha)f'(\gamma)$.

3. Prove that $f(x)$ has a maximum on any closed finite interval, $[\alpha, \beta]$.

4. (Budan's theorem). Let $f(x)$ have degree n and assume $\alpha < \beta$ in Φ are not roots of $f(x)$. Let W_γ denote the number of variations in sign in the sequences:

$f(\gamma), f'(\gamma), \cdots, f^{(n)}(\gamma)$. Prove that $W_\alpha - W_\beta$ exceeds the number of roots of $f(x)$ in Φ in (α, β) *counting the multiplicities* of these roots by a non-negative even integer.

5. Deduce from ex. 4 *Descartes' rule:* Let $f(x) = \alpha_0 x^n + \alpha_1 x^{n-1} + \cdots + \alpha_l x^{n-l}$, $\alpha_0 \neq 0, \alpha_l \neq 0, \alpha_i \varepsilon \Phi$. Let P denote the number of variations in sign in the sequence $(\alpha_0, \alpha_1, \cdots, \alpha_l)$. Show that P exceeds the number of positive roots of $f(x)$, counting multiplicities, by a non-negative even integer.

4. Real closure of an ordered field. We have seen that every formally real field can be imbedded in a real closed field. In particular, this applies to ordered fields. We shall now show that, if Φ is an ordered field, then there exists a real closed algebraic extension field Δ of Φ whose (unique) ordering is an extension of that of Φ. Moreover, we shall see that Δ is essentially unique. To prove the existence of Δ we need the following

Lemma. *Let Φ be an ordered field, Ω an algebraic closure of Φ and let* E *be the subfield of Ω/Φ obtained by adjoining to Φ the square roots of the positive elements of Φ. Then* E *is formally real.*

Proof. Suppose we have a relation $\Sigma\xi_i^2 = 0$ in E. Then the ξ_i are contained in a finite dimensional extension field of the form $\Phi(\sqrt{\beta_1}, \sqrt{\beta_2}, \cdots, \sqrt{\beta_r})$ where the β_i are positive elements of Φ. Hence it suffices to show that every subfield $\Phi(\sqrt{\beta_1}, \sqrt{\beta_2}, \cdots, \sqrt{\beta_r})$, $\beta_i > 0$, of E is formally real. We prove this by induction on the dimensionality of the subfield and for this it is convenient to prove the apparently stronger statement that, if $\Sigma\gamma_i\xi_i^2 = 0$ for $\gamma_i > 0$ in Φ and ξ_i in $\Phi(\sqrt{\beta_1}, \sqrt{\beta_2}, \cdots, \sqrt{\beta_r})$, then every $\xi_i = 0$. This is clear for Φ since this is an ordered field. Suppose it holds for subfields of the indicated form of lower dimensionality than that of $\Gamma = \Phi(\sqrt{\beta_1}, \cdots, \sqrt{\beta_r})$. We may assume that $\Gamma \supset \mathrm{H} = \Phi(\sqrt{\beta_1}, \cdots, \sqrt{\beta_{r-1}})$, so the result holds for H. Now assume $\Sigma\gamma_i\xi_i^2 = 0$, $\xi_i \varepsilon \Gamma$, $\gamma_i > 0$ in Φ. Write $\xi_i = \eta_i + \zeta_i\sqrt{\beta_r}$, $\eta_i, \zeta_i \varepsilon \mathrm{H}$. Then $\Sigma\gamma_i\eta_i^2 + \Sigma\beta_r\gamma_i\zeta_i^2 + 2(\Sigma\gamma_i\eta_i\zeta_i)\sqrt{\beta_r} = 0$. Since $\sqrt{\beta_r} \notin \mathrm{H}$, $\Sigma\gamma_i\eta_i\zeta_i = 0$, so $\Sigma\gamma_i\eta_i^2 + \Sigma\beta_r\gamma_i\zeta_i^2 = 0$. Since $\gamma_i, \beta_r\gamma_i \varepsilon \Phi$, $\gamma_i > 0$, $\beta_r\gamma_i > 0$, and $\eta_i, \zeta_i \varepsilon \mathrm{H}$, every η_i and $\zeta_i = 0$. Then every $\xi_i = 0$ and the result is valid for Γ.

Definition 4. *Let Φ be an ordered field. Then an extension field Δ of Φ is called a* real closure *of Φ if* (1) *Δ is real closed,* (2) *Δ is algebraic over Φ,* (3) *the ordering of Δ is an extension of that of Φ.*

We can now prove the following basic result.

Theorem 8. *Every ordered field Φ has a real closure. If Φ_1 and Φ_2 are ordered fields with the real closures Δ_1 and Δ_2, respectively, then any order isomorphism of Φ_1 onto Φ_2 has a unique extension to an isomorphism of Δ_1 onto Δ_2. The extension is an order isomorphism.*

Proof. Let Φ be an ordered field, Ω an algebraic closure of Φ. Let E be the subfield of Ω obtained by adjoining to Φ the square roots of all the positive elements of Φ. Then E is formally real and Ω is an algebraic closure of E. We have seen that there exists a real closed subfield Δ of Ω/E (Th. 3). Suppose $\beta \in \Phi$ and $\beta > 0$. Then $\beta = \rho^2$, $\rho \in \Delta$. Hence $\beta > 0$ in Δ so the ordering in Δ is an extension of that of Φ. Hence Δ is a real closure of Φ.

Next let Φ_i, $i = 1, 2$, be ordered fields, Δ_i a real closure of Φ_i and let $\alpha \rightarrow \bar{\alpha}$ be an order isomorphism of Φ_1 onto Φ_2. We wish to extend the given isomorphism to an isomorphism of Δ_1 onto Δ_2. We note first that, if $f(x) \in \Phi_1[x]$, then $f(x)$ and its image $\bar{f}(x)$ under $\alpha \rightarrow \bar{\alpha}$ have the same number of roots in Δ_1 and Δ_2 respectively. We have seen that there exists a $\mu > 0$ in Φ_1 such that every root of $f(x)$ in Δ_1 is contained in $(-\mu, \mu)$. Moreover, by Sturm's Theorem, the number of roots of $f(x)$ in Δ_1 in the interval $(-\mu, \mu)$, hence the total number of roots of $f(x)$ in Δ_1, is given by $V_{-\mu} - V_\mu$ where V_γ is the number of variations in sign of the standard sequence (7) for f at γ. Since the standard sequence of f is contained in $\Phi_1[x]$, all of this carries over to $\bar{f}(x)$ and Δ_2. Hence the number of roots of $\bar{f}(x)$ in Δ_2 is the same as the number of roots of $f(x)$ in Δ_1. We note next that, if $F = \{\rho_1, \rho_2, \cdots, \rho_n\}$ is a finite subset of Δ_1, then there exists a subfield Γ_1 of Δ_1/Φ containing F and an isomorphism τ of Γ_1 into Δ_2 which extends $\alpha \rightarrow \bar{\alpha}$ and is such that, if $\rho_1 < \rho_2 < \cdots < \rho_n$, then $\rho_1^\tau < \rho_2^\tau < \cdots < \rho_n^\tau$. For this purpose let $f(x)$ be a polynomial in $\Phi_1[x]$ which has the elements ρ_i, $1 \leq i \leq n$, $\sigma_j = \sqrt{\rho_{j+1} - \rho_j}$, $1 \leq j \leq n - 1$, among its roots. We note that the $\sigma_j \in \Delta_1$ since Δ_1 is real closed and $\rho_{j+1} - \rho_j > 0$ (proof of Th. 2). Let Γ_1 be the finite dimensional extension of Φ_1 generated by the roots of $f(x)$ in Δ_1. Then $\Gamma_1 = \Phi_1(\theta_1)$ and, if $g(x)$ is the minimum polynomial of θ_1 over Φ_1, $\bar{g}(x)$ has a root θ_2 in Δ_2. We have an isomorphism τ of Γ_1 onto $\Phi_2(\theta_2)$ such that $\alpha^\tau = \bar{\alpha}$, $\alpha \in \Phi_1$, and $\theta_1^\tau = \theta_2$. Then $\rho_{j+1}^\tau - \rho_j^\tau = (\rho_{j+1} - \rho_j)^\tau = (\sigma_j^\tau)^2 > 0$. Hence $\rho_1^\tau < \rho_2^\tau < \cdots < \rho_n^\tau$ in Δ_2 as required. We shall now define a mapping η of Δ_1 into Δ_2 in the following way: Let ρ be an element of Δ_1 and let $h(x)$ be its minimum polynomial over Φ_1. Let the roots of $h(x)$ in Δ_1 be $\rho_1 < \rho_2 < \cdots < \rho_m$ and suppose $\rho_k = \rho$. Then $\bar{h}(x)$ has exactly m roots $\rho_1' < \rho_2' < \cdots < \rho_m'$ in Δ_2 and we now set $\rho^\eta = \rho_k'$. Evidently $\alpha^\eta = \bar{\alpha}$, $\alpha \in \Phi_1$, and it is easy to see that η is 1–1 and surjective. We assert that, if $\rho, \sigma \in \Delta_1$, then $(\rho + \sigma)^\eta = \rho^\eta + \sigma^\eta$, $(\rho\sigma)^\eta = \rho^\eta\sigma^\eta$ so that η is an isomorphism of Δ_1 into Δ_2 extending $\alpha \to \bar{\alpha}$. Let F be a finite subset of Δ_1 which includes the roots in Δ_1 of the minimum polynomials over Φ_1 of ρ, σ, $\rho + \sigma$ and $\rho\sigma$. Then we have seen that there exist a subfield Γ_1 of Δ_1 over Φ_1 containing F and an isomorphism τ of Γ_1 in Δ_2 extending $\alpha \to \bar{\alpha}$ such that τ preserves the order of the elements of F. As before, let $h(x)$ be the minimum polynomial of ρ over Φ_1 and let $\rho_1 < \rho_2 < \cdots < \rho_m$ be the roots of $h(x)$ contained in Δ_1. Then $\rho_i \in F$ and $\rho_1^\tau < \rho_2^\tau < \cdots < \rho_m^\tau$. We have $\bar{h}(\rho_i^\tau) = 0$ and it follows from the definition of η that $\rho^\eta = \rho^\tau$. Similarly, we see that $\sigma^\eta = \sigma^\tau$, $(\rho + \sigma)^\eta = (\rho + \sigma)^\tau$, $(\rho\sigma)^\eta = (\rho\sigma)^\tau$. Since τ is an isomorphism, this implies that $(\rho + \sigma)^\eta = \rho^\eta + \sigma^\eta$, $(\rho\sigma)^\eta = \rho^\eta\sigma^\eta$. Hence η is an isomorphism of Δ_1 onto Δ_2 extending the given isomorphism of Φ_1 onto Φ_2. Now let η' be any isomorphism of Δ_1 onto Δ_2. Since η' maps squares into squares it is clear that η' is an order isomorphism. Suppose also that η' extends the mapping $\alpha \to \bar{\alpha}$. Let $\rho \in \Delta_1$ and let $\rho_1 < \rho_2 < \cdots < \rho_m$ be the roots in Δ_1 of the minimum polynomial $h(x)$ of ρ over Φ_1. Then $\rho_1^{\eta'} < \rho_2^{\eta'} < \cdots < \rho_m^{\eta'}$ are the roots in Δ_2 of $\bar{h}(x)$. It follows that $\rho^{\eta'} = \rho^\eta$. Hence the extension η is unique. This completes the proof of the theorem.

If Δ_1 and Δ_2 are two real closures of a given ordered field Φ, then the identity mapping on Φ can be extended to an order isomorphism of Δ_1 onto Δ_2. In this sense real closures are equivalent and we may therefore speak of *the* real closure of Φ.

EXERCISES

1. Let Φ be an ordered field, Λ an extension field such that the only relations of the form $\Sigma\gamma_i\xi_i^2 = 0$ with γ_i positive in Φ and ξ_i in Λ are those in which every $\xi_i = 0$. Show that Λ can be ordered in such a way that its ordering is an extension of that of Φ.

2. Let Φ be an ordered field, Λ a real closed extension field whose order is an extension of that of Φ. Show that Λ contains a real closure of Φ.

5. Real algebraic numbers. We have seen that the field R_0 of rational numbers has a unique ordering (ex. 1, § 1). This ordered field has a real closure Δ_0 which is determined up to isomorphism. We shall call any real closure Δ_0 of R_0 *the field of real algebraic numbers*. Clearly $\Omega_0 = \Delta_0(\sqrt{-1})$ is an algebraic closure of R_0, and we shall call this field *the field of algebraic numbers*.

Now let $\Gamma = R_0(\theta)$ be a finite dimensional extension field of the rationals. Then if $n = [\Gamma:R_0]$, we have n distinct isomorphisms of Γ/R_0 into Ω_0/R_0. These are determined by mapping $\theta \rightarrow \theta_i$, $1 \leq i \leq n$, where $\{\theta_1, \theta_2, \cdots, \theta_n\}$ is the set of roots of the minimum polynomial $g(x)$ of θ in Ω_0. Let $\theta_1, \theta_2, \cdots, \theta_r$ be those θ_i which belong to Δ_0. We shall call these the *real conjugates* of θ. We agree to set $r = 0$ if θ has no real conjugates. Let τ_i, $1 \leq i \leq r$, be the isomorphism of Γ/R_0 into Δ_0/R_0 such that $\theta^{\tau_i} = \theta_i$. Then the ordering of $R_0(\theta_i) \subseteq \Delta_0$ imposed by the unique ordering in Δ_0 provides an ordering of Γ: We define $\rho > 0$ for $\rho \in \Gamma$ if and only if $\rho^{\tau_i} > 0$. We shall refer to this ordering of Γ as the ordering determined by τ_i. Now suppose we have any ordering of Γ and let Δ be a real closure of Γ relative to this ordering. Since Γ is algebraic over R_0, it is clear that Δ is a real closure of R_0. Consequently, we have an order isomorphism τ of Δ/R_0 onto Δ_0/R_0. The restriction of τ to Γ coincides with one of the τ_i and it is clear that the given ordering of Γ is the same as the ordering determined by τ_i. Finally, suppose τ_i and τ_j provide the same ordering of Γ. Then we have an order preserving isomorphism of $R_0(\theta_i)$ onto $R_0(\theta_j)$ such that $\theta_i \rightarrow \theta_j$. Since Δ_0 is a real closure of $R_0(\theta_i)$ and of $R_0(\theta_j)$, by Theorem 8, we have an automorphism of Δ_0 (over R_0) sending θ_i into θ_j. On the other hand, since Δ_0 is a real closure of R_0, Theorem 8 shows also that the identity is the only automorphism of Δ_0 over R_0. Hence we must have $\theta_j = \theta_i$. These results establish the following

Theorem 9. Let Γ *be a finite dimensional extension of the field of rational numbers. Then the number of distinct orderings of* Γ *is the same as the number of isomorphisms of* Γ/R_0 *into the field* Δ_0/R_0 *of real algebraic numbers.*

In particular, this number cannot exceed $[\Gamma:R_0]$ and there are no orderings of $\Gamma = R_0(\theta)$ if and only if the minimum polynomial of θ over R_0 has no real roots, that is, no roots in Δ_0.

We shall now apply this result to obtain a theorem of Hilbert and Landau which gives a necessary and sufficient condition that an element of $\Gamma = R_0(\theta)$, θ algebraic, is a sum of squares in this field. First, let Φ be any field of characteristic $\neq 2$ and, as in § 1, let $\Sigma(\Phi)$ be the subset of Φ of elements of the form $\Sigma\alpha_i^2$, $\alpha_i \in \Phi$. Next, we introduce the following definition.

Definition 5. *An element ρ of a field is called* totally positive *if $\rho > 0$ in every ordering of the field.*

In particular, it will be understood that in a field which has no ordering, then every element is totally positive. Thus every element of a field which is not formally real is totally positive. We have the following general criterion

Theorem 10. *Let Φ be a field of characteristic $\neq 2$. Then an element $\rho \neq 0$ in Φ is totally positive in Φ if and only if ρ is a sum of squares of elements of Φ.*

Proof. If $0 \neq \rho = \Sigma\alpha_i^2$, then clearly $\rho > 0$ in every ordering of Φ. Conversely, assume $\rho \neq 0$ is not a sum of squares in Φ. Let Ω be an algebraic closure of Φ and consider the collection of subfields E of Ω/Φ in which ρ is not a sum of squares. This collection contains Φ and is inductive; hence it contains a maximal element P. Now P is formally real; otherwise, the lemma of § 1 shows that every element of P is a sum of squares, but we know that ρ is not a sum of squares in P. Then P can be ordered. We note next that $-\rho$ is a square in P. Otherwise, we have the field $P(\sqrt{-\rho})$ in Ω and this properly contains P. Hence in this field we must have $\rho = \Sigma(\xi_i + \eta_i\sqrt{-\rho})^2$, ξ_i, η_i in P. This gives $\rho = \Sigma\xi_i^2 - \rho\Sigma\eta_i^2 + 2(\Sigma\xi_i\eta_i\sqrt{-\rho})$. It follows that $\Sigma\xi_i\eta_i = 0$ so $\rho(1 + \Sigma\eta_i^2) = \Sigma\xi_i^2$. Then $1 + \Sigma\eta_i^2 \neq 0$ by the formal reality of P; hence $\rho = (\Sigma\xi_i^2)(1 + \Sigma\eta_i^2)^{-1}$ is a sum of squares in P by the lemma of § 1. This contradicts the choice of P. Hence we see that $-\rho = \beta^2$, $\beta \in P$. This implies that $-\rho > 0$ and $\rho < 0$ in every ordering of P. Since P can be ordered, the induced ordering in Φ gives an ordering of Φ in which $\rho < 0$. Thus ρ is not totally positive.

This criterion and the result we obtained before on the form of the orderings of a finite dimensional extension of the rationals evidently imply the following

Theorem 11 (Hilbert-Landau). *Let Γ be a finite dimensional extension field of rationals and let $\tau_1, \tau_2, \cdots, \tau_r$ $(r \geq 0)$ be the different isomorphisms of Γ/R_0 into the field of real algebraic numbers. Then an element $\rho \neq 0$ of Γ is a sum of squares in Γ if and only if $\rho^{\tau_i} > 0$ for $i = 1, 2, \cdots, r$.*

EXERCISES

1. Let Φ be an ordered field and Λ a formally real field over Φ. Let ρ be an element of Λ which cannot be written in the form

$$\Sigma\beta_i\xi_i^2, \quad \beta_i \geq 0 \text{ in } \Phi, \tag{10}$$

ξ_i in the larger field. Show that there exists an algebraic extension P of Λ such that ρ is not of the form (10) in P (ξ_i in P) but ρ has this form in every proper algebraic extension P′ of P. Show that every positive element of Φ is a square in P and hence that ρ is a sum of squares in any $P' \supset P$, P′ algebraic over P. Prove that P is real closed and that the ordering in P is an extension of that of Φ. Prove that $\rho < 0$ in the ordering of P. Hence prove the following theorem: A necessary and sufficient condition that an element ρ of Λ have the form (10) in Λ is that $\rho \geq 0$ in every ordering of Λ which extends the ordering of Φ.

2. Let Φ be an ordered field, Δ the real closure of Φ and Γ a finite dimensional extension of Φ. Prove the following generalization of Theorem 9: If r is the number of isomorphisms of Γ/Φ into Δ/Φ, then r is the number of ways of extending the ordering of Φ to an ordering of Γ.

6. Positive definite rational functions. One of the problems proposed by Hilbert in his address to the 1900 Paris Congress of Mathematicians was the following: Let Q be a rational function of n variables with rational coefficients such that $Q(\xi_1, \cdots, \xi_n) \geq 0$ for all real $(\xi_1, \cdots, \xi_n)$ for which Q is defined. Then is Q necessarily a sum of squares of rational functions with rational coefficients? * By a rational function with rational coefficients we mean a mapping $(\xi_1, \cdots, \xi_n) \rightarrow Q(\xi_1, \cdots, \xi_n)$ where $Q(x_1, \cdots, x_n)$ is a rational expression in indeterminates x_i with rational coefficients. The domain of definition of Q is the set of real n-tuples $(\xi_1, \cdots, \xi_n)$ for which the denominator of $Q(x_1, \cdots, x_n)$ is not 0. In 1927 Artin gave an affirmative answer to Hilbert's question and proved the following stronger result.

Theorem 12 (Artin). *Let Φ be a field of real numbers (that is, a subfield of the field R of ordinary real numbers) which has a unique*

* This is known as Hilbert's 17th problem. See D. Hilbert, *Mathematische Probleme*, Göttinger Nachrichten, 1900, p. 284 or *Gosammelte Abhandlungen*, Vol. 3, p. 317.

ordering and let Q be a rational function with coefficients in Φ which is rationally definite in the sense that $Q(\xi_1, \cdots, \xi_n) \geq 0$ for all rational (ξ_i) for which Q is defined. Then Q is a sum of squares of rational functions with coefficients in Φ.

Instances of fields Φ which satisfy the condition are: the rational field, any real closed subfield of real numbers, the field of all real numbers. If we take Φ to be the first of these, then Artin's theorem gives a stronger result than that suggested by Hilbert. Let Φ be as in the theorem and consider the field $\Phi(x_i) \equiv \Phi(x_1, x_2, \cdots, x_n)$ in indeterminates x_i with coefficients in Φ. This field is formally real (ex. 4, § 1). According to Theorem 10, $Q(x_1, \cdots, x_n) \neq 0$ in $\Phi(x_i)$ is a sum of squares in this field if and only if $Q > 0$ in every ordering of $\Phi(x_i)$. Hence Theorem 12 will follow if we can show that, if $Q \neq 0$ is rationally definite, then $Q > 0$ in every ordering of $\Phi(x_i)$. This will follow from the following

Theorem 13. *Let Φ be a field of real numbers, $\Phi(x_i) \equiv \Phi(x_1, \cdots, x_n)$ the field of rational expressions in n indeterminates x_i with coefficients in Φ and suppose an ordering has been given to $\Phi(x_i)$ which extends the ordering of Φ as a subfield of the field of real numbers. Suppose $f_1(x_1, \cdots, x_n), \cdots, f_k(x_1, \cdots, x_n)$ is a finite set of elements of $\Phi(x_i)$. Then there exists a rational n-tuple $(a_1, \cdots, a_n)$ such that for every j, $1 \leq j \leq k$, $f_j(x_1, \cdots, x_n)$ is defined at (a_i) and f_j and $f_j(a_1, \cdots, a_n)$ have the same sign in the sense that $f_j(a_1, \cdots, a_n) \gtreqless 0$ according as $f_j \gtreqless 0$ in the given ordering of $\Phi(x_i)$.*

Suppose that this result holds and let Φ be as in Theorem 12. Let $Q \neq 0$ be an element of $\Phi(x_i)$ which is not a sum of squares in $\Phi(x_i)$. Then we know that there exists an ordering of $\Phi(x_i)$ for which $Q < 0$. Since Φ has only one ordering, the ordering of $\Phi(x_i)$ is an extension of that of Φ. Hence Theorem 13 gives a set (a_i), a_i rational, such that $Q(a_i) < 0$. Then Q is not rationally definite. Hence we see that, if Q is rationally definite, then it is a sum of squares of elements of $\Phi(x_i)$ and this is Artin's theorem.

We shall prove Theorem 13—after some necessary preliminaries—by induction on the number n of x_i. The result is clear if $n = 0$ since in this case $\Phi(x_i) = \Phi$, so the functions are just constant functions. It remains to prove the inductive step, so we assume the result for $\Phi(x_1, \cdots, x_n)$ and we shall prove it for $\Phi(x_1, \cdots, x_n, y)$, where y is an extra indeterminate. We shall see also that it is, in essence, sufficient to consider polynomials in y with coefficients in $\Phi(x_i)$. Let $F_1(x_i, y), \cdots, F_k(x_i, y) \in \Phi(x_i)[y]$. Then we shall call a property P of this set of polynomials in y *rationally specializable* if there exists a set of elements $\psi_1(x_i), \cdots, \psi_h(x_i)$ in $\Phi(x_i)$ such that, if $(a_1, \cdots, a_n)$ is any rational n-tuple for which $\psi_1, \cdots, \psi_h$ is defined and $\psi_l(a_i)$, $1 \leq l \leq h$, has the same sign as ψ_l (in a given ordering of $\Phi(x_i)$), then the coefficients of all the $F_j(x_i, y)$ are defined at $(a_1, \cdots, a_n)$ and the polynomials $F_1(a_i, y), \cdots, F_k(a_i, y)$ have the property P. We shall require two results on specializable properties which we state as lemmas.

Lemma 1. *The property that* $F(x_i, y) = y^m + \varphi_1(x_i)y^{m-1} + \cdots + \varphi_m(x_i)$ *has precisely* r *roots in the real closure of* $\Phi(x_i)$ *is rationally specializable.*

Proof. We are assuming that Φ is a subfield of the field of real numbers and Φ is ordered by the ordering of the field of real numbers. The subfield of the latter of elements which are algebraic over Φ is a real closure Δ of Φ (Cor. to Th. 6 and ex. 2, § 4). The assertion of the lemma is that there exist $\psi_1, \psi_2, \cdots, \psi_h$ in $\Phi(x_i)$ such that if $(a_1, \cdots, a_n)$ is any rational n-tuple such that every ψ_l is defined at $(a_1, \cdots, a_n)$ and $\psi_l(a_1, \cdots, a_n)$ has the same sign as ψ_l, then $F(a_i, y)$ is defined and the number of its real roots (or roots in Δ) is r. Let $F_0 = F(x_i, y), F_1, \cdots, F_s$ be the standard sequence for $F(x_i, y)$ (as in (7)). If $(a_1, \cdots, a_n)$ is a rational n-tuple for which the non-zero coefficients of the F_j and of the quotients Q_j as in (7) are defined and have non-zero values in Φ, then $F_0(a_i, y), \cdots, F_s(a_i, y)$ is the standard sequence for $F(a_i, y) = F_0(a_i, y)$. Let $\eta(x_i) = \sum_{j=1}^{m} \varphi_j(x_i)^2 + (m + 1)$. Then we have seen (p. 283) that the r roots of $F(x_i, y)$ in the real closure of $\Phi(x_i)$ lie in the interval $(-\eta, \eta)$. By Sturm's theorem the difference in the variations in sign between the two sequences $F_0(x_i, -\eta), F_1(x_i, -\eta), \cdots, F_s(x_i, -\eta)$ and $F_0(x_i, \eta), F_1(x_i, \eta), \cdots, F_s(x_i, \eta)$ is r. Now let $\{\psi_1(x_i), \cdots, \psi_h(x_i)\}$ be the set of elements of $\Phi(x_i)$ consisting of the coefficients of the standard sequence for F and of the quotients Q_j in this sequence, and the elements $F_j(x_i, -\eta(x_1, \cdots, x_n))$, $F_j(x_i, \eta(x_1, \cdots, x_n))$, $0 \leq j \leq s$. Then it is clear from Sturm's theorem that, if $(a_1, \cdots, a_n)$ is a rational n-tuple such that the ψ's are defined at $(a_1, \cdots, a_n)$ and every $\psi_l(a_1, \cdots, a_n)$ has the same sign as ψ_l, then $F(a_i, y)$ is defined and has exactly r roots in Δ in $(-\eta(a_1, \cdots, a_n), \eta(a_1, \cdots, a_n))$. If we refer to the result on bounds for the roots again we see that there are no real roots of $F(a_i, y)$ outside of the indicated interval. Hence r is the number of real roots of $F(a_i, y)$.

Lemma 2. *Let $\{F_1(x_i, y), \cdots, F_t(x_i, y)\}$ be a sequence of polynomials (not necessarily distinct) belonging to $\Phi(x_i)[y]$. Assume the leading coefficients are* 1. *The property that $F_j(x_i, y)$ has a root ρ_j in the real closure* P *of $\Phi(x_i)$ and $\rho_1 < \rho_2 < \cdots < \rho_t$ is rationally specializable.*

Proof. The elements ρ_k and $(\rho_{j+1} - \rho_j)^{1/2}$, $1 \leq j \leq t - 1$, are contained in P and these generate a finite dimensional extension field Λ of $\Phi(x_i)$ which has a primitive element θ. Let $g(x_i, y)$ be the minimum polynomial of θ over $\Phi(x_i)$. We have $\rho_k = \varphi_k(x_i, \theta)$, $(\rho_{j+1} - \rho_j)^{1/2} = \sigma_j(x_i, \theta)$ where $\varphi_k(x_i, y), \sigma_j(x_i, y) \varepsilon \Phi(x_i)[y]$. Since $F_k(x_i, \rho_k) = 0$, $F_k(x_i, \varphi_k(x_{i'}, y))$ has θ as root and since $g(x_i, y)$ is the minimum polynomial of θ, we have

(11) $$F_k(x_i, \varphi_k(x_{i'}, y)) = G_k(x_i, y)g(x_i, y), \quad 1 \leq k \leq t.$$

Similarly the relation $\rho_{j+1} - \rho_j = \sigma_j(x_i, \theta)^2$ or $\varphi_{j+1}(x_i, \theta) - \varphi_j(x_i, \theta) = \sigma_j(x_i, \theta)^2$ gives a relation

(12) $$\varphi_{j+1}(x_i, y) - \varphi_j(x_i, y) - \sigma_j(x_i, y)^2 = H_j(x_i, y)g(x_i, y), \quad 1 \leq j \leq t - 1,$$

in $\Phi(x_i)[y]$. Since $\sigma_j(x_i, \theta) \neq 0$ it has an inverse $\tau_j(x_i, \theta)$ in Λ and so we have relations of the form

(13) $$\sigma_j(x_i, y)\tau_j(x_i, y) - 1 = k_j(x_i, y)g(x_i, y)$$

in $\Phi(x_i)[y]$. Let $\{\psi_l(x_1, \cdots, x_n)\}$ be a finite set of elements of $\Phi(x_i)$ which includes the coefficients of the $F_k(x_i, y)$, all the coefficients of the polynomials in y appearing in (11), (12), and (13) and a set of elements given in Lemma 1 to insure that $g(a_i, y)$ has a real root γ. Moreover, if the a_i are chosen so that every $\psi_l(a_i)$ is defined, then substitution of γ for y in every polynomial which occurs in (11), (12), and (13) is permissible. Substituting $y = \gamma$ in (11) we see that $F_k(a_i, y)$ has the root $\beta_k \equiv \varphi_k(a_i, \gamma)$. Substituting $y = \gamma$ in (12) we see that $\beta_{j+1} - \beta_j = \varphi_{j+1}(a_i, \gamma) - \varphi_j(a_i, \gamma) = \sigma_j(a_i, \gamma)^2 \geq 0$. By (13), we have $\sigma_j(a_i, \gamma)\tau_j(a_i, \gamma) = 1$. Hence $\sigma_j(a_i, \gamma) \neq 0$ and $\beta_{j+1} > \beta_j$. Thus we see that $F_j(a_i, y)$ has the real root β_j and $\beta_1 < \beta_2 < \cdots < \beta_t$ as required.

We can now give the

Proof of Theorem 13. As we have seen before, it suffices to prove that, if the theorem holds for $\Phi(x_1, \cdots, x_n)$, then it holds for $\Phi(x_1, \cdots, x_n, y)$, y an extra indeterminate. Let P′ be a real closure of the ordered field $\Phi(x_i, y)$, P the real closure of $\Phi(x_i)$ contained in P′. We are given a finite set of elements $F(x_i, y)$ of $\Phi(x_i, y)$ and we have to show that rational a_i and b can be chosen so that $F(a_i, b)$ is defined and has the same sign as $F(x_i, y)$ in the ordering in P′ and this holds for every F in the given set. We can write $F = \varphi(x_1, \cdots, x_n)P_1(x_i, y)^{e_1} \cdots P_h(x_i, y)^{e_h}$ where $\varphi(x_1, \cdots, x_n) \in \Phi(x_i)$, e_j is an integer, $P_j(x_i, y)$ is irreducible in $\Phi(x_i)[y]$ and has leading coefficient 1, and the $P_j(x_i, y)$ are distinct. If a_i, b have the property that $\varphi(a_1, \cdots, a_n)$, $P_j(a_i, b)$ are defined and have the same sign as φ and P_j, $1 \leq j \leq h$, then $F(a_1, \cdots, a_n, b)$ is defined and has the same sign as F. This remark shows that we may as well suppose that the given set consists of elements $\varphi \in \Phi(x_i)$ and $F \in \Phi(x_i)[y]$ such that every F is irreducible in $\Phi(x_i)[y]$ and has leading coefficient 1. Let $\rho_1 < \rho_2 < \cdots < \rho_t$ be the roots in P of the given set $\{F\}$ of polynomials in y. We can form a sequence $F_1, F_2, \cdots, F_t$ whose terms are in $\{F\}$ so that ρ_j is a root of F_j. Since the F are irreducible and the field is of characteristic 0, the roots of F are distinct. Also distinct F's are relatively prime. Hence if $G(x_i, y)$ is the product of the distinct F's, then G has distinct roots. By Lemma 1, we can find elements $\psi_1, \cdots, \psi_h$ in $\Phi(x_i)$ such that, if $a_1, \cdots, a_n$ are rational and every ψ_l is defined at $(a_1, \cdots, a_n)$ and $\psi_l(a_1, \cdots, a_n)$ has the same sign as ψ_l, $1 \leq l \leq h$, then $G(a_i, y)$ is defined and has t real roots. By Lemma 2, we have elements $\psi_{h+1}, \cdots, \psi_k$ so that, if the a's are rational and $\psi_m(a_1, \cdots, a_n)$ is defined and has the same sign as ψ_m, $h + 1 \leq m \leq k$, then $F_j(a_i, y)$ is defined and has a real root β_j so that $\beta_1 < \beta_2 < \cdots < \beta_t$. We now add to the ψ's already given all the elements φ of the set given initially and the discriminant δ of $G(x_i, y)$, which is different from 0, since G has distinct roots. By the induction hypothesis, we can choose rational a_i so that all the conditions given by the ψ's, the φ's, and δ are satisfied. Now in P[y] we have the factorization

$$(14) \quad F_j(x_i, y) = (y - \rho_{j_1})(y - \rho_{j_2}) \cdots (y - \rho_{j_{t_j}})Q_1(y) \cdots Q_{s_j}(y)$$

where the Q's are irreducible quadratics with leading coefficients 1 and $\{j_1, j_2, \cdots, j_{t_j}\}$ are distinct and are a subset of $\{1, 2, \cdots, t\}$. Our choice of the a's insures that in the field of real numbers we have

$$(15) \quad F_j(a_i, y) = (y - \beta_{j_1})(y - \beta_{j_2}) \cdots (y - \beta_{j_{t_j}})\, S_1(y) \cdots S_{s_j}(y)$$

where the S's are irreducible quadratics with leading coefficients 1. Since y is transcendental over $\Phi(x_i)$ and the ρ_j are algebraic over this field, it is clear that y is contained in one of the following open intervals in P$'$: $(-\infty, \rho_1)$, (ρ_1, ρ_2), $\cdots$, (ρ_{t-1}, ρ_t), (ρ_t, ∞). Also we have seen that an irreducible quadratic with coefficients in a real closed field and leading coefficient 1 has the form $(y - \gamma)^2 + \delta^2$, $\delta \neq 0$ (see (3)). This implies that every $Q(y) > 0$ in P$'$, and for any real number b, $S(b) > 0$ for the S's in (15). It now follows from (14) and (15) that, if y is in the k-th interval of the sequence $(-\infty, \rho_1)$, (ρ_1, ρ_2), $\cdots$, (ρ_t, ∞) and b is any real number contained in the k-th interval $(-\infty, \beta_1)$, (β_1, β_2), $\cdots$, (β_t, ∞), then $F_j(a_i, b)$ and $F_j(x_i, y)$ have the same sign and this will hold for every j. Now it follows from the archimedean property of the real field that every open real interval contains a rational number. Hence we can choose a rational b so that $F_j(a_i, b)$ has the same sign as $F_j(x_i, y)$ for all the j. This completes the proof.

Remark. It is natural to ask if a result like Artin's holds for polynomials with coefficients in a field Φ such as in Artin's theorem. In view of this theorem one can formulate the question in the following way: Let $P(x_1, \cdots, x_n) \in \Phi[x_1, \cdots, x_n]$ such that $P = \Sigma R_i(x_1, \cdots, x_n)^2$ where the R_i are rational expressions in the x's with coefficients in Φ. Does this imply that $P = \Sigma P_j(x_1, \cdots, x_n)^2$, $P_j \in \Phi[x_1, \cdots, x_n]$. Artin has shown that this is correct if $n = 1$ and Φ is any field of real numbers. On the other hand, some examples due to Hilbert show that the result is false for $n \geq 2$ even for Φ the field of real numbers.

EXERCISE

1. Let Φ be an ordered field and $\Phi(x_1, \cdots, x_n)$ the field of rational expressions in indeterminates $x_1, \cdots, x_n$ over Φ. Suppose $Q \in \Phi(x_i)$ satisfies $Q(\xi_1, \cdots, \xi_m) \geq 0$ for all ξ_i in a real closure Δ of Φ for which $Q(\xi_1, \cdots, \xi_n)$ is defined. Prove that $Q = \Sigma\beta_j F_j(x_1, \cdots, x_n)^2$ where the β_j are non-negative elements of Φ and the $F_j \in \Phi(x_i)$. (Hint: See ex. 1 of § 5 and prove a suitable analogue of Th. 13.)

7. Formalization of Sturm's theorem. Resultants. In the next few sections we shall develop an algorithm, due to Tarski and to Seidenberg, for testing the solvability in a real closed field of a finite system of polynomial equations and inequalities (in several variables). The ultimate test (Th. 16) will consist in the verification of a finite system of polynomial equations and inequalities in the coefficients of the given system. In this section we shall consider first a reformulation of Sturm's theorem in this manner. We shall develop also an elimination method based on resultants which will be essential in the sequel.

To obtain the formalized version of Sturm's theorem it is convenient to begin with the ring $\mathfrak{A}[x]$ where $\mathfrak{A} = R_0[t_1, \cdots, t_r]$, x and the t_i are indeterminates and R_0 is the field of rational numbers. Let $F(t_1, \cdots, t_r; x) \in \mathfrak{A}[x]$, so $F(t_i; x) = a_n x^n + a_{n-1}x^{n-1} + \cdots + a_0$ where the $a_i \in \mathfrak{A}$ and $a_n \neq 0$. If the t_i are specialized to $t_i = \tau_i \in \Phi$, then we obtain the polynomial $f(x) \equiv F(\tau_i; x) \in \Phi[x]$ where $\deg f(x) \leq n$. We shall now obtain a number of sequences E: $\{F_0(t_i; x), F_1(t_i; x), \cdots, F_s(t_i; x)\}$ such that for any τ_i, $1 \leq i \leq r$, in Φ there exists one of these sequences E such that the specialized sequence $\{F_0(\tau_i; x), F_1(\tau_i; x), \cdots, F_s(\tau_i; x)\}$ is essentially the standard Sturm sequence (7) for $f(x)$.

Our choice for $F_0(t_i; x)$ is any one of the polynomials $a_m x^m + a_{m-1}x^{m-1} + \cdots + a_0$, $a_m \neq 0$, $m \leq n$ obtained by dropping leading terms $a_n x^n + \cdots + a_{m+1}x^{m+1}$ of $F(t_i; x)$. Next we take $F_1(t_i; x) = F_0'(t_i; x)$ the formal derivative of F_0 considered as a polynomial in x. Suppose we have already defined $F_0, F_1, \cdots, F_k$. If $F_k = 0$, we break off the sequence with $F_0, F_1, \cdots, F_s = F_{k-1}$. Otherwise, let $F_{k-1} = b_p x^p + \cdots + b_0$, $F_k = c_q x^q + \cdots + c_0$ where $b_p \neq 0$, $c_q \neq 0$, and $p > q$. The usual division process shows that we can find polynomials $Q(x)$, $R(x)$ in $\mathfrak{A}[x]$ such that $c_q^{p-q+1}F_{k-1} = QF_k - R$ where $\deg_x R(t_i; x) < \deg_x F_k(t_i; x)$.

For our purposes it is preferable to replace $p - q + 1$ by the smallest even integer $e \geq p - q + 1$. Thus, changing notation, we write $c_q{}^e F_{k-1} = QF_k - R$ and we call Q and R the *quotient* and *remainder* on dividing F_{k-1} by F_k. It is clear that these are unique. We now take F_{k+1} to be R or one of the polynomials obtained from R by dropping leading terms in the manner that $F_0(t_i; x)$ was obtained from $F(t_i; x)$. The sequences $\{F_0, F_1, \cdots, F_s\}$ obtained in this way will be called *generic standard sequences* for $F(t_i; x)$. Clearly the number of these is finite.

Let $E: \{F_0, F_1, \cdots, F_s\}$ be one of the generic standard sequences for F. Then we associate with E two finite subsets $\delta(E)$ and $\lambda(E)$ of $\mathfrak{A} = R_0[t_i]$. The set $\delta(E)$ is the set of coefficients of the terms dropped in forming the sequence. Thus, $\delta(E)$ consists of the coefficients (of the powers of x) in $F(t_i; x) - F_0(t_i; x)$ and those of $R - F_{k+1}$, $k \geq 1$, as above. We let $\lambda(E)$ be the set consisting of the leading coefficients of the F_i in the sequence.

Now let $\tau_i \in \Phi$, $1 \leq i \leq r$, and consider the polynomial $f(x) = F(\tau_i; x)$. We assume $f(x) \neq 0$. Suppose $\deg f(x) = m \leq n$. Then we shall take $F_0(t_i; x) = a_m x^m + \cdots + a_0$ and we have $F_0(\tau_i; x) = f(x)$, $a_m(\tau_i) \neq 0$. The condition $F_0(\tau_i; x) = f(x)$ gives $a_n(\tau_i) = \cdots = a_{m+1}(\tau_i) = 0$. One sees easily by an inductive argument that there exists a generic standard sequence $\{F_0, F_1, \cdots, F_s\}$ for F such that, if l_k is the leading coefficient of F_k, then

$$
\begin{aligned}
&l_k(\tau_i) \neq 0, \quad 0 \leq k \leq s \\
(16) \qquad &F_0(\tau_i; x) = f(x) \\
&l_k(\tau_i)^{e_k} F_{k-1}(\tau_i; x) = F_k(\tau_i; x) Q_k(\tau_i; x) - F_{k+1}(\tau_i; x),
\end{aligned}
$$

where $0 \leq k \leq s$, $F_{s+1} = 0$ and e_k is an even integer, and Q_k is the quotient on dividing F_{k-1} by F_k. Since $F_0(\tau_i; x) = f(x)$, $F_1(\tau_i; x) = f'(x)$, $l_k(\tau_i)^{e_k} > 0$ and the degrees of the $F_k(\tau_i; x)$ are decreasing, it is clear that the terms of $\{F_0(\tau_i; x), F_1(\tau_i; x), \cdots, F_s(\tau_i; x)\}$ differ from those of the standard sequence (7) for $f(x)$ by positive multipliers in Φ. Hence the sequence $\{F_0(\tau_i; x), \cdots\}$ can be substituted for the standard sequence in applying Sturm's theorem.

We shall now formalize the conditions given in this theorem by considering the finite collection of systems of equations and inequalities made up of relations of the following types:

$$
(17)\qquad
\begin{aligned}
F_k(t_i; y)F_l(t_i; y) &< 0,\\
F_{k+1}(t_i; y) = \cdots = F_{l-1}(t_i; y) &= 0, \quad k < l\\
F_p(t_i; z)F_q(t_i; z) &> 0,\\
F_{p+1}(t_i; z) = \cdots = F_{q-1}(t_i; z) &= 0, \quad p < q,
\end{aligned}
$$

where y and z are indeterminates and we require that the pairs $(k, l), \cdots, (p, q), \cdots$ used in the two sets of relations are such that, if $\epsilon_k = 0, 1, -1$, $\eta_p = 0, 1, -1$ satisfy the same conditions, namely, $\epsilon_k\epsilon_l < 0$, $\epsilon_{k+1} = \cdots = \epsilon_{l-1} = 0, \cdots, \eta_p\eta_q > 0$, $\eta_{p+1} = \cdots = \eta_{q-1} = 0, \cdots$, then the number of variations of sign of $\{\epsilon_0, \epsilon_1, \cdots, \epsilon_s\}$ exceeds that of $\{\eta_0, \eta_1, \cdots, \eta_s\}$. Now let (β, γ) be an interval in Φ, $\beta < \gamma$, such that $f(\beta) \neq 0$, $f(\gamma) \neq 0$. Then it is clear from Sturm's theorem that $f(x)$ has a root in (β, γ) if and only if $t_i = \tau_i$, $y = \beta$, $z = \gamma$ satisfies one of the systems of relations (17).

If we take into account all the generic sequences E and observe that (16) is equivalent to the conditions $l(\tau_i) \neq 0$, $d(\tau_i) = 0$ for all $l \in \lambda(E)$ and $d \in \delta(E)$, we see that (16) and (17) give a finite collection of conditions $\{G_1, G_2, \cdots, G_h\}$, where each G_j is a finite set of polynomial equations and inequalities with rational coefficients, such that $f(x)$ has a root in (β, γ) if and only if $t_i = \tau_i$, $y = \beta$, $z = \gamma$ satisfies all the conditions of one of the systems G_j.

Now let $f(x) = \alpha_m x^m + \alpha_{m-1}x^{m-1} + \cdots + \alpha_0$ where $\alpha_m \neq 0$. Then we know that all the roots in Φ of $f(x)$ are contained in the interval $(-\eta, \eta)$ where $\eta = m + 1 + \sum_0^{m-1} \alpha_j^2\alpha_m^{-2}$. We have $F_0(t_i; x) = a_m x^m + a_{m-1}x^{m-1} + \cdots + a_0$ where $a_m = a_m(t_1, \cdots, t_r) \neq 0$. If we substitute $y = -(m + 1) - \sum_0^{m-1} a_j^2 a_m^{-2}$ in $F_k(t_i; y)$ and $z = m + 1 + \sum_0^{m-1} a_j^2 a_m^{-2}$ in $F_k(t_i; z)$ and clear of fractions by multiplying by a suitable even power of a_m, we obtain polynomial relations like those in (17), for the t_i alone. In this way one obtains a finite collection $\{G_1, G_2, \cdots, G_h\}$ where each G_j is a finite system of polynomial equations and inequalities with rational coefficients in the t_i alone such that the statement that $f(x)$ has a root in Φ is equivalent to the validity of one of the systems G_j for $t_i = \tau_i$.

We shall consider next a classical determinant criterion for the existence of a common factor of positive degree for two polynomials. We consider the polynomials $f(x) = \alpha_n x^n + \alpha_{n-1}x^{n-1} + \cdots + \alpha_0$, $g(x) = \beta_m x^m + \beta_{m-1}x^{m-1} + \cdots + \beta_0$ in $\Phi[x]$ where Φ is an arbitrary field. We assume $m > 0$, $n > 0$, but we shall allow $\alpha_n = 0$ or $\beta_m = 0$. The result we require is the following

Theorem 14. *Let* $f(x) = \alpha_n x^n + \alpha_{n-1}x^{n-1} + \cdots + \alpha_0$, $g(x) = \beta_m x^m + \beta_{m-1}x^{m-1} + \cdots + \beta_0$ *where* $m, n > 0$ *and put*

$$(18)\quad R(f,g) = \begin{vmatrix} \alpha_n & \alpha_{n-1} & \cdots & \alpha_0 & & & \\ & \alpha_n & \alpha_{n-1} & \cdots & \alpha_0 & & \\ \cdot & \cdot & \cdot & \cdot & \cdot & \cdot & \cdot \\ & & & \alpha_n & \alpha_{n-1} & \cdots & \alpha_0 \\ \beta_m & \beta_{m-1} & \cdots & \beta_0 & & & \\ & \beta_m & \beta_{m-1} & \cdots & \beta_0 & & \\ \cdot & \cdot & \cdot & \cdot & \cdot & \cdot & \cdot \\ & & & \beta_m & \beta_{m-1} & \cdots & \beta_0 \end{vmatrix} \begin{matrix} \left.\vphantom{\begin{matrix}a\\a\\a\\a\end{matrix}}\right\} m \text{ rows} \\ \left.\vphantom{\begin{matrix}a\\a\\a\\a\end{matrix}}\right\} n \text{ rows} \end{matrix}$$

Then $R(f,g) = 0$ *if and only if either* $\alpha_n = 0 = \beta_m$ *or* $f(x)$ *and* $g(x)$ *have a common factor of positive degree in* x.

Proof. If $\alpha_n = 0 = \beta_m$, then the first column of the determinant is 0. Hence $R(f,g) = 0$. Next assume $f(x)$ and $g(x)$ have a common factor $h(x)$ of positive degree and that either $\alpha_n \neq 0$ or $\beta_m \neq 0$. Then $f(x) = f_1(x)h(x)$, $g(x) = g_1(x)h(x)$ and either $f_1(x) \neq 0$ or $g_1(x) \neq 0$, according as $\alpha_n \neq 0$ or $\beta_m \neq 0$. By symmetry, we may assume $\alpha_n \neq 0$, $f_1(x) \neq 0$. We have $f(x)g_1(x) = g(x)f_1(x)$, $f(x) = f_1(x)h(x) \neq 0$. If $\deg h(x) = r$, then $\deg f_1 = n - r$. If $g(x) = 0$, we have $g_1(x) = 0$; otherwise the relation $f(x)g_1(x) = g(x)f_1(x)$ gives $\deg g_1(x) \leq m - r$. Hence we may write $f_1(x) = -\gamma_{n-1}x^{n-1} - \gamma_{n-2}x^{n-2} - \cdots - \gamma_0$, $g_1(x) = \delta_{m-1}x^{m-1} + \delta_{m-2}x^{m-2} + \cdots + \delta_0$, so that we have

$$(19)\quad (\alpha_n x^n + \cdots + \alpha_0)(\delta_{m-1}x^{m-1} + \cdots + \delta_0) + (\beta_m x^m + \cdots + \beta_0)(\gamma_{n-1}x^{n-1} + \cdots + \gamma_0) = 0.$$

If we equate to 0 the coefficients of $x^{m+n-1}, x^{m+n-2}, \cdots, 1$ in (19), we obtain the following equations:

$$
\begin{aligned}
\alpha_n\delta_{m-1} + \beta_m\gamma_{n-1} &= 0 \\
\alpha_n\delta_{m-2} + \alpha_{n-1}\delta_{m-1} + \beta_m\gamma_{n-2} + \beta_{m-1}\gamma_{n-1} &= 0 \\
&\cdot \\
&\cdot \\
&\cdot \\
\alpha_0\delta_0 + \beta_0\gamma_0 &= 0.
\end{aligned}
\tag{20}
$$

We consider this as a system of homogeneous equations in the γ's and δ's taken in the order $\delta_{m-1}, \delta_{m-2}, \cdots, \delta_0, \gamma_{n-1}, \cdots, \gamma_0$. Since not all the γ's are 0, the determinant of the coefficients of the γ's and δ's is 0. If we take the transpose of this determinant obtained by ordering the γ's and δ's as indicated, we obtain (18). Hence $R(f, g) = 0$. Conversely, assume $R(f, g) = 0$. Then we can re-trace the steps through (20) and (19) and conclude that there exist $f_1(x)$, $g_1(x)$ such that $f(x)g_1(x) = g(x)f_1(x)$ where $\deg f_1 \leq n - 1$, $\deg g_1 \leq m - 1$, and either $f_1 \neq 0$ or $g_1 \neq 0$. Assume $f_1 \neq 0$. If $g_1 = 0$, then $g = 0$, $\beta_m = 0$, and either $f(x)$ is a non-zero common factor of f and g or $\alpha_n = 0$. If $g_1 \neq 0$ and $g = 0$ the foregoing argument applies. Now suppose $g_1 \neq 0$ and $g \neq 0$. Then the relations $f(x)g_1(x) = g(x)f_1(x)$, $f_1 \neq 0, g_1 \neq 0$, $g \neq 0$ imply $f \neq 0$. Either $\alpha_n = 0 = \beta_m$ or we may assume $\alpha_n \neq 0$ which implies that $\deg f(x) = n$. Since $\deg f_1(x) \leq n - 1$, the relation $f(x)g_1(x) = g(x)f_1(x)$ and the factorization of the non-zero polynomials f, f_1, g, g_1 into irreducible factors implies that $f(x)$ and $g(x)$ have a common factor of positive degree.

We shall call $R(f, g)$ the *resultant* of f and g (relative to x). If either highest coefficient of f or of g is not 0, then the vanishing of $R(f, g)$ is a polynomial relation on the α_i, β_j with integer coefficients which is equivalent to the statement that f and g have a common factor of positive degree.

EXERCISE

1. Prove that, if $f(x) = x^n + \alpha_{n-1}x^{n-1} + \cdots + \alpha_0$, then $R(f, f')$ is the discriminant of $f(x)$ (cf. § 3.1).

8. Decision method for an algebraic curve. In this section we shall give Seidenberg's method for deciding the solvability of an equation $f(x, y) = 0$ in a real closed field. This will be based on the result which we shall now establish, that if $f(x, y) = 0$ has a solution in Φ, then the equations $f(x, y) = 0$, $(y - \delta)\dfrac{\partial f}{\partial x} - (x - \gamma)\dfrac{\partial f}{\partial y} = 0$, have a common solution in Φ for any γ, δ in Φ. The geometric idea underlying this result will be clear from the two lemmas which are used to prove it.

Lemma 1. *Let $f(x, y) \varepsilon \Phi[x, y]$, x, y indeterminates, Φ a real closed field. Then if $f(x, y) = 0$ has a solution in Φ, it has a solution (α, β) nearest the origin.*

Proof. We consider the intersection in the space $\Phi^{(2)}$ of pairs (ξ, η), $\xi, \eta \varepsilon \Phi$, of the curve C: $f(x, y) = 0$ with the circle $x^2 + y^2 = \gamma^2$, $\gamma \geq 0$. Our hypothesis implies that there exist γ for which this intersection is not vacuous, and we have to show that the set S of $\gamma \geq 0$ such that C meets $x^2 + y^2 = \gamma^2$ in $\Phi^{(2)}$ has a minimum. We now consider the polynomials $f(x, y)$ and $x^2 + y^2 - c^2$ as polynomials in y with coefficients in $\Phi(c, x)$, where c and x are regarded as indeterminates, and we form the resultant $g(c, x)$ of these two polynomials. The formula (18) shows that $g(c, x)$ is a polynomial in c and x with coefficients in Φ. If (α, β) is a point of intersection of the circle $x^2 + y^2 = \gamma^2$ and the curve C, then $f(\alpha, y)$ and $y^2 + \alpha^2 - \gamma^2$ have a common factor $y - \beta$. Hence $g(\gamma, \alpha) = 0$ and $g(\gamma, x)$ has the root $\alpha \varepsilon \Phi$. Moreover, $-\gamma \leq \alpha \leq \gamma$. Conversely, assume that for $\gamma \geq 0$, $g(\gamma, x)$ has a root α in Φ, $-\gamma \leq \alpha \leq \gamma$. Since the leading coefficient of y in $y^2 + \alpha^2 - \gamma^2$ is 1, it follows from Theorem 14 that $y^2 + \alpha^2 - \gamma^2$ and $f(\alpha, y)$ have a common factor in $\Phi[y]$. Since the factors of $y^2 + \alpha^2 - \gamma^2$ are $y \pm \beta$ where $\beta = (\gamma^2 - \alpha^2)^{1/2}$, it follows that (α, β) or $(\alpha, -\beta)$ is a point of intersection of the two curves. Hence we see that the set S of $\gamma \geq 0$ such that C meets $x^2 + y^2 = \gamma^2$ in $\Phi^{(2)}$ is the same as the set of $\gamma \geq 0$ such that $g(\gamma, x) = 0$ has a root $x = \alpha \varepsilon \Phi$, $-\gamma \leq \alpha \leq \gamma$. Let S' be the subset of S of the γ such that $g(\gamma, \pm\gamma) \neq 0$. For these the condition is that $g(\gamma, x)$ has a root in $(-\gamma, \gamma)$. It is clear that we can obtain $g(\gamma, x)$ by a suitable specialization of a polynomial with rational coefficients such that one of the parameters is specialized to γ. Hence we can apply the result obtained in the last section to conclude that S' is the union of a finite number of sets defined by polynomial equations and inequalities of the form $p(c) = 0$, $q(c) = 0$, $r(c) \gtrless 0$ where p, q, r are polynomials with coefficients in Φ. One sees easily that such a set is a union of a finite number of intervals which may be open, closed, half open, single points or extend to infinity. Since the set of γ such that $g(\gamma, \pm\gamma) = 0$ is either finite or all $\gamma \geq 0$, it is clear that S has the same structure as S'. The result will now follow by showing that the complement of S in the set of non-negative elements is the union of open intervals; for, this implies that S is the union of a finite number of closed intervals and hence has a minimal element. Thus, let $\delta \geq 0$, $\delta \notin S$. Then $g(\delta, x) = 0$, $-\delta \leq x \leq \delta$ has no solution x in Φ. Write $g(c, x) = g_0(x) + g_1(x)(c - \delta) + \cdots + g_m(x)(c - \delta)^m$ where the $g_i(x)$ are polynomials in x. Then $g_0(x) \neq 0$ in $-\delta \leq x \leq \delta$. It follows that there exists a $\delta' > \delta$ such that $g_0(x) \neq 0$ in $-\delta' \leq x \leq \delta'$. Then there exist $b > 0$, $B > 0$ such that $|g_0(x)| \geq b$, $|g_i(x)| \leq B$ for all x in $[-\delta', \delta']$ (ex. 3, § 3). Then, if $|c - \delta| < \frac{1}{2}$ and $|c - \delta| < b/4B$ and $x \in [-\delta', \delta]$,

$$|g(c, x)| \geq |g_0(x)| - |g_1(x)(c - \delta) + \cdots + g_m(x)(c - \delta)^m|$$

$$\geq b - 2B|c - \delta| > b - \frac{b}{2} = \frac{b}{2}.$$

This implies that every δ'' satisfying $\delta'' \leq \delta'$, $\delta'' < \delta + \frac{1}{2}$, $\delta'' < \delta + b/4B$ is contained in the complement of S. Hence this complement contains an open interval containing δ and the proof is complete.

As in the classical case of the field of real numbers, a point (α, β) on $f(x, y) = 0$ is called a *simple point* if

$$\left(\left(\frac{\partial f}{\partial x}\right)_{(\alpha,\beta)}, \left(\frac{\partial f}{\partial y}\right)_{(\alpha,\beta)}\right) \neq (0, 0).$$

Then the normal vector at (α, β) is $\left(\left(\frac{\partial f}{\partial x}\right)_{(\alpha,\beta)}, \left(\frac{\partial f}{\partial y}\right)_{(\alpha,\beta)}\right)$ and the *tangent line* to the curve at (α, β) is defined by the equation

$$\left(\frac{\partial f}{\partial x}\right)_{(\alpha,\beta)}(x-\alpha)+\left(\frac{\partial f}{\partial y}\right)_{(\alpha,\beta)}(y-\beta)=0.$$

Now let (α, β) be a point on C: $f(x, y) = 0$ in $\Phi^{(2)}$ nearest the origin. We wish to show that $\beta\left(\frac{\partial f}{\partial x}\right)_{(\alpha,\beta)} - \alpha\left(\frac{\partial f}{\partial y}\right)_{(\alpha,\beta)} = 0.$ This is clear if $(\alpha, \beta) = (0, 0)$ or (α, β) is not a simple point. Otherwise, the equation states that the vector joining $(0, 0)$ to (α, β) and the normal vector are linearly dependent. Hence C and the circle with center at the origin and radius $(\alpha^2 + \beta^2)^{1/2}$ have the same tangent at (α, β). If this is not the case, then the tangent to C at (α, β) contains interior points of the circle whereas C itself does not. The result will therefore follow from

Lemma 2. *Let p be a point of intersection (coordinates in Φ) of a circle and a curve C: $f(x, y) = 0$, $f(x, y) \in \Phi[x, y]$. Assume p is a simple point and the tangent at p to C has points interior to the circle. Then C itself has points interior to the circle.*

Proof. We take $p = (0, 0)$ and the tangent to C at p to be the x-axis. Then $f(0, 0) = 0$ and $\left(\frac{\partial f}{\partial x}\right) = 0$, and we may suppose that $\left(\frac{\partial f}{\partial y}\right)(0, 0) = 1$. The center of the circle is not on the x-axis, so we may denote it as (a, b) with $a \neq 0$. We have $f(x, y) = f(0, 0) + \left(\frac{\partial f}{\partial x}\right)_0 x + \left(\frac{\partial f}{\partial y}\right)_0 y + \frac{1}{2!}\left[\left(\frac{\partial^2 f}{\partial x^2}\right)_0 x^2 + 2\left(\frac{\partial^2 f}{\partial x \partial y}\right)_0 xy + \left(\frac{\partial^2 f}{\partial y^2}\right)_0 y^2\right] + \cdots$, so taking into account the conditions on f we see that we can write $f(x, y) = y(1 + h(x, y)) + g(x)$ where $h(0, 0) = 0$ and $g(x)$ is a polynomial in x divisible by x^2. Since $h(0, 0) = 0$ we may choose a $\delta > 0$ such that $|h(x, y)| \leq \frac{1}{2}$ if $|x| \leq \delta$ and $|y| \leq \delta$. Then $\frac{1}{2} \leq 1 + h(x, y) \leq \frac{3}{2}$ and $\delta(1 + h(x, \delta))$ is between $\frac{1}{2}\delta$ and $\frac{3}{2}\delta$ while $-\delta(1 + h(x, -\delta))$ is between $-\frac{1}{2}\delta$ and $-\frac{3}{2}\delta$ for all x satisfying $|x| \leq \delta$. Since $g(0) = 0$ there exists a δ', $0 < \delta' \leq \delta$ such that $f(x, \delta) = \delta(1 + h(x, \delta)) + g(x) > 0$ and $f(x, -\delta) < 0$ if $|x| \leq \delta'$. Then for every x_0,

$|x_0| \leq \delta'$ there exists a $y_0 \in [-\delta, \delta]$ such that $f(x_0, y_0) = 0$. Then $y_0 = -g(x_0)(1 + h(x_0, y_0))^{-1}$ and

$$(a - x_0)^2 + (b - y_0)^2$$
$$= (a - x_0)^2 + \left(b + \frac{g(x_0)}{1 + h(x_0, y_0)}\right)^2$$
$$= a^2 + b^2 - 2ax_0 + x_0{}^2 + \frac{2bg(x_0)}{1 + h(x_0, y_0)} + \frac{(g(x_0))^2}{(1 + h(x_0, y_0))^2}.$$

Since $g(x_0)$ is divisible by $x_0{}^2$, it is clear that, if we take x_0 sufficiently small so that $ax_0 > 0$, then $(a - x_0)^2 + (b - y_0)^2 < a^2 + b^2$. Hence (x_0, y_0) is a point on C interior to the given circle.

Our results now show that, if $C: f(x, y) = 0$ has a solution in Φ, then there exists a solution in Φ which is also in $y \dfrac{\partial f}{\partial x} - x \dfrac{\partial f}{\partial y} = 0$. If we replace the origin by (γ, δ) where $\gamma, \delta \in \Phi$, then we see in the same way that the intersection of C and the curve D: $(y - \delta) \dfrac{\partial f}{\partial x} - (x - \gamma) \dfrac{\partial f}{\partial y} = 0$ contains a point in $\Phi^{(2)}$.

We shall now apply this to obtain Seidenberg's procedure for deciding the solvability in Φ of $f(x, y) = 0$. First, we can obtain the highest common factor of the coefficients of the powers of y in $f(x, y)$ and put $f(x, y) = d(x) f_1(x, y)$ where $f_1(x, y)$ is not divisible by a polynomial of positive degree in x alone. Evidently $f(x, y) = 0$ has a solution in Φ if and only if either $d(x) = 0$ or $f_1(x, y) = 0$ has such a solution. This reduces the discussion to polynomials which are not divisible by polynomials of positive degree in x alone. Next we can compute a highest common factor in $\Phi(x)[y]$ of $f(x, y)$ and $\dfrac{\partial}{\partial y} f(x, y)$ by using the Euclidean algorithm. We may assume this belongs to $\Phi[x, y]$ and is not divisible by polynomials of positive degree in x alone. Then we can divide out by this highest common factor and obtain a polynomial $g(x, y)$ which is a factor of $f(x, y)$, has the same irreducible factors as $f(x, y)$ and has no multiple factors in $\Phi[x, y]$. Clearly, $f(x, y) = 0$ is solvable in Φ if and only if $g(x, y) = 0$ is solvable in Φ.

If we replace f by g and change the notation back to f we may suppose that $f(x, y)$ has no multiple factors of positive degree and no factors of positive degree in x alone. The first of these conditions implies that $f(x, y)$ and $\frac{\partial}{\partial y} f(x, y)$ have no common factor contained in $\Phi(x)[y]$ of positive degree in y.

We now consider the polynomial $g(x, y) = y \frac{\partial f}{\partial x} - (x - \gamma) \frac{\partial f}{\partial y}$ where γ is any element of Φ. We know that, if the curve $C: f(x, y) = 0$ contains a point in $\Phi^{(2)}$, then the intersection of C and $D: g(x, y) = 0$ contains such a point. Before we can make use of this it is necessary to arrange matters, by choosing a suitable γ so that the intersection of C and D is a finite set. To do this we introduce another indeterminate c and we consider the polynomial $g(c; x, y) = y \frac{\partial f}{\partial x} - (x - c) \frac{\partial f}{\partial y}$. Let $R(c; x)$ be the resultant relative to y (that is, considering the polynomials as polynomials in y) of $f(x, y)$ and $g(c; x, y)$. We claim that $R(c; x) \neq 0$. Otherwise, $R(\gamma; x) = 0$ for all γ. Now, Theorem 14 shows that if γ has this property, then $g(\gamma; x, y)$ and $f(x, y)$ have a common factor in $\Phi(x)[y]$ and, consequently, in $\Phi[x, y]$ of positive degree in y (see Vol. I, p. 125). Hence if $R(\gamma; x) = 0$ for all γ, then there exist distinct γ, say, γ_1 and γ_2 such that $f(x, y)$, $g(\gamma_1; x, y)$ and $g(\gamma_2; x, y)$ all have a common factor of positive degree in y. This follows from the fact that to within associates $f(x, y)$ has only a finite number of different factors in $\Phi[x, y]$. We can then conclude that $f(x, y)$ and $(\gamma_1 - \gamma_2) \frac{\partial}{\partial y} f(x, y) = g(\gamma_1; x, y) - g(\gamma_2; x, y)$ have a common factor of positive degree in y and this contradicts the fact that $f(x, y)$ and $\frac{\partial}{\partial y} f(x, y)$ have no such common factor. This proves that $R(c; x) \neq 0$.

We can now choose a $\gamma \in \Phi$ such that $R(x) \equiv R(\gamma; x) \neq 0$. Set $g(x, y) = g(\gamma; x, y)$. Then $f(x, y)$ and $g(x, y)$ have no common factor of positive degree in y and no common factor of positive degree in x alone; hence they have no common factors except units in $\Phi[x, y]$. It follows that the resultant $Q(y)$ of $f(x, y)$ and $g(x, y)$

relative to x is not 0. Let V be the intersection in $\Omega^{(2)}$, where $\Omega = \Phi(\sqrt{-1})$ is the algebraic closure of Φ, of $C: f(x, y) = 0$ and $D: g(x, y) = 0$. If $(\rho, \sigma) \in V$, $f(\rho, \sigma) = 0 = g(\rho, \sigma)$ imply that $R(\rho) = 0$ and $Q(\sigma) = 0$. Since $R(x) \neq 0$, $Q(y) \neq 0$, this gives only a finite number of possibilities. Hence V is a finite set. We know that, if C contains a point in $\Phi^{(2)}$, then V has such a point and, consequently, $R(x)$ has a root in Φ. Conversely, suppose $R(x)$ has a root α in Φ. If α is not a root of the polynomial in x which is the coefficient of the highest power of y in $f(x, y)$, then $R(\alpha) = 0$ implies the existence of a $\sigma \in \Omega$ such that $(\alpha, \sigma) \in V$. If $\sigma = \beta \in \Phi$, then we have the desired result that V, and hence C, has a point in $\Phi^{(2)}$. Otherwise, $(\alpha, \bar{\sigma}) \in V$ where $\bar{\sigma} \neq \sigma$ is the conjugate of σ under the automorphism $\neq 1$ of Ω/Φ. Then we have two points in V: (α, σ) and $(\alpha, \bar{\sigma})$ with the same abscissa.

We can easily overcome—by a suitable choice of axes—the two difficulties which we have noted which may prevent concluding that V, and hence C, has a point in $\Phi^{(2)}$ from the fact that $R(x)$ has a root in Φ. We shall change to an x', y'-system where $x = \mu(x' + y')$, $y = y'$ and $\mu \neq 0$ will be chosen suitably in Φ. The equation of C in the x', y'-system is $f(\mu(x' + y'), y') = 0$. Let $f_n(x, y)$ be the homogeneous part of highest degree n (>0) in x and y in the polynomial $f(x, y)$. Then the coefficient of $(y')^n$ in $f(\mu(x' + y'), y')$ is $f_n(\mu, 1)$. Since $f_n(x, 1) \neq 0$, we can choose $\mu \in \Phi$ so that $f_n(\mu, 1) \neq 0$. Since the total degree of $f(x, y)$ is n, it will follow that the constant $f_n(\mu, 1) \neq 0$ is the polynomial in x' which is the coefficient of the highest power of y' in $f(\mu(x' + y'), y')$. This will take care of one of the difficulties. To take care of the other we compute, by using the Euclidean algorithm, applied to $R(x)$ and $R'(x)$, a polynomial $r(x)$ which has simple roots that are the same as those of $R(x)$. Similarly, we compute a polynomial $q(y)$ having simple roots the same as those of $Q(y)$. We note next that we can compute a polynomial $s(x)$ whose roots are $(\rho_i - \rho_{i'})(\sigma_j - \sigma_{j'})^{-1}$ where $\rho_1, \cdots, \rho_s$ are the roots of $r(x)$ and $\sigma_1, \cdots, \sigma_t$ are those of $q(y)$, $i \neq i', j \neq j'$. For this we introduce indeterminates ξ_i, $1 \leq i \leq s$, η_j, $1 \leq j \leq t$, and we consider the polynomial

$$\prod_{\substack{i \neq i' \\ j \neq j'}} [(\eta_j - \eta_{j'})x - (\xi_i - \xi_{i'})].$$

This is invariant under all permutations of the ξ's and η's so the coefficients of the powers of x are polynomials with integer coefficients in the elementary symmetric polynomials of the ξ's and the η's (Vol. I, p. 109). If we replace these elementary symmetric polynomials by the corresponding coefficients of $r(x)$ and $q(y)$ normalized to have leading coefficients 1, we obtain a polynomial $s(x)$ whose roots are $(\rho_i - \rho_{i'})(\sigma_j - \sigma_{j'})^{-1}, i \neq i', j \neq j'$. Assume now that μ is not a root of $s(x)$ (as well as not a root of $f_n(x, 1)$) and consider the set of points (ρ_i, σ_j). This contains V and no two distinct points in this set have the same abscissa in the x', y'-system since (ρ, σ) is the point $(\mu^{-1}\rho - \sigma, \sigma)$ in the x', y'-system. Hence $\mu^{-1}\rho_i - \sigma_j \neq \mu^{-1}\rho_{i'} - \sigma_{j'}$ if $(i, j) \neq (i', j')$.

We now choose μ as indicated and we replace $f(x, y), g(x, y)$ by $h(x, y) = f(\mu(x + y), y)$ and $k(x, y) = g(\mu(x + y), y)$. Let $f(x)$ be the resultant relative to y of $h(x, y)$ and $k(x, y)$. Then the argument shows that $f(x, y) = 0$ is solvable in Φ if and only if $f(x)$ has a root in Φ. The latter problem can be decided by Sturm's theorem.

In order to carry this over to more than two variables it is necessary to consider polynomials involving parameters and to apply an inductive procedure. This necessitates an extension of the decision method we have just given to take care of an equation $f(x, y) = 0$ restricted by an inequality $g(x) \neq 0$. To handle this we first obtain a highest common factor $d(x)$ of $g(x)$ and the coefficients of the powers of y in $f(x, y)$ by the Euclidean algorithm. Write $f(x, y) = d(x)f_1(x, y)$, $g(x) = d(x)g_1(x)$. Then the pair of conditions $f(\alpha, \beta) = 0$, $g(\alpha) \neq 0$ is equivalent to the pair: $f_1(\alpha, \beta) = 0$, $g(\alpha) \neq 0$. This remark permits us to reduce the consideration to the case in which $g(x)$ and $f(x, y)$ have no common factor of positive degree. To avoid considering trivial cases we assume also that $\deg g(x) > 0$ and $\deg_x f(x, y) > 0$. Let $T(y)$ be the resultant relative to x of $f(x, y)$ and $g(x)$. Then $T(y) \neq 0$ since, otherwise, $f(x, y)$ and $g(x)$ have a common factor of positive degree in x in $\Phi[x, y]$ contrary to our arrangement. We now choose τ in Φ so that $T(\tau) \neq 0$ and we replace $f(x, y)$ by $h(x, y) = f(x, y + \tau)$. Then the resultant relative to x of $h(x, y)$ and $g(x)$ is $T(y + \tau)$ which is not 0 for $y = 0$. This implies that $g(x)$ and $h(x, 0)$ are relatively prime. Clearly we can replace the pair $f(x, y)$, $g(x)$ by the pair $h(x, y)$, $g(x)$ for the problem of testing the existence of a solution in Φ of $f(x, y) = 0, g(x) \neq 0$. Now let $k(x, y) = h(x, g(x)y)$. Then if (α, β) satisfies $h(\alpha, \beta) = 0$, $g(\alpha) \neq 0$, we have $k(\alpha, \gamma) = 0$ for $\gamma = \beta g(\alpha)^{-1}$. On the other hand, if $k(\alpha, \gamma) = 0$, $h(\alpha, g(\alpha)\gamma) = 0$, so $g(\alpha) \neq 0$ since $h(x, 0)$ and $g(x)$ are relatively prime. Hence α and $\beta = g(\alpha)\gamma$ satisfy $h(\alpha, \beta) = 0, g(\alpha) \neq 0$. This shows that $f(x, y) = 0, g(x) \neq 0$ has a solution (α, β) in $\Phi^{(2)}$ if and only if $k(x, y) = 0$ has a solution in $\Phi^{(2)}$ and this is the situation we handled before.

9. Equations with parameters. If one attempts to extend the method which we have given in the last section to more than two variables, one is led to treat all but two of the variables as parameters and to seek a reduction of the number of variables by means of the method. This leads to the consideration of polynomials involving parameters. Since the parameters will be allowed to take on any values in the real closed field, there is no loss in generality in assuming that the coefficients of the polynomials are rational numbers. Moreover, the result one obtains in this way will be applicable impartially to all real closed fields, and this can be used to establish an important principle due to Tarski which states that any elementary statement of algebra (this has to be made precise) which is valid for one real closed field is valid for all real closed fields. The main result in Seidenberg's method for treating these questions is the following

Theorem 15. *Let* $F(t_i; x, y) \in R_0[t_1, \cdots, t_r; x, y]$, $G(t_i; x) \in R_0[t_1, \cdots, t_r; x]$, t_i, x, y *indeterminates,* R_0 *the field of rational numbers. Then one can determine in a finite number of steps a finite set of pairs of polynomials* $(F_j(t_i; x), G_j(t_i))$, $F_j \in R_0[t_i; x]$, $G_j \in R_0[t_i]$, $j = 1, 2, \cdots, h$, *such that, if* Φ *is any real closed field, then* $\tau_i \in \Phi$, $1 \leq i \leq r$, *has the property that*

$$F(\tau_i; x, y) = 0, \quad G(\tau_i; x) \neq 0 \tag{21}$$

is solvable for $x, y \in \Phi$ *if and only if one of the conditions:*

$$G_j(\tau_i) \neq 0 \quad and \quad F_j(\tau_i; x) = 0 \tag{22}$$

is solvable for x *in* Φ, *is satisfied.*

The proof of this theorem is essentially a formalization of the decision method of the last section. We consider first some necessary preliminary notions.

We shall call the set $\Phi^{(r)}$ of r-tuples $(\tau_1, \tau_2, \cdots, \tau_r)$, $\tau_i \varepsilon \Phi$, the *parameter space*. A finite set of pairs of finite subsets (δ_j, λ_j) of $\mathfrak{A} = R_0[t_i]$, $j = 1, 2, \cdots, h$, will be called a *rational cover* if for any Φ of characteristic 0, $\Phi^{(r)}$ is the union of the sets S_j where S_j is *defined* by (δ_j, λ_j) in the sense that it is the set of $(\tau_i) \varepsilon \Phi^{(r)}$ satisfying $d(\tau_i) = 0$, $d \varepsilon \delta_j$, $l(\tau_i) \neq 0$, $l \varepsilon \lambda_j$. If (δ_j', λ_j'), $j = 1, \cdots, h$, $(\delta_k'', \lambda_k'')$, $k = 1, \cdots, q$ are rational covers, then so is $(\delta_j' \cup \delta_k'', \lambda_j' \cup \lambda_k'')$, $j = 1, \cdots, h$, $k = 1, \cdots, q$. The corresponding sets are the intersections of those of the two given rational covers. We shall call this a *refinement* of the two rational covers.

We have noted that, if $F = a_n x^n + \cdots + a_0$, $G = b_m x^m + \cdots + b_0$, $a_j, b_j \varepsilon R_0[t_i]$, $a_n \neq 0$, $b_m \neq 0$, $n \geq m$, then we have a uniquely determined division algorithm which yields an even integer $e \geq n - m + 1$ and a quotient Q and remainder R in $R_0[t_i; x] = \mathfrak{A}[x]$, $\mathfrak{A} = R_0[t_i]$, such that $b_m^e F = QG - R$ where $\deg_x R < \deg_x G$. This can be extended to the case $n < m$ or $F = 0$ by taking $e = 0$, $Q = 0$, $R = -F$. We now associate with the pair (F, G) a number of *generic Euclidean sequences* $F_0, F_1, \cdots, F_s$ determined by the following rules F_0 and F_1 are F and G or are obtained from these respectively by dropping leading terms. Thus, $F_0 = a_p x^p + \cdots + a_0$ where $0 \leq p \leq n$, $F_1 = b_q x^q + \cdots + b_0$, $0 \leq q \leq m$. If $F_1 = 0$ we take $s = 0$ and let the sequence consist of F_0 alone. Otherwise, we divide F_0 by F_1 and we let F_2 be the remainder or a polynomial obtained from the remainder by dropping leading terms. If $F_2 = 0$, we stop with F_0, F_1; otherwise, we repeat the process. Clearly, this process breaks off in a finite number of steps and, since we have only a finite number of choices for every F_k, we obtain a finite number of generic Euclidean sequences E for (F, G). Set $D(t_i; x) = F_s(t_i; x)$ the last term in the sequence E. Then $D \neq 0$ unless $F_0 = F_1 = 0$ and, except in this case, we can divide F and G by D obtaining $m(t_i)^e F = F^{(1)} D - R^{(1)}$, $m(t_i)^f G = G^{(1)} D - S^{(1)}$ where $m(t_i)$ is the leading coefficient of D, e and f are the even integers, and $F^{(1)}$, $G^{(1)}$ the quotients, $R^{(1)}$, $S^{(1)}$ the remainders obtained in the division. With each E we associate also the pair of subsets $(\delta(E)$,

$\lambda(E))$ of $R_0[t_i]$ where $\delta(E)$ is the set of coefficients of the dropped terms in the process of forming E (e.g., the coefficients of $F - F_0$ and $G - G_0$) and $\lambda(E)$ is the set of leading coefficients of the F_k.

Now let Φ be any field of characteristic 0, let $(\tau_i) \in \Phi^{(r)}$ and set $f(x) = F(\tau_i; x), g(x) = G(\tau_i; x)$. It is easily seen that there exists a generic Euclidean sequence E for (F, G) such that $d(\tau_i) = 0$ for all $d \in \delta(E)$, $l(\tau_i) \neq 0$ for all $l \in \lambda(E)$. Hence the set of pairs $(\delta(E), \lambda(E))$ for all generic E is a rational cover. If E is chosen as indicated for (τ_i), then $d(x) = D(\tau_i; x)$ is a highest common factor in $\Phi[x]$ of $f(x)$ and $g(x)$ and, if $D(t_i; x) \neq 0$, we have the polynomials $F^{(1)}(t_i; x), G^{(1)}(t_i; x)$ such that $m(\tau_i)^e f(x) = d(x) f_1(x)$, $m(\tau_i)^f g(x) = d(x) g_1(x)$ where $f_1(x) = F^{(1)}(\tau_i; x), g_1(x) = G^{(1)}(\tau_i; x)$ and $m(t_i)$ is the leading coefficient of $D(t_i; x)$. We have $m(\tau_i) \neq 0$ since $m(t_i) \in \lambda(E)$.

The procedure we have just indicated can be extended in an obvious way to any finite set of polynomials. We shall need the process also for polynomials in two indeterminates x, y (besides the t_i). Here we begin with $F(t_i; x, y)$ and $G(t_i; x, y)$ in $\mathfrak{A}[x, y] = R_0[t_i; x, y]$ and we treat x like one of the t_i. The division algorithm with respect to y gives $l(t_i; x)^e F = QG - R$ where $\deg_y R < \deg_y G$. If we observe that a relation $d(\tau_i; x) = 0$ for $d(t_i; x) \in R_0[t_i; x]$ is equivalent to $l_k(\tau_i) = 0$ for all the coefficients $d_k(t_i)$ of $d(t_i; x)$ and $l(\tau_i; x) \neq 0$, $l(t_i, x) \in R_0[t_i; x]$, holds if and only if $l_k(\tau_i) \neq 0$ for one of the coefficients l_k, we see that we can determine a rational cover (δ_j, λ_j), $j = 1, 2, \cdots, h$, and polynomials $D_j(t_i; x, y)$ and $F_j^{(1)}(t_i; x, y)$, $G_j^{(1)}(t_i; x, y)$ if $D_j \neq 0$, such that if (τ_i) is in the subset S_j defined by (δ_j, λ_j), then $d(x, y) = D_j(\tau_i; x, y)$ is a highest common factor in $\Phi(x)[y]$ of $f(x, y) = F(\tau_i; x, y)$ and $g(x, y) = G(\tau_i; x, y)$. Moreover, if $D(t_i; x, y) \neq 0$ and $m(t_i; x)$ is its leading coefficient regarding D as a polynomial in y, then $m(x) = m(t_i; x) \neq 0$ and $m(x)^e f(x, y) = d(x, y) f_1(x, y)$, $m(x)^f g(x, y) = d(x, y) g_1(x, y)$ where $f_1(x, y) = F_j^{(1)}(\tau_i; x, y)$, $g_1(x, y) = G_j^{(1)}(\tau_i; x, y)$.

There is one more device we shall need which will take the place of the step in the decision method of choosing an element γ in Φ such that for a given polynomial $f(x) \neq 0$ one has $f(\gamma) \neq 0$. Let $F(t_i; x) = F_q(t_i)x^q + \cdots + F_0(t_i)$ where $F_q(t_i) \neq 0$. Assume first that (τ_i) in $\Phi^{(r)}$ satisfies $F_q(\tau_i) \neq 0$. If we recall the bound for the roots in Φ of a polynomial given in § 3 we see that $\eta = (q+1) + \sum_0^{q-1} F_k(\tau_i)^2 F_q(\tau_i)^{-2}$ is not a root of $F(\tau_i; x)$. Hence if we set $Q(t_i) = (q+1)F_q(t_i)^2 + \sum_0^{q-1} F_k(t_i)$, $P(t_i) = F_q(t_i)^2$, then $P(\tau_i) \neq 0$, $Q(\tau_i) \neq 0$ for all (τ_i) satisfying $F_q(\tau_i) \neq 0$ and $\eta = Q(\tau_i)P(\tau_i)^{-1}$ is not a root of $F(\tau_i; x)$. Next assume $F_q(\tau_i) = 0$ and $F_p(\tau_i) \neq 0$ for the first non-zero coefficient $F_p(t_i)$ after $F_q(t_i)$. Then we can repeat the argument with p replacing q. Continuing in this way we obtain a rational cover (δ_j, λ_j), $j = 1, 2, \cdots, h$, such that $F(\tau_i; x) = 0$ for $(\tau_i) \in S_h$ and for $j < h$ we have $P_j(t_i)$, $Q_j(t_i)$ such that $P_j(\tau_i) \neq 0$, $Q_j(\tau_i) \neq 0$ and $F(\tau_i; Q_j(\tau_i)P_j(\tau_i)^{-1}) \neq 0$ for $(\tau_i) \in S_j$.

We are now ready to give the

Proof of Theorem 15. We note first that it is sufficient to give a rational cover (δ_k, λ_k), $k = 1, \cdots, m$, such that for each k one defines a finite set of pairs of polynomials $G_{kj}(t_i) \in R_0[t_i]$, $F_{kj}(t_i; x) \in R_0[t_i; x]$ having the property that, if $(\tau_i) \in S_k$, the subset of $\Phi^{(r)}$ defined by (δ_k, λ_k), then $F(\tau_i; x, y) = 0$, $G(\tau_i; x) \neq 0$ is solvable in Φ if and only if one of the conditions: $G_{kj}(\tau_i) \neq 0$ and $F_{kj}(\tau_i; x) = 0$ is solvable in Φ, is satisfied. If we have this situation, we put $F_{kj}{}^*(t_i; x) = F_{kj}(t_i; x)^2 + \sum_{d \in \delta_k} d(t_i)^2$, $G_{kj}{}^*(t_i) = G_{kj}(t_i) \prod_{l \in \lambda_k} l(t_i)$. Then the finite set of pairs $(F_{kj}{}^*(t_i; x), G_{kj}{}^*(t_i))$ satisfies the condition for the set of pairs $(F_j(t_i; x), G_j(t_i))$ in the statement of the theorem.

We consider next the reduction of the theorem from the pair of conditions $F(t_i; x, y) = 0$, $G(t_i; x) \neq 0$ to a single condition $F(t_i; x, y) = 0$. (This corresponds to the second half of the argument given in the last section.) We shall use an induction on $\deg_x F$ and we note that the result is trivial if F does not involve x. Then we can take $F(t_i; x)$ to be the polynomial obtained by replacing y by the missing x and take $G(t_i)$ to be the sum of the squares of the coefficients of $G(t_i; x)$. We now assume $\deg_x F(t_i; x, y) > 0$ and we apply the considerations on highest common factors to $G(t_i; x)$ and the coefficients of the powers of y in $F(t_i; x, y)$. Accordingly, we obtain a rational cover such that for each member (δ, λ) of the cover we can determine polynomials

$m(t_i)$, $D(t_i; x)$, $F^{(1)}(t_i; x, y)$, $G^{(1)}(t_i; x)$ with rational coefficients such that $D(\tau_i; x)$ is a highest common factor of $G(\tau_i; x)$ and the coefficients of the y terms in $F(\tau_i; x, y)$ and $m(\tau_i) \neq 0$,

$$m(\tau_i)^e F(\tau_i; x, y) = D(\tau_i; x) F^{(1)}(\tau_i; x, y),$$
$$m(\tau_i)^f G(\tau_i; x) = D(\tau_i; x) G^{(1)}(\tau_i; x)$$

for all (τ_i) in the set S defined by (δ, λ). We can replace the pair $F(t_i; x, y)$, $G(t_i; x)$ by the pair $F^{(1)}(t_i; x, y)$, $G^{(1)}(t_i; x)$ in the set S so if $\deg_x F^{(1)} < \deg_x F$ the induction can be used. Hence we may assume equality of the degrees indicated, which means that we have $\deg_x D = 0$. Then $D(t_i; x) = m(t_i)$, and $G(\tau_i; x)$ and the coefficients of $F(\tau_i; x, y)$ are relatively prime. Now let $T(t_i; y)$ be the resultant relative to x of $F(t_i; x, y)$ and $0x + G(t_i; x)$. Then $T(\tau_i; y) \neq 0$ for all $(\tau_i) \varepsilon S$ and by passing to a refinement of the rational cover we may assume also that we can find $P(t_i)$, $Q(t_i) \varepsilon R_0[t_i]$ such that $P(\tau_i) \neq 0$, $Q(\tau_i) \neq 0$, and $T(\tau_i; Q(\tau_i) P(\tau_i)^{-1}) \neq 0$ for (τ_i) in S. We replace $F(t_i; x, y)$ by $H(t_i; x, y) = P(t_i)^f F(t_i; x, y + Q(t_i) P(t_i)^{-1})$ where $f = \deg_y F(t_i; x, y)$. The resultant of $H(t_i; x, y)$ and $G(t_i; x)$ relative to x has the form $P(t_i)^g T(t_i; y + Q(t_i) P(t_i)^{-1})$ and this is not 0 for $(\tau_i) \varepsilon S$, $y = 0$. It follows that $H(\tau_i; x, y) = 0$, $G(\tau_i; x) \neq 0$ is solvable in Φ if and only if $K(\tau_i; x, y) = 0$ is solvable in Φ for $K(t_i; x, y) = H(t_i; x, G(t_i; y) y)$.

We now consider a single equation $F(t_i; x, y) = 0$. By considering the highest common factor of the coefficients of the powers of y of F we reduce the consideration to subsets S defined by a rational cover and polynomials $F(t_i; x, y)$ such that $F(\tau_i; x, y)$ is not divisible by a polynomial of positive degree in x for $(\tau_i) \varepsilon S$. Next we consider the highest common factor of F and $\dfrac{\partial F}{\partial y}$ and after a refinement we may assume that we have determined polynomials $m(t_i; x)$, $D(t_i; x, y)$, $F_1(t_i; x, y)$ with rational coefficients such that $D(\tau_i; x, y)$ is a highest common factor in $\Phi(x)[y]$ of $F(\tau_i; x, y)$ and $\dfrac{\partial}{\partial y} F(\tau_i; x, y)$, $m(\tau_i; x) \neq 0$ and $m(\tau_i; x)^e F(\tau_i; x, y) = D(\tau_i; x, y) F_1(\tau_i; x, y)$. Then $F_1(\tau_i; x, y)$ has no multiple factors of positive degree in y and F and F_1 have the same irreducible factors of positive degree in y in $\Phi[x, y]$. Again we can determine

$k(t_i)$, $L(t_i; x)$, $F_2(t_i; x, y)$ such that $k(\tau_i)^f F_1(\tau_i; x, y) = L(\tau_i; x)F_2(\tau_i; x, y)$ where $F_2(\tau_i; x, y)$ is not divisible by a polynomial of positive degree in x. Then it is clear that we may replace F by F_2 and so we may assume that for $(\tau_i) \varepsilon S$, $F(\tau_i; x, y)$ has no multiple factors of positive degree in y and no factor of positive degree in y alone. Then $F(\tau_i; x, y)$ and $\frac{\partial}{\partial y} F(\tau_i; x, y)$ have no common factors of positive degree. Set $G(t_i, c; x, y) = y \frac{\partial F}{\partial x} - (x - c) \frac{\partial F}{\partial y}$ where c is another indeterminate and let $R(t_i, c; x)$ be the resultant relative to y of $G(t_i, c; x, y)$ and $F(t_i; x, y)$. Then one can argue as in the decision method itself that $R(\tau_i, c; x) \neq 0$. By going to a refinement of the rational cover we can obtain $P(t_i)$, $Q(t_i) \varepsilon R_0[t_i]$ such that $P(\tau_i) \neq 0$, $Q(\tau_i) \neq 0$, $R(\tau_i, Q(\tau_i) P(\tau_i)^{-1}; x) \neq 0$. If we replace $G(t_i, c; x, y)$ by $G(t_i; x, y) \equiv P(t_i)G(t_i, Q(t_i)P(t_i)^{-1}; x, y)$, we see that the resultant $R(t_i; x)$ of $F(t_i; x, y)$ and $G(t_i; x, y)$ relative to y satisfies $R(\tau_i; x) \neq 0$, $(\tau_i) \varepsilon S$. As before, we can argue that also the resultant $Q(t_i; y)$ of F and G relative to x satisfies $Q(\tau_i; y) \neq 0$. The remainder of the proof can be made along the lines of the decision method itself. We leave it to the reader to carry this out.

10. Generalized Sturm's theorem. Applications. We can now prove the following generalization of Sturm's theorem which is due to Tarski.

Theorem 16. *Let φ be a finite set of polynomial equations and inequalities of the form $F(t_1, \cdots, t_r; x_1, \cdots, x_n) = 0$, $G(t_1, \cdots, t_r; x_1, \cdots, x_n) \neq 0$ or $H(t_1, \cdots, t_r; x_1, \cdots, x_n) > 0$ where $F, G, H \varepsilon R_0[t_1, \cdots, t_r; x_1, \cdots, x_n]$. Then one can determine in a finite number of steps a finite collection of finite sets ψ_j of polynomial equations and inequalities of the same type in the parameters t_i alone such that, if Φ is any real closed field, then the set φ has a solution for the x's in Φ for $t_i = \tau_i$, $1 \leq i \leq r$, if and only if the τ_i satisfy all the conditions of one of the sets ψ_j.*

Proof. We show first that we can reduce the system φ to a single equation of the form $F(t_i; x_j) = 0$ where the number of x's may have to be increased. First it is clear that an inequality

$G \neq 0$ is equivalent to $G^2 > 0$. Next we can replace an inequality $H > 0$ by the equivalent equation $z^2H - 1 = 0$ where z is an extra indeterminate. Finally, a number of equations $F_i = 0$ can be replaced by the single equation $\Sigma F_i^2 = 0$. These observations prove the assertion, so we take φ to be a single equation $F(t_i; x_j) = 0$. We show first by induction on the number n of x's that we can determine a finite number of sets of equations of the form $F_k(t_i; x) = 0$, $G_k(t_i) \neq 0$ such that a set $\tau_1, \cdots, \tau_r$, τ_i in Φ, has the property that $F(\tau_i; x_j) = 0$ is solvable for x's in Φ if and only if for some k one has $G_k(\tau_i) \neq 0$ and $F_k(\tau_i; x) = 0$ is solvable in Φ. This is trivial for $n = 1$ and it is a consequence of Theorem 15 if $n = 2$. Assume it holds for $n - 1 \geq 2$. Then treating x_n as one of the parameters we conclude that we can determine a finite number of pairs of polynomials $(F_k(t_i, x_n; x), G_k(t_i, x_n))$ with rational coefficients such that, if the τ_i and $\xi_n \varepsilon \Phi$, then $F(\tau_i; x_1, \cdots, x_{n-1}, \xi_n) = 0$ is solvable for $x_1, \cdots, x_{n-1}$ in Φ if and only if, for some k, $G_k(\tau_i, \xi_n) \neq 0$ and $F_k(\tau_i, \xi_n; y) = 0$ is solvable in Φ. By Theorem 15, for each k one can find a finite set of pairs of polynomials $(F_{kj}(t_i; x), G_{kj}(t_i))$ with rational coefficients such that $F_k(\tau_i, x; y) = 0$, $G_k(\tau_i; x) \neq 0$ is solvable in Φ if and only if for some j we have $G_{kj}(\tau_i) \neq 0$ and $F_{kj}(\tau_i; x) = 0$ is solvable in Φ. It follows that the set of pairs $(F_{kj}(t_i; x), G_{kj}(t_i))$ satisfies the required condition for $F(t_i; x_1, \cdots, x_n)$. We now denote these pairs as $(F_j(t_i; x), G_j(t_i))$. For each $F_j(t_i; x)$ the version of Sturm's theorem we considered in § 7 shows that a finite set of polynomial equations and inequalities with rational coefficients in the t_i can be found such that these are satisfied by $t_i = \tau_i \varepsilon \Phi$ if and only if $F_j(\tau_i; x)$ is solvable in Φ. If we add to each set the inequality $G_j(\tau_i) \neq 0$ we obtain the sets ψ satisfying the requirement of the theorem.

Suppose now that we have a system of equations and inequalities with rational coefficients which have a solution in one real closed field Φ_1. It is clear that we can introduce parameters and change our assertion to one that a certain system with parameters and rational coefficients has a solution in Φ_1 for certain rational values of the parameters. Then Theorem 16 implies that these rational numbers satisfy one of a certain set of rational equations and inequalities. Then if Φ is any other real closed field we can apply Theorem 16 again in the reverse direction and conclude that the original system has a solution in Φ.

Again, suppose we have a system of equations and inequalities with rational coefficients involving parameters and suppose that for one real closed field Φ_1 it is true that the system has a solution in Φ_1 for all choices of the parameters in Φ_1. Then one concludes from Theorem 16 that this is equivalent to the statement that every set of values for the parameters in Φ_1 satisfies one of a certain finite collection of finite sets of equations and inequalities. It is easy to see that this in turn is equivalent to the statement that there are no solutions in Φ_1 of any one of another finite collection of finite sets of rational equations and inequalities for the parameters. The foregoing result shows that this carries over to every real closed field Φ. Hence we see that the original system has a solution in Φ for all choices of the parameters in Φ where Φ is any real closed field.

We shall now consider an application of these results to an important theorem on division algebras.

A long time ago, before real closed fields were invented, Frobenius proved the following theorem: The only finite dimensional division algebras over the field R of real numbers are: (1) R itself, (2) $R(\sqrt{-1})$, (3) Hamilton's quaternion algebra over R. The known proofs of this theorem are algebraic and give the same result for any real closed field. The reader may refer to Dickson's *Algebras and Their Arithmetics*, p. 62, for an elementary proof of this type. We now drop the assumption of associativity which we have made throughout this book and consider non-associative algebras. These are defined to be vector spaces over a base field Φ in which a multiplication xy is defined satisfying the distributive laws and the rule $\alpha(xy) = (\alpha x)y = x(\alpha y)$, $\alpha \in \Phi$. Such an algebra which is finite dimensional is called a division algebra if it has no zero divisors: $xy = 0$ implies either $x = 0$ or $y = 0$ in the algebra. Besides the examples noted above there is one other important example of a non-associative division algebra, namely, an algebra of eight dimensions of octonions which was discovered by Cayley and by Graves. The known examples of finite dimensional non-associative division algebras over the field of real numbers have dimensions 1, 2, 4, and 8. It was conjectured for a long time that these are the only possible dimensions and this was finally established by deep topological considerations by Bott and Milnor. It would be a hardy task to attempt to carry over the proof to the case of real closed fields. Moreover, this is unnecessary since it is quite easy to conclude the result for arbitrary real closed fields from its validity for the field of real numbers. Assuming the Bott-Milnor result for the field of real numbers, we shall prove that, if $n \neq 1, 2, 4, 8, \infty$ and Φ is a real closed field, then there exists no n dimensional non-associative division algebra over Φ. To prove this let $\mathfrak{A}$ be a non-associative algebra with the basis $(u_1, u_2, \cdots, u_n)$ over Φ and suppose $u_iu_j = \Sigma\gamma_{ijk}u_k$ where the $\gamma_{ijk} \varepsilon \Phi$. If $x = \Sigma\xi_iu_i$, $\xi_i \varepsilon \Phi$, then the mapping $y \to xy$ in $\mathfrak{A}$ is a linear one whose matrix relative to the basis $(u_1, \cdots, u_n)$ is (ρ_{jk}) where $\rho_{jk} = \sum_i \xi_i\gamma_{ijk}$. The existence of a $y \neq 0$ such that $xy = 0$ is equivalent to the statement that $y \to xy$ is a singular linear transformation and this is the case if and only if $F(\gamma_{ijk}; \xi_i) \equiv \det(\rho_{jk}) = 0$. To show that $\mathfrak{A}$ is not a division algebra we have to show that there exists an $x \neq 0$ such that $F(\gamma_{ijk}; \xi_i) = 0$. We now see that our assertion is equivalent to the following: Let $F(t_{ijk}; x_i) = \det\left(\sum_i x_it_{ijk}\right)$ which can be considered as a polynomial in indeterminates t_{ijk}, x_i with rational coefficients. Then for all choices $t_{ijk} = \gamma_{ijk} \varepsilon \Phi$ there exists a solution $x_i = \xi_i$ in Φ of the system $F(\gamma_{ijk}; x_i) = 0$, $\Sigma x_i^2 \neq 0$. Now by the Bott-Milnor theorem this holds for $\Phi = R$ the field of real numbers. Hence our results show that it holds for every real closed field.

Another example of the same type is a theorem of Hopf's which states that the only possible finite dimensionalities for real non-associative commutative division algebras are $n = 1, 2$. Commutativity, of $\mathfrak{A}$ is equivalent to the condition $\gamma_{ijk} = \gamma_{jik}$ for all i, j. Hence in the foregoing argument we consider indeterminates t_{ijk} for $i \leq j$ and define $t_{jik} = t_{ijk}$ for $j > i$. Then $\det\left(\sum_i x_it_{ijk}\right)$ is a polynomial with rational coefficients in the indeterminates t_{ijk}, $i \leq j$. The rest of the argument carries over and shows that Hopf's theorem is valid for all real closed fields.

There is a general class of statements on real closed fields which can be treated in the foregoing manner. These are the so-called elementary sentences of algebra. We shall not attempt to give the precise definition for these but refer the reader to the literature (see the bibliographic notes on this chapter). The results we have considered are special cases of the general *principle of Tarski* that any elementary sentence of algebra is either true for all real closed fields or is false for all real closed fields.

EXERCISES

1. Assuming the result for the field of real numbers prove that, if Φ is any real closed field and $F_1(x_1, \cdots, x_n) = 0, \cdots, F_k(x_1, \cdots, x_n) = 0$ where the F's ε $\Phi[x_1, \cdots, x_n]$ has a solution $x_i = \xi_i \varepsilon \Phi$, then it has a solution nearest the origin.
2. Prove the analogue of Theorem 16 for algebraically closed fields Φ of characteristic 0 and finite sets of equations $F(t_1, \cdots, t_r; x_1, \cdots, x_n) = 0$ and inequalities $G(t_1, \cdots, t_r; x_1, \cdots, x_n) \neq 0$ where the $F, G \varepsilon R_0[t_i; x_j]$. (Hint: A simple proof of this can be based on the generic Euclidean sequences and the following simple observation due to Tarski: if $f(x), g(x) \varepsilon \Phi[x]$ and $\deg f > 0$, $\deg g > 0$, then $f(x) = 0$, $g(x) \neq 0$ has a solution in Φ if and only if $f(x)$ is not a divisor of $g(x)^{\deg f(x)}$).
3. Prove the result of ex. 2 also for Φ of characteristic $p \neq 0$ by developing the corresponding results on generic Euclidean sequences of $I_p[t_i; x]$, $I_p = I/(p)$.

11. Artin-Schreier characterization of real closed fields. We shall complete our discussion of real closed fields by proving a beautiful characterization of real closed fields which is due to Artin and Schreier. We recall that, if Φ is a field not containing $\sqrt{-1}$ and $\Phi(\sqrt{-1})$ is algebraically closed, then Φ is real closed (Th. 6). We shall now prove

Theorem 17. *Let Ω be an algebraically closed field and Φ a proper subfield which is of finite co-dimension in Ω. Then Φ is real closed and $\Omega = \Phi(\sqrt{-1})$.*

Proof. Let $\Phi' = \Phi(\sqrt{-1}) \subseteq \Omega$. The theorem will follow from the result quoted if we can show that $\Phi' = \Omega$. Hence we suppose that $\Omega \supset \Phi'$. Let E be an algebraic extension of Φ'. Then E is isomorphic to a subfield of Ω over Φ' and so $[E:\Phi'] \leq [\Omega:\Phi']$. Hence the dimensionalities of algebraic extensions of Φ' are bounded. This implies that Φ' is perfect. Otherwise, the characteristic is $p \neq 0$ and there exists a $\beta \varepsilon \Phi'$ which is not a p-th power. Then for every $e > 0$, $x^{p^e} - \beta$ is irreducible in $\Phi'[x]$ (ex. 1, § 1.6) and this provides an algebraic extension of p^e dimensions over Φ'.

Since e is arbitrary, this contradicts what we proved. Thus Φ' is perfect and so Ω is separable over Φ'. Since Ω is algebraically closed it is Galois over Φ' and its Galois group G over Φ' is $\neq 1$ because $\Omega \supset \Phi'$. Hence G contains a cyclic subgroup of prime order q and consequently there exists a subfield $E \supset \Phi'$ such that Ω is cyclic of q dimensions over E. Since Ω is an algebraic closure of E and $[\Omega: E] = q$, it is clear that Ω and E are the only algebraic extensions of E. It now follows that the characteristic of Φ is not q. Otherwise, Ω is a cyclic q-extension of E, and the existence of such an extension of E implies the existence of cyclic q^m-extensions of E for every m (Th. 3.16). This has been ruled out and so the characteristic is not q. This implies that Ω, which is algebraically closed, contains q distinct roots of 1. Since these are roots of $(x - 1)(x^{q-1} + x^{q-2} + \cdots + 1)$ and since the irreducible polynomials in $E[x]$ have degrees 1 or q, all the q-th roots of 1 are contained in E. Since Ω is cyclic q-dimensional over E, $\Omega = E(\sqrt[q]{\alpha})$ where $\alpha \in E$ and is not a q-th power in E (Th. 2.5). Consider the polynomial $g(x) = \prod_1^{q^2} (x - \zeta^i\rho)$ where ζ is a primitive q^2-root of 1 and ρ is an element of Ω such that $\rho^{q^2} = \alpha$. Since the inclusion $\zeta^i\rho \in E$ implies that E contains an element $(\zeta^i\rho)^q = \beta$ such that $\beta^q = \alpha$, we see that no $\zeta^i\rho \in E$. Since $g(x) = x^{q^2} - \alpha \in E[x]$, it follows that all of its irreducible factors in $E[x]$ are of degree q. If β is the constant term of one of these, then $\beta = \rho^q\eta$ where η is a power of ζ. Since $(\rho^q)^q = \alpha$ and $\Omega = E(\sqrt[q]{\alpha})$, $\rho^q \notin E$ and $\Omega = E(\rho^q) = E(\beta\rho^{-q}) = E(\eta)$. Since E contains all the q-th roots of 1, we see that η is a primitive q^2-root of 1. Let Φ_0 be the prime field of Ω and now consider the subfield $\Phi_0(\eta)$ of Ω. If Φ_0 is the field R_0 of rational numbers we know that the dimensionality of the field of q^r-th roots of 1 is $\varphi(q^r)$ (Th. 3.2) and this goes to infinity with r. If Φ_0 has characteristic $p \neq q$, then the field of the q^r-th roots of 1 over Φ_0 contains at least q^r elements, so again the dimensionality of this field over Φ_0 approaches infinity. In any case it now follows that there exists a positive integer r such that $\Phi_0(\eta)$ contains a primitive q^r-th root of 1 but no primitive q^{r+1}-st root of 1. Since η is a primitive q^2-root of 1, $r \geq 2$. The field Ω contains a primitive q^{r+1}-st root of 1, say ξ. Let $h(x)$ be the minimum polynomial of ξ over E. Since $\eta \notin$ E, $\xi \notin$ E, so deg $h(x)$ $= q$. Also $h(x)$ is a factor of $x^{q^{r+1}} - 1 = \prod_{1}^{q^{r+1}} (x - \xi^i)$, so the coefficients of $h(x)$ are contained in $\Phi_0(\xi)$; hence they are contained in the field $\Gamma = \Phi_0(\xi) \cap \mathrm{E}$. It follows that $[\Phi_0(\xi):\Gamma] = q$. Next we consider the subfield $\Gamma' = \Phi_0(\gamma)$, $\gamma = \xi^q$, of $\Phi_0(\xi)$. Evidently γ is a primitive q^r-th root of 1, so Γ' contains q distinct q-th roots of 1. On the other hand, $\Phi_0(\xi) = \Gamma'(\xi)$ where $\xi^q = \gamma \in \Gamma'$, so either $\Phi_0(\xi) = \Gamma'$ or $\Phi_0(\xi)$ is cyclic of q dimensions over Γ'. If $\Phi_0(\xi) = \Gamma' = \Phi_0(\gamma)$, we have $\Phi_0(\xi) \subset \Phi_0(\eta)$ since $\Phi_0(\eta)$ contains all the q^r-th roots of 1. Then $\Phi_0(\eta)$ contains ξ, a primitive q^{r+1}-st root of 1, contrary to hypothesis. Thus we have $[\Phi_0(\xi):\Gamma'] = q$. Now $\Gamma' \neq \Gamma$. Otherwise, Γ contains a primitive q^r-th root of 1, so Γ and E contain η contrary to $\Omega = \mathrm{E}(\eta) \supset \mathrm{E}$. We have therefore proved that the field $\Phi_0(\xi)$ of the q^{r+1}-st roots of 1 over the prime field contains two distinct subfields Γ and Γ' over which it is q-dimensional. It follows that the Galois group of $\Phi_0(\xi)$ over Φ_0 is not cyclic. By Lemma 1 of § 1.13 and Theorem 3.5, this is the case only if the characteristic is 0 and $q = 2$. Then the element η considered before is a primitive 4-th (q^2 with $q = 2$) root of 1. On the other hand, E contains Φ' which contains $\sqrt{-1}$ and this is a primitive 4-th root of 1. Hence we have $\Omega = \mathrm{E}(\eta) = \mathrm{E}$ contrary to $\Omega \supset \mathrm{E}$. This contradiction shows that $\Phi' = \Phi(\sqrt{-1}) = \Omega$ and Φ is real closed.

SUGGESTIONS FOR FURTHER READING

Chapter I. The classical Galois correspondence between groups of automorphisms and subfields has been extended in a number of different directions. First, one has Krull's Galois theory of infinite dimensional extensions which is considered in Chapter VI. Next one has the Galois theory of division rings which is due (independently) to H. Cartan and the present author. An account of this can be found in the author's *Structure of Rings*, A.M.S. Colloquium Vol. 37 (1956), Chapter VII. (Our development of the Galois theory in Chapter I is based on the methods which were developed originally to handle the non-commutative theory.) A Galois theory of finite dimensional separable extensions based on the notion of a self-representation of a field is due to Kaloujnine. This is contained in a more general theory given by the present author in two papers in *Am. J. Math.*, Vol. 66 (1944), pp. 1–29 and pp. 636–644. See also two papers by Hochschild and by Dieudonné in the same journal, Vol. 71 (1949), pp. 443–460 and Vol. 73 (1951), pp. 14–24.

Quite recently a Galois theory of automorphisms of commutative rings has been developed jointly by S. U. Chase, D. K. Harrison, and A. Rosenberg. This paper will appear in *Transactions A.M.S.*

A general cohomology theory of fields has been given by Amitsur in *Trans. A.M.S.*, Vol. 90 (1959), pp. 73–112. See also the paper by Rosenberg and Zelinsky on this subject in *Trans. A.M.S.*, Vol. 97 (1960), pp. 327–356, and Amitsur's paper in *J. Math. Soc. Japan*, Vol. 14 (1962), pp 1–25.

Chapter II. We have indicated in the text the unsolved problem of the existence for a given field Φ and a given finite group G of a Galois extension P/Φ whose Galois group is isomorphic to G. A closely related question is that of the existence of an equation with coefficients in Φ having a given subgroup of S_n as group. These problems have been studied extensively for Φ the field of rational numbers and more generally for algebraic number fields (finite dimensional extensions of the rationals). Two methods have been developed for this problem: one based on arithmetic properties of number fields, and a second more elementary method based on an irreducibility criterion due to Hilbert. The deepest results thus far obtained in the arithmetic theory are due to Šafarevič. A summary of his results is given in *Math. Reviews*, Vol. 16 (1955), pp. 571–572.

The Hilbert method (which was used by Hilbert to prove the existence of rational equations with S_n as Galois group) has two stages. Given a field Φ one requires first a purely transcendental extension field

$\Phi(t_1, \cdots, t_r)$ and a Galois extension P of $\Phi(t_i)$ with Galois group isomorphic to the given group G. This problem is still open except for special cases (S_n, alternating group and some others). Next one needs to know that Φ is a Hilbertian field in the sense that Hilbert's irreducibility theorem holds for Φ. (For example, the rational field is Hilbertian; the field of p-adic numbers and finite fields are not.) A discussion of this theorem and its relation to Galois theory is given in S. Lang's book *Diophantine Geometry*, New York, 1962, Chapter VIII.

An interesting aspect of the classical Galois theory of equations is Klein's theory of form problems. A development of this from the point of view of algebras, particularly crossed products, is due to R. Brauer in *Math. Annalen*, Vol. 110 (1934), pp. 437–500. Reference to the classical works on the subject is given in this paper.

A general reference book for Galois theory of equations is Tschebotaröw's *Gurndzüge der Galois'schen Theorie*, Groningen, 1950 (translated from Russian by Schwerdtfeger).

Chapter III. D. K. Harrison has given a general theory of abelian extension fields in *Trans. A.M.S.*, Vol. 106 (1963), pp. 230–235.

Chapter IV. Some of the deeper results of this chapter have been developed to meet the needs of algebraic geometry. The reader may consult S. Lang's *Introduction to Algebraic Geometry*, 1958, or A. Weil's *Foundations of Algebraic Geometry*, A.M.S. Colloquium Vol. 29, Providence, 1st. Ed., 1946, 2nd. Ed., 1962, for these connections.

Chapter V. There are several directions that one may take in pursuing the subject matter of this chapter. First, one can study the general theory of valuations as given in Zariski-Samuel's *Commutative Algebra* Vol. II, D. Van Nostrand Co., Inc., Princeton, 1960, Chapter VI. Secondly, this chapter leads to the arithmetic theory of number fields and fields of algebraic functions of one variable. For this the reader may consult Chevalley's book *Algebraic Functions of One Variable*, Princeton, 1951, Artin's book *Theory of Algebraic Numbers*, Göttingen, 1959, and E. Weiss' book *Algebraic Number Theory*, New York, 1963. A third direction which one can take after studying Chapter V is local class field theory. For this the reader may consult Serre's book *Corps Locaux*, Paris, 1962.

Chapter VI. The original Artin-Schreier theory is given in papers by Artin and Schreier and by Artin in the *Hamburg Abhandl.*, Vol. 5 (1927). Our exposition follows these papers rather closely. Seidenberg's work is in *Annals of Math.*, Vol. 60 (1954), pp. 365–374. This contains also a statement of Tarski's principle and, of course, a reference to Tarski's earlier paper. Much of the present chapter can be developed also as a part of mathematical logic, more exactly, as an aspect of the theory of models. The reader may consult A. Robinson's book, *Model Theory*, Amsterdam, 1963, particularly Chapter VIII. Also references to the literature are given in this book.

INDEX

GPSR Compliance

The European Union's (EU) General Product Safety Regulation (GPSR) is a set of rules that requires consumer products to be safe and our obligations to ensure this.

If you have any concerns about our products, you can contact us on ProductSafety@springernature.com

In case Publisher is established outside the EU, the EU authorized representative is:

Springer Nature Customer Service Center GmbH
Europaplatz 3
69115 Heidelberg, Germany

Batch number: 10682323

Printed by Printforce, the Netherlands